DISSOCIATIVE RECOMBINATION OF MOLECULAR IONS WITH ELECTRONS

DISSOCIATIVE RECOMBINATION OF MOLECULAR IONS WITH ELECTRONS

Edited by

Steven L. Guberman
Institute for Scientific Research
Winchester, Massachusetts

Kluwer Academic / Plenum Publishers
New York, Boston, Dordrecht, London, Moscow

Library of Congress Cataloging-in-Publication Data

Dissociative recombination of molecular ions with electrons/edited by Steven L. Guberman.
p. cm.
Proceedings of an American Chemical Society symposium held in Chicago, Aug. 26–30, 2001.
Includes bibliographical references and index.
ISBN 0-306-47765-3
1. Dissociation—Congresses. 2. Ions—Congresses. 3. Electrons—Congresses. 4. Molecular structure—Congresses. I. Guberman, S. (Steven).

QD517.G83 2003
541.3′64—dc21

2003047449

Proceedings of the American Chemical Society symposium "Dissociative Recombination of Molecules with Electrons," held August 26–30, 2001, in Chicago, Illinois.

ISBN 0-306-47765-3

233 Spring Street, New York, New York 10013

http://www.wkap.nl/

10 9 8 7 6 5 4 3 2 1

A C.I.P. record for this book is available from the Library of Congress

Printed in the United States of America

Preface

This volume is based upon a symposium entitled "Dissociative Recombination of Molecules with Electrons," which was sponsored by the Division of Physical Chemistry at the American Chemical Society national meeting in Chicago, Aug. 26-30, 2001. The four day symposium brought together the leading contributors to modern dissociative recombination research. A total of 43 papers were presented, covering all aspects of dissociative recombination research including recent theoretical and experimental developments and applications to astronomy and ionospheric studies. Several papers were also presented in the closely related area of dissociative attachment. A brief summary of the symposium has appeared.*

Unfortunately, less than two weeks before the start of the meeting, one of the pioneers of modern chemical physics and an invited speaker, Sheldon Datz, died. In the past decade, Sheldon had made invaluable and innovative contributions to the establishment of storage rings as major instruments in the study of dissociative recombination. I will fondly remember him for his generosity in praising the work of others. The symposium was dedicated to his memory.

A photograph of the participants can be found at http://www.sci.org/meet/acs.htm.

Steven L. Guberman

Institute for Scientific Research

* S. L. Guberman, Science **294**, 1474 (2001).

Acknowledgements

Thanks are due to the session chairpersons: Ray Flannery, Al Wagner, Eric Herbst, Nigel Adams, Rainer Johnsen, and Ann Orel. I also thank Richard Stratt, Marc Gordon, and Kristin Hinders for their assistance with the logistics and financial details. I am indebted to each of the referees who kindly agreed to review the chapters. Their voluntary efforts have provided important contributions to this volume.

The conference was supported by the U. S. National Science Foundation and the Petroleum Research Fund of the American Chemical Society.

The symposium would not have taken place if it had not been suggested by Susan L. Greenblatt. I am deeply grateful to her for encouragement and support during the meeting organization and during the preparation of this volume.

Recombination Verse

Ions Mine

Air: "Clementine."

In the dusty lab'ratory,
  'Mid the coils and wax and twine,
There the atoms in their glory
  Ionise and recombine.

Chorus. Oh my darlings! Oh my darlings!
    Oh my darling ions mine!
  You are lost and gone for ever
    When just once you recombine!

From Poft-Prandial Proceedings of the Cavendifh Society, 6th edition (Bowes and Bowes, Cambridge, 1926).

Far beyond the flight of airmen,
High above the smoke and grime,
There the atoms in their glory,
Ionize and Recombine.

Chorus. Oh my darlings! Oh my darlings!
  Oh my darlings ions mine!
  You are lost and gone forever
  If but once you recombine!

From University College London as quoted in D. R. Bates, Case Studies in Atomic Physics **4**, 57 (1974).

High above the acid rainfall, Where the sun must sometimes shine
There the atoms in their glory, Ionize and recombine

Oh my darlings, Oh my darlings, Oh my darlings ions mine
Thou art lost and gone forever if but once you recombine

In a cavity, in a discharge, Back in nineteen forty- nine
FRED BIONDI waved his wand-y And the ions recombine

In the labs of PEART and DOLDER, By the banks of coaly TYNE,
There's inclined beams, intertwined beams, There are beams of either sign

At the plasma physics meetings, Mid the crates of beer and wine
NORMAN BARDSLEY, TOM O'MALLEY, Dissociate and recombine.

Tokamaks and Stellerators,, Eating Taxes, yours and mine,
And all that they are heating Is tungsten twenty - nine.

Air-Clementine (Traditional)

From Johannes Hasted in *Physics of Ion-Ion and Electron-Ion Collisions*, ed. by F. Brouillard and J. W. McGowan (Plenum Press, New York, 1983), p.vi.

Contents

Storage Rings

General Theory

H_3^+ Theory and Experiment*

* Additional discussion of H_3^+ can be found in the chapters starting on p.87 and 177.

Attachment and Detachment

Dissociative Recombination Mechanisms

An Introduction and Overview

Steven L. Guberman
Institute for Scientific Research, 22 Bonad Road, Winchester, MA 01890
slg@sci.org

1. INTRODUCTION

This chapter presents an introduction to dissociative recombination (DR) and to the papers in this volume by reviewing how DR works. In DR a molecular ion, AB^+, captures an electron, e^-, and breaks up,

$$AB^+ + e^- \rightarrow A + B. \quad (1)$$

Process (1) is dissociative in the exit channel and recombinative in the entrance channel. The Process involves the interaction of two continua, an electron in a nearly coulomb field on the left side of (1) with a nuclear continuum, i.e. "free" heavy particles on the right side of (1). AB^+ can be a diatomic or polyatomic molecular ion. The product fragments are often neutral and if energetically allowed, they can also be excited. Indeed, if the electron energy is high enough, both positively and negatively charged fragments, i.e. ion pairs, can be produced. Because Reaction (1) is driven by the coulomb attraction between the ion and the electron, it often proceeds without a barrier to electron capture. In DR, energy must be released in order to conserve energy and momentum. The release is often accomplished rapidly by the molecular dissociation, which finalizes the process since after dissociation, the electron can no longer be emitted (autoionization). The absence of a barrier and the rapid fragmentation allows DR rate coefficients to often (but not always) be high, i.e. about 1 x 10^{-7}cm^3/sec.

The theoretical study of DR is both a dynamics problem and a bound state problem involving many current areas of research in chemical physics. For the dynamics, it is necessary to describe the scattering of the incident electron. Is the electron elastically scattered without capture? Is the electron temporarily captured and emitted before dissociation? How is the electron captured? At energies just above the lowest vibrational level of an ion, many superexcited states are accessible. These include states having more than a single electron excited from the ground state neutral configuration in addition to those in which a single inner shell electron is excited. States with a single excited outer shell electron may also be found in this energetic region if they are vibrationally excited. If the electron is captured into these states, how do they decay? Do the different decay mechanisms interfere with each other? Does the interference affect the quantum yields of the products? Can the electron be captured only by its interaction with the other electrons or can it be captured by its interaction with the vibrating nuclei? If the electron is emitted (autoionization) does it leave behind a molecular ion that has been rovibrationally relaxed or excited? If the electron is permanently captured leading to dissociation to neutrals, the motion of the neutrals on potential curves or surfaces needs to be described. For polyatomic ions, the product neutral molecules can be both electronically and rovibrationally excited. Highly accurate potential curves and surfaces are needed if accurate rate coefficients are to be calculated.

Considerable progress has been made by both experiment and theory in answering many of these questions and that progress is described in the papers in this volume. Below, several of the questions raised above are answered and the mechanisms that drive DR are elucidated.

2. THE DIRECT MECHANISM

The possibility that molecular ions could undergo DR was suggested by Bates and Massey[1] 55 years ago. They were exploring a chemical model of the Earth's ionosphere and needed a mechanism for removing the electrons at night, which photoionization by the Sun's rays had produced during the day. They recognized the possibility that DR could play an important role, but they were uneasy with the mass mismatch between the electron and the molecular nuclei. How could an electron, several thousand times smaller than the nuclei, cause the molecule to fall apart? The explanation was given by Bates in a one page paper.[2] His mechanism is shown in Fig. 1 for the capture of an electron with energy, ε. The incoming electron is captured through its interaction with the other electrons and not through its interaction with the nuclei. Capture occurs into a repulsive state upon which the nuclei can fly apart following the solid arrow in Fig. 1. After capture and prior to dissociation, the electron can be emitted (autoionization). Once the internuclear distance is greater than the distance at which the ion and neutral curves cross, autoionization is unlikely, and DR is completed. Note that the

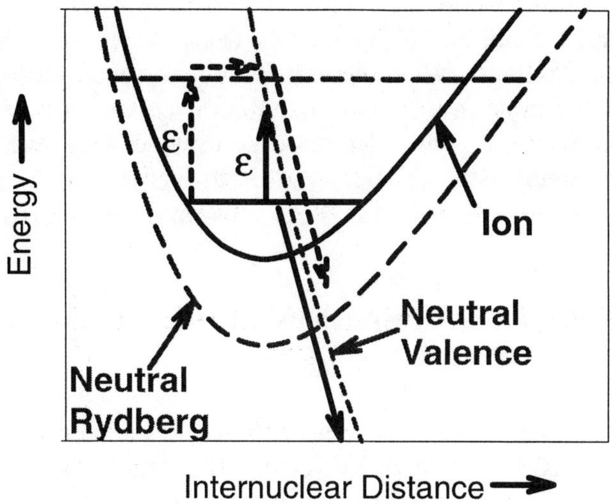

Figure 1. DR mechanisms. ε and ε' are electron energies.

mechanism requires a neutral potential curve to cross the ion curve. By the non-crossing rule, this situation is forbidden for a fully optimized state since it must cross through an infinite number of Rydberg states of the same symmetry. The crossing state must therefore be a diabatic state, i.e. a state that does not optimize the full electronic Hamiltonian. In practice, these states are taken to be fully optimized valence states that omit Rydberg character. Nevertheless, DR can also be described by fully optimized states, with Rydberg character, which do not cross the ion. This adiabatic picture is described below. For now, however, we will use the diabatic approach.

The diabatic process of Bates,[2] shown by the solid arrows in Fig. 1, is known as direct DR. In this case, the rate coefficient is approximately proportional to the square of a matrix element between the vibrational wave functions for the ion and neutral states. A high direct DR rate coefficient is often possible if the repulsive state crosses the ion curve between the turning points of the ion vibrational level, giving a high vibrational wave function overlap. The matrix element describing capture is given by

$$< \Psi_{ie}(x,R)\, X_v(R) \mid H(R) \mid \Psi_d(x,R)\, F_d(R) > \qquad (2)$$

which can be rewritten as

$$<X_v(R) \mid (\Gamma_{ie,d}(R)/2\pi)^{1/2} \mid F_d(R) >. \qquad (3)$$

Here H is the electronic Hamiltonian, R is the internuclear distance, x represents the electron coordinates, Ψ_{ie} is the electronic wave function for the ion plus a "free" electron, Ψ_d is the electronic wave function for the dissociative state, $X_v(R)$ is the ion bound vibrational wave function, $F_d(R)$ is the continuum vibrational wave function in the dissociative state and $\Gamma_{ie,d}(R)$ is the electronic width. A small electron capture width will lead to a small probability for capture while a large width will increase the possibility of autoionization. A favorable width for DR is between these two extremes.

3. THE BORN-OPPENHEIMER BREAKDOWN INDIRECT MECHANISM

The dashed arrows of Fig. 1 show the pathway for indirect recombination, introduced by Bardsley,[3] in which an electron of energy ε' is captured into a neutral, vibrationally excited Rydberg state. This state is predissociated by the repulsive state of the direct DR mechanism. Capture occurs by a Born-Oppenheimer breakdown mechanism [Bardsley, 1968] involving the interaction between the electronic and nuclear motion. The electrostatic matrix element, analogous to (2) but with a Rydberg state replacing the dissociative state, is zero. The appropriate matrix element is given by

$$< \Psi_{ie}(x,R)\, X_v(R) \mid T_N(R) \mid \Psi_{ryd}(x,R)\, X_{v'}(R) > \tag{4}$$

where T_N is the nuclear kinetic energy operator. Alternatively, the matrix element in (4) can be determined from the variation of the quantum defect, μ with R. The quantum defect is determined from the energy difference between the ion and the neutral potential curves, $\mu(R) = n - (2E)^{-1/2}$, where n is the principal quantum number. Once capture occurs, the electron can autoionize via the coupling in (4) or the Rydberg state can predissociate via the electrostatic coupling in (2) with $\Psi_{ryd}(x,R)$ replacing $\Psi_{ie}(x,R)$ on the left side of the matrix element. Once in the dissociative state, autoionization can occur by this electrostatic mechanism or dissociation can take place along the route shown by the dashed arrow in Fig. 1.

Note that for Born-Oppenheimer breakdown capture, the propensity rule, $\Delta v = 1$ applies.[4] For electron capture by an ion in v=0, this coupling is largest if the electron enters the v=1 level of a Rydberg state.

The Rydberg states are the source of structure in the DR cross sections. The resultant cross section shape near a Rydberg resonance is due to the interference between direct and indirect recombination. The interference induced shape can be understood by calculating the Fano profile index from the above matrix elements[5] using the same approach than Fano introduced for understanding atomic photoionization cross sections near resonances.[6] This approach is successful if the resonances are isolated and not overlapping.

4. THE ELECTRONIC INDIRECT MECHANISM

In this mechanism, the same states are excited as in the Bardsley indirect mechanism except that the nuclear kinetic energy operator is not involved. In the electronic indirect mechanism, the electron is captured first into the dissociative state, which then "leaks" into the vibrationally excited Rydberg state by an electrostatic mechanism. The flux eventually returns to the dissociative state and the process is completed. This process has been discussed previously[7,8] and has been included in the Multichannel Quantum Defect Theory (MQDT) approach for calculating DR cross sections.[5] The matrix element takes the form[5]

$$\iint X_v(R)\langle\Psi_{ie}|H(R)|\Psi_d\rangle F_d(R_<)G_d(R_>)\times\langle\Psi_d|H(R')|\Psi_{ryd}\rangle X_{v'}(R')\, dR\, dR' \quad (5)$$

where $R_<$ and $R_>$ are the lesser and greater of R and R' respectively and G_d is the irregular continuum dissociative wave function. Because the electronic Hamiltonian acts twice in (5), this mechanism is often referred to as the second order indirect mechanism. The magnitude of the matrix element in (5) often exceeds that for the indirect Born-Oppenheimer breakdown mechanism.[5]

A very important aspect of the electronic indirect mechanism is that the propensity rule for the Bardsley indirect mechanism no longer applies. Indeed transitions with $\Delta v \gg 1$ occur often. In these cases, the energetics are such that transitions to high v states with low n are very important. Resonances with low n can have large widths and can cover a large section of the cross section versus electron energy plot.

The second order electronic mechanism is now included in most MQDT calculations of DR cross sections where its interference with direct recombination and the Born-Oppenheimer breakdown indirect mechanism is fully taken into account. Approaches that calculate the full K matrix, account for higher order indirect electronic coupling.

5. CORE EXCITED STATES

In the indirect mechanisms discussed above, the Rydberg states of interest have as the core, the ion state undergoing recombination. Each rovibrational state of the ion is the limit of an infinite series of these Rydberg states. But there are other Rydberg states that can also play a role. These are states having excited ion states as the core.[9] For ions with excited states lying less than 4 eV above the ground state, Rydberg states with an excited core can affect the DR of the lowest ground state ion vibrational levels. However, these states differ in a fundamental way from the Rydberg states discussed in earlier sections. These states differ by at least a double excitation from the ground state ion plus a free electron. In this situation the Born-Oppenheimer breakdown matrix elements (4) are zero. Therefore the coupling matrix elements for populating these states are electrostatic and are given by

$$< \Psi_{ie}(x,R)\, X_v(R) \mid H(R) \mid \Psi_{ryd}^{ex\,core}(x,R)\; X_{v'}(R) > \qquad (6)$$

This is the same matrix element as in (2) except that the excited ion core Rydberg state replaces the dissociative state on the right side. These Rydberg levels will cause structure in the cross section, as do the levels discussed in the prior Sections. However, for Rydberg states with the ground state ion core, the v = m resonances cannot appear in DR of the v = m level of the ion since these must lie below the ion level. However, for excited core states, only the energetics determine which excited core Rydberg vibrational levels lay above the recombining ion level. These excited core states have already been included in MQDT cross section calculations. For N_2^+, a partial inclusion of the $A\,^2\Pi_u$ Rydberg states (i.e. only $^3\Pi_u$ Rydberg states were included) increases the rate coefficient at room temperature by about 10%.[10] The assignment of cross section structure to core excited states has also been discussed in storage ring studies of OH^+[11] and CD^+[12] and excited core states have been included in a theoretical study of CH^+ and CD^+ DR.[13]

6. DISSOCIATIVE STATE MIXING

If there are more than two dissociative routes of the same symmetry, the dissociative routes can mix via the Rydberg states of the same symmetry. These dissociative routes predissociate the same symmetry Rydberg states and as a result, they are all connected together in the region of the ion. This leads to the interesting situation in which capture into a favorable crossing dissociative state can lead to partial dissociation along another dissociative route of the same electronic symmetry but with an ion curve crossing that is unfavorable to direct recombination. Another interesting effect, which has been demonstrated for N_2^+,[14] occurs in the case where one dissociative route has a favorable ion intersection and another of the same symmetry has an unfavorable intersection. Without the presence of the favorable crossing route, the cross section structure for DR along the unfavorable route consists of narrow resonances, which are dips. However, with the favorable route included in the calculations, the dips change to peaks since in the energy region of the resonance, the unfavorable dissociative route mixes in some of the favorable route via the Rydberg resonances.

A similar situation can occur if multiple routes of different electronic symmetries drive DR if a nonzero coupling is present which couples the intermediate Rydberg states together. This situation occurs in O_2^+ where spin-orbit coupling couples together $^1\Sigma_u^+$ and $^3\Sigma_u^-$ dissociative states by coupling together the Rydberg states of these symmetries.[15] The $^1\Sigma_u^+$ state crosses the ion between v=1 and v=2 and dissociates to $O(^1S) + O(^1D)$. Therefore, the quantum yield of $O(^1S)$ from DR of v=0 along $^1\Sigma_u^+$ is very small. The $^1\Sigma_u^+$ state is the only state that generates $O(^1S)$ from the low ion levels. The $^3\Sigma_u^-$ state has a favorable intersection with v=0 and dissociates to $O(^3P) + O(^1D)$. However, after favorable capture of the electron into the $^3\Sigma_u^-$ dissociative state, some of the flux transfers into the $^3\Sigma_u^-$ Rydberg states, which are connected by spin-orbit coupling to the $^1\Sigma_u^+$ Rydberg states. The latter are predissociated by the $^1\Sigma_u^+$ dissociative state leading to the $O(^1S) + O(^1D)$ products. This spin-orbit coupling mechanism causes over an order of magnitude increase in the quantum yield for $O(^1S)$.[15] This mechanism will, of course, occur in other molecular ions but it requires that the molecular ion have nonzero spin and angular momentum projection upon the internuclear axis.

7. DR IN THE ABSENCE OF A CURVE CROSSING

For many molecules, there are no diabatic states that cross though the ion potential. Examples include HeH^+ [16-18] and H_3^+.[19] At first glance, one might therefore expect that the DR rate coefficients would be very small. However, it is known from studies of DR in several molecular ions that when direct recombination is slow, the Rydberg resonance states often appear as peaks in DR cross sections. DR occurs by electron capture into a Rydberg resonance, which is predissociated by a curve that does not intersect the ion curve (see Fig. 2). Therefore, in the vicinity of each resonance, DR suddenly becomes possible and the cross section increases, yielding a peak. The question to be answered in these cases is: are there enough of these peaks and are they high and broad enough to give a high rate coefficient?

In the case of HeH^+, the matrix element that describes direct DR is given by

$$< \Psi_{ie}(x,R)\, X_v(R) \mid T_N \mid \Psi_{ryd}(x,R)\, F_d(R) > . \tag{7}$$

For HeH^+, there are no valence dissociative routes that cross the ion and dissociation takes place on the repulsive wall of Rydberg states with asymptotes that lie below the ion v=0 level.[16,17] Fig. 2 shows two Rydberg curves, one of which is dashed and the other has dashes and dots. Direct

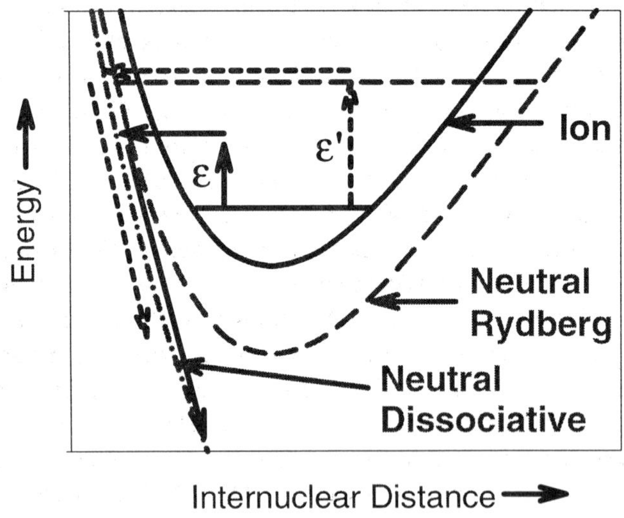

Figure 2. Dissociative recombination without a curve crossing.

recombination is shown for an electron with energy, ε. By Born-Oppenheimer breakdown coupling, the electron is captured into the dash-dot potential curve. The dash-dot state is a member of the Rydberg series that has the ion as its limit. The electrostatic coupling between these states is zero. The ion plus a free electron and the Rydberg state are connected by Born-Oppenheimer breakdown coupling. The direct recombination cross section calculated with the matrix element in (7) is small. DR is primarily driven by capture of an electron (with energy ε′ in Fig. 2) into a vibrationally excited Rydberg state followed by dissociation upon the dash-dot Rydberg state. Both Rydberg states are members of the same series and are connected by the Born-Oppenheimer breakdown coupling. For HeH^+, there is a strong coupling between these Rydberg states and the rate coefficient for HeH^+ DR is not negligible. The theoretically calculated rate at room temperature is 2.6×10^{-8} cm^3/sec[16,17] for ^3HeH and the experimentally derived rate coefficient is 3×10^{-8} cm^3/sec for ^4HeH.[20] Note that the theory did not include rotational effects and isotope effects are large in these systems. So the agreement between theory and experiment may be fortuitous. The same mechanism described here for HeH^+ also drives the DR of H_3^+. However, DR of the latter is, of course, considerably more complicated because of the need to account for more than one nuclear motion coordinate. The magnitude of the DR rate coefficient for H_3^+ is controversial and is the subject of several papers in this volume.

8. OTHER SOURCES OF CROSS SECTION STRUCTURE

Rydberg vibrational levels are not the only sources of structure in DR cross sections. As the electron energy increases, the cross section drops abruptly as new autoionization channels (i.e. ion vibrational levels) become accessible. Broad cross section structure can also arise from the variation of the Franck-Condon factor between ion and continuum vibrational wave functions as the electron energy changes.[21]

A very interesting contribution to cross section structure has been described recently in studies of HD^+ DR to $H^+ + H^-$.[22-24] In H_2, DR occurs on states that are mostly doubly excited near the ion.[25] The lowest dissociative state, $Q_1\,{}^1\Sigma_g^+$, crosses the ion near the outer turning point of the v=1 vibrational level and has a well at R=4-5a_o. The outer wall of this potential is mostly due to the $H^+ + H^-$ coulomb attraction. Several Rydberg states of H_2 are crossed twice by this Q_1 potential so that there is more than one way to dissociate to $H^+ + H^-$. An electron can be captured into the Q_1 state followed by transfer of some of the dissociating flux to a Rydberg state at the first crossing. At the second crossing with the Q_1 state, the flux partially returns to the Q_1 state and dissociation continues to $H^+ + H^-$. Alternatively, the flux can diabatically follow the Q_1 state out to the asymptotic limit. The interference resulting from these different paths causes ripples in the theoretically[24] determined cross section, which compares well to the experimentally derived cross section.[22] This type of interference will be important in the DR of many other molecular ions.

9. CONCLUSIONS

The mechanisms discussed here will take place in the DR of many molecular ions that have not yet been studied. However, the list of mechanisms described here is undoubtedly incomplete; it is likely that as more ions are studied, the detailed energetics and interactions involved in each case will reveal still more mechanisms. It is important to keep in mind, in future studies, that very small interactions can play a large role in DR. The indirect mechanisms discussed here often involve Rydberg states with high n. These Rydberg states, of different electronic symmetries, can be separated in energy by only a few wave numbers. In these cases, small interactions can completely mix these states, as discussed above, for the spin-orbit coupling in O_2.

The mechanisms discussed here also occur in the DR of polyatomic ions, but they are more complicated due to the additional degrees of freedom in the nuclear motion. For example, intramolecular energy transfer can compete with dissociation in polyatomic ions so that DR may not be the most likely event to occur after electron capture. For polyatomics, rules are needed to describe how the excess energy is distributed. Under what conditions and for what molecular structures will recombination lead to rovibrational excitation or dissociation? Can capture of an electron in one part of a molecule lead to dissociation in another part? How do the polyatomic superexcited states decay? Clearly, much remains to be learned about DR.

ACKNOWLEDGEMENTS

Several of the mechanisms described above were initially studied in this laboratory. The author gratefully acknowledges support from NASA Grants NAG5-11428, NAG5-12220, NAG5-11827, and from the U. S. National Science Foundation under Grants ATM-9812034 and ATM-0225256.

REFERENCES

1 D. R. Bates and H. S. W. Massey, Proc. Roy. Soc. A **192**, 1 (1947).
2 D. R. Bates, Phys. Rev. **78**, 492 (1950).
3 J. N. Bardsley, J. Phys. B **1**, 365 (1968).
4 G. Herzberg and Ch. Jungen, J. Mol. Spectrosc. **41**, 425 (1972).
5 S. L. Guberman and A. Giusti-Suzor, J. Chem. Phys. **95**, 2602 (1991).
6 U. Fano, Phys. Rev. **124**, 1866 (1961).
7 T. F. O'Malley, J. Phys. B **14**, 1229 (1981).
8 A. P. Hickman, J. Phys. B **20**, 2091 (1987).
9 S. L. Guberman, Ab Initio Studies of Dissociative Recombination, in *Dissociative Recombination: Theory, Experiment, and Applications*, ed. by J. B. A. Mitchell and S. L. Guberman (World Scientific, Singapore, 1989), p. 45.
10 S. L. Guberman, New Mechanisms for Dissociative Recombination, in *The Physics of Electronic and Atomic Collisions, XIX International Conference, Book of Invited Papers*, ed. by L. J. Dube et al., (American Institute of Physics Press, New York, 1995), p.307.
11 Z. Amitay, D. Zajfman, P. Forck, T. Heupel, M. Grieser, D. Habs, R. Repnow, D. Schwalm, A. Wolf, and S. L. Guberman, Phys. Rev. A **53**, R644 (1996).
12 P. Forck, C. Broude, M. Grieser, D. Habs, J. Kenntner, J. Liebmann, R. Repnow, D. Schwalm, A. Wolf, Z. Amitay and D. Zajfman, Phys. Rev. Lett. **72**, 2002 (1994).
13 L. Carata, A. E. Orel, M. Raoult, I. F. Schneider and A. Suzor-Weiner, Phys. Rev. A **62**, 052711 (2000).
14 S. L. Guberman, The Role of Rydberg States in Dissociative Recombination, in *Dissociative Recombination: Theory, Experiment, and Applications III*, ed. by D. Zajfman, J. B. A. Mitchell, D. Schwalm and B. R. Rowe (World Scientific, Singapore, 1996), p. 64.
15 S. L. Guberman, Science **278**, 1276 (1997).
16 S. L. Guberman, Recent Developments in Dissociative Recombination, in *Atomic Collisions: A Symposium in Honor of Christopher Bottcher (1945-1993)*, (American Institute of Physics Press, New York, 1994), p.88.
17 S. L. Guberman, Phys. Rev. A **49**, R4277 (1994).

18 B. K. Sarpal, J. Tennyson and L. A. Morgan, J. Phys. B **27**, 5943 (1994).
19 H. H. Michels and R. H. Hobbs, Ap. J. **286**, L27 (1984).
20 C. Strömholm, J. Semaniak, S. Rosén, H. Danared, S. Datz, W. van der Zande, and M. Larsson, Phys. Rev. A **54**, 3086 (1996).
21 S. L. Guberman, Theoretical Studies of Dissociative Recombination, in *Thermophysical Aspects of Re-entry Flows*, ed. by J. N. Moss and C. D. Scott, (American Institute of Aeronautics and Astronautics, New York, 1986) pp.225-242; S. L. Guberman, Can. J. Phys. **64**, 1621 (1986).
22 W. Zong, G. H. Dunn, N. Djurić, M. Larsson, C. H. Greene, A. Al-Khalilil, A. Neau, A.M. Derkatch, L. Vikor, W. Shi, A. Le-Padellec, S. Rosén, H. Danared, and M. af Ugglas, Phys. Rev. Lett. **83**, 951 (1999).
23 Å. Larson, C. H. Greene, A. E. Orel, I. F. Schneider and A. Suzor-Weiner in *Dissociative Recombination: Theory, Experiment, and Applications IV*, ed. by M. Larsson, J. B. A. Mitchell and I. F. Schneider (World Scientific, Singapore, 2000), p.101.
24 Å. Larson and A. E. Orel, Phys. Rev. A **64**, 062701 (2001).
25 S. L. Guberman, J. Chem. Phys. **78**, 1404 (1983).

Dissociative Recombination of Electrons and Ions: The Early Experiments

Manfred A. Biondi
Dept. of Physics and Astronomy, University of Pittsburgh, Pittsburgh, Pennsylvania, 15260; biondi+@pitt.edu

1. INTRODUCTION

An account is given of laboratory studies made over 50 years ago that revealed a very fast electron-ion recombination process and the subsequent experiments that established the underlying mechanism. These studies were facilitated by development during World War II of major new technologies such as microwaves and fast electronics. Many of these developments took place at MIT in its classified Radiation Laboratory. In 1946, these extensive facilities were converted to the Research Laboratory for Electronics, where Physics Professors W. P. Allis and S. C. Brown formed the Microwave Discharge Group, initially with four graduate student experimenters.

Of the four students, three were assigned topics aimed at elucidating the mechanism of microwave breakdown, i.e., how ionization of the gas took place. The fourth, the author, was assigned as a thesis topic, "Electron Removal Processes in Ionized Gases". In the course of this and the follow-on research, his colleagues were S. C. Brown at MIT, T. D. Holstein and W. A. Rogers at Westinghouse Research Laboratories, and T. R. Connor and L. Frommhold at the University of Pittsburgh.

2. EXPERIMENTS

The approach adopted for these studies involved creating a discharge plasma using a microwave pulsed magnetron and then making observations in the post-discharge (afterglow) period. In order to deduce the electron removal processes, a reliable method for determining the electron density was required. At microwave frequencies and low gas pressures (< 30 Torr) the electrons in the plasma behave like a slightly lossy dielectric. If such a plasma is placed within a microwave cavity, its resonant frequency f_0 is perturbed in direct proportion to an appropriate average of the electron density n_e, i.e., $\Delta f/f_0 = C[n_e]_{av}$, where C is calculated from the spatial distributions of the microwave field and the electron density in the cavity.[1]

A minimal representation of the Microwave Afterglow Apparatus is shown in Fig. 1. The cavity, resonant at ~3 GHz, contains a quartz cylinder connected to a high vacuum gas handling system. A low pressure of noble gas, e.g., helium, is introduced to the cylinder and ionized by a pulse (~0.5 ms duration) of power from the high power magnetron that is repeated periodically (~20 ms). The electron density builds up during the discharge pulse and decays in the afterglow period when the magnetron is off. The electron density variation during the afterglow is determined by probing the cavity with low power from a tunable klystron. By measuring the resonant frequency of the cavity at various times in the afterglow, the variation of the electron density, $n_e(t)$, is determined.

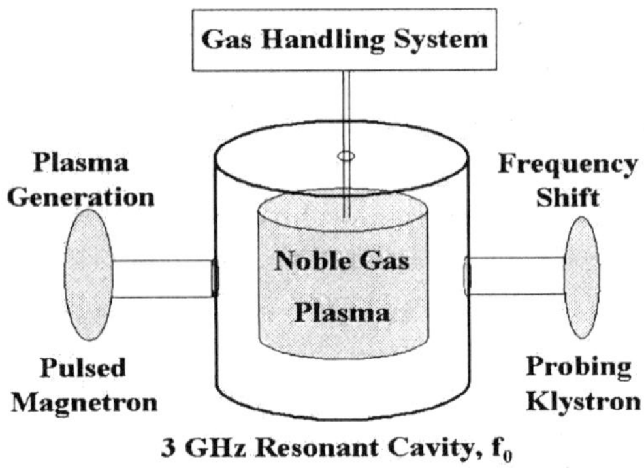

Figure 1. Schematic representation of the Microwave Afterglow Apparatus.

The electron continuity equation provides a means of determining the relevant atomic collision processes from the electron density variations. There are two limiting cases of interest; in the first case the only significant electron loss process is ambipolar diffusion of the electrons and ions. The continuity equation for the afterglow reduces to $dn_e/dt = D_a\nabla^2 n_e$, where D_a is the ambipolar diffusion coefficient. The solution of this equation for lowest mode diffusion loss is[2]

$$n_e(t) = n_e(0)\exp(-t/\tau) , \tag{1}$$

where the decay time $\tau = \Lambda^2/D_a$, with Λ the characteristic length of the plasma container. In the second limiting case, electron-ion recombination dominates, so that $dn_e/dt = -\alpha n_e n_+$ and, with $n_e \equiv n_+$, the decay is given by[3]

$$1/n_e(t) = 1/n_e(0) + \alpha t \,. \tag{2}$$

In the 1940s, it was generally believed that noble gas ions were atomic in nature (X^+, where X = He, Ne, ...). Thus, in low pressure, weakly ionized noble gases, the electrons and ions diffused together ambipolarly, they recombined very slowly (radiatively), and the electrons could not attach to the atoms to form stable negative ions. Therefore, it was concluded that ambipolar diffusion would be the only significant electron loss process in noble gas afterglows.

Helium was chosen for the initial afterglow studies because the gas samples were the purest obtainable, and it was thought to be the simplest gas to understand. The observed afterglow electron densities approximately obeyed the predicted ambipolar diffusion time decay (Eq. 1), with the decay time increasing linearly with gas density. The deduced D_a values agreed both with the values obtained from quantum calculations of the He^+/He resonant charge transfer and with those deduced from some ion mobility measurements.

Following this success with helium, neon afterglows were investigated. Here, the measured electron density decays differed completely from our expectations; the initial slopes were very steep, they did not follow an exponential time decay, and the decay curves did not vary appreciably with changes in gas pressure from ~5 – 30 Torr. Radiative electron-ion recombination was far too slow (time scale > 10 s) to explain the results. Months were spent repeating the experiments and attempting to reduce gas sample impurities, with no change.

Finally, we tried a "recombination plot" (Eq. 2) of the data, with the results[3] shown in Fig.2a. Recombination loss seemed clearly indicated, but the deduced recombination coefficient α exceeded 10^{-7} cm^3/s, some 10^5 times the radiative value. We next proceeded to argon, with similar results, as indicated in Fig. 2b. Also, upon re-examining the helium diffusion data, we found that the more rapid electron loss seen at high n_e could be explained by inclusion of recombination with $\alpha \sim 10^{-8}$ cm^3/s, an order of magnitude smaller than for neon and argon, but large compared to radiative recombination. When these very large recombination coefficients were discussed at the first two Gaseous Electronics Conferences in 1948 and 1949, the results were greeted with considerable scepticism. We therefore carried out experiments to demonstrate the reality of this very large electron-ion recombination loss.

One person who was not sceptical of our recombination results was D. R.

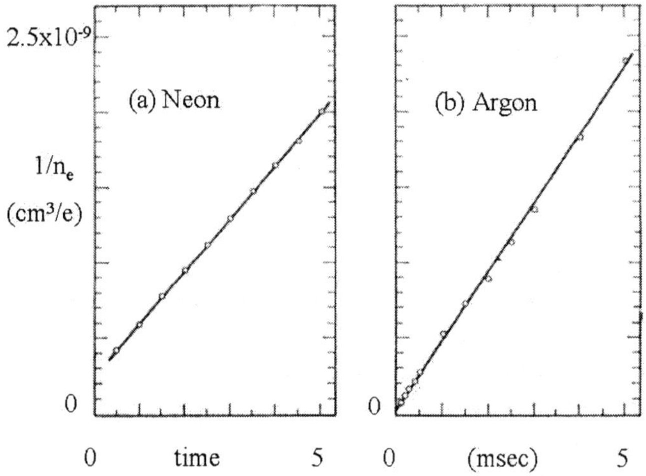

Figure 2. "Recombination loss" plots of electron density decays at T=300K in: a) neon at 24 Torr pressure and b) argon at 14 Torr.

Bates, a former student of H. S. W. Massey's. For many years Massey's group had been carrying out quantum calculations of a variety of fundamental atomic collision processes such as resonant charge transfer, radiative recombination, and electron attachment. They also were interested in the ionized layers in Earth's upper atmosphere that had recently been discovered through ionosonde determinations of atmospheric electron densities. They were puzzled by the rapid decrease in electron density in the E-layer after the sun set, cutting off the photo-ionization of the gas. Various attempts to explain the observations were advanced, e.g., electron attachment, but were subsequently rejected. Bates and Massey also considered dissociative recombination, but they thought it unlikely because of the weak direct coupling of the electronic and nuclear motions. The process was sufficiently speculative that Bates failed to mention it in a 1949 review of upper atmosphere processes. Owing to rather poor intercontinental and interdisciplinary communication in those days, we were as unaware of Bates's work as he was of ours.

Upon publication in June 1949 of our helium afterglow work[2] Bates responded[4] rapidly with a paper in which he suggested that the ions under study were He_2^+ and not He^+ and that the observed recombination was dissociative recombination. (Six months later, our December 1949 paper on recombination studies in neon, argon, etc.[3] made a much stronger case for his suggestion, which he further amplified.)[5] A. V. Phelps completely rebuilt the MIT microwave afterglow apparatus to include a small, magnetic mass spectrometer. He showed that the principal afterglow ions in helium initially were He^+ but that they converted to He_2^+ via 3-body collisions with two He atoms.[6]

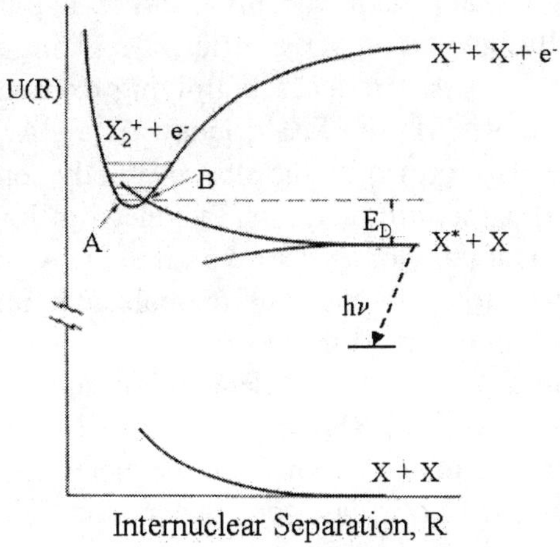

Figure 3. Simplified potential energy curves for the dissociative recombination process.

The dissociative recombination (DR) reaction proposed by Bates can be represented by the equation

$$X_2^+ + e^- \rightleftharpoons (X_2^{**})_{repulsive} \rightarrow X^* + X \quad (+E_D), \tag{3}$$

where X = He, Ne, etc. and E_D is the dissociation energy. A schematic representation of the process is given in Figure 3. A repulsive curve of a doubly excited state, X^{**}, crosses the molecular ion curve near its v = 0 vibration state (A). When an electron is captured by the molecular ion there is competition between autoionization and dissociation, but once the nuclei have separated beyond point (B), autoionization can no longer occur. In this case an excited and a ground state atom are the end products.

Experiments were devised to test Bates's proposal, based on the properties of the DR process, namely:

1) **molecular** ions are required,
2) diatomic ions lead to excited **atoms,**
3) the excited atoms have **kinetic energy** of dissociation.

The second property was established in spectroscopic studies of noble gas afterglows at Harvard by R. B. Holt and his students[7,8] and in work at Westinghouse Research Laboratories.[9] These studies showed that under conditions favoring molecular ions in the afterglow, only line spectra of the noble gas atoms were observed and their intensities varied as n_e^2, as expected for a recombination origin.

To demonstrate that **molecular** ions are required, studies were carried out first in pure argon and then in dilute argon-helium mixtures. Since discharges normally involve the most efficient ionizing path, in pure argon, fast-electron impact on Ar produces high-lying excited states $Ar^{*'}$, which then associatively ionize via the fast reaction $Ar^{*'} + Ar \rightarrow Ar_2^+ + e^-$. (The molecular ions are also formed by the slower 3-body conversion.) However, in dilute ($1{:}10^3$) argon-helium mixtures, fast-electron impact creates chiefly He^M, which then Penning ionizes Ar via $He^M + Ar \rightarrow He + Ar^+ + e^-$. Converson of these atomic argon ions to molecular ions should be slow, owing to the small argon concentration.

The results in pure argon, where molecular ions are expected, revealed the large recombination loss, as shown in Figure 2b. In the dilute Ar-He mixtures, where atomic ions are expected, the electron density decays were as shown[9] in Figure 4. In this case the final decays followed an ambipolar diffusion law (the initial rise in electron density was the result of Penning ionization by He^M left over from the discharge). The decay time exhibited the proper variation with gas pressure, and the deduced ambipolar diffusion coefficient was appropriate for Ar^+ ions moving in helium. No e-i recombination loss was discernable.

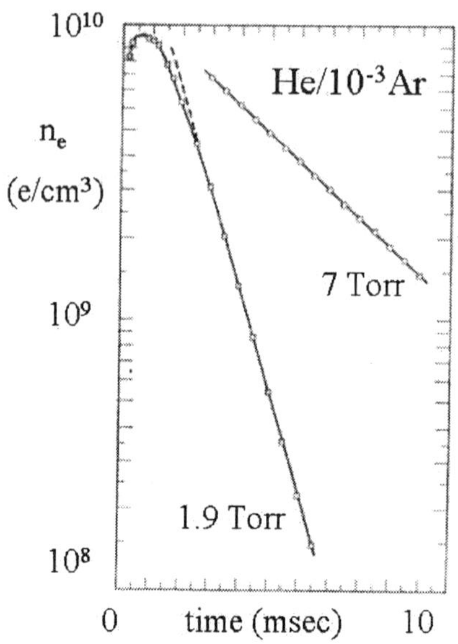

Figure 4. Electron density decays in dilute ($1{:}10^3$) Ar-He mixtures at T=300K.

The third property offers the most convincing test of whether DR is the operative process, i.e., can one detect kinetic energy of dissociation in the recombination products? If E_D is the order of 1 eV, then the radiating excited atoms should emit a line distinctly different in shape and broader than a thermal Gaussian profile.

We selected a Fabry-Perot interferometer (FPI) for the afterglow line profile determinations in order to provide both very high resolution and large optical throughput. A photomultiplier was employed as detector to obtain optimum sensitivity. A schematic representation of the FPI instrument[10] is shown in Figure 5. The wavelength λ of the central spot of the concentric-ring interference pattern is given by $N\lambda = 2\mu t$, where N is the interference order (typically ~30,000 in our experiments), μ is the refractive index of the medium between the etalon plates, and t is their spacing. Rather than "tuning" the wavelength by changing the plate spacing while maintaining extremely accurate parallelism, t is held constant while the refractive index μ is varied by changing the density of the gas between the plates. The etalon plates are mounted in a pressure housing, and argon gas from a high pressure tank passes through a needle valve at a constant rate, resulting in a linear change of λ with time.

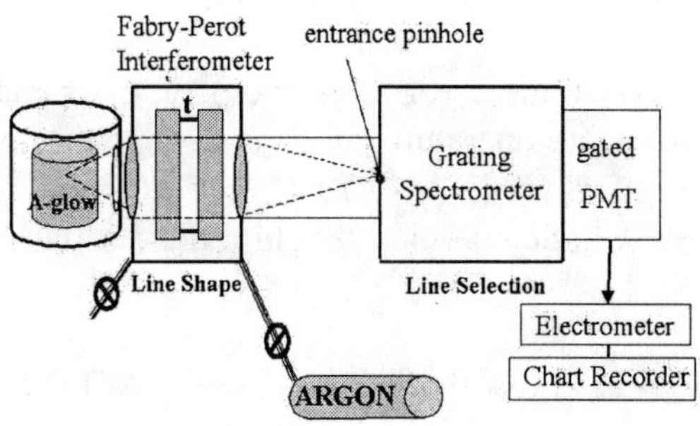

Figure 5. Simplified diagram of the refractive-index-tuned Fabry-Perot interferometer.

Radiation from the afterglow is collimated and passes through the etalon, then is focused on an entrance pinhole at the center of the interference pattern. In the early experiments, where only one or two spectral lines were studied, the desired line was selected by an interference filter placed between the afterglow cavity and the FPI housing. In the later work, where many (>20) lines were studied, a grating spectrometer of modest resolution and large throughput was used to select the desired line. The PMT detector was electronically gated to detect only the afterglow radiation; its analogue output was fed to an electrometer and displayed on a chart recorder.

The interferometric studies were carried out in three phases. The first phase involved studies of helium afterglows using an FPI of modest resolution (moderately flat plates, ~λ/30, and silver reflective coatings). The resolving power was sufficient to determine the line width but not its detailed shape. The helium 587.6 nm ($3\,^3D - 2\,^3P$) line was found to exhibit a modest broadening of the afterglow line compared to the discharge's near-thermal width.[11]

Since recombination in helium was not well established, the second phase of the work focused on neon, where the large electron-ion recombination rate and correlated excited atom emission had been well demonstrated. A new microwave afterglow apparatus with a greatly improved FPI (~λ/100 plate flatness and high reflectivity/low loss dielectric coatings) was employed in studies of the strong afterglow neon 585.2 nm ($2p_1 - 1s_2$) line at T = 300K. With this instrument it was possible to determine the detailed line shapes.

Rather different discharge and afterglow line profiles were expected. The short microwave pulse causes only modest heating of the neon gas in the cavity, so that the radiation from the excited atoms produced by fast-electron impact should exhibit a near-thermal (Gaussian) profile of the form

$$I_{th}(\nu) = I_0 \exp(-a\nu^2/T), \tag{4}$$

where ν is the wave number (or frequency) of the radiation (= 0 at line center), T is the atoms' temperature, and a is a group of physical constants.

During the afterglow, for $E_D >> kT$, it can be shown[11] that excited atoms having dissociation kinetic energies should radiate a flat-topped line profile of the form

$$I_{diss}(\nu) = I_0(\nu)\,[\mathrm{erf}\,(b\nu/T + cE_D) - \mathrm{erf}\,(b\nu/T - cE_D)], \tag{5}$$

where b and c are groups of physical constants. In practice, the afterglow line shape may be complicated by the fact that some of the fast excited atoms will transfer their excitation to slow atoms before they can radiate, i.e.,

$$X_f^* \rightarrow X_f^{*'} + h\nu \quad \text{(radiation)} \tag{6a}$$

$$X_f^* + X_{th} \rightarrow X_f + X_{th}^* \quad \text{(transfer).} \tag{6b}$$

As a result the afterglow line will be a composite of a fast and a slow contribution, as shown in Figure 6. The first interferometric studies of neon were carried out at T = 300K[12] and yielded 585.2 nm afterglow line profiles that exhibited the composite form suggested by Eqs. 6a and 6b.

The third phase of the interferometric studies involved a further refinement of the apparatus to permit studies of 22 neon lines and 5 argon lines emitted from gases that could be cooled to 77K.[13] Lowering the gas temperature reduced the thermal broadening effects, yielding sharper details

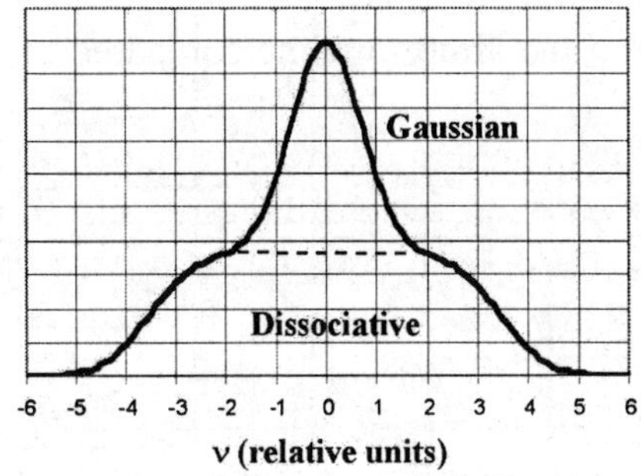

Figure 6. Theoretical spectral line profile for the dissociative recombination process.

of the line structure. Again, the neon transition most studied was the 585.2 nm line, since from the estimated binding energy of Ne_2^+ and the known Ne atomic levels, E_D was expected be in excess of 1 eV for this transition.

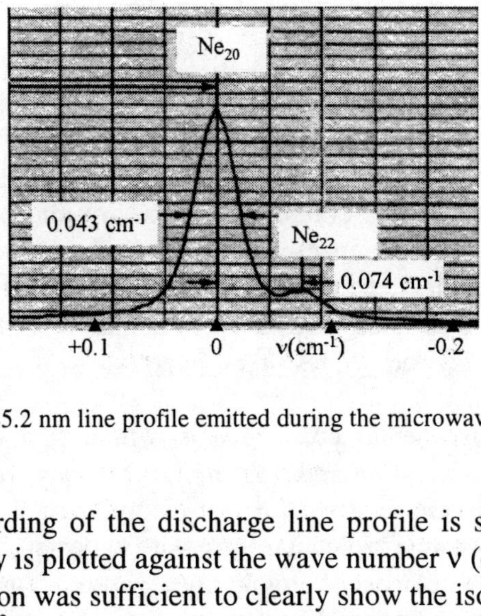

Figure 7. The neon 585.2 nm line profile emitted during the microwave discharge at T = 77K and p = 1 Torr.

The chart recording of the discharge line profile is shown in Figure 7, where the intensity is plotted against the wave number ν (defined by $\lambda\nu = 1$). The FPI's resolution was sufficient to clearly show the isotope shift between the ^{20}Ne and the ^{22}Ne atoms in natural neon, and the width of the line was consistent with a gas temperature well below room temperature.

During the afterglow the line profile shown in Fig. 8 was observed. The composite line shape indicated in the theoretical profile of Figure 6 is clearly seen when the distortion introduced by the 9% ^{22}Ne isotope is taken into account. The lowest practicable neon pressure (0.5 Torr) was used to minimize the excitation transfer to thermal neon atoms, with the result that only ~30% of the radiated profile was thermal.

While the 585.2 nm line profile represented one of the best examples of the theoretical dissociative shape (Fig. 6), all of the 22 neon and the 5 argon

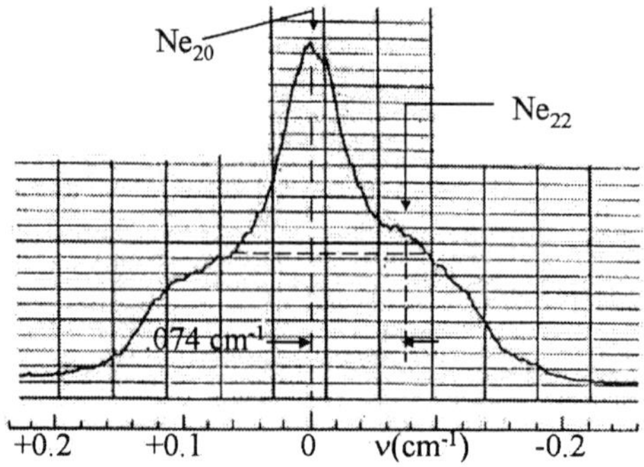

Figure 8. Afterglow 585.2 nm line profile at T = 77K and p = 0.5 Torr.

afterglow lines studied exhibited various degrees of broadening. Thus, these studies confirmed conclusively the third, most definitive property of the proposed dissociative recombination process.

3. SUMMARY

The introduction in the late 1940s of microwave afterglow techniques for studying fundamental atomic collision processes in ionized gases led to the discovery of a very fast electron-ion recombination process in noble gas afterglows. These results led to the proposal by D. R. Bates in 1950 that the operative process was dissociative recombination (DR). As a result, experiments were devised to test this hypothesis by examining several features of DR.

By creating experimental conditions in which first atomic ions were predominant in the afterglow and then molecular ions were, it was shown that molecular ions were required for fast recombination to occur. Next, in afterglows where diatomic molecular ions were expected, the recombination radiation was found to consist of atomic lines whose intensities varied as n_e^2, consistent with a recombination origin of the radiation following dissociation of the excited molecule. Finally, using high-resolution interferometry to determine the afterglow line profiles, a characteristic DR spectral line shape was observed. From these results, we concluded that DR was indeed the process responsible for the rapid electron loss.

Today, DR is seen as ubiquitous, playing a significant role in interstellar and interplanetary environments, in planetary upper atmospheres, in laboratory plasmas, and in various engineering applications. In response, new techniques - ranging from flowing afterglows to storage rings - have been added to the experimental methods applied to studies of the many facets of the DR process.

ACKNOWLEDGEMENTS

The author is greatly indebted to R. Johnsen and many other colleagues who have been of great assistance to him during the course of the researches described here. The preparation of this paper has been supported, in part, by the National Science Foundation under grant ATM-9906408.

REFERENCES

1 M. A. Biondi, Rev. Sci. Instr. 22, 500 (1951).
2 M. A. Biondi and S. C. Brown, Phys. Rev. **75**, 1700 (1949).
3 M. A. Biondi and S. C. Brown, Phys. Rev. **76**, 1697 (1949).
4 D. R. Bates, Phys. Rev., **77**, 718 (1950).
5 D. R. Bates, Phys. Rev., **78**, 492 (1950).
6 A. V. Phelps and S. C. Brown, Phys. Rev., **86**, 102 (1952).
7 R. B. Holt, J. M. Richardson, B. Howland, and B. T. McClure, Phys. Rev. **77**, 239 (1950).
8 A. Redfield and R. B. Holt, Phys. Rev. **82**,874 (1951).
9 M. A. Biondi, Phys. Rev. **129**, 1181 (1963).
10 M. A. Biondi, Rev. Sci. Instr. **27**, 36 (1956).
11 W. A. Rogers and M. A. Biondi, Phys. Rev. **134A**, 1215 (1964).
12 T. R. Connor and M. A. Biondi, Phys. Rev. **140A**, 778 (1965).
13 L. Frommhold and M. A. Biondi, Phys. Rev. **185**, 244 (1969).

Optical Spectroscopy of Recombining Ions in Flowing Afterglow Plasmas

R. Johnsen, M. Skrzypkowski, T. Gougousi, R. Rosati, and M. F. Golde[1]
Dept. of Physics and Astronomy and Dept. of Chemistry[1]*, University of Pittsburgh, Pittsburgh, Pennsylvania, 15260 USA*
rj@vms.cis.pitt.edu

1. INTRODUCTION

Spectroscopy has been an important tool in studies of dissociative recombination (DR) for many years. The early work by Biondi[1] and co-workers established the dissociative nature of the process by demonstrating the Doppler broadening of spectral lines due to the momentum that is imparted to the separating recombination fragments. In more recent work, spectroscopic methods have been used to identify products of DR of diatomic and polyatomic ions and to quantitatively measure yields for specified states of radiating recombination products. The formation of products in radiating states is of potential interest for the interpretation of planetary spectra and perhaps those observed from interstellar clouds, and of course, experimenters hope that identification of product states will lead to new insights into the mechanisms of recombination.

Both stationary or flowing afterglow plasmas are excellent media for studying emissions from recombining molecular ions. A good introduction and references to the relevant methods can be found in the review article by Adams[2] who carried out a survey of emission spectra from several recombining ions. Quantitative spectroscopy is considerably more difficult and hence data on quantitative yields of radiating species are still quite sparse. As was pointed out by Herbst,[3] in this meeting, absolute yields are needed for the interpretation of spectra from interstellar clouds. Such measurements are being carried out in our group at the University of

Pittsburgh and they are the principal subject of this contribution to the symposium.

It is acknowledged that, compared to spectroscopic afterglow methods, ion-storage rings (ISR)[4] are a more powerful tool for measuring complete product yields into all channels, but at this time it is not possible to carry out optical spectroscopy in ion storage rings due to the low concentration of ions and electrons. Thus, we view afterglow methods and the ISR techniques as complementary rather than as competing with each other.

In the following, we will use the term "spectroscopic yield" to denote the fraction of recombination events that results in formation of a specified electronic state. This yield is not necessarily identical to the nascent recombination yield of this product state since the state may also be populated by a radiative transition from a higher state that is formed in recombination. The nascent yield may be obtained by measuring and subtracting the cascading contribution, if any, from higher states. The procedure usually requires comparing photon count rates that are observed at widely different wavelengths, which in turn requires a precise knowledge of the wavelength-dependence of the photometric sensitivity of the optical detection system. The finer points will be discussed later.

2. STATIONARY AND FLOWING AFTERGLOWS

In a stationary (or static) afterglow system,[5] a plasma is generated by exposing a suitable gas mixture, generally a rare gas with small admixtures of other gases, to a pulsed microwave or other electrical discharge, or to an intense flux of ultraviolet radiation, or perhaps a high-energy electron beam. The microwave discharge is used most often. Upon termination of the discharge, the electron density decay is measured by using Langmuir probes, by measuring the frequency shift of a resonant microwave cavity mode, or by measuring the real part of the plasma conductivity using a radio-frequency method. Other diagnostic tools typically include a mass spectrometer that records the evolution of the relative ion concentrations as a function of afterglow time.

The spectroscopic measurements most often employ monochromators, sometimes interference filters, and, for line-shape measurements, Fabry-Perot interferometers. Photon counting techniques are used almost exclusively. The analysis of the plasma decay in stationary afterglows is simple when only a single-ion species is present and when ambipolar diffusion is very slow. In general, numerical solutions are required. Since the intensity of emissions that result from short-lived radiating recombination products is proportional to the square of the electron density, n_e, a simple but very useful test of the consistency of experimental observations consists of comparing the rates of decay of the electron density, the square root of the emission intensity, and the mass selected ion currents that are detected by the mass spectrometer (where available).

The stationary afterglow method has some strengths but also some weaknesses. One problem is that the molecular additives that are required to make the desired ions are exposed to an intensive discharge, which may result in dissociation, electronic and vibrational excitation, and formation of new chemical species. Vibrational excitation of the neutral gases can persist into the afterglow and produce heating of the electron gas and persistent vibrational excitation of the ions under study.[6] In pure rare gases, of course, such problems do not arise.

The flowing afterglow, to be described in detail later, has the virtue of allowing far better control over the ion-chemistry than is possible in the stationary afterglow. Here, the molecular gases are added downstream from the discharge region and they are never directly exposed to the discharge that typically operates in pure helium. Energetic particles, mainly metastable helium atoms, are safely converted to argon ions before molecular gases are added.

A brief note of warning to flowing-afterglow users: While helium metastables are indeed removed by Penning ionization with argon, the resulting atomic argon ions recombine with electrons by three-body recombination and form metastable argon atoms (with concentrations of 10 to 20% of the ion density) and those can lead to unexpected reactions. Also, there is some evidence that radiatively trapped ultraviolet photons from the discharge enter the reaction region. This seems to happen in spite of the usual precaution of putting a right-angle bend into the discharge inlet such that there is no line-of-sight path to the reaction region.

Modeling of the ion-kinetic processes in flowing afterglow plasmas is almost always required. In the work in Pittsburgh, we use computer models that include the relevant kinetic processes, ambipolar diffusion in gas flowing with the proper parabolic dependence of flow velocity on the radial coordinate, and an approximate treatment of the gas mixing downstream from the gas injection port.

2.1 Stationary Afterglow Data on Molecular Dimer Ions

We mention this subject here, in part, to draw attention to the problem of accounting for the anomalously large recombination coefficients of cluster and dimer ions. It is striking that weakly bound molecular dimer ions, such as $N_2^+ \cdot N_2$ and $CO^+ \cdot CO$ recombine much faster than the monomers. Bates[7] postulated that such ions undergo what he called "super dissociative recombination" and he ascribed the high recombination rate to the availability of dissociation channels into excited states of the molecular fragments. Bates proposed that the super-dissociative mechanism is also responsible for the high recombination rates of the weakly bound rare gas dimer ions (except for He_2^+), which were the first ions studied by Biondi.[1]

Partly in response to Bates's suggestion, Cao and Johnsen[8] carried out measurements on $N_2^+ \bullet N_2$ recombination in a high-pressure (several hundred Torr) stationary afterglow system that was well suited for this work because the high pressure promotes rapid conversion of the monomer to dimers ions. In addition, $CO \bullet CO^+$ ions were studied (unpublished). The recombination rates indeed were found to be very large (~2.5 x 10^{-6} cm^3/s) and both exhibited strong emissions of molecular bands. In the case of N_4^+, the second positive band system of N_2 ($C^3\Pi_u$ – $B^3\Pi_g$) was observed while three band systems were found to arise from recombination of $CO \bullet CO^+$, the third positive band ($b^3\Sigma$ –$a^3\Pi$), the Ångström band ($B^1\Sigma$-$A^1\Pi$), and the Herman band ($e^3\Sigma$-$a^3\Pi$). The origin of the emissions was clearly recombination, as evidenced by their disappearance on adding an electron scavenger (SF_6) and the observed scaling of intensities with n_e^2.

The observation of the molecular bands supports the conjecture of Bates, but since absolute yields were not determined it has not been shown that the observed dissociation channels are dominant and responsible for the large recombination coefficient. The same statement applies to the recombination of rare gas dimer ions; we still do not know if the spectroscopically observed states are indeed dominant. It would be interesting to perform measurements on dimer ions in an ion storage ring experiment to see if (a) similar high rate coefficients are found and (b) if other product channels exist.

3. ABSOLUTE SPECTROSCOPY IN FLOWING AFTERGLOWS

We now turn our attention to measurements of absolute yields. The flowing afterglow system[9] that we used has the configuration shown in Fig. 1. Metastable helium atoms, produced in a microwave discharge in helium, are converted by Penning ionization to Ar^+ ions by downstream addition of argon (density ~ 15% of that of helium). Ion and electron densities at this point are ~ 4×10^{10} cm^{-3}. The reagent gases that initiate formation of the desired molecular ions are added through movable and fixed reagent inlets.

The diagnostic tools include a quadrupole mass filter that analyzes the ion composition of the plasma at a point downstream from the reagent inlet, and a movable Langmuir probe (diameter = 25μm, length = 3-4 mm) that measures electron densities. The Langmuir probes are used in the electron collecting mode since a series of tests[10] showed that ion collecting probes produce misleading results.

Emission spectroscopy of the plasma is carried out by a movable (along the direction of the flow) monochromator/phototube system (McPherson 0.3m monochromator, Hamamatsu R943-02 phototube) that views the plasma through a long quartz window. The system covers the spectral range from 200 to 900 nm. An in-situ absolute calibration of the optical systems was performed by measuring the intensity of the $N_2(C^3\Pi_u\text{-}B^3\Pi_g)(0,0)$ band at 337 nm, excited by the reaction

$$Ar^* + N_2 \rightarrow Ar + N_2(C^3\Pi_u) \quad \text{(yield } 0.80\pm0.20\text{))}.$$ [11,12]

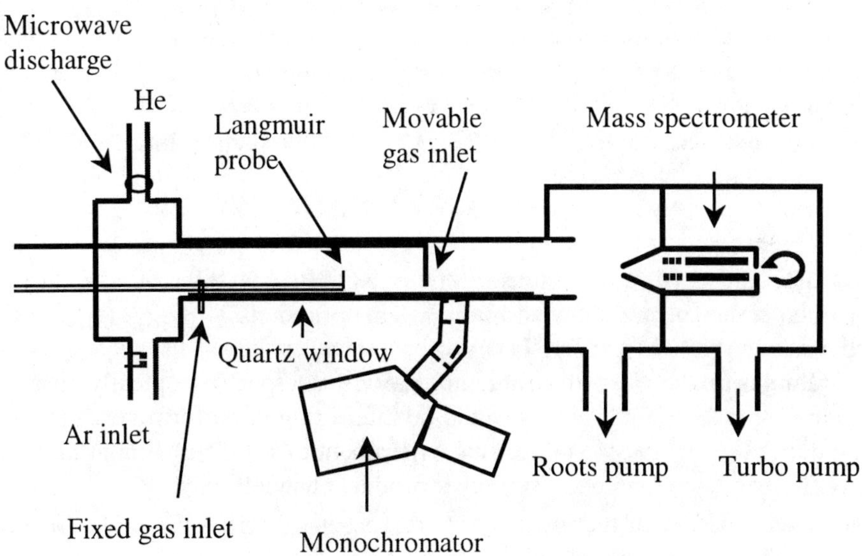

Figure 1. Schematic diagram of the flowing-afterglow tube.

The flux of Ar* was determined by converting the Ar* to NO^+,

$$Ar^* + NO \rightarrow Ar + NO^+ + e \quad \text{(yield } 0.35\pm0.10\text{)}$$ [13,14],

and using the Langmuir probe to measure the electron density increase due to this reaction. The spectral response of the system at other wavelengths was measured using a calibrated tungsten lamp in the red, a deuterium lamp in the UV to visible range, and joining the two response curves in the overlapping spectral range.

To distinguish emissions from recombining ions from those of extraneous sources, an electron scavenging gas (SF_6) was added.

The methods of analyzing flow-tube data are very well known and will not be described here. We use computer models that allow accurate modeling of diffusion, gas mixing, ion-molecule reactions, recombination, and product formation. Some details of these methods have been presented in a recent publication.[9]

3.1 Emission spectra and yields from DR of N_2O^+, HCO_2^+, N_2OH^+, HCO^+/COH^+, and CO_2^+

In this section, we give a few examples of recent and ongoing flowing afterglow measurements of spectroscopic yields from the recombination of several ions. Due to space limitations, we cannot reproduce Figures of all spectral scans. Some may be found in a recent book chapter.[15]

Many recombining ion species were found to produce measurable spectral emissions but the absolute yields of the radiating states were often found to be quite small.

For instance, the yield for $NO(B^2\Pi_i)$ from the recombination

$$N_2O^+ + e \rightarrow NO(B^2\Pi_i) + N(^4S),$$

as determined from the intensity of the NO ($B^2\Pi$-X $^2\Pi$) band, is only 0.6 percent. The spectra show additional weak emissions of the N_2 ($C^3\Pi$ - $B^3\Pi$) bands and the NO ($A^2\Sigma^+$- $X^2\Pi$) bands.

Similarly, the yield of the recombination branch,

$$HCO_2^+ + e \rightarrow OH(A^2\Sigma^+) + CO(X^1\Sigma^+),$$

as determined from the intensity of the observed $OH(A^2\Sigma^+\text{-}X^2\Pi_i)$ (v=0 and v=1) band system, is also quite small (1.5%).

An interesting result of this work, which still awaits a complete analysis, is the finding that both NH ($A^3\Pi_i$-$X^3\Sigma$) and OH ($A^2\Sigma^+$-$X^2\Pi_i$) spectra are emitted in recombining plasmas containing N_2OH^+ ions. This suggests that the emissions arise from recombination of the two isomeric species $HNNO^+$ and $NNOH^+$:

$$NNOH^+ + e \rightarrow OH(A^2\Sigma^+) + N_2(X^1\Sigma_g^+) \quad (2\ \%)$$
$$HNNO^+ + e \rightarrow NH(A^3\Pi_i) + NO(X^2\Pi_r) \quad (1\text{-}5\ \%).$$

The intensity of the NH band decayed far more rapidly with axial position than that of OH. This observation supports the assumptions that the bands arise from two distinct ions species, while ruling out the alternative explanation that recombination involves a major rearrangement of the molecule. The yields of both radiating states are again quite small, 2% for $OH(A^2\Sigma^+)$ and 1-5% for $NH(A^3\Pi_i)$.

Recombination of HCO^+ and its isomeric form COH^+ are interesting processes for astrophysical applications. In a still ongoing experiment, we are studying the emission spectra that arise from the dissociative recombination of HCO^+/COH^+ by producing a mixture of those ions using a reaction scheme that starts from

$$Ar^+ + H_2 \rightarrow ArH^+ + H$$

and

$$ArH^+ + H_2 \rightarrow H_3^+ + H.$$

By adding CO and H_2 the subsequent reactions,[16]

$$\begin{aligned} H_3^+ + CO &\rightarrow HCO^+ + H_2 \qquad (94\%) \\ &\rightarrow COH^+ + H_2, \qquad (6\%) \end{aligned}$$

and the reactions

$$\begin{aligned} ArH^+ + CO &\rightarrow Ar + HCO^+ \\ &\rightarrow Ar + COH^+ \\ Ar^+ + CO &\rightarrow Ar + CO^+ \end{aligned}$$

and

$$\begin{aligned} CO^+ + H_2 &\rightarrow HCO^+ + H \qquad (\sim 50\%) \\ &\rightarrow COH^+ + H \qquad (\sim 50\%) \end{aligned}$$

produce both isomers.[17] Kinetic modeling, based on currently available branching fractions, shows that the fractional concentration of the more energetic COH^+ isomer should remain small (several %) under our conditions. Conversion of COH^+ to HCO^+ occurs by the reactions[17]

$$COH^+ + CO \rightarrow HCO^+ + CO$$

and

$$COH^+ + H_2 \rightarrow HCO^+ + H_2.$$

The intensity of the wavelength integrated UV spectra show that recombination of HCO^+/COH^+ results in $CO(a^3\Pi_r)$ with an effective yield (including cascading from higher states) of (26±10)% for a mixture of the two isomeric forms. Since HCO^+ is far more abundant, the $CO(a^3\Pi_r)$ product must predominantly arise from HCO^+. However, several intense visible emissions (e-a, d-a, a'-a) from higher-lying states of CO are observed also. It is energetically possible to populate the e and d states in recombination of COH^+ but not in recombination of HCO^+.

Fig. 2 shows a spectrum of the overlapping visible emissions from the higher-lying states of CO. Clearly, the spectrum contains much information but it is difficult to identify the origin of particular features. It helps to investigate how the spectrum changes with the axial position in the flow tube ('afterglow time'). Measurements show that emissions from the d state and from a'(v>4) levels decay much faster than those originating from the a' (v≤4) levels. This seems to indicate that the d state is populated by

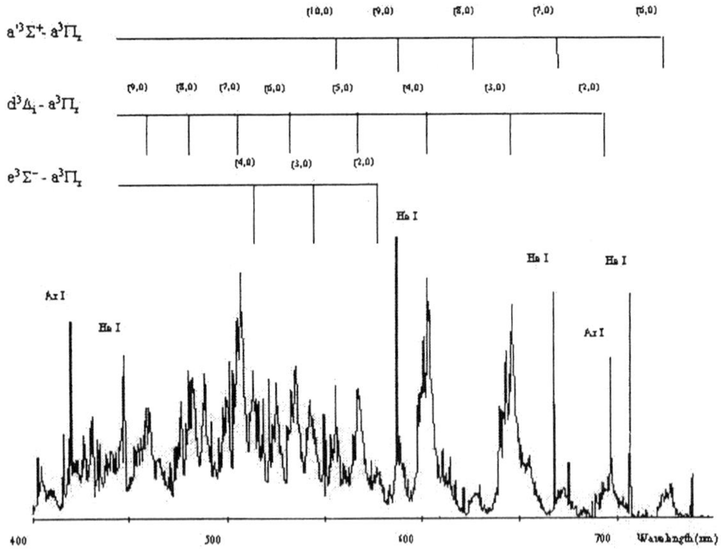

Figure 2. Spectral scan from 400 to 700 nm of CO(e-a, d-a, a'-a) band emissions from recombination of a mixture of HCO^+/COH^+ ions. The atomic helium and argon lines are due to scattered light from the discharge and upstream region of the plasma.

recombination of COH^+ while different vibrational levels of a' may be populated by recombination of HCO^+ and COH^+. A faster decay of COH^+ is to be expected since this isomer is destroyed (converted to HCO^+) in the reactions with H_2 and CO. The data available at the time of this writing are compatible with the assumption that both isomers recombine with the same rate, but more data are needed for different conditions before a firm conclusion can be reached.

Our earlier published work on CO_2^+ recombination[18] indicated substantial emissions from the long-lived $a^3\Pi_r$ state. This confirmed earlier work.[19,20] It was concluded that this state is populated directly by the recombination channel

$$CO_2^+ + e \rightarrow CO(a^3\Pi_r) + O(^3P)$$

and, in addition, by cascading from higher states ($a'^3\Sigma^+$, d $^3\Delta_i$, $e^3\Sigma^-$) but the magnitude of the cascading contribution was not determined at that time. The spectroscopic yield (~29%) of $CO(a^3\Pi_r)$ was only about one half of that obtained in the earlier work (55%).[19,20] After a more thorough calibration of the optical system and additional model calculations, we concluded that cascading contributes from 50 to 100 % to the long-lived CO(a). The

combined yield for the states a', d, and e was found to be (22±7)%. Using the relative yields measured by Tsuji et al.,[21] the absolute yields would be 13 % for a', 8% for d, and 1.7% for e. The uncertainty in the cascading contribution to CO(a) is still rather large. One problem is that our computer models do not accurately reproduce the z-dependence of the measured intensities in the visible range, especially in the vicinity of the reagent inlet. Recent modifications of the gas inlet and a reduction of light scattering in the flow tube have improved the situation and we hope to be able to arrive at a more accurate determination of the cascading contributions.
(*Note added in proof*: Improved measurements, carried out during the year after the original submission of this paper, showed that the cascading contribution to CO(a) formation is only (13±3)%. The partial yields for the a', d, and e states given above should be reduced by a factor 0.6.)

4. NO EMISSIONS FROM DR OF H_3^+?

We had expected to observe emissions from H_3^+ recombination since some experimental work suggested that such emissions should be present (see also discussion and references in Gougousi et al.[22]). The very recent work by Glosik and coworkers in Prague[23] strongly suggests that the recombination of H_3^+ proceeds via a long-lived H_3 complex which can reasonably be expected to undergo radiative transitions, as had been suggested earlier.[24] However, no spectral emissions that could be ascribed to dissociative recombination of H_3^+ were found by us in the wavelength range from 200 to 900 nm. Adams[24] reported a similar negative finding. It seems ironic that the first observations of the H_3 spectrum by Herzberg and coworkers (see e.g. Dabrowski and Herzberg[25]) were ascribed to recombination. At this time, the most likely explanation of the origin of the spectra observed by Herzberg and coworkers may be that the process responsible was dissociative recombination of H_5^+ ions rather than of H_3^+. In this context, the observation of line broadening in some H_3 rotational lines by Miderski and Gellene[26] is very interesting. The authors ascribed the broadening to the energy released in dissociative recombination of H_5^+ into H_2 and H_3*. It would be interesting to repeat these experiments in a cooled flowing afterglow experiment and with higher-resolution interferometric techniques.

5. CONCLUSIONS

The spectroscopy of recombining afterglow plasmas provides a rich and important tool for investigating dissociative recombination process. However, the interpretation of observed spectra is subtle and complicated, more so if one seeks to obtain absolute yields for particular states. In many cases, the yields for radiating states are small. That does not make them uninteresting since the observed spectral features may still be useful for diagnostic purposes in laboratory and space plasmas. One also quickly discovers that the recombining plasma may contain mixtures of isomeric forms of the same ion and that those can give rise to different spectra. While this complicates the interpretation further, it also allows studies of the different recombination behavior of different isomeric forms, which are difficult to do by other methods.

The recombination mechanism of H_3^+ ions received much attention during this symposium but it remains an 'enigma.' The absence of measurable emissions may contain some hints but we do not know in which direction they point.

ACKNOWLEDGMENT

This work was, in part, supported by the NASA Planetary Atmospheres Program.

REFERENCES

1 M. A. Biondi, this volume, p. 13.
2 N. G. Adams, Spectroscopic Determination of the Products of Electron-Ion Recombination in *Advances in Gas Phase Ion Chemistry*, Vol. 1 (JAI Press, 1992), p. 271-310.
3 E. Herbst, elsewhere in this volume.
4 S. Datz, G. Sundström, C. Biederman, L. Broström, H. Danared, Phys. Rev. Lett. **77**, 4891 (1995).
5 M.A. Biondi, this volume, p. 13.
6 R. Johnsen, Int. J. Mass Spectr. and Ion Proc. **81**, 67 (1987).
7 D. R. Bates, J. Phys. B **24**, 703 (1991).
8 Y.S. Cao and R. Johnsen, J. Chem. Phys. **95**, 7356 (1991) and Y.S. Cao, PhD thesis, University of Pittsburgh (1991).
9 T. Gougousi, R. Johnsen, M. F. Golde, J. Chem. Phys. **107**, 2430 (1997).
10 R. Johnsen, E.V. Shun'ko, T. Gougousi, and M.F. Golde, Phys. Rev. E **50**, 3994 (1994).
11 J. H. Kolts, H. C. Brashears, and D. W. Setser, J. Chem. Phys. **67**, 2931 (1977).
12 N. Sadeghi and D. W. Setser, Chem. Phys. Lett. **82**, 44 (1981).
13 M. F. Golde, Y.-S. Ho, and H. Ogura, J. Chem. Phys. **76**, 3535 (1982).
14 M. T. Jones, T. D. Dreiling, D. W. Setser, and R. N. McDonald, J. Phys. Chem. **89**, 4501 (1985).
15 R. Johnsen, M. Skrzypkowsi, T. Gougousi, and M. F. Golde, Spectroscopic Emissions from the Recombination of N_2O^+, N_2OH^+/HN_2O^+, CO_2^+, CO_2H^+, HCO^+/COH^+, H_2O^+, NO_2^+, HNO^+ and LiF Measurements of the H Atom Yield from H_3^+ in *Dissociative*

Recombination, Theory, Experiment and Applications ed. by M. Larsson, J. B. A. Mitchell, I. F. Schneider (World Scientific. Singapore, 2000), p. 200.

16 A. J. Illies, M. F. Jarrold, and M. T. Bowers, J. Am. Chem. Soc. **105**, 2562, 1982).

17 C. G. Freeman, J. S. Knight, J. G. Love, M. J. McEwan, Int. J. Mass Spec. and Ion Proc. **80**, 255 (1987).

18 M. P. Skrzypkowski, T. Gougousi, R. Johnsen, M. F. Golde, J. Chem. Phys. **108**, 8400 (1998).

19 T. S. Wauchop and H. P. Broida, J. Chem. Phys. **56**, 330 (1972).

20 G. Taieb and H. P Broida, Chem. Phys. **21**, 313 (1977).

21 M. Tsuji, K. Kobarai, H. Kouno, H. Obase, and Y. Nishimura, J. Chem. Phys. **94**, 1127 (1991).

22 T. Gougousi, R. Johnsen, and M. F. Golde, Int. J. Mass Spec. and Ion Proc. **149/150**, 131 (1995).

23 R. Plašil et al., this volume, p. 249.

24 N. G. Adams, private communication.

25 I. Dabrowski and G. Herzberg, Can. J. Phys. **58**, 1238 (1980).

26 C. A. Miderski and G. I. Gellene, J. Chem. Phys. **88**, 5331 (1988).

Technique for Distinguishing and Determining the Origin of Photon Emissions from He^+/Ar^+ Plasmas

Recombination Emissions with the Addition of OCS, CS_2 and H_2S

Nigel G. Adams, Toufik Mostefaoui, and Lucia M. Babcock
Department of Chemistry, University of Georgia, Athens, GA 30602
adams@chem.uga.edu

1. INTRODUCTION

Experimental studies of electron-ion recombination have traditionally been made in three ways: (1) the determination of recombination rate coefficients, some as a function of temperature, (2) the identification of ground state neutral products and (3) the detection of excited state products. Initial studies of (1) were made in stationary afterglows, predominantly by Biondi and co-workers[1] and Smith and co-workers[2] and were followed by investigations using the more versatile flowing afterglow.[3] In parallel with the latter studies, emission spectroscopy was used primarily for detecting excited state products in ion- and metastable-neutral reactions,[4-6] but with some studies in which excited products of recombination were detected. Studies were later made by Adams and co-workers[7,8] and by Johnsen and co-workers[9] to detect the ground state products of recombination, a more difficult problem but where substantial progress has been made. At the same time, determination of product distributions was also starting in storage rings.[10,11] In this area, storage ring studies have been more numerous, due predominantly to the degree of effort put into the technique by a number of workers. Also, with this method, it is possible to inject a single ion species into the ring, and for heteronuclear ions to radiationally cool the vibrations.[12] In the flowing afterglow, it is more complicated to create a single ion plasma, ensure that the ions are vibrationally relaxed and to detect all of the products simultaneously. Where the same systems have been studied using both techniques, there is a measure of agreement, but with some discrepancies.[13]

Where the flowing afterglow has been more effective than storage rings has been in the detection of excitation (electronic and vibrational) in the recombination products. However, again, as the ion species studied become more complicated, there is a greater complication in creating a single ion plasma and being able to distinguish between recombination emissions and emissions generated from the ion-molecule reactions producing the recombining ions and those from the ionisation source. Some identification of the emission source has been possible by using electron-attaching gases (i.e., SF_6 and CCl_4) to rapidly remove electrons from the plasma and quench electron-ion recombination emissions. Examples of this are the detection of $OH(A^2\Sigma^+)$ from O_2H^+ and N_2OH^+ + e,[14] $NH(A^3\Pi_i)$ from N_2OH^+ + e,[14] $CO(a^3\Pi_r)$ from HCO^+ + e,[15,16] $N_2(B^3\Pi_g)$ from N_2H^+ + e[15,16] and $CO(a^3\Pi_r)$ from CO_2^+ + e.[17] Such studies require a plasma containing only one recombining ion and exceptional stability in the apparatus was required over spectral scans with and without the attaching gases added. Any variations in the background emissions between the two scans, which can take more than an hour, would result in spurious signals. In any event, such an approach limits the minimum recombination signal that can be detected.

To reduce these limiting effects, a technique has been developed in which reactive gases are pulsed into the flowing afterglow. Pulsing with even a low repetition rate of ~5 Hz means that the effects of drifts over a time scale of > 40ms will be eliminated. Also, by judicious choice of the reactive gas, emissions from specific reactions or reaction sequences can be isolated. The technique has been initially tested in a He^+/Ar^+ plasma to isolate emissions from reactions originating from an Ar^+ precursor, by pulsing H_2, which is very reactive with Ar^+ but unreactive with He^+. Addition of OCS, CS_2 and H_2S generating emissions downstream of the H_2 injection point, provided stringent tests of the technique. Thus, the difference in the emissions between H_2 absent and present (Ar^+ present and absent) would eliminate a multitude of emissions from an Ar^+ precursor. A further benefit of the technique is that the Ar^+ reaction with H_2 results in the production of H_3^+. Thus, additional emissions resulting from this precursor would register as a negative signal in the difference spectrum and be separately identifiable.

2. PULSED GAS TECHNIQUE

Details of the technique and the pulsed valve operation have been discussed elsewhere,[18,19] so only an overview of the salient features is given here. A simplified schematic of the apparatus is given in Figure 1. He is passed though a microwave cavity discharge producing He^+ and He^m, which

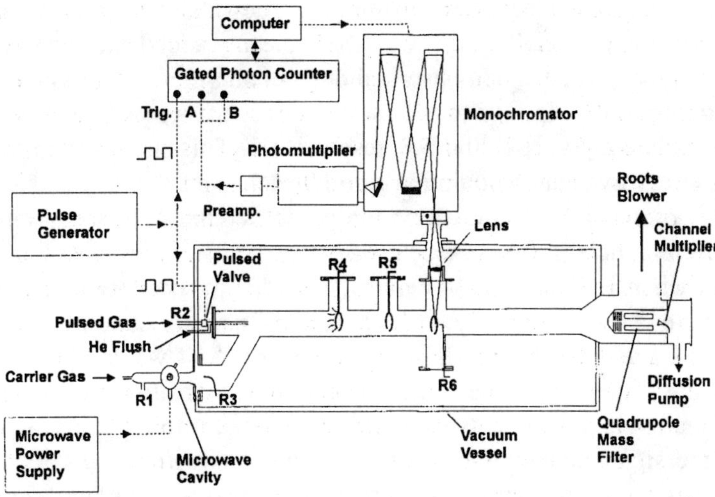

Figure 1. Simplified schematic of the flowing afterglow with pulsed gas and spectroscopic capabilities (R1 to R6 are reactant gas inlet ports).

move along the flow tube under the action of a Roots blower pump. Ions are detected downstream by sampling through a small orifice in a nosecone, mass analysed and counted after amplification with a channel multiplier and a pre-amplifier. He^+ slowly associates to give He_2^+ and, after Ar addition, He^m and He_2^+ react to produce Ar^+ whilst He^+ remains unaffected. Further downstream, in the viewing region of an optical monochromator, a gas ABC is added (variously OCS, CS_2 or H_2S) which reacts with He^+ and Ar^+ giving a variety of ionic products, some of which are likely to recombine with electrons. In any of these processes, where energetically possible, photon emission from excited states may occur. The emissions generated in the viewing region are dispersed in the monochromator and counted into both channels, A and B, of a dual channel gated photon counter, after amplification in a cooler photomultiplier. The initial composition of the reacting plasma is modified by pulsed addition of H_2 which converts Ar^+ to H_3^+ with the He^+ being largely unchanged by virtue of the very small rate coefficient for reaction with H_2 (1(-13) cm^3s^{-1}).[20] Because of the high reactivity of Ar^+ and the product ions in the reaction sequence with H_2, only a very small flow of H_2 is required to convert all of the Ar^+ to H_3^+ before the ABC addition point is reached. This flow is insufficient to cause significant reaction with H_2 in the viewing region of the product ions of the ABC reaction.

In order to get a clear demarcation between H_2 being absent and present in the flow tube, a He flush gas is used to transport the H_2 from the calibrated capillary (used to measure the average flow rate of H_2) and provide sufficient pressure for the H_2 to move efficiently through the pulsed valve. A second

He flush gas is introduced immediately after the valve to rapidly remove the H_2 after closing the valve. In this way, H_2 is cleanly added and removed. That this is possible, has been demonstrated by Williams et al.[19] by monitoring the modulation of He_2^+, H_3O^+ and SF_6^- at the downstream mass spectrometer by pulse modulating Ar, H_2O and SF_6 respectively. This showed that, even for sticky gases, very clean modulation could be achieved.

The two channels, A and B, of the photon counter are separately gated with different delays from the pulsed valve actuation, so that they count emission when H_2 is absent (A) and present (B) in the flow tube. The gate widths were chosen to be 70% of the on and off periods and these were made equal. This avoided counting in regions where the H_2 was rapidly changing and ensured that the data acquisition was insensitive to gate delay. Discrimination levels in the two counters were made closely equal at –10 mV so that the signal sensitivities were the same. Any difference will result in apparent signal even when the count rates in the two gating periods are equal. Also, any differences in the gate widths will have a similar effect. These two effects limit the maximum sensitivity of the technique. At each wavelength in a spectral scan, counts in channels A and B are accumulated until a statistically significant count in the difference, S, is achieved. The count period at each wavelength is chosen to be that necessary to achieve a pre-determined S/dS, where dS is the statistical noise, and this is given by

$$\frac{S}{dS} = \frac{N_A - N_B}{\sqrt{N_A + N_B}}\sqrt{\tau} \qquad (1)$$

where N_A and N_B are the count rates into channels A and B respectively and τ is the count period.[18] This ensures that the statistical noise on individual signal peaks is the same fraction of the peak height independent of the absolute magnitude of the S signal. Distinct situations are possible. If N_A, N_B, and S are similar, only a short count period is required. However, since the τ to achieve a pre-determined S/dS increases as N_A and N_B become larger and S becomes smaller, the times to accumulate data for very small, and thus less significant, features will become very long. Therefore, on data collection, a decision has to be made to determine whether there is a significant signal peak or not. This can be done by determining S/dS as a function of time, since statistically it can be shown[18] that

$$\frac{S}{dS}(t) = \frac{x}{\sqrt{\tau}}\sqrt{t}. \qquad (2)$$

Thus, by extrapolating this curve, it can be determined whether the pre-determined value of S/dS can be achieved in an acceptable time T. If $\tau > T$, it

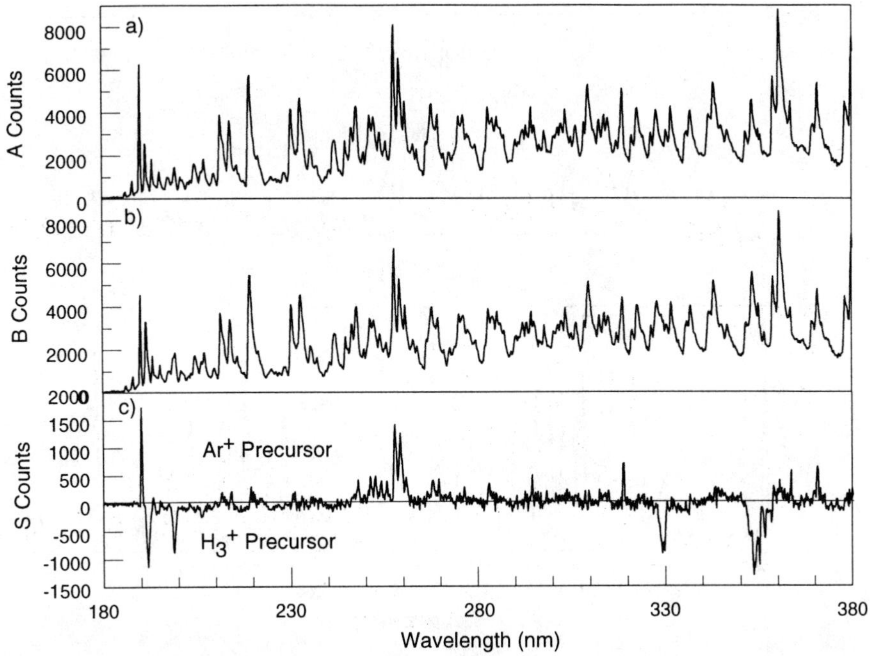

Figure 2. Spectral scan of the A, B and S counts for a $He^+/Ar^+/OCS$ plasma with H_2 modulation at 5 Hz. The signal, S, spectrum separately shows features resulting from Ar^+ and H_3^+ ion precursors.

is concluded that there is no significant signal and the count is terminated and the approximate noise count rate determined. If $\tau < T$, then the count is continued until the pre-determined value of S/dS is obtained and the count rate again determined (but in this case it is larger and statistically significant). The wavelength is then stepped and the procedure repeated for the next value of wavelength in the scan. In this way, spectra such as that given in Figure 2 are obtained.

3. SPECTRAL ANALYSIS

The spectra in Figure 2 dramatically illustrate how the signal from reactions resulting from a single precursor can be recovered using the pulsing technique. The signal S can be obtained when it is both comparable to A and B and when it is much smaller. Noise levels in the signal spectrum are achieved even when A and B are large and the level of the noise bears little relationship to their magnitudes. Figure 3 shows a magnified region of the S spectrum together with the S/dS values with good correlation both where there is signal and where signal is absent, i.e., only noise. For this spectrum, the reactions that can contribute to the emissions are

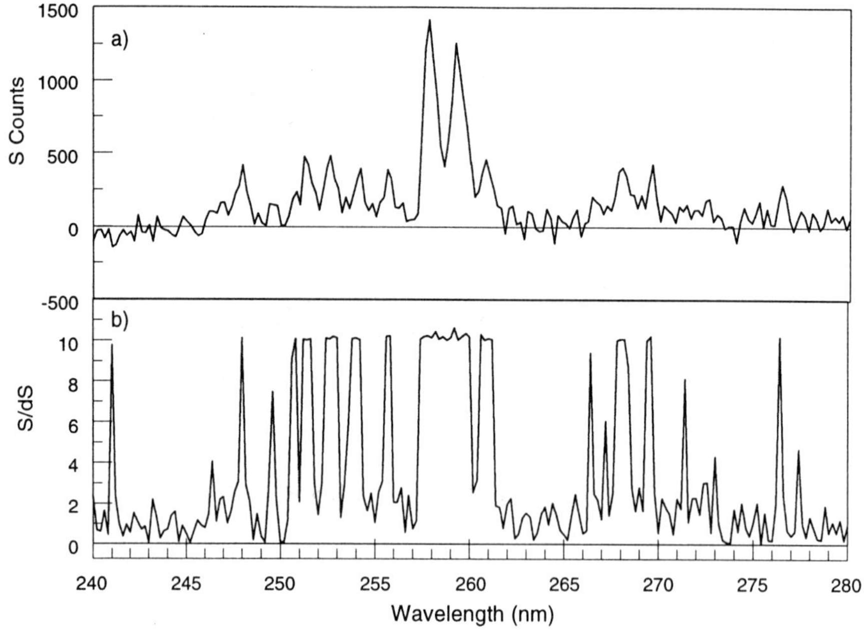

Figure 3. Magnification of a section of the signal, S, spectral scan in Figure 2, together with the corresponding values of S/dS. When significant signal is present, the pre-determined value of S/dS (= 10 in this case) is achieved.

$$Ar^+ + OCS \rightarrow S^+ + CO + Ar - 0.054\ eV \quad 95\% \tag{3a}$$
$$\rightarrow OCS^+ + Ar + 4.58\ eV \quad 5\% \tag{3b}$$

with a rate coefficient of 1.3(-9) cm^3s^{-1}. All kinetic data are taken from the compilation by Anicich[20] unless otherwise stated and energetics are calculated using data from the NIST Webbook.[21] This is followed by

$$OCS^+ + e \rightarrow S + CO + 8.018\ eV \tag{4a}$$
$$\rightarrow SO + C + 2.266\ eV \tag{4b}$$
$$\rightarrow CS + O + 4.258\ eV \tag{4c}$$

for the Ar^+ precursor (note that all energetically possible channels are shown for the OCS^+ and other recombination reactions and are not necessarily significant channels). With H_2 present, the reactions are

$$H_3^+ + OCS \rightarrow HOCS^+ + H_2 + 2.138\ eV \quad > 90\% \tag{5a}$$
$$\rightarrow HCO^+ + H_2S + 1.706\ eV \quad < 10\% \tag{5b}$$

with k = 1.9(-9) cm^3s^{-1},

$$HCO^+ + e \rightarrow H + CO + 7.441\ eV \quad (6)$$

and

$$HOCS^+ + e \rightarrow H + OCS + 7.086\ eV \quad (7a)$$
$$\rightarrow OH + CS + 4.601\ eV \quad (7b)$$
$$\rightarrow SH + CO + 7.612\ eV \quad (7c)$$
$$\rightarrow CH + SO + 1.701\ eV \quad (7d)$$
$$\rightarrow O + HCS + 2.219\ eV \quad (7e)$$
$$\rightarrow C + HSO + ? \quad (7f)$$
$$\rightarrow S + HCO + 4.587\ eV \quad (7g)$$

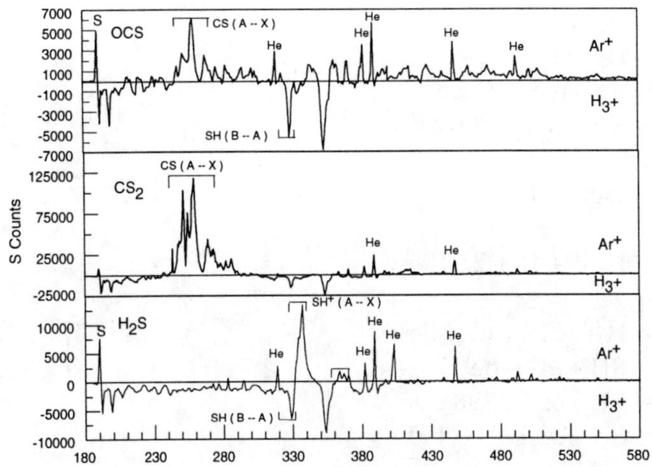

Figure 4. Signal, S, spectral scans for He^+/Ar^+ plasmas with OCS, CS_2 and H_2S addition. Many of the emissions are identified with some being common between the spectra, whilst others are not (see text).

for the H_3^+ precursor. This spectrum is compared with those for the $He^+/Ar^+/CS_2$ and $He^+/Ar^+/H_2S$ systems in Figure 4 with many of the peaks identified. Note that the spectral region above 580 nm is not included since there is very little activity in this region. The reactions that can contribute to the production of these features for CS_2 are

$$Ar^+ + CS_2 \rightarrow S^+ + CS + Ar + 0.833\ eV \quad 88\% \quad (8a)$$
$$\rightarrow CS_2^+ + Ar + 5.686\ eV \quad 12\% \quad (8b)$$

$$CS_2^+ + e \rightarrow S + CS + 5.507 eV \quad (9a)$$
$$\rightarrow C + S_2 + 2.568\ eV \quad (9b)$$

for an Ar^+ precursor and

$H_3^+ + CS_2 \rightarrow HCS_2^+ + H_2 + 2.693$ eV $k = 2.0(-9)$ cm^{3}s^{-1} (10)

$HCS_2^+ + e \rightarrow H + CS_2 + 6.531$ eV   (11a)
$\rightarrow SH + CS + 7.346$ eV   (11b)
$\rightarrow CH + S_2 + 2.512$ eV   (11c)
$\rightarrow S + HCS + 4.021$ eV   (11d)
$\rightarrow C + HS_2 + ?$ (11e)

for an H_3^+ precursor. For H_2S, the corresponding reactions are

$Ar^+ + H_2S \rightarrow H_2S^+ + Ar + 5.302$ eV $k = 1.3(-9)$ cm^{3}s^{-1} (12)

$H_2S^+ + e \rightarrow H + SH + 6.541$ eV   (13a)
$\rightarrow S + H_2 + 7.372$ eV   (13b)
$\rightarrow S + 2H + 3.066$ eV   (13c)

for the Ar^+ precursor and

$H_3^+ + H_2S \rightarrow H_3S^+ + H_2 + 2.910$ eV $k = 3.7(-9)$ cm^{3}s^{-1} (14)

$H_3S^+ + e \rightarrow H + H2S + 6.314$ eV   (15a)
$\rightarrow SH + H_2 + 6.917$ eV   (15b)
$\rightarrow SH + 2H\ 2.398$ eV   (15c)
$\rightarrow S + H + H_2 - 0.175$ eV   (15d)

for the H_3^+ precursor.

There are several features obvious in the spectra in Figure 4. The $S(4^5S^o - 3^3D^o)$ peak in the CS_2 case is very small whereas it is substantial in the other cases. This is to be expected since $S(4^5S^o)$ production is endothermic for CS_2. Also, the $CS(A^1\Pi - X^1\Sigma^+)$ bands are absent for H_2S since no C is available in this system. These peaks are positive in the S spectrum showing that the precursor is Ar^+. Here, the reaction kinetics point to production in reactions (4a) and (9a) for $S(4^5S^o)$ and CS(A) production also in reaction (9a). However, CS(A) production is not energetically possible for the OCS system (equation (4c)) unless the recombining OCS^+ is internally excited by at least 0.553 eV. SF_6 has been added upstream in the plasma to attach the electrons and quench and detect recombination emissions. This is illustrated in Figure 5, where the CS(A – X) emissions at 257 nm, the (0,0) transition, are plotted versus CS_2 flow for zero SF_6 and a small flow of SF_6 sufficient to attach all of the electrons. In the absence of SF_6, the signal increases with CS_2 flow as expected for CS_2^+ production from reaction (8b), but decreases dramatically with SF_6 addition as expected for a product of CS_2^+ recombination with electrons. Note that, in Figure 4 in addition to the positive going features

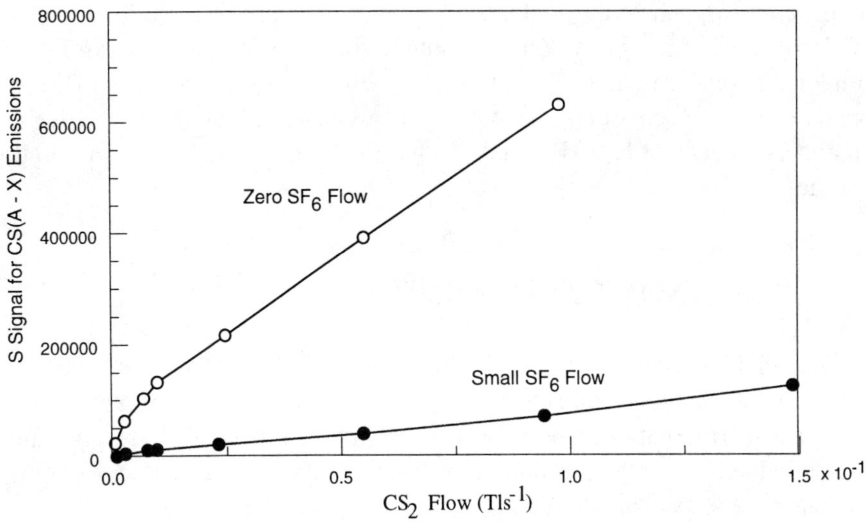

Figure 5. Variation with CS_2 flow of CS(A - X) emission in the (0,0) vibrational band at 257 nm for $He^+/Ar^+/CS_2$ plasma without and with SF_6 addition to attach electrons.

from an Ar^+ precursor, there are negative going features from an H_3^+ precursor. The two features, $SH(B^2\Sigma - A^2\Sigma^+)$ and an, as yet, unidentified emission at 354 nm, occur in all of the spectra; the SH(B) is energetically accessible in all three cases. The variation in these emissions versus SF_6 flow for the H_2S system, shows that they are almost totally quenched by electron attachment and thus result from recombination of H_3S^+ in reaction (15b).

4. CONCLUSIONS

A flowing afterglow plasma technique has been developed which uses pulsing of reactive gases to extract photon emission spectra for individual reactions and sequences of reactions where there are extensive emissions from the ionization source and from other interfering reactions. It has been successfully used to investigate electron recombination emissions from He^+/Ar^+ plasmas to which OCS, CS_2 and H_2S have been added. Studies are continuing to investigate the variation of emission intensities with both reactant, ABC, and SF_6 flows, in order to definitively establish the origins of all of the emissions. In particular, variation of vibrationally resolved band emissions are underway to determine whether the relative intensities of the individual vibrational bands vary with ABC flow indicating quenching of excitation in the precursor ion before reaction or of the emitting species when it has a long radiative lifetime. Once completed, the vibrational distributions of the emitting molecular states and thus of the reaction products can be determined. This has been done previously for other systems[15] using Franck Condon factors calculated with a program developed by Kent Ervin[22] together with spectroscopic constants from standard sources.[23] Different pulsing sequences of H_2, ABC and SF_6 will be used to further isolate specific emissions.

ACKNOWLEDGEMENTS

Financial support of the ACS-PRF (Grant 33892-AC6) for technique development and ion-molecule reaction studies and NASA (Grant NAG5-8951) for instrument development and recombination studies are gratefully acknowledged. Thanks are due to Jason McLain and Louis Atwood for assistance with the data analysis.

REFERENCES

1 R. Johnsen, Int. J. Mass Spectrom. Ion Proc. **81**, 67 (1987).
2 I. C. Plumb, D. Smith, and N. G. Adams, J. Phys. B **5**, 1762 (1972).
3 N. G. Adams and D. Smith, in Dissociative Recombination: *Theory, Experiment and Applications*, ed. by J. B. A. Mitchell and S. L. Guberman (World Scientific, Singapore, 1989), p. 124.
4 J. A. Coxon, P. J. Marcoux and D. W. Setser, Chem. Phys. **17**, 403 (1976).
5 M. Tsuji, in *Techniques for the Study of Ion-Molecule Reactions*, ed. by J. M. Farrar and W. H. Saunders (Wiley, New York, 1988), p. 489.
6 M. Tsuji, Adv. Gas Phase Ion Chem., Vol. **4**, in press (2002).
7 N. G. Adams, Adv. Gas Phase Ion Chem. **1**, 271 (1992).
8 T. L. Williams, L. M. Babcock and N. G. Adams, Mon. Not. Roy. Astro. Soc. **282**, 413 (1996).
9 T. Gougousi, R. Johnsen and M. F. Golde, J. Chem. Phys. **107**, 2430, (1997); *ibid*, p. 2440.
10 M. Larsson, in *Adv. Ser. Phys. Chem.: Photoionization and Photodetachment*, ed. by C.-Y. Ng (World Scientific, Singapore, 2000), p. 693.
11 M. Larsson, Adv. Gas Phase Ion Chem. **4**, in press (2002).
12 M. Larsson, Ann. Rev. Phys. Chem. **48**, 151 (1997).
13 N. G. Adams and L. M. Babcock, in *Dissociative Recombination: Theory, Experiment and Applications IV*, ed. by M. Larsson, J. B. A Mitchell and I. F. Schneider (World Scientific, Singapore, 1999), p. 190.
14 B. L. Foley, N. G. Adams, and H. S. Lee, J. Phys. Chem. **97**, 5218 (1993).
15 N. G. Adams and L. M. Babcock, J. Phys. Chem. **98**, 4564 (1994).
16 J. M. Butler, L. M. Babcock and N. G. Adams Mol. Phys. **91**, 81 (1997).
17 M. P. Skrzypkowski and R. Johnsen, J. Chem. Phys. **108**, 8400 (1998).
18 T. Mostefaoui, N. G. Adams and L. M. Babcock, Rev. Sci. Instrum. submitted.
19 T. L. Williams, B. K Decker, L. M. Babcock and N. G. Adams Rev. Sci. Instrum. **71**, 2169 (2000).

20 V. G. Anicich, J. Phys. Chem. Ref. Data **22**, 1469 (1993).
21 NIST Standard Reference Database, Number 69, July 2001 Release, http://webbook.nist.gov.
22 K. M. Ervin and W. C Lineberger, Adv. Gas Phase Ion Chem. **1**, 121 (1992).
23 K. P. Huber and G. Herzberg, *Molecular Spectra and Molecular Structure: IV Constants of Diatomic Molecules* (Van Nostrand Reinhold, New York, 1979).

Reactions of Electrons with Hydrocarbon Cations: From Linear Alkanes to Aromatic Species

M. Hassouna, J. L. Le Garrec, C. Rebrion-Rowe, D. Travers, and B. R. Rowe
Equipe d'astrochimie expérimentale, laboratoire P.A.L.M.S., UMR. 6627 du C.N.R.S., Université de Rennes I, 35042 Rennes Cedex, France
christiane.rebrion@univ-rennes1.fr

1. INTRODUCTION

In the framework of our systematic study of the recombination of hydrocarbon ions, we have already obtained results for classical aliphatic carbocations,[1] small unsaturated rings[2] and methyl substituted benzenic rings.[3] The recombination rate constants of these ions range from 5 to 12 x 10^{-7} cm^3s^{-1} and are independent of the number of carbon atoms or structure.

In this paper we report results on polycyclic aromatic ions, namely ions formed from naphthalene, acenaphthene, phenanthrene and fluoranthene. These ions react with electrons, at a rate that is slightly higher than observed for aliphatic or monocyclic cations.

2. EXPERIMENTAL

2.1 The Apparatus

The technique is an extension of a classic FALP technique that includes a moveable mass spectrometer, which is used to identify the ions and to measure their relative densities as a function of distance along the flow. When this measurement is possible, combined with that of the electron density, it allows the recombination of one particular ion to be studied, even if other ions in the flow also undergo recombination

In the apparatus, an afterglow forms in a helium buffer gas. This afterglow contains neutral ground state and metastable excited state helium atoms, He^M (2^3S), He^+ and He_2^+ ions. The measurements reported here are performed at a helium density of $\approx 10^{17}$ cm^{-3}. Under these conditions, the hydrodynamic measuring time is 1 ms with a flow velocity $v = 10^4$ cm/s and the initial electron density is varied from 10^9 to 10^{10} cm^{-3}. N_2 is used as a precursor gas, and is introduced at a second entry port, downstream from the microwave cavity (Fig. 1). N_2 reacts by ion-molecule reactions[4] with He^+ and He_2^+ and the metastable helium atoms He^M (2^3S) are destroyed by Penning ionization[5] so as to produce a plasma dominated by N^+ and N_2^+. These ions are called the "precursor ions" and are denoted P^+. N_2^+ is at least partially removed by dissociative recombination[6] ($\alpha = 2.6 x 10^{-7}$ $cm^3 s^{-1}$).

The parent vapors of the ions to be studied are introduced through the needle entry port, located further downstream as shown in Fig. 1. At this point, the plasma contains no He^+ or He_2^+, and consists of a mixture of N^+ and N_2^+. The ratio, N_2^+/N^+, depends upon the initial electron density.

2.2 Features Specific to the Use of Heavy Hydrocarbons

2.2.1 The injection of the neutral hydrocarbons

As the mass of the neutral parent increases, its injection as a vapor becomes more difficult because its vapor pressure at room temperature decreases, reaching a few micro-Torr for fluoranthene. All neutrals which are solid at room temperature were vaporized upstream of the needle injector in an oven, whose temperature was maintained sufficiently high and constant. The vapors were flushed into the needles by means of a helium flow, which together with the total pressure in the oven, could be adjusted to ensure the total conversion of the precursor ion into the hydrocarbon ions. This total conversion could not be obtained for fluoranthene and will be discussed later. All the gas lines between the oven and the needles were heated, the upstream parts being cooler than the downstream ones, and the needles themselves were ohmically heated.

2.2.2 The Langmuir probe and the sampling orifice of the mass spectrometer

Once injected, a non-negligible part of the hydrocarbon stays in its neutral form, and as the experiment chamber is at room temperature, redeposits on all

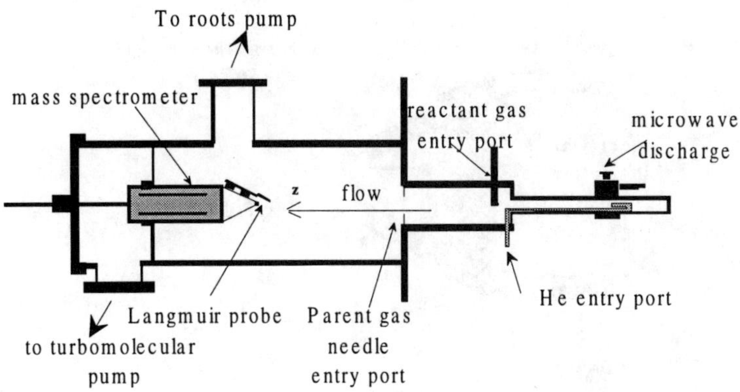

Figure 1. The FALP-MS apparatus.

surfaces. In order to keep our measurement tools operational, we had to continuously heat them to remove surface deposits.

The classical Langmuir probe consists of a single, very thin, gold plated wire that collects electrons from the plasma, provided a suitable bias voltage is applied. Our modified Langmuir probe consists of a loop of the same wire. A continuous current of 100-150 mA heats the wire, and is superimposed upon the current due to electron collection. The whole device has been described in detail elsewhere[7] and has been used in studies involving neutrals heavier than toluene.

The other surface that had to be protected was the sampling orifice of the mass spectrometer, which is a fraction of a millimeter wide. In order to avoid deposition onto it of solid material, an electrical wire was wrapped around the entrance cone of the mass spectrometer and a temperature probe was used to check that the cone temperature was about 150 °C. A circulating water jacket protected the remainder of the mass spectrometer from this heating, as shown on Fig. 2.

2.3 Determination of the recombination rate coefficient

Let M_i^+ be the DR ion of interest. It is produced by the reaction of the precursor ion with the neutral parent or via secondary ion-molecule reactions. It decays by reaction with electrons (DR, rate α_i), or with neutrals C_n (rate k_n), which can be impurities and/or the parent itself, and also by ambipolar diffusion.

The equations for the decay of $[M_i^+]$ and [e] *vs* distance z along the flow are:

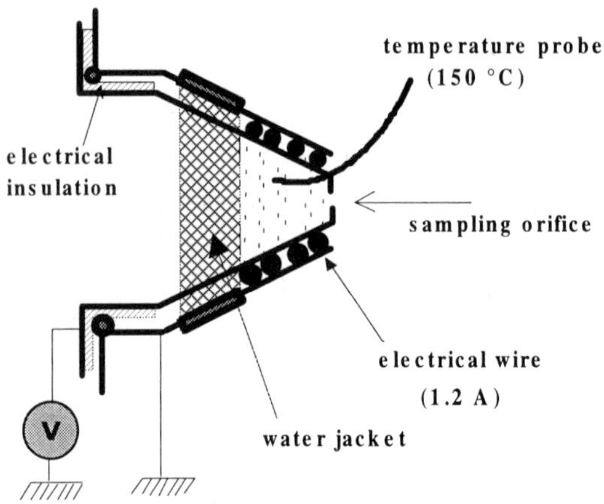

Figure 2. The entrance cone of the mass spectrometer.

$$v\frac{d[M_i^+]}{dz} = \underbrace{-\alpha_i[M_i^+][e]}_{DR} - \underbrace{\sum_n k_n[M_i^+][C_n]}_{\text{ion-mol reactions}} + \underbrace{\sum_j k_j[P^+][Hyd]}_{\text{production}} - \underbrace{\frac{D_A}{\Lambda^2}[M_i^+]}_{\text{diffusion}}$$

and $$v\frac{d[e]}{dz} = \underbrace{-\sum_i \alpha_i [M_i^+][e]}_{DR} - \underbrace{\sum_j \beta_i [N_j][e]}_{\text{attachment}} - \frac{D_A}{\Lambda^2}[e].$$

Provided that M_i^+ is not produced in the recombination zone and that there is no electron attachment we have:

$$\text{Ln}\frac{[M_i^+]_z}{[M_i^+]_{z_0}} = -\frac{\alpha}{v}\int_{z_0}^{z}[e]dz - \frac{1}{v}\left(k_n[C_n] + \frac{D_A}{\Lambda^2}\right)(z - z_0)$$

If ion-molecule reactions and diffusion can be neglected, the plot of $\text{Ln}\frac{[M_i^+]_z}{[M_i^+]_{z_0}}$ *vs* $\frac{1}{v}\int_{z_0}^{z}[e]dz$ yields α_I (Fig. 3). If not, this plot will give an upper limit to the DR rate, and the actual rate can be obtained by plotting $\text{Ln}\frac{[M^+]_{z_1}}{[M^+]_{z_0}}$ vs $\frac{1}{v}\int_{z_0}^{z_1}[e]dz$ at a fixed position, z_1 for different initial electron densities. The term representing other losses than DR is a constant in this case.

Under the same conditions:

$$\frac{1}{[e]_z} - \frac{1}{[e]_{z_0}} = \sum_i x_i\alpha_i \frac{z - z_0}{v}$$

where $x_i = \dfrac{[M_i^+]}{\sum_i [M_i^+]}$ and the slope of the plot of 1/[e] vs z yields the mean recombination rate for all ions present in the flow. The uncertainty in the results is 30%.

3. RESULTS

3.1 $C_8H_{10}^+$

This ion is produced from naphthalene, via the reaction:

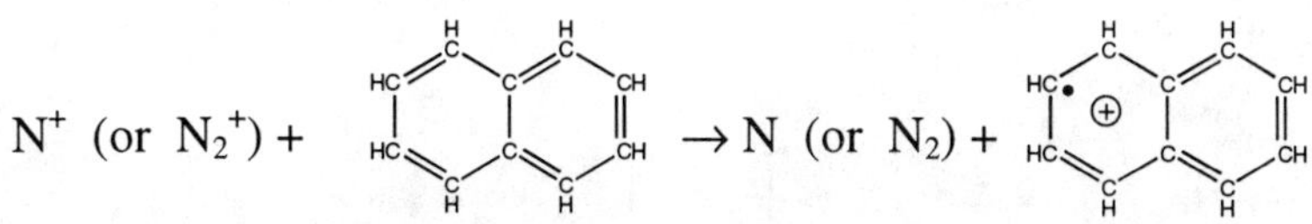

As reported in a recent ion-molecule reaction study,[8] the high resolution mass spectrum shows no hydrogen loss from naphthalene, and little C_2H_2 loss. The azulene cation is another less stable bicyclic isomeric form of $C_8H_{10}^+$, but if both ions pass through common structures prior to fragmentation,[9] isomerization of naphthalene$^+$ into azulene$^+$ is not observed.[10] A value of $8 \pm 2 \times 10^{-7}$ cm^{3}s^{-1} was obtained for the recombination of this ion.

3.2 $C_{12}H_{10}^+$ and $C_{12}H_9^+$

These ions are the main products of the reaction of N^+ and N_2^+ with acenaphthene (Fig. 4). Acenaphthene can be seen as a naphthalene molecule that is bridged by C_2H_4, and as naphthalene exhibited no H-loss after reaction with the precursor ions, it is likely that the H-loss on acenaphthene occurs on the aliphatic part of the molecule, leaving the bicyclic aromatic part unaltered. Unfortunately, to our knowledge, nothing is known concerning the fragmentation pathways of this ion.

$C_{12}H_{10}^+$ and $C_{12}H_9^+$ are produced in nearly equal amounts. The rate constants for the recombination of both ions were obtained by recording the mass spectrum first at low resolution, in order to determine the ionic fraction of both ions together, then at high resolution, in order to determine the ionic fraction of the separated ions. The recombination rates are 1.5×10^{-6} cm^{3}s^{-1} for $C_{12}H_{10}^+$ and 1.1×10^{-6} cm^{3}s^{-1} for $C_{12}H_9^+$.

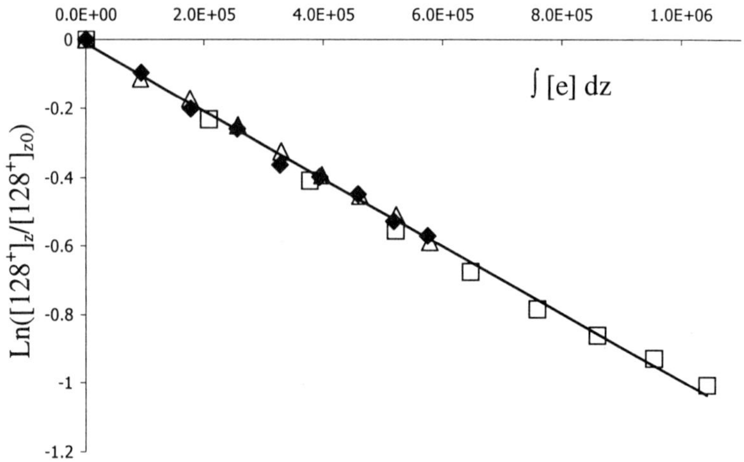

Figure 3. Determination of the rate constant for $C_{10}H_8^+$ + e: the slope gives $\alpha = 9.8 \times 10^{-7} cm^3 s^{-1}$.

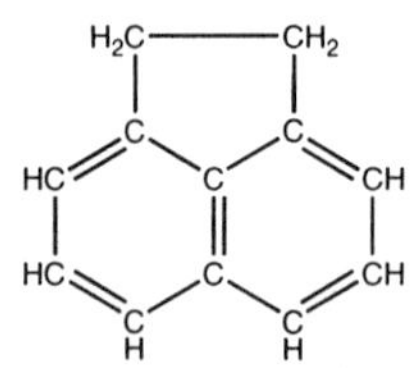

Figure 4. Acenaphthene

3.3 $C_{14}H_{10}^+$

This ion is one of the (numerous) reaction products of N^+ and N_2^+ with phenanthrene. The fragmentation patterns of $C_{14}H_{10}^+$ have been experimentally studied by time-resolved photoionization in the VUV,[11] but some of the fragments are not observed in our experiment, in particular $C_{14}H_9^+$, although its formation is exothermic by 2.72 eV. A high resolution mass scan indicated that the peak around mass 178 consists in fact of a single ion, with its ^{13}C isotope at mass 179.

This ion is the phenanthrene cation, represented in Fig. 5. Of course, the positive charge can be located anywhere in the molecule. One important fact is that this ion does not isomerize to the anthracene cation[11] so that our measurements concern only the phenanthrene cation. Its recombination rate constant is $1.7 \times 10^{-6} cm^3 s^{-1}$.

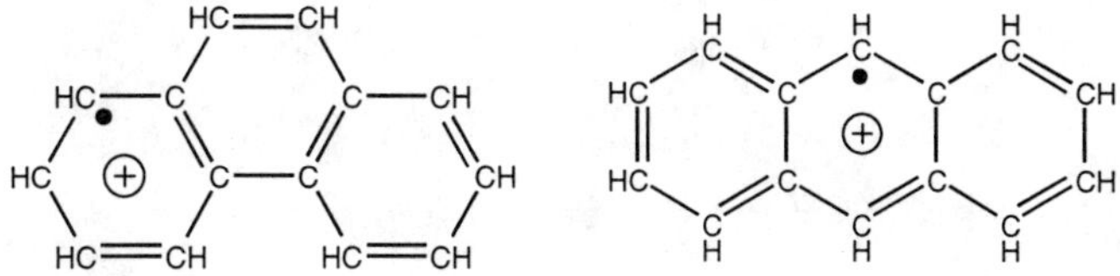

Figure 5. The phenanthrene cation (left) and the anthracene cation (right).

3.4 $C_{16}H_{10}^+$

This ion was one of the ions formed from the reaction of the precursor ions with fluoranthene (Fig. 6)

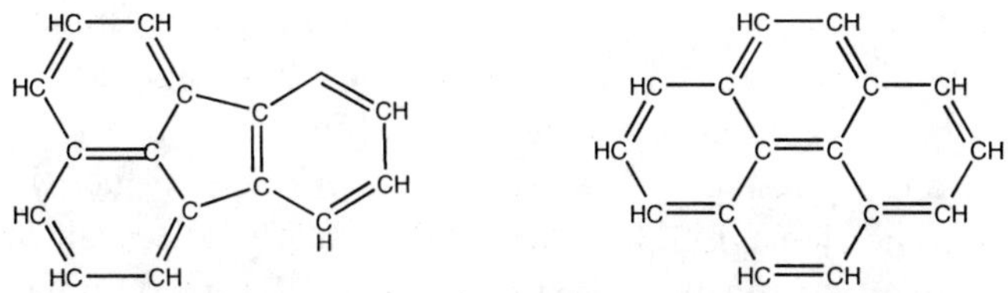

Figure 6. Fluoranthene (left) and pyrene (right).

Its recombination rate could not be obtained in the usual way for several reasons. Firstly, no reliable measurement of the relative density of $C_{16}H_{10}^+$ was possible, because of mass discrimination. This phenomenon is well known to mass-spectrometer users, and manifest by a distorted representation of the peak heights in the mass spectrum, according to the mass of the ions. This is directly linked to the mass range that is explored, and in the case of the ions created from fluoranthene, ions ranged up to mass 404. Secondly, once the precursor ions were destroyed by ion-molecule reactions, the ions of interest were also almost totally converted into the dimer ($C_{16}H_{10}^+$. $C_{16}H_{10}$). Conversely, when enough $C_{16}H_{10}^+$ was kept in the flow, the precursor ions were also present. Thirdly, pyrene (Fig. 6), which attaches an electron very efficiently, was present as an impurity at a 2% concentration. For all these reasons, the usual mathematical treatment was no longer valid, and the chemistry had to be modeled.

Because the electron density was the parameter measured in the most reliable way, it was modeled according to the network represented in Fig. 7. $k_1[C_{16}H_{10}]$ and $k_2[C_{16}H_{10}]$ were measured. k_3 was calculated from k_1. α_3 is known and ambipolar diffusion was taken into account. The adjustable

$$N^+ + C_{16}H_{10} \xrightarrow{k_1} C_{16}H_{10}^+ + N$$

$$N_2^+ + C_{16}H_{10} \xrightarrow{k_2} C_{16}H_{10}^+ + N_2$$

$$C_{16}H_{10}^+ + C_{16}H_{10} + M \xrightarrow{k_3} (C_{16}H_{10})_2^+ + M$$

$$C_{16}H_{10}^+ + e \xrightarrow{\alpha_1} \text{products}$$

$$(C_{16}H_{10})_2^+ + e \xrightarrow{\alpha_2} \text{products}$$

$$N_2^+ + e \xrightarrow{\alpha_3} N + N$$

$$C_{16}H_{10}\ \text{(pyrene)} + e \xrightarrow{\beta} C_{16}H_{10}^-$$

Figure 7. The chemical reactions involved in the model.

parameters were α_1, α_2 and β. Nine series of measurements were made, with 8.5×10^8 cm^{-3} < $[e]_0$ < 8.5×10^9 cm^{-3} and $[C_{16}H_{10}]$ ranging between 7.11×10^{10} cm^{-3} and 2.1×10^{11} cm^{-3}. The best agreement between measurements and the model was achieved for $\alpha_1 = 3 \times 10^{-6}$ cm^{3}s^{-1}, $\alpha_2 = 1 \times 10^{-5}$ cm^{3}s^{-1} and $\beta = 1 \times 10^{-9}$ cm^{3}s^{-1}. These values have to be considered as order of magnitude results rather than precise values since the network used for modeling is very simplified and the recombination of the other ions present in the flow has not been taken into account.

4. THE "DIMERS"

The phenanthrene and fluoranthene cations easily react by ion-molecule reactions with their neutral parent to form dimers by termolecular association. Our experiment does not allow the recombination rate of such ions to be measured because their parents, which are the phenanthrene or fluoranthene cations, are present in the flow. The model for fluoranthene on one hand, and the increase of the mean recombination rate as a function of distance, z, that is observed for phenanthrene on the other, give strong indications however that the recombination of such ions is faster than that of their monomers.

5. CONCLUSIONS

Several technical improvements of the FALP-MS apparatus have made possible the measurement of the recombination rate of polycyclic hydrocarbon ions with electrons. A slight increase of the recombination rate is observed, compared to what was obtained for non-cyclic and mono-cyclic hydrocarbon ions.

Table 1. Recombination rate summary.

Type of ion		Recombination rate (in 10^{-7} cm^3 s^{-1})	remarks
Non-cyclic		5 to 10	Mainly tertiary carbocations
Mono-cyclic		3 to 11	3 to 7 carbon atoms in the cycle
Poly-cyclic	$C_{10}H_8^+$	8 ± 2	measured
	$C_{12}H_9^+$	11.5 ± 3.5	measured
	$C_{12}H_{10}^+$	15 ± 4.5	measured
	$C_{14}H_{10}^+$	17 ± 5	measured
	$C_{16}H_{10}^+$	30 ± 9	modeled

Although these measurements are promising, it seems difficult to go ahead with heavier molecules with the FALP-MS apparatus. The first reason is that the heavier the hydrocarbon, the less is known about its chemical or physical properties, like electron attachment or vapor pressure. The second reason is that our experimental design requires large amounts of solid hydrocarbons, of high isomeric purity, whose price increases exponentially with size. The third reason is that the reactions of the ion with its neutral parent produce other ions whose recombination mask the recombination of the ion of interest.

For all these reasons, a new experiment has to be designed, in which the ions of interest can be injected directly into the helium plasma.

REFERENCES

1 L. Lehfaoui , C. Rebrion-Rowe , S. Laubé , J.B.A. Mitchell and B. R. Rowe, J. Chem. Phys. **106**, 5406 (1997).

2 C. Rebrion-Rowe, L. Lehfaoui, B.R. Rowe and J.B.A. Mitchell, J. Chem. Phys. **108,** 7185 (1998).

3 C. Rebrion-Rowe, T. Mostefaoui, S. Laubé and J.B.A. Mitchell, J. Chem. Phys. **113**, 3039 (2000).

4 Y. Ikezoe, S. Matsuoka, M. Takabe and A. Viggiano, "Gas Phase Ion-Molecule Reaction Rates through 1986", (Maruzen Co., Japan, 1987).

5 T. P. Parr, D. M. Parr and R. M. Martin, J. Chem. Phys. **62**, 3590 (1982).

6 A. Canosa, J. C. Gomet, B. R. Rowe and J. L. Queffelec, J. Chem. Phys. **94**, 7159 (1991).

7 S. Laubé, T. Mostefaoui and B. R. Rowe, Rev. Sci. Instrum. **71**, 519 (2000).

8 A. J. Midey, S. Williams, S. T. Arnold, I. Dotan, R. A. Morris and A. A. Viggiano, Int.J.Mass Spectrom. **195/196**, 327 (2000).

9 H. W. Jochims, H. Rasekh, E. Rühl, H. Baumgärtel and S. Leach, Chem. Phys. **168**, 159 (1992).

10 W. Cui, B. Hadas, B. Cao and C. Lifshitz, J. Phys. Chem. A **104**, 6339 (2000).

11 Y. Ling and C. Lifshitz, J. Phys. Chem. A **102**, 708 (1998).

Dissociative Recombination of Xe_2^+ and XeH^+

J. B. A. Mitchell,[1] R. H. Lipson[2], and B. K. Sarpal[3]
[1]P.A.L.M.S. UMR No. 6627 du C.N.R.S., Université de Rennes I, 35042 Rennes, France: [2]Dept. Of Chemistry, University of Western Ontario, London, ON, Canada N6A 5B7: [3]British Telecommunications plc, 81 Newgate Street, London, EC1A 7AJ
Brian.Mitchell@univ-rennes1.fr

1. INTRODUCTION

When excited in a plasma, xenon has useful emissions in the visible and ultraviolet. Indeed, its visible spectrum resembles closely the solar spectrum making it very attractive as an illumination light source. Xenon is used in discharge lamps, lasers and in plasma display devices. Being inert, it is non-toxic and therefore represents an attractive replacement for mercury in fluorescent lamps. Given this interest in its spectral response, it is not surprising therefore, that there have been many spectroscopic studies of xenon. During one such study, the authors came upon an unusual phenomenon that owes its existence to associative ionization and dissociative recombination processes. The recombination rates of homonuclear rare gas dimers (with the exception of He_2^+) are large due to the fact that they have suitably placed curve crossings that allow the capture of an electron by the ion to be followed rapidly by dissociation along a neutral state. This dissociation process results in the conversion of potential energy into kinetic energy, and thus, the potential energy of the electron plus ion system drops below the ionization potential, rendering re-ionization impossible. The free atoms resulting from this dissociative recombination process can be formed in a variety of different energy levels, and information on the branching ratios for the formation of these states is very important for the modeling of the light emission from plasmas containing such molecular ions.

2. Xe_2^+ : ASSOCIATIVE IONIZATION AND RECOMBINATION

A multiphoton excitation experiment[1] was performed with the goal of studying the spectra of n=8-11 p and n=4-8 f levels of xenon atoms. This experiment used frequency doubled photons from a neodymium/YAG pumped dye laser, and the lines of interest were excited by two photon absorptions. A pulsed valve, fed by a 10%Xe/90%He mixture at a backing pressure of ≈ 114 psi was used to produce the free jet gas target which was located inside a vacuum chamber maintained at ≈ 10^{-6} Torr. The excitation of the lines was observed using a resonance enhanced multi-photon ionization (REMPI) technique. This method involves forming the state being examined via a two photon absorption and then absorbing a third photon so that the system enters the ionization continuum, thus ejecting an electron that is then detected. The excitation spectrum thus obtained is illustrated in Fig. 1. The lines displayed are labeled using tabulated energy levels published elsewhere.[2]

The apparatus used is very powerful analytically as it incorporates both a photoelectron spectrometer and a time-of-flight mass spectrometer that can be used to identify the states giving rise to the absorption. A typical photoelectron spectrum for an np state is shown in Fig. 2 and it is seen that one line appears corresponding to the kinetic energy of the departing electron (following the third photon absorption). The energy of this electron is close to 5.5 eV, which is exactly as expected since the third photon carries the system well into the ionization continuum. Fig. 2 also shows a corresponding photoelectron spectrum for an nf state, and here it is seen that the situation is quite different. The peak near 5.5 eV is now quite weak while lower energy peaks predominate. Examination of the energy levels indicates that these peaks must arise due to excitation of primarily 5d and 6p states that lie *below* the nf state, though as seen from Fig. 1, the two-photon absorption occurred exactly where the nf state was expected. This means that rather than undergoing direct excitation into the continuum as occurred for the np states during absorption of a third photon, the nf state relaxes down to a lower state which then undergoes ionization into the continuum, releasing the photoelectron. This rapid relaxation process is thus the central feature of this experiment, and was examined in more detail.

A time-of-flight mass spectrum was taken of the ions produced as a result of (2+1) REMPI, and while for the np states, these yielded peaks corresponding to Xe^+ ions, an additional feature due to Xe_2^+ was also seen in the case of the nf states. The origin of this peak must be associative ionization, i.e.:

$$Xe\,(nf) + Xe\,(^1S_0) \rightarrow Xe_2^+ + e\,.$$

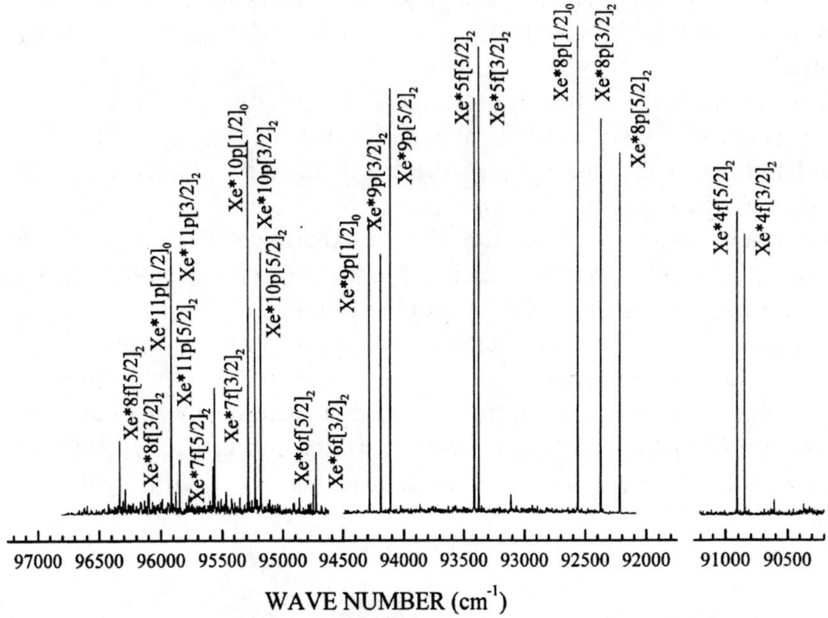

Figure 1. (2+1) REMPI spectrum of ^{132}Xe between 90,200 and 96,800 cm^{-1}.

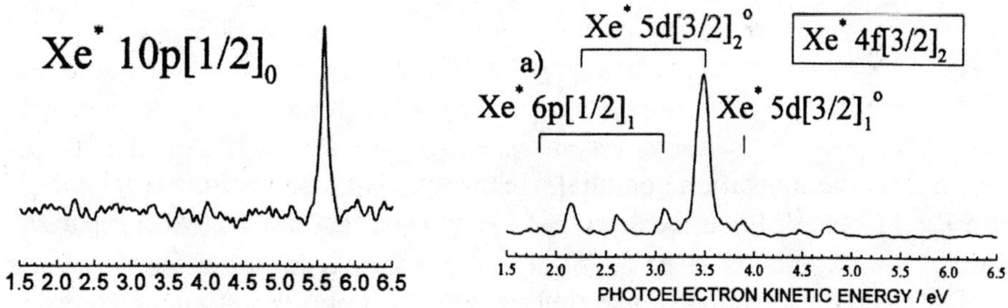

Figure 2. Photoelectron spectra arising from absorptions to 10p $[1/2]_0$ and $4f[3/2]_2$ states.

The peak does not arise due to ionization of dimers formed in the jet as no such peak is seen for the np states under the same operating conditions. (Only at higher backing pressures used to form the jet were dimer ions seen in the mass spectra taken for both np and nf absorptions).

Following this associative ionization process, the Xe_2^+ dimer ions rapidly undergo dissociative recombination with electrons to form predominantly 5d and 6p states that are then ionized by the third photon to yield the photoelectron spectra, similar to that shown in Fig. 2.

In summary therefore, two main details emerge from this experiment:

a) nf states readily lead to associative ionization while np states are directly ionized.
b) 5d and 6p states are primarily produced as a result of the dissociative recombination of the Xe_2^+ dimer ions so formed.

The latter is not unexpected given that the dissociating states with 5d and 6p asymptotes pass through the Xe_2^+ potential well close to the v=0 level. The observation is also in close accord with experimental findings[3] that 6p states are a dominant product of Xe_2^+ dissociative recombination (though their apparatus was not capable of detecting 5d atoms).

The finding that associative ionization is so state selective is less easy to explain. In their original paper, Xu et al.[1] discounted the possibility that orbital size was the determining factor. Subsequent reflection however, has led us to believe that this conclusion may in fact be wrong. Table 1 indicates the mean radii for n between 4 and 11, calculated using the hydrogenic formula:

$$< r_{n,l} >= \frac{n^{*2}}{z}\left\{1+\frac{1}{2}\left[1-\frac{l(l+1)}{n^{*2}}\right]\right\}a_0$$

where n* = n-δ, δ is the quantum defect and z is the effective nuclear charge (assumed = 1 for Table 1). For a given atomic level, n* can be calculated using the formula:

$$E_{n,l} = IP - \frac{Ry}{n^{*2}}$$

where IP is the ionization potential of the atom, $E_{n,l}$ is the atomic level energy and Ry =13.6 eV. It can be seen that f- and p-states do not have comparable radii.

Unfortunately, such a simple representation ignores important physics that is relevant for both the p and f levels of Xenon. In the radial Schrödinger equation, the form of the potential in which the electrons move is:

$$V_{eff}(r) = V_{Coulomb}(r) + \frac{l(l+1)\hbar^2}{2 m_e r^2}$$

and for higher l values the form of this potential is a double-well structure such as that illustrated in Fig. 3. The effect of this potential on atomic structure and on the resulting spectroscopy of heavy atoms has been reviewed by Connerade.[4]

n	p[1/2]	p[3/2]	p[5/2]	f[3/2]	f[5/2]
4				9.29	9.40
5				16.31	16.47
6				24.93	25.14
7				35.15	35.40
8	15.95	15.40	14.98	46.95	47.24
9	24.00	23.36	22.85		
10	33.64	32.89	32.29		
11	44.84	44.00	43.31		

Table 1: Mean radii (Å) for p and f states in Xenon with z = 1 (see text).

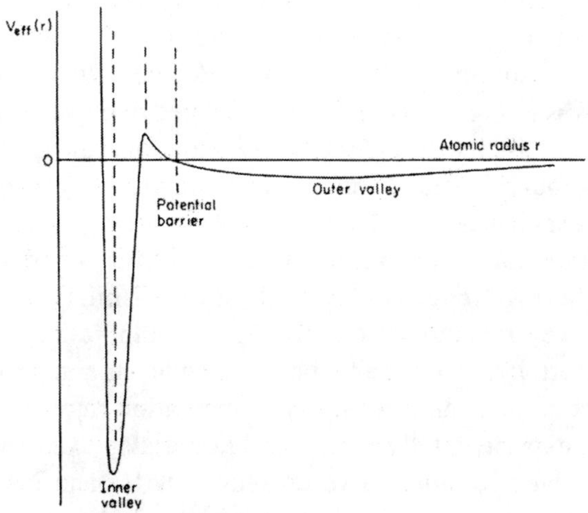

Figure 3. Double well potential.

For atoms such as xenon, the centrifugal barrier prevents electrons in f orbitals from entering the core, and their maximum wave function amplitude will be found entirely within the outer well. With increasing z, the wave function amplitude of such orbitals will increase in the inner well as the Coulomb part of $V_{eff}(r)$ begins to dominate over the repulsive centrifugal component. For Xe, the inner electrons also effectively screen the excited f-electrons as the core is relativistically contracted. Calculations indicate that for these orbitals, z is ~unity[5] and therefore, the hydrogen formula for $<r_{n,l}>$ may in fact be a good approximation. Conversely, p-electrons are more penetrating than f-electrons since the centrifugal barrier is smaller. As a result, the effective z "felt" by the p-electrons should be > 1.

The relevance of this to the present problem is that for comparable energies, f electrons in xenon are probably located much further out with larger mean orbital radii than p electrons. This should make the atoms in f orbitals much more chemically reactive than the corresponding p orbital atoms and therefore, one should see a much greater tendency for associative ionization for the f states than for the p states, for a given n value. This qualitative picture still needs to be reinforced with calculations of actual mean radii. These will be presented in a subsequent publication.

3. XeH^+ RECOMBINATION

The dissociative recombination of rare gas hydride ions is also a subject of great importance having applications in astrophysics and nuclear technology (for HeH^+) and for plasma physics and laser technology for the heavier hydrides. The dissociative recombination of HeH^+ has been the subject of much experimental[6-8] and theoretical work[9,10] following the original measurement of Yousif and Mitchell[11] that the recombination cross section was not small as had been expected due to the lack of a suitable curve crossing through which direct recombination could proceed. This work has had a profound influence on our understanding of dissociative recombination having revealed the mechanism of non-crossing or tunneling recombination in which the ionic system can capture an electron and decay via a coupling to a neighboring repulsive neutral state. While this mechanism permits dissociative recombination to occur, the rate coefficient for HeH^+ recombination at room temperature is small in comparison with typical rates for diatomic molecular ions.[12] Recombination rates for NeH^+ and ArH^+ have not been experimentally measured but are also expected to be small as these ions also lack suitable curve crossings. KrH^+ and XeH^+ recombination has been studied using afterglow techniques[13,14] and here a surprise was found. While the rate for KrH^+ was too small ($<1.15 \times 10^{-8}$ cm^3s^{-1}) to be accurately determined using the afterglow method, that for XeH^+ was found by Le Padellec et al.[14] to have the rather large rate of 8.3×10^{-8} cm^3s^{-1}. Clearly there is a fundamental difference in the character of this ion that induces this phenomenon. A clue to this difference can be found by considering the ionization potentials for rare gas atoms. While the IPs for helium, neon and argon are large (24.6, 21.6 and 15.8 eV respectively), that for krypton (14 eV) is similar to that for atomic hydrogen. The IP for xenon (12.13 eV) is smaller. The ground states of the rare gas hydride ions have $R + H^+$ as their asymptotes except for the xenon hydride ion which has $Xe^+ + H$. It has been shown[15] that the $1^1\Sigma$ state leading to this asymptote crosses the ion ground state near 5.5 Bohr. Neutral Rydberg states with this ion state as their limit will cross the ion state at smaller internuclear distances. All of the rare gas hydrides have repulsive ground states but their adiabatic excited states are bound.[16,17] Calculations of the excited potential energy curves for the lighter rare gas hydrides reveal a progression of curves that is very Rydberg in structure, closely mirroring the ion ground state. Calculations[18-20] for ArH and XeH reveal structure in the Rydberg series suggesting that in a diabatic

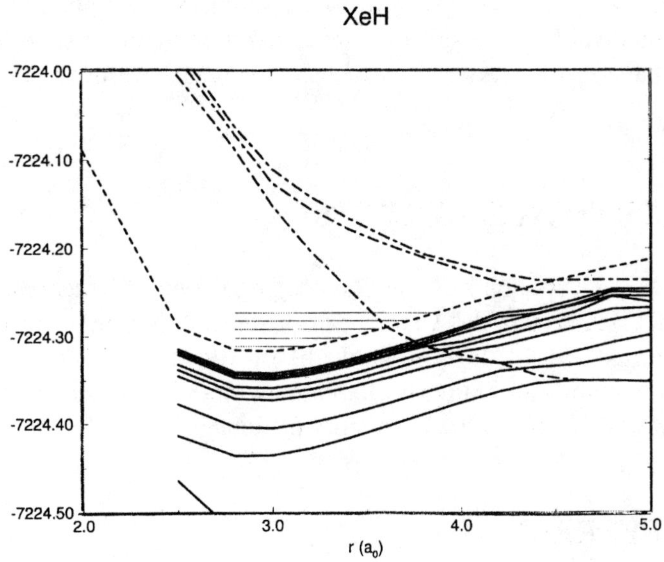

Figure 4. Potential energy curves for XeH^+ (dashed curve), XeH $^2\Sigma$ bound states (solid curves) and XeH $^2\Sigma$ resonance states (chain curves).

representation, the ground state of the ions might be crossed by a neutral state. For ArH, these crossings would appear to be far from the minimum of the ion ground state but for XeH, the perturbed region is much closer and this could explain the rather large recombination rate, typically found when the direct mechanism is permitted.

Using an R-matrix approach,[21] one of the authors (BKS) has performed a calculation of both resonant and bound states involved in the scattering of electrons from XeH^+. The results of this calculation are shown in Fig. 4. It is seen that there is in fact a curve crossing between the $^2\Sigma$ resonance and the ionic ground state of XeH^+ that occurs in the vicinity of the v=2 level. In fact the calculation also revealed another crossing between the ion state and a $^2\Pi$ resonance that occurred closer to the ion potential minimum but the width of this resonance was so narrow that its contribution to the dissociative recombination cross section is likely to be small compared to that of the $^2\Sigma$ state. Certainly the presence of such a curve crossing will enhance the recombination probability and a calculation involving only the direct mechanism indicated 300K rates in the range of a few times $10^{-8}cm^3s^{-1}$ and a few times 10^{-7} cm^3s^{-1} for the v=1 and v=2 levels respectively. A full calculation taking account of both the direct and the indirect mechanism should be performed however, to fully understand this process.

Interestingly, a curve crossing was also found for the case of KrH^+ but this occurred near the v=5 level and calculations based upon the direct mechanism predicted very small rate coefficients for ground state ions that would not have been possible to measure using the flowing afterglow method.

ACKNOWLEDGMENTS

This work was supported by the National Engineering Research Council (NSERC) of Canada and the Centre Nationale de Recherche Scientifique (CNRS) of France. R. H. Lipson would like to thank CNRS for a visiting scientist grant that allowed him to collaborate with the Equipe d'Astrochimie of the PALMS laboratory at the University of Rennes I.

REFERENCES

1 X. K. Xu, J. B. A. Mitchell, and R. H. Lipson, Phys. Rev. A **62**, 052712 (2000).
2 C.E. Moore, *Atomic Energy Levels*, Natl. Bur. Stand. (U.S.) Circ. No. 467 (U.S.G.P.O, Washington, DC, 1971), Vol. III.
3 Y. J. Shiu, M. A. Biondi, and D. P. Sipler, Phys. Rev. A **15** 494, (1977).
4 J. P. Connerade, Contemp. Phys. **19,** 415 (1978).
5 K. T. Cheng, and C. Froese Fischer, Phys. Rev. A **28**, 2811 (1983).
6. F. B. Yousif, J. B. A. Mitchell, M. Rogelstad, A. Le Padellec, A. Canosa, and M.I. Chibisov, Phys. Rev. A **49**, 4610 (1994).
7 G. Sundstrom, S. Datz, J. R. Mowat, S. Mannervik, L. Brostrom, M. Carlson, H. Danared, and M. Larsson, Phys. Rev. A **50**, R2806 (1994).
8 J. R. Mowat, H. Danared, G. Sundstrom, M. Carlson, L. H. Andersen, L.H. Vejby-Christensen, M. af Ugglas, and M. Larsson, Phys. Rev. Lett. **74**, 50 (1995).
9 S. L. Guberman, Phys. Rev. A **49**, R4277 (1994).
10 B. K. Sarpal, and J. Tennyson, J. Phys. B **27**, 5943 (1994).
11 F. B. Yousif, and J. B. A. Mitchell, Phys. Rev. A **40**, 4318 (1989).
12 J. B. A. Mitchell, Phys. Rep. **186**, 215 (1990).
13 M. Geoghegan, N. G. Adams, and D. Smith, J. Phys. B **24**, 2589 (1991).
14 A. Le Padellec, S. Laubé, O. Sidko, B.R. Rowe, B. K. Sarpal, and J. B. A. Mitchell, J. Phys. B **30**, 963 (1997).
15 C. Kubach, and V. Sidis, J. Phys. B **6**, L289 (1973).
16 R. H. Lipson, Chem. Phys. Lett. **129**, 82 (1986).
17 I. Dabrowski, G. Herzberg, and R. H. Lipson, Mol. Phys. **63**, 289 (1988).
18 G. Theodorakopoulos, I.D. Petsalakis, and R.J. Buenker, Mol. Phys. **71**, 1055 (1990).
19 I. D. Petsalakis, and G. J. Theodorakopoulos, J. Phys. B **27**, 4483 (1994).
20 I. D. Petsalakis, G. J. Theodorakopoulos, and S. Costa, Mol. Phys. **75**, 805 (1992).
21 I. Rabadan and J. Tennyson, J. Phys. B **27**, 1975 (1997).

Dissociative Recombination of Protonated Dimer Ions

J. Glosík, R. Plašil, P. Zakouřil, and V. Poterya
Charles University Prague, Mathematics and Physics Faculty, Czech Republic
glosik@plasma.troja.mff.cz

1. INTRODUCTION

Clusters play an important role in the positive ion chemistry of the terrestrial atmosphere. The combined water/aldehyde and water/acetonitrile ionic clusters are amongst the most abundant positive ions in the troposphere of the Earth.[1,2] The thermodynamic data on these clusters are known,[3] but there are no available data on recombination of these clusters with electrons.[4] These data can now easily be obtained using our High Pressure Flowing Afterglow (HPFA) apparatus[5-9] that operates at prcssurcs up to 20 Torr. To facilitate further discussion of the kinetics of these processes we will designate these clusters as $H^+ \cdot (\ldots)_n$, $n = 1$-3, without representing their actual molecular structure, e. g. $H^+ \cdot (HCOH)_2$.

In a flowing afterglow plasma used for a study of recombination of large cluster ions, a sequence of ion molecule reactions has to proceed to form cluster ions prior to their recombination with electrons. The HPFA is very suitable for these studies because the high pressure of a buffer gas enhances association reactions.[5-7] In the HPFA at pressures over 5 Torr, cluster ions prior to their recombination undergo many collisions and we can even assume that temperature of the ions is equal to the temperature of the helium ($T_+ = T_g = 300$ K). The situation is not so simple with the electron temperature. The final electron temperature depends on the balance between energy gain and loss in collisions with He buffer. Actual temperature varies along the flow tube and has to be measured in the experiment.

Dissociative Recombination of Molecular Ions with Electrons
Edited by S. Guberman, Kluwer Academic/Plenum Publishers, 2003

2. EXPERIMENT

The High Pressure Flowing Afterglow (HPFA) was described in detail in our previous papers.[5-7] The schematic diagram in Fig. 1 outlines the essential parts of this apparatus including the axially movable Langmuir probe. The fundamental features of HPFA are similar to a conventional FALP apparatus,[10] but the operational pressure in the HPFA can be increased up to 20 Torr. The flow tube is a stainless steel tube 40 cm long, 1.8 cm inner diameter. The afterglow plasma is created in the upstream glass part of the flow tube in a microwave discharge in He (30 W). Reactant gases are introduced via several entry ports.

The cylindrical axially movable Langmuir probe (tungsten wire 14 micrometers in diameter and 4 mm long) is used to measure the electron number densities and the electron energy distribution functions (EEDF) along the flow tube.

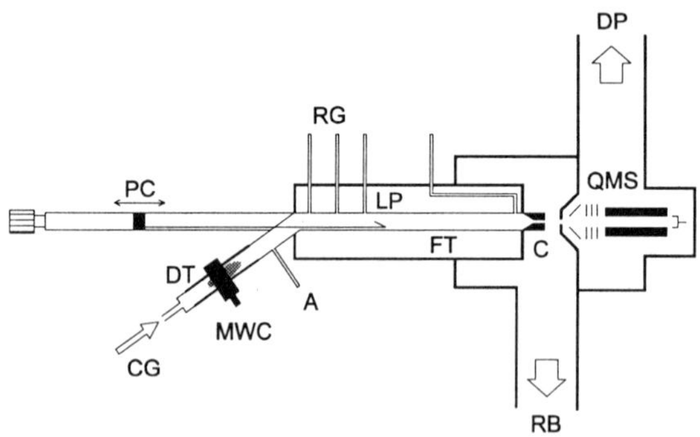

Figure 1. The High Pressure Flow Tube. CG: flow of He carrier gas, A: Ar entry port, RG - reactant gas entry port, MWC: microwave cavity, FT: flow tube, C: capillary, QMS: quadruple mass spectrometer, LP: Langmuir probe.

The very small diameter of the probe is essential for work at higher pressures. To obtain absolute values of electron number densities required for determination of recombination rate coefficients the probe is calibrated by measuring the well established recombination rate coefficient of O_2^+ with electrons (for details see Refs. 5, 11).

3. RESULTS AND DISCUSSION

3.1 Ion–Molecule Reactions And Kinetics Of Formation Of Protonated Dimer Ions

All studied dimer ions are formed in a sequence of ion molecule reactions. The kinetics of formation of the dimer ions have to be carefully studied because the results of this part of study are used in the formation of the plasma with the dominance of the particular dimer ion in the recombination region of the flow tube.

By adding traces of water to plasma dominated by Ar^+ ions we start this sequence by formation of H_3O^+ in the post discharge region. After the addition of the reactant gas further downstream the sequence continues by fast proton transfer reactions. The chain is terminated by three-body association, which is enhanced by high pressure of He.

We will describe this processes on formation of $H^+\cdot(CH_3CN)_2$. Ions H_3O^+ react with the acetonitrile in the fast proton transfer:

$$H_3O^+ + CH_3CN \rightarrow CH_3CNH^+ + H_2O. \tag{1}$$

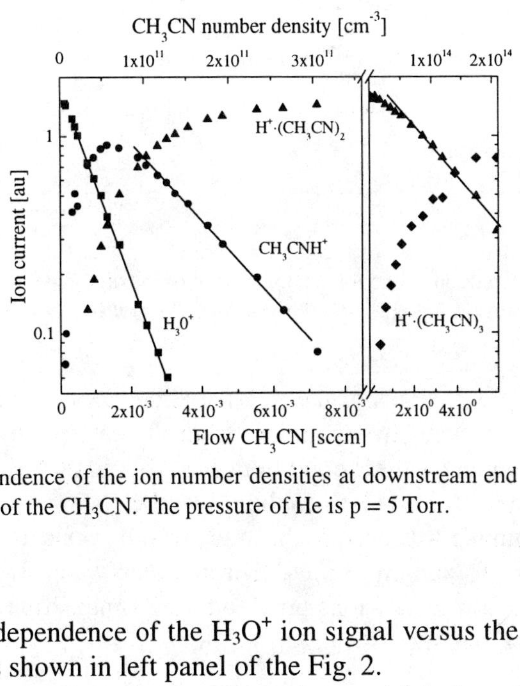

Figure 2. The dependence of the ion number densities at downstream end of the flow tube on the number density of the CH_3CN. The pressure of He is p = 5 Torr.

The typical dependence of the H_3O^+ ion signal versus the number density of acetonitrile is shown in left panel of the Fig. 2.

The obtained reaction rate coefficient is very high; k_1 at the pressures 3-5 Torr = 5.9×10^{-9} cm^3 s^{-1}. The formed CH_3CNH^+ ions further react with present acetonitrile molecules in three-body association:

$$CH_3CNH^+ + CH_3CN + He \rightarrow H^+\cdot(CH_3CN)_2 + He \tag{2}$$

The obtained effective binary reaction rate coefficient is $k_{2eff} = (2.1 \pm 0.7)\times10^{-9}$ cm^3 s^{-1}. If the flow rate of CH_3CN is high enough then the $H^+\cdot(CH_3CN)_2$ cluster ions further associate with CH_3CN:

$$H^+\cdot(CH_3CN)_2 + CH_3CN + He \rightarrow H^+\cdot(CH_3CN)_3 + He \tag{3}$$

The corresponding decay curve is plotted in the right panel of the Fig. 2. This reaction is very slow, $k_{3eff} = (1.1 \pm 0.3) \times 10^{-12}\ cm^3\ s^{-1}$.

We have taken advantage of the large difference between the flows of the acetonitrile required to form $H^+ \cdot (CH_3CN)_2$ and $H^+ \cdot (CH_3CN)_3$ to create such conditions in the flow tube that $H^+ \cdot (CH_3CN)_2$ were the dominant ions in the recombination controlled region.

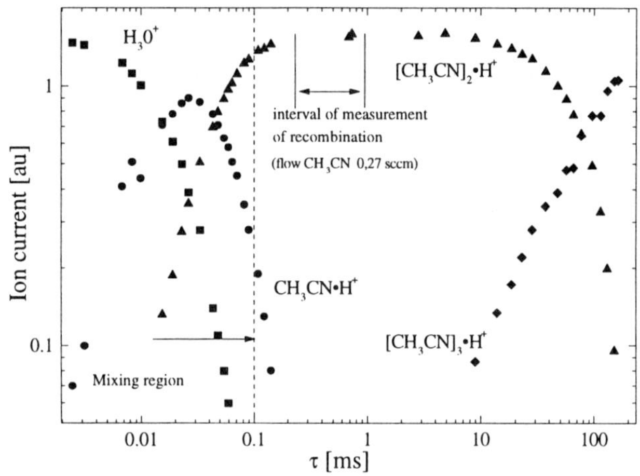

Figure 3. The time evolution of the ionic composition of the afterglow plasma modeled using the dependencies plotted in Fig. 2 for the flow rate of CH_3CN $\Phi_{CH3CN} = 0.27$ sccm.

Fig. 3 shows the changes of the relative abundance of the ions along the flow tube (with reaction time) obtained from the data plotted in Fig. 2. The model predicts that under these conditions the $H^+ \cdot (CH_3CN)_2$ ions become the dominant ions after 0.12 ms (1 cm in the flow tube).

A similar technique was used in the study of other ions. First we measured the kinetics of the formation of the dimer and than we estimated the flow of the parent gas to form dimer ions prior to the recombination region (0.12 ms or better to say 1 cm in the flow tube).

3.2 Electron–Ion Recombination

Once the kinetics of the clustering reactions have been established we can discuss the results of the study of the recombination of the dimer ions with electrons. It should be emphasized again that the formation of the dimer ions is very fast and their removal by ion molecule reactions is very slow (if there is any) in comparison with the rate of the recombination.

3.2.1 Recombination of $H^{+}\cdot (CH_3COH)_2$, $H^{+}\cdot (HCOH)_2$, $H^{+}\cdot (CH_3CN)_2$, Ions with Electrons

We will discuss the procedure for studying recombination of $H^{+}\cdot (CH_3COH)_2$. For other ions just results will be given.

By adding the calculated flow of acetaldehyde (0.19 sccm), the majority of ions entering the reaction region are converted to $H^{+}\cdot (CH_3COH)_2$ cluster ions within a few millimeters from the entry port. Traces of ions of mass 59 and 103 were also present in the mass spectrum. Because of the large recombination rate coefficient of the cluster ions, the recombination starts to be the dominant loss process in the plasma within 0.15 ms. The recombination rate coefficient was calculated by the conventional FALP method.[12]

When the recombination and diffusion are the dominant loss processes in the flowing afterglow the analytical solution of the balance equation for electrons[13] is:

$$n_e = \left[\alpha \frac{\exp(\nu_D \tau) - 1}{\nu_D} + \frac{1}{n_{e0}} \exp(\nu_D \tau) \right]^{-1} \qquad (4)$$

Where n_{e0} is the electron number density at the beginning of the recombination controlled region, v_D characterizes the ambipolar diffusion losses and α is the recombination rate coefficient. For v_D approaching zero, the last equation converts to the simple form, $1/n_e = 1/n_0 + \alpha t$, and is the equation that is usually used in a recombination dominated plasma with negligible diffusion losses. Equation (4) was used to fit the data and to calculate α. The recombination rate coefficient, measured at 6 to 8.2 Torr, is $\alpha = (1.4 \pm 0.5)\ 10^{-6}\ cm^3\ s^{-1}$. The "apparent" electron temperature is ~ (620 ± 180) K. The instabilities of the plasma potential, the fluctuation of the probe potential (noise on the probe) and the inhomogeneity of the probe surface (work function) can increase the "apparent" value of the electron temperature. We assume that actual electron temperature in the recombination controlled region is at the lower end of the given interval. This will give T_e ~ 500 K in the recombination dominated region.

Recombination rate coefficients for the ions $H^{+}\cdot (HCOH)_2$, $H^{+}\cdot (CH_3CN)_2$, were measured in a similar manner and we obtained: $(2.8 \pm 1)\times 10^{-6}\ cm^3\ s^{-1}$ for $H^{+}\cdot (CH_3CN)_2$ and $(3.3 \pm 1)\times 10^{-6}\ cm^3\ s^{-1}$ for $H^{+}\cdot (HCOH)_2$ at (400 ± 100) K and 450 K respectively. The study of acetonitrile was complicated by rapid plasma enhanced deposition on the probe.

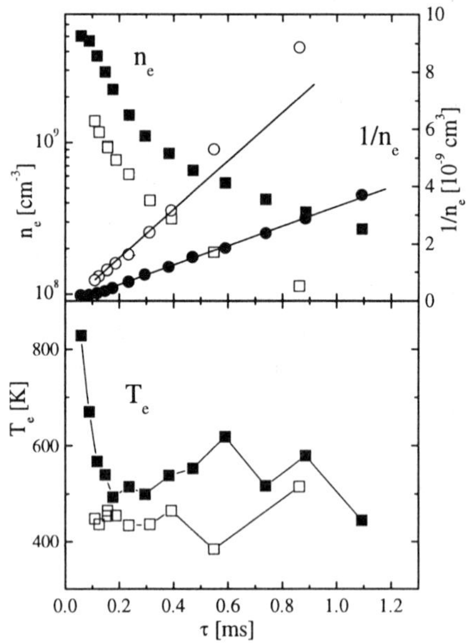

Figure 4. Upper panel: "Time evolutions" of the electron number densities n_e and reciprocal electron number densities ($1/n_e$) in the afterglow dominated by the recombination of $H^{+}\cdot (CH_3COCH_3)_2$ with electrons. The open symbols indicate "high" temperature data and closed symbols indicate "low" temperature. Lower panel: "Time evolutions" of the electron temperatures. T_e are obtained from the measured electron energy distribution functions, EEDF.

3.2.2 Recombination of $H^{+}\cdot (CH_3COCH_3)_2$ ions with electrons

We measured the recombination of $H^{+}\cdot (CH_3CHCH_3)_2$ in two different regimes of the afterglow. For the first set of measurements, we obtained a recombination rate coefficient of $(3.4 \pm 1.5)\times10^{-6}$ cm^3 s^{-1}. The electron temperatures obtained for the same set of data were nearly constant $T_e = (580 \pm 150)$ K.

In another set we measured the recombination rate coefficients at several pressures in a plasma with a temperature of $T_e = (450 \pm 100)$ K and we obtained $\alpha = (7 \pm 2.5)\times10^{-6}$ cm^3 s^{-1}. This recombination rate coefficient is very large and is considerably higher than that obtained at $T_e = (580 \pm 150)$ K. The evolution of n_e and $1/n_e$ versus recombination time for both sets are plotted in the upper panel of Figure 4. The corresponding evolution of T_e is plotted in the lower panel of Figure 4 showing the obvious differences. The recombination is faster in the plasma with the lower electron temperature.

The main conclusion of this part of the study is that the recombination coefficient is decreasing with increasing electron temperature. The negative temperature dependence of α seems to be very pronounced.

4. CONCLUSION

We studied formation of the protonated dimer ions, $H^+\cdot (CH_3CN)_2$, $H^+\cdot (HCOH)_2$, $H^+\cdot (CH_3COH)_2$ and $H^+\cdot (CH_3COCH_3)_2$, and their subsequent dissociative recombination with electrons. The measured recombination rate coefficients are plotted in Figure 5 versus helium pressure.

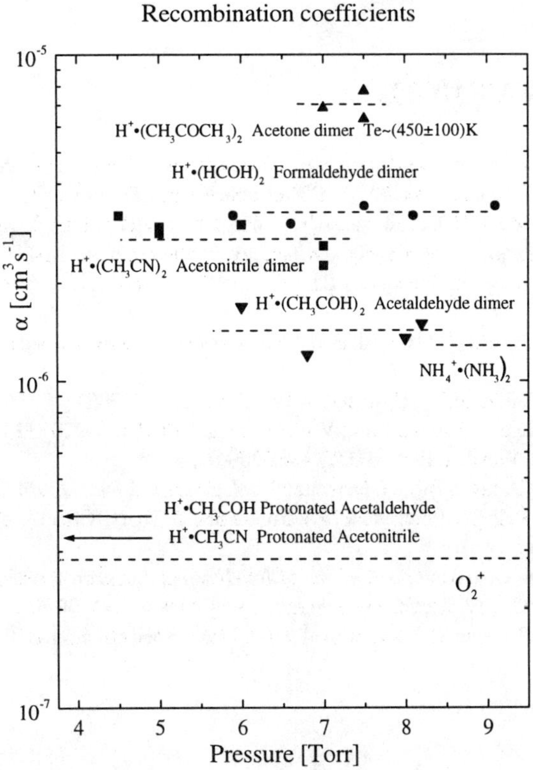

Figure 5. The measured recombination rate coefficients of the recombination of protonated dimers with electrons with indication of corresponding T_e. Included are also recombination coefficients of protonated acetaldehyde, protonated acetonitrile and O_2^+.

The evolutions of the electron energy distribution function and the electron temperature in the recombination-dominated flowing afterglow plasma were also measured.

Usually in FALP studies of recombination it was assumed that electron temperature is equal to gas temperature. This assumption is based on estimations of rate of loss of electron energy in collisions with He atoms. We measured actual electron temperature of the recombining electrons and we obtained that T_e is by 150-200 degrees higher than He temperature.

ACKNOWLEDGEMENTS

Thanks for financial support are due to GACR (202/00/1689, 202/99/D061 and 202/99/M004), GAUK (146/2000 and 171/2000 B FYZ MFF) and MSM 1132000002. The experiments were carried out with support of EU in the frame of the ETR network (HPRN-CT-2000-00142) and with support from EURATOM.

REFERENCES

1 R. P. Wayne, *Chemistry of Atmospheres, Clarendon*, (Oxford Press, 1985).
2 D. Smith, N. G. Adams and E. Alge, Planet Space Sci. **29**, 449 (1980).
3 R. G. Keesee and A. W. Castleman, Jr., J. Phys. Chem. Ref. Data **15**, 1011 (1986).
4 J.B.A. Mitchell and C. Rebrion-Rowe, Int. Rev. in Phys. Chem. **16**, 201 (1997).
5 J. Glosík, G. Bánó, R. Plašil, A. Luca, and P. Zakouřil, Int. J. Mass Spec. **189**, 103 (1999).
6 J. Glosík, P. Zakouřil, V. Hanzal, and V. Skalský, Int. J. Mass Spec. Ion Proc. **149/150**, 187 (1995).
7 R. Plašil, J. Glosík, and P. Zakouřil, J. Phys. B **32**, 3575 (1999).
8 J. Glosík, R. Plašil, P. Zakouřil and V. Poteriya, J. Phys.B **34**, 2781 (2000).
9 J. Glosík and R. Plašil, J. Phys.B **33**, 4483 (2000).
10 D. Smith, N. G. Adams, A. G. Dean and M. J. Church, J. Phys. D. **8**, 141 (1975).
11 J. Glosík, R. Plašil, V. Poteriya, P. Kudrna, and M. Tichý, Chem. Phys. Lett. **331**, 209 (2000).
12 D. Smith and N. G. Adams, in *Physics of Ion-Ion and Electron-Ion Collisions*, edited by F. Brouillard and J. W. McGowan (Plenum Press, New York, 1983).
13 D. Smith and P. Španěl, International J. of Mass Spectr. and Ion Processes **129**, 163 (1993).

An Electrostatic Storage Ring for Studying Electron-Ion Collisions at KEK

Tetsumi Tanabe,[1] Katsuhisa Chida,[1] Kouji Noda,[2] Ikuo Watanabe[3]
[1] *High Energy Accelerator Research Organization (KEK),Oho, Tsukuba 305-0801, Japan;*
[2] *National Institute of Radiological Sciences, Anagawa, Chiba 263-8555, Japan;*
[3] *Toshiba Corporation, Keihin Product Operations, Yokohama 230-0045, Japan*
Tetsumi.Tanabe@kek.jp

1. INTRODUCTION

Since the late 1980s, many storage rings with electron coolers have been operational, some of which have been used for studying atomic physics. Many new phenomena have been found, especially concerning the dissociative recombination of molecular ions with electrons.[1] The object of the present research is directed toward heavy molecular ions. One of the difficulties in studying low-charge and heavy molecular ions with the usual magnetic storage rings is the increasing magnetic rigidity with ion mass. An electrostatic storage ring is more suitable for such ions because the electrostatic rigidity is independent of the ion mass at a given energy. There are also many advantages of electrostatic storage rings, such as compactness, ease of construction and good reproducibility. The ring is expected to be very useful for the research of atomic and molecular science in the future. Pioneering work for the electrostatic storage ring was performed by S. P. Møller (the ring ELISA in Aarhus[2]), where spherical deflectors were used to balance the horizontal and vertical focusing forces. This focusing resulted in very narrow waists in the middle of the deflectors. However, it was observed that the beam lifetime decreased greatly with increasing beam intensity.[3] In order to avoid such a difficulty, we designed a ring using cylindrical deflectors instead of spherical deflectors.[4] The lattices using the cylindrical deflectors allow both solutions, with and without waists, in the middle of the deflectors. It is very attractive to expand the research object from light molecular ions mainly studied so far to heavy macro-molecular ions. In order to study the feasibility of the storage of bio-molecular ions in the ring, an electrospray ion source accompanied by an ion trap was constructed. An electron-beam target is indispensable for studying electron-ion collisions. In magnetic storage rings, electron coolers have been used to compress the phase space occupied by beams and at the same time act as an electron target for ion-electron collision research. On the other hand, in the electrostatic ring, the electron cooling device can function mainly as an electron-beam target, because the ion velocity is not necessarily high enough to cool the heavy-ion beam in the longitudinal directions. An electron-beam target, which is a compact electron cooler, has been designed. In this paper, we report on the design of the ring and the results of beam tests for the light ions and bio-molecular ions produced by an electron cyclotron resonance (ECR) ion source and an electrospray ion (ESI) source, respectively.

2. LAYOUT OF THE EQUIPMENT

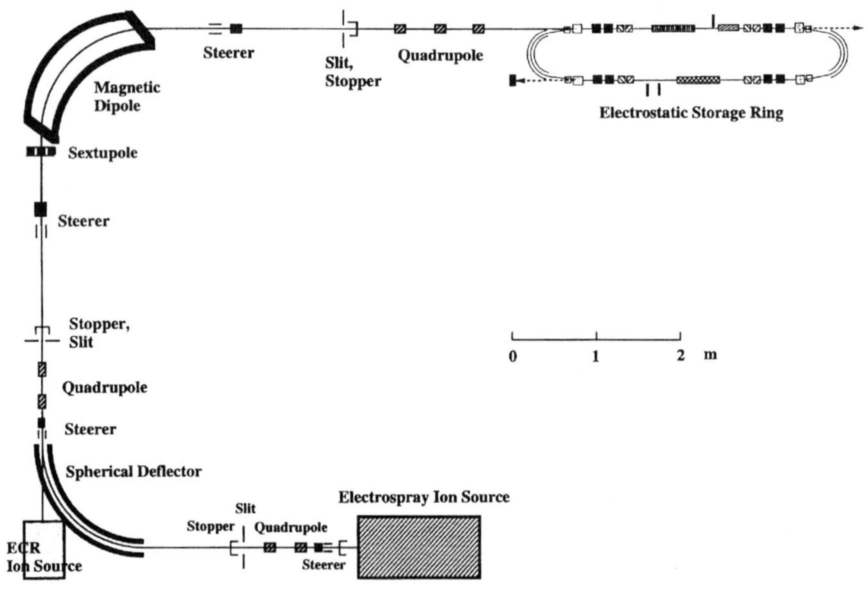

Figure 1. Entire layout of the ion sources, the mass spectrometer and the electrostatic storage ring.

Light atomic and molecular ions are produced in a compact ECR ion source (see Fig. 1) with an acceleration voltage of 20 kV, and are injected into a beam-analyzing dipole magnet. A momentum-analyzed beam is then injected into the ring through a matching section consisting of an electrostatic quadrupole triplet. On the other hand, bio-molecular ions with an energy of

30 keV/charge are produced by the ESI source and injected into the ring after being bunched and mass analyzed. The mass analyzer consists of an electrostatic spherical deflector, a sextupole magnet and a dipole magnet that also serves to analyze the momentum of ion-beams from the ECR source. The ECR and ESI sources can be used alternatively without breaking the vacuum in the merging section.

3. ION SOURCES

3.1 Electron Cyclotron Resonance Ion Source

The design of the compact ECR source, which is driven by a 2.45 GHz microwave power source, is based on a straightforward application of the method adopted for conventional ECR ion sources for multiply charged ions.[5] Gaseous materials are fed into the source and ionized. The source can produce molecular ions as well as low and highly charged atomic ions with moderate intensities.

3.2 Electrospray Ion Source

The ESI source (JEOL, model MS-API15, Japan) was installed on a high-voltage terminal. A sample solution at flow rates of between 1 and 10 $\mu l/m$ enters the ESI source chamber through a capillary. The capillary is electrically biased at a few kilovolts relative to a counter electrode. The resulting field at the capillary tip charges the surface of the emerging liquid, dispersing it by Coulomb forces into a fine spray of charge droplets. Driven by the electric field, the droplets move towards the counter electrode. A flow of nitrogen gas at a flow rate of about 5 l/m helps to nebulize the solution and hastens evaporation of the solvent from each droplet, decreasing its diameter as it drifts towards the orifice of the counter electrode. Consequently, the charge density on the droplet surface increases until the Coulomb repulsion exceeds the surface tension, resulting in producing charged daughter droplets and ions by Coulomb explosion. A syringe pump controls the flow of the solution, typically a water-methanol mixture, containing the sample.

Ions created by the ESI source are then conveyed to the entrance of a linear octupole RF ion trap by way of a quadrupole RF ion-guide. The ion trap accumulates ions initially, and then ejects the trapped ions quickly in order to transfer them to the storage ring as a bunch. Trap electrodes are electrically isolated into five pieces, in order to independently apply a DC voltage superimposed on the RF voltage to each electrode. This makes it possible to form an axial confinement field during accumulation, and also to form an axial gradient field during ejection, in collaboration with the entrance and exit electrodes. The ions trapped into the potential well are thermalized through collisions with He gas fed into the trap. After accumulation for a period of a few seconds, ions are released by abruptly changing each DC bias with a switching time of about 0.4 μs and transferred to another quadrupole RF ion-guide. The thus-obtained bunched beam is then accelerated up to 30 keV/charge by an accelerator tube.

The ion trap and the beam bunching system dramatically increased the pulse beam current approximately by three orders of magnitude in comparison with DC operation.

4. ELECTROSTATIC STORAGE RING

The layout of the ring is given in Fig. 2 and the main parameters are listed in Table 1. A racetrack lattice consists of two 160° deflectors, four 10° deflectors, which allow beam injection and extraction, and four quadrupole doublets. For the 160° deflectors, a cylindrical shape has been adopted. In this ring, there are four types of stable regions with and without waists of the beam envelope in the middle of the deflectors.

The type of beam injection is single-turn injection. During injection, the high voltage for the first 10° deflector is turned off, and after filling the beam in the ring it is turned on. The rising and falling times of the high voltage power supplies used in the injection are about 0.1 μs, which is sufficiently short compared with the circulation time of ions.

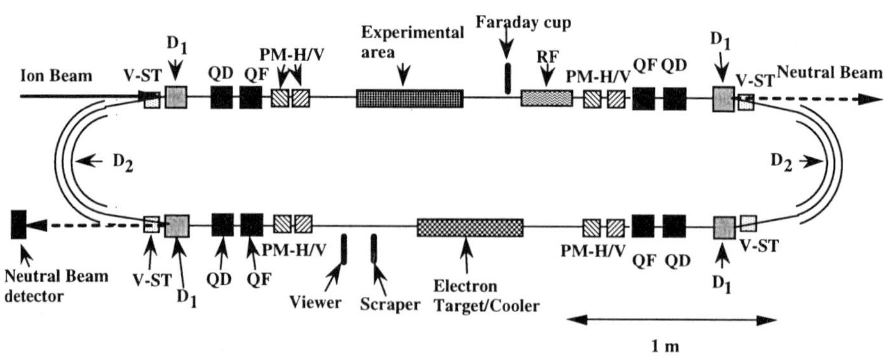

Figure 2. Layout of the electrostatic storage ring. D_1 and D_2 indicate 10° and 160° deflectors, QF and QD the horizontally focusing and defocusing electrostatic quadrupoles, PM-H and PM-V the horizontal and vertical position monitors, RF the drift-tube RF system and V-ST the vertical steerer.

Table 1. Design parameters for the electrostatic storage ring.

Maximum energy	30 keV/charge
Circumference	8.136 m
Revolution time	14 μs (O^+, 30 keV)
160° cylindrical deflectors	
Central radius	250 mm
Gap	30 mm
Nominal voltage	± 3.6 kV
10° parallel plate deflectors	
Gap	60 mm
Length	92 mm
Nominal voltage	± 2.6 kV
Quadrupoles	
Inscribed radius	25 mm
Length	100 mm
Nominal voltage	± 0.75 kV
RF	
Drift tube length	203 mm
Voltage	<10 V

Several kinds of beam diagnostic elements are installed in the straight sections, as shown in Fig. 2. The injected beam profile is observed destructively with a viewer consisting of an alumina (doped with oxide chromium) screen and a view port. The surface of the screen is covered by a tungsten mesh. This allows a measurement of the beam current hitting the screen. The injected beam current is also more accurately measured by a Faraday cup. Four horizontal and vertical electrostatic position monitors provide position information concerning bunched beams nondestructively. A beam scraper defines the maximum beam size both horizontally and vertically, which can also be used for the beam position and profile measurements.

A vacuum pumping system is composed of four ion pumps and six titanium sublimation pumps. The entire ring is bakeable at a temperature of less than 300 °C. The vacuum pressure of the ring is on the order of 10^{-11} Torr.

A neutral beam produced by collisions with the residual gas is measured by a micro-channel plate with a phosphor anode (32 mm in diameter) installed in a vacuum extension 1.25-m downstream of the 10° deflector, as shown in Fig. 2. The count rate of the neutral beam is proportional to the number of stored ions at a constant residual-gas pressure, which gives exact information about the beam lifetime. Furthermore, a projection of the circulating beam can be observed from the neutral beam profile on the phosphor screen. The position and size of the neutral beam profile are useful for adjusting the ring parameters.

5. STORAGE OF ATOMIC AND MOLECULAR IONS

The 20-keV beams from the ECR ion source were momentum-analyzed (typical momentum resolution: $\Delta p/p \sim 10^{-3}$) and injected into the ring after being chopped. The injected beam size observed by the viewer is about 5 mm in diameter.

The change in the number of stored atomic ions with time is expressed by $N = N_0 \exp(-t/\tau)$, where τ is the lifetime. If the ring is operated at a stable working point, the finite lifetime of ions in the ring is mainly determined by interactions with the residual gas. First, ion beams are multiply scattered by the Coulomb field of the residual gas. The ions moving in a direction that exceeds the acceptance angle of the ring are lost. The 1/e lifetime, (τ_{MS}), for losses due to multiple scattering is proportional to Θ^2_{max}, where Θ_{max} is a maximum acceptance angle.[6] Second, ions which change their charge state by capturing or stripping electrons deviate from the central orbit and are lost. The lifetimes by capture, (τ_{cap}), and stripping, (τ_{strip}), can be expressed by the cross sections (σ_{cap} and σ_{strip}) as $\tau_{cap} = 1/\rho\upsilon\sigma_{cap}$ and $\tau_{strip} = 1/\rho\upsilon\sigma_{strip}$ (ρ, density of the residual gas atoms; υ, the velocity of ions), respectively. The cross sections for the residual gas interactions of atomic ions are almost energy-independent at these low energies. Therefore, the lifetimes are inversely proportional to the velocity and increase with a decrease in the ion velocity. Unlike in the case of multiple scattering, the lifetimes due to the charge-changing process do not depend on the acceptance angle. The entire lifetime (τ) is related to the lifetimes due to multiple scattering, and those due to capture and stripping by $1/\tau = 1/\tau_{MS} + 1/\tau_{cap} + 1/\tau_{strip}$.

There are four stable regions with different tune values for betatron oscillations. We studied the characteristics of a stored beam at a working point in each stable region for 20-keV O^+. The lifetimes depend greatly on the working points. The difference in the lifetimes has been deduced to come from the difference in the multiple-scattering lifetimes, which depend on the maximum acceptance angle. The dependence of the coasting-beam lifetime on the injected beam current was measured for a 20-keV O^+ beam. At low currents, beams decay almost exponentially with a lifetime of about 12 s, while at higher currents, the beams decay faster initially. The lifetime does not depend much on the intensity at a current of less than about 100 nA, which corresponds to 1×10^7 injected particles. This current is tolerable for practical use, like in atomic collision experiments. A rapid decrease in the beam intensity observed at high currents, which does not obey an exponential decay law, seems to be caused by intra-beam scattering, which is the scattering between beam particles. In the ELISA ring in Aarhus, spherical deflectors were recently replaced by cylindrical deflectors, and a similar lifetime dependence on beam intensity has been reported.[7]

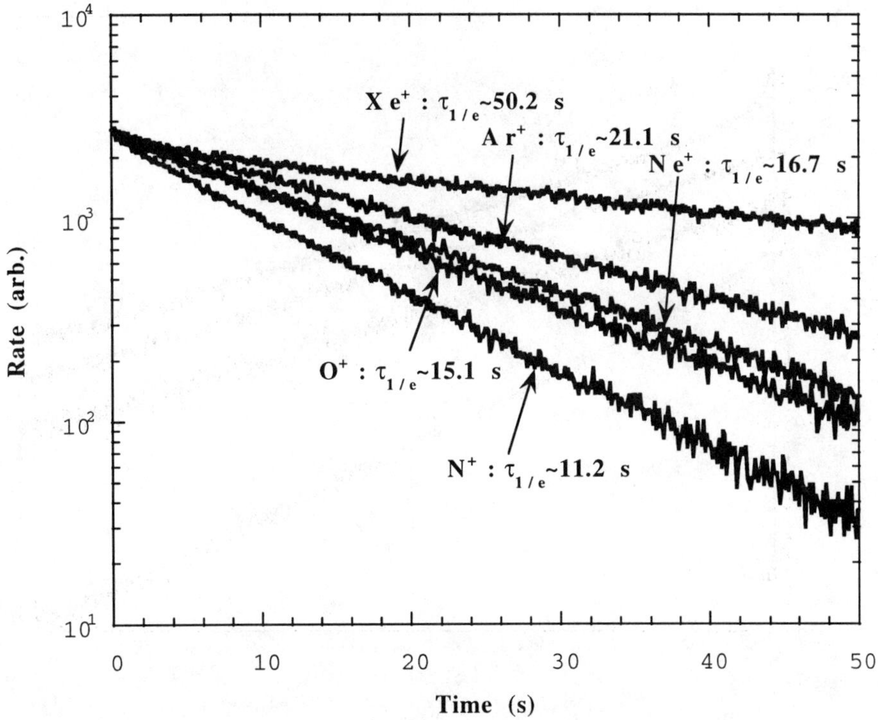

Figure 3. Time dependence of the neutral beam production rate for 20-keV atomic ions at an injected current of 10~20 nA. The vacuum pressure was about 3×10^{-11} Torr. The decay curves are arbitrarily scaled in magnitude to facilitate a comparison.[4]

So far, we have stored atomic and molecular ions of N^+, O^+, Ne^+, Ar^+, Xe^+, N_2^+, NO^+, O_2^+, and H_2O^+ at 20 keV. These ions are stored at the same voltages for the deflectors and quadrupoles. The decays of the beam intensity for atomic ions are shown in Fig. 3.

The lifetimes increase with the mass number (A). This is partly due to the fact that the lifetimes (τ_{MS}, τ_{cap} and τ_{strip}) are all proportional to $\sqrt{A}$ for a fixed energy, although they also depend on the cross sections. For molecular ions, the beams initially consists of many vibrationally excited states. The cross sections for the residual gas collisions are different for each excited state. Furthermore, the molecular ions with dipole moments disintegrate into the ground state during storage by spontaneous emission. Thus, the decay of the beam is not necessarily expressed by a single exponential, but is given by a multi-exponential. The decay of the intensities of molecular ions is shown in Fig. 4.

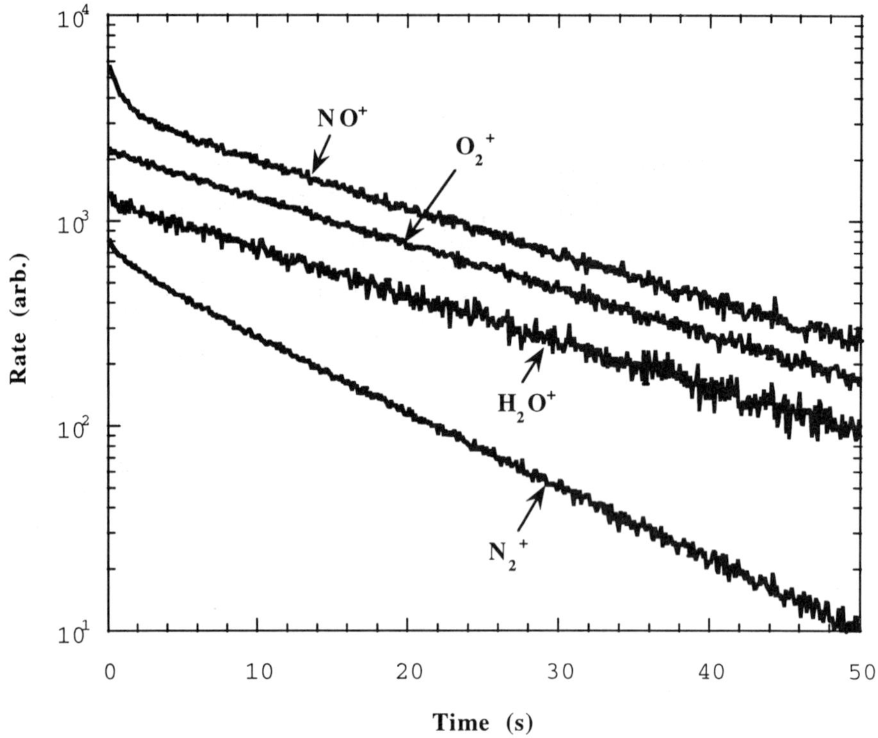

Figure 4. Time dependence of the neutral beam production rate for 20-keV molecular ions at an injected current of 10~20 nA. The vacuum pressure was about 4×10^{-11} Torr. The decay curves are arbitrarily scaled in magnitude to facilitate a comparison.[4]

6. STORAGE OF BIO-MOLECULAR IONS

The ions from the ESI source were accumulated in the trap for a period of a few seconds. The ions were then ejected in a bunch and accelerated to an energy of 20 keV/charge. After being mass analyzed, the ions were injected into the ring. Fig. 5 shows the neutral particle production rate as a function of the storage time for the singly and multiply charged positive ions of the arginine dimer, reserpine, angiotensin I, insulin chain B and insulin from bovine pancreas which have molecular weights of about 331, 609, 1298 3499 and 5738, respectively. As can be seen in the Figure, the beams have lifetimes of 12 s - 20 s, which are surprisingly long because large bio-molecules have usually been assumed to be fragile, resulting in a short surviving time in collisions with residual gas. The cross sections for the collision with residual gas deduced from the beam lifetime are on the order of 10^{-15} cm^2.

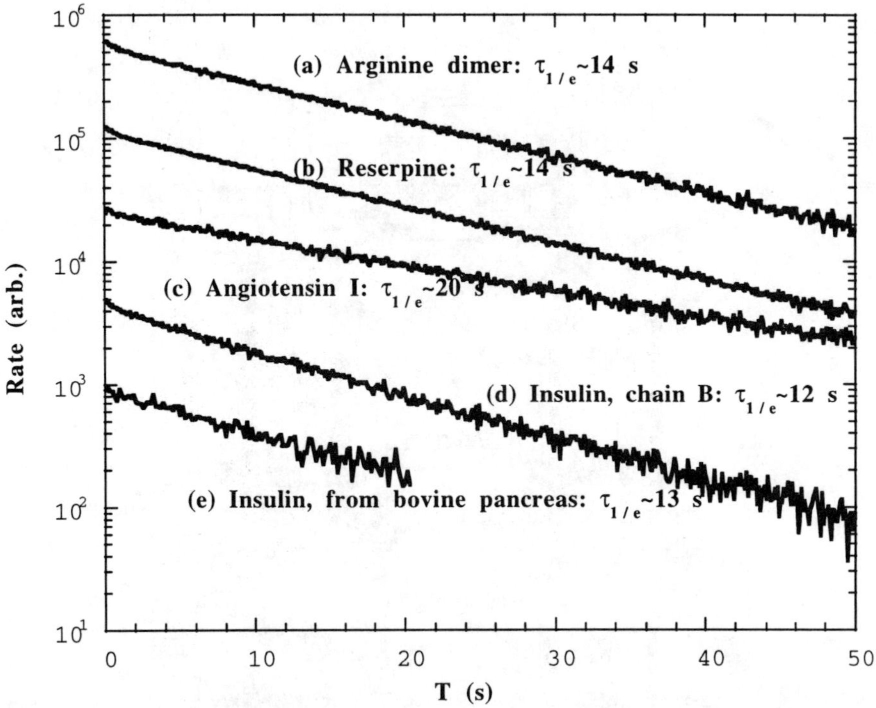

Figure 5. Time dependence of the neutral beam production rates for 20-keV/charge bio-molecular ions: (a) arginine dimer (charge, 1+; M.W., ~331), (b) reserpine (1+; ~609), (c) angiotensin I (1+; ~1298), (d) insulin chain B (3+; ~3499) and (e) insulin from bovine pancreas (5+; ~5738). The 1/e lifetimes are indicated in the Figure. The vacuum pressure was about 4×10^{-11} Torr. The decay curves are arbitrarily scaled in magnitude to facilitate a comparison.

7. DESIGN OF AN ELECTRON BEAM TARGET

Fig. 6 shows an electron beam target with an interaction range of 20 cm. The structure is almost the same as an ordinary electron cooling device.[8] Electrons are emitted from a thermo-cathode with a diameter of 20 mm and guided in a uniform solenoid field after acceleration. The main design parameters are: maximum energy, 100 eV; maximum current, 2 mA; and maximum solenoid field, 120 Gauss. This will be installed in one of the straight sections of the ring (see Fig. 2). Steering of the ion beam at the toroidal field is corrected by a pair of electrostatic horizontal steerers at the entrance and the exit of the electron beam device.

Another choice is an adiabatic-expansion-type electron beam target[9] in which electrons are emitted from a cathode with a diameter of 5 mm in a

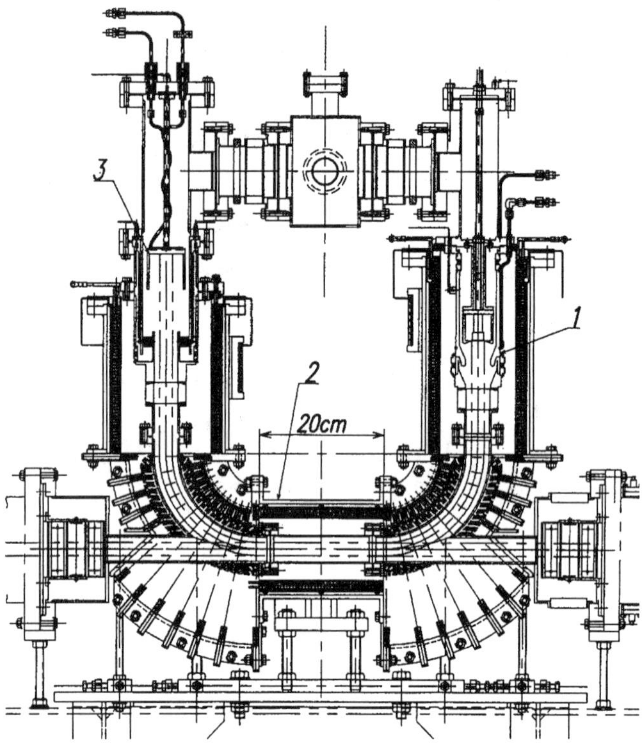

Figure 6. Schematic drawing of the electron beam target with (1) electron gun, (2) ion-electron interaction region, and (3) electron collector.

solenoid field of 1 kG and expanded to 20mm. In this case, the electron beam target can also function as a cooler for light ions.

8. CONCLUSION

It is shown here that an electrostatic storage ring can be used to store atomic and molecular ions, including bio-molecular ions, at sufficiently long lifetimes with moderate intensities. The lifetimes are long enough to cool vibrationally excited ions for those molecules with dipole moments. The ring can store ion beams independent of their masses. Therefore, it is especially suitable for studying macromolecules, like cluster or bio-molecular ions. If we introduce electrons into the ring, they can open up a way to conduct research on atomic collisions. The dissociative recombination for bio-molecular ions is one of the most interesting research items. Thus, the ring is very attractive for studying chemistry and biology as well as physics.

ACKNOWLEDGMENTS

The authors thank R. Arakawa for valuable suggestions on the design of the ESI source, M. Wada, I. Katayama and K. Okada for discussions on the design of the ion trap, M. Yoshizawa for analyzer controlling and Y. Ito for preparing a sample solution of bio-molecules. This work was performed under a Grant-in-Aid for Scientific Research (A) from the Ministry of Education, Culture, Sports, Science and Technology.

REFERENCES

1 T. Tanabe, I. Katayama, N. Inoue, K. Chida, Y. Arakaki, T. Watanabe, M. Yoshizawa, S. Ohtani, and K. Noda, Phys. Rev. Lett. **70**, 422 (1993).
2 S. P. Møller, Nucl. Instr. and Meth. **A 394**, 281 (1997).
3 S. P. Møller, U.V. Pedersen, Proceedings of the 1999 Particle Accelerator Conference, New York, 1999, p. 2295.
4 T. Tanabe K. Chida, K. Noda and I. Watanabe, Nucl. Instr. and Meth. **A** , **482**, 595 (2002).
5 M. Sekiguchi, H. Matsushita, I. Jonoshita, M. Kabasawa, Rev. Sci. Instrum. **69**, 837 (1998).
6 D. Habs et al., Nucl. Instr. and Meth. **B43**, 390 (1989).
7 S. P. Møller et al., Proceedings of the 7th European Particle Accelerator Conference, EPAC2000, Vienna, 2000, p. 788.
8 T. Tanabe, K. Noda, T. Honma, M. Kodaira, K. Chida. T. Watanabe, A. Noda, S. Watanabe, A. Mizobuchi, M. Yoshizawa. T. Katayama, H. Muto and A. Ando, Nucl. Instr. and Meth. **A307**, 7 (1991).
9 T. Tanabe, K. Noda, T. Yosiyuki, M. Hosino, I. Watanabe, K. Chida, T. Watanabe, T. Honma, I. Katayama, Y. Arakaki et al., Nucl. Instr. and Meth. **A441**, 326 (2000).

Studies of Dissociative Recombination in CRYRING

M. Larsson[1], N. Djuric[2], G.H. Dunn[2], A. Neau[1], A.M. Derkatch[1], F. Hellberg[1], S. Kalhori[1], D.B. Popovic[2], J. Semaniak[3], Å. Larson[1], and R. Thomas[1]
[1]Department of Physics, SCFAB, Stockholm University, SE-106 91 Stockholm, Sweden[2]:JILA, University of Colorado and National Institute of Standards and Technology, Boulder, Colorado 80309-0440:[3]Institute of Physics, Pedagogical University, 25-406 Kielce, Poland

1. INTRODUCTION

This article describes some of the most recent experiments at the ion storage ring CRYRING at the Manne Siegbahn Laboratory, Stockholm University. A comprehensive review of recent work at ion storage rings has recently been published,[1] and there are several other reviews of recent date,[2,3] as well as the proceedings from the previous dissociative recombination (DR) meeting at Nässlingen in Sweden.[4] Finally, a brief review of three-body breakup dynamics has recently been published.[5]

2. EXPERIMENTS AT CRYRING

The ion storage ring CRYRING is operated by the Manne Siegbahn Laboratory, which serves as a national user facility. It is located on the Stockholm University campus, and formally the laboratory is a unit within the university. The funding is partly from the Swedish Science Council, and partly from Stockholm University. In order to obtain beam time at CRYRING, a formal proposal must be submitted. Proposals are evaluated by an independent, external, program committee, and the beam time is distributed based on the recommendations by the program committee.

CRYRING experiments not covered in this volume include cross sections and branching ratios for cluster ions of atmospheric interest, such as $H_5O_2^+$, $H_7O_3^+$, $NO^+(H_2O)_n$, and $H^+(NH_3)_n$. The Principal Investigator is Jan

Pettersson from Göteborg University in Sweden. The results are not yet published. In addition, resonant ion pair formation experiments have been performed for HD^+,[6,7] OH^+,[7] NO^+,[8] and HF^+,[9] as well as H_3^+. The latter is briefly described in this article. Lastly, branching ratios for the polyatomic molecular ions H_3O^+ [10] and $HCNH^+$[11] have been measured.

3. BEAM EXPERIMENTS IN A STORAGE RING

There are several advantages with using an ion storage ring for the study of dissociative recombination (DR). The long storage time, which can be tens of seconds, allows metastable components to be removed and vibrationally excited levels to be relaxed. The high beam energy, which is in the MeV range, is an advantage in several respects. First, the charge transfer cross section in collisions between the stored ion beam and residual gas molecules is small at high beam energy, which significantly reduces the background. Second, the use of energy resolving surface barrier detectors is facilitated by the high beam energy. This is a very important aspect, not least in measurements of product branching ratios for polyatomic molecular ions. Third, the high beam energy allows merged-beams studies of fairly heavy ions. In CRYRING, merged-beams experiments with ions up to m = 73 amu have been carried out, but there is no reason to believe that this is an upper limit. Fourth, one can use special tricks that make use of the fact that the neutral particles arrive at the detector at high speed; for example, the particles can penetrate through a foil (which is exploited in experiments on three-particle dissociation dynamics).[12]

In the ion storage rings, the electron beam is merged with the ion beam over a distance of the order of a meter (it varies from ring to ring). The use of merged-beams in atomic and molecular physics has recently been reviewed.[13] In merged-beams DR studies, one of the beams is composed of particles very much lighter than the particles in the other beam. The resolution in measurements of cross sections is determined by the energy spread in the beam of the light particles (electrons).

The electron-velocity distribution in the ion's rest frame is given by

$$f(\mathbf{v}_e) = \frac{m_e}{2\pi kT_\perp} \frac{m_e}{2\pi kT_{||}} \times \exp -\frac{m_e v_\perp^2}{2kT_\perp} - \frac{m_e(v_{||} - v_d)^2}{2kT_{||}} \qquad (1)$$

where v_d is the detuning velocity (shown in Fig. 1), and $v_\perp$ and $v_{||}$ are the longitudinal and transverse electron velocities, respectively. The dissociative recombination rate coefficient in the electron cooler is given by

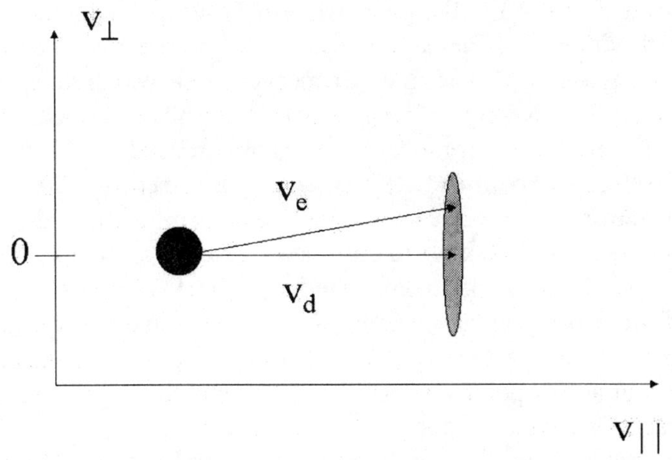

Figure 1. An ion colliding with electrons with flattened velocity distribution. The coordinate system is chosen to be the ion's rest frame.

$$\langle \sigma v \rangle = \int \sigma(v) v \, f(\mathbf{v}_e) \, d^3\mathbf{v}_e. \tag{2}$$

The resolution in an experiment with merging electron and ion beams is determined by the electron velocity spread. Owing to the invention of the adiabatically expanded electron beam,[14] it has been possible to reduce the transverse electron temperature, $kT_\perp$, from 100 meV to about 1 meV in the CRYRING electron cooler. This has allowed for the observation of structures in the DR cross sections for several molecules. One example is HeH^+ for which the cross section has been measured at different electron temperatures in CRYRING.[15, 16]

Resonances in the low-energy part of the H_3^+ DR cross section have never been observed. We decided to use the ultra-cold CRYRING electron beam to search for these resonances.

4. RECENT MERGED-BEAMS STUDIES OF H_3^+

The dissociative recombination of H_3^+ has been the most debated subject in the field of DR, and the present conference is no exception. The experimental and theoretical situation at the time of the *Royal Society Discussion Meeting* in London in February 2000 has been reviewed.[17,18] More recent experimental and theoretical results are presented at the present conference. In terms of H_3^+ experiments at ion storage rings, recent cross section results from ASTRID give a thermal rate constant at 300 K of $(1.0\pm0.2)\times10^{-7}$ cm^3s^{-1}.[19] Kinematic correlations have been studied by means of imaging techniques,[20, 21] and in experiments at TSR which make use of the Coulomb explosion imaging technique it has been shown that H_3^+ stored in a ring relaxes to its lowest vibrational level in a few seconds.[22]

In June 2001, an experiment with H_3^+ was performed in CRYRING. The ultra-cold electron beam was used in this experiment, and for the first time structures in the cross section were observed, as can be seen in Fig. 2. This is a breakthrough in the experimental study of the DR of H_3^+ and will be important for the detailed comparison of experiment and theory. In Fig. 2 a relative cross section is given. The absolute cross section was measured in the experiment, but at this point we prefer to show only the relative results since not all corrections have been applied to the data yet.

The cross section was also measured for different ion source conditions. A preliminary analysis shows that the cross section is not independent of the source conditions, indicating that the internal state distribution affects the cross section. Since only the zeroth vibrational level remains populated after storage for a few seconds, it must be the rotational distribution that is critical. An interesting observation is that the high-energy peak at 10 eV is unaffected by the source conditions and thus serves as a normalisation peak. The ion-pair channel $H_3^+ + e \rightarrow H^- + H_2^+$ opens at an electron energy of about 5.4 eV. It has been studied by Peart et al.[23] and by Mitchell and co-workers.[24] Fig. 3 shows the results from the CRYRING experiment in June 2001. The absolute cross section was measured, but since the data analysis has not been completed, we give a relative cross section. The data of Ref. 24 are absolute, but since they are compared with relative cross sections, we have labelled the scale as "arbitrary units". The shapes of the cross sections are quite different. Whereas the cross section from CRYRING shows no distinct threshold, the data of Mitchell and co-workers, which were claimed to be due to H_3^+ ions in their zeroth vibrational levels, show a threshold. The cross section from CRYRING peaks at a higher electron energy close to that of the high-energy DR peak. The question is why the data from the two experiments are quite different, although there are arguments that both pertain to vibrational ground state ions. In order to shed some light on this, we performed wave packet calculations using the potential curves of Michels and Hobbs.[25]

The results are shown in Fig. 4 and are quite illustrative. A single Gaussian-like peak results from the calculations, which assume population of $v = 0$ in H_3^+. A double peak is obtained when $v = 1$ in H_3^+ is populated. A comparison of Fig.s 3 and 4 makes us believe that the results of Mitchell and co-workers were obtained for an H_3^+ beam containing a mixture of $v = 0$ and 1, whereas the experiment in CRYRING was performed with vibrational ground state ions.

While further work with the data analysis of the experiment in June 2001

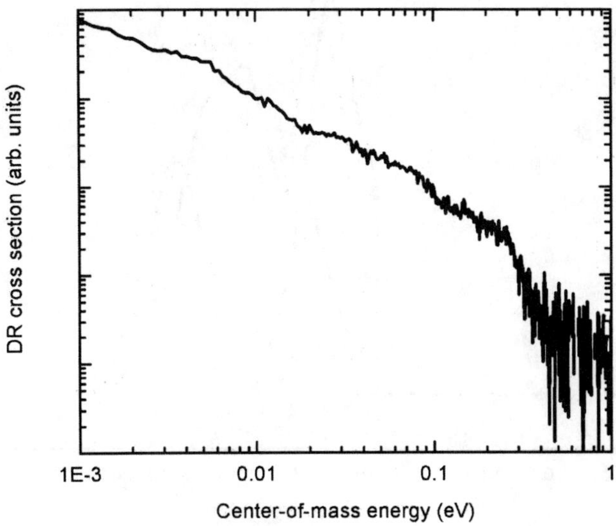

Figure 2. Relative cross section for dissociative recombination of H_3^+ measured in CRYRING.

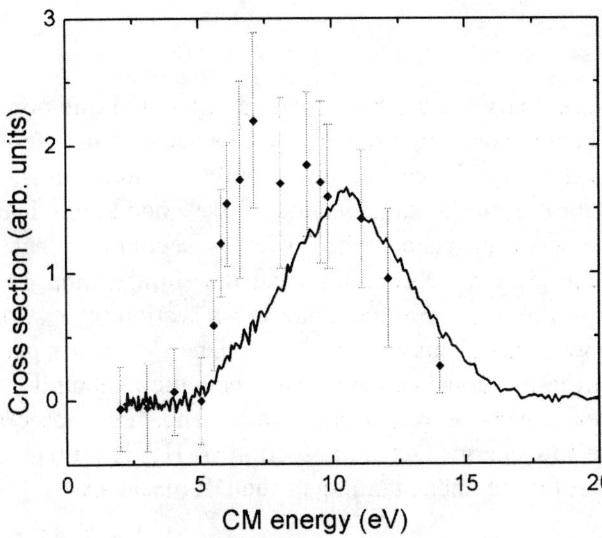

Figure 3. The new data from CRYRING for the process $H_3^+ + e \rightarrow H^- + H_2^+$ are represented by the full drawn curve; the black diamonds show the results from Ref. 24 when the ion source was operated at high pressure to produce vibrationally quenched ions.

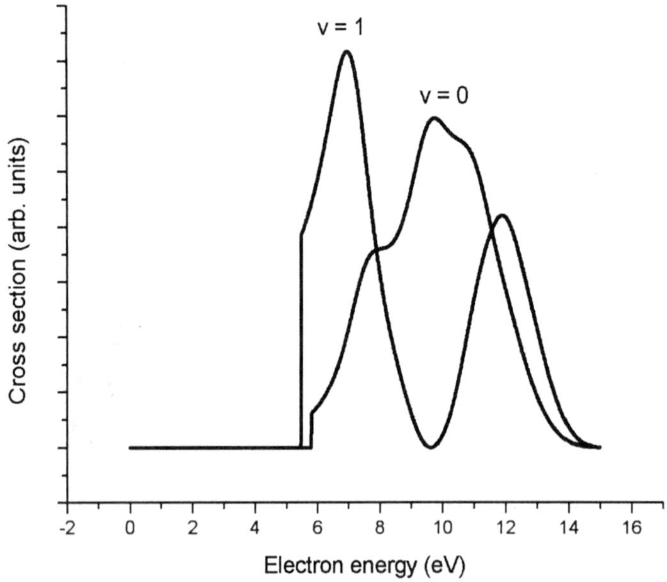

Figure 4. Calculated cross sections (in arbitrary units) for production of $H^- + H_2^+$. The calculations were performed with a wave packet technique and potential curves from Ref. 25. The two curves are obtained assuming H_3^+ populating v = 0 and 1, respectively.

is needed, a preliminary picture has emerged already at this point. The ion-pair formation data in combination with wave packet calculations suggest that the results are valid for H_3^+ ions in their ground vibrational level. This means that the DR data are also obtained under these conditions. The rotational distribution seems to influence the DR cross section so that it changes depending on the ion source condition and the storage time. Long storage times (20–40 s) gave in general a smaller cross section than when a storage time of 4–5 s was used. This is not well understood at this point, and the hypothesis that the rotational distribution affects the magnitude of the cross section requires further experimental work. The first observation of a structure in the low-energy DR cross section of H_3^+ is a breakthrough and will be important for the understanding of the DR mechanism.

ACKNOWLEDGEMENT

The authors are grateful to the staff members of the Manne Siegbahn Laboratory for their assistance and help in the experiment. This work was supported by the Swedish Research Council, the Swedish Foundation for International Cooperation in Research and Higher Education, and the Fifth Framework Programme of the EU (HPRN-CT-2000-00142).

REFERENCES

1 M. Larsson, in *Gas Phase Ion Chemistry, Vol. 4*, eds. N.G. Adams and L. Babcock (JAI Press Inc.: Greenwich, CT, 2002), p. 179.

2 M. Larsson, in *Adv. Ser. Phys. Chem., Vol. 10: Photoionization and Photodetachment* (World Scientific: Singapore, 2000), p. 693.

3 M. Larsson, in *The Physics of Electronic and Atomic Collisions: XXI International Conference, AIP Conference Proceedings 500*, eds. Y. Itikawa, K. Okuno, H. Tanaka, and M. Matsuzawa (American Institute of Physics: New York, 2000), p. 44.

4 M. Larsson, J. B. A. Mitchell, and I. F. Schneider (eds.), *Dissociative Recombination: Theory, Experiment and Applications IV* (World Scientific, Singapore, 2000).

5 M. Larsson and R. Thomas, Phys. Chem. Chem. Phys. **3**, 4471 (2001).

6 W. Zong, G. H. Dunn, N. Djuric, M. Larsson, C. H. Greene, A. Al-Khalili, A. Neau, A. M. Derkatch, L. Vikor, W. Shi, A. Le Padellec, S. Rosén, H. Danared, and M. af Ugglas, Phys. Rev. Lett. **83**, 951 (1999).

7 Å. Larson, N. Djuric, W. Zong, C. H. Greene, A. E. Orel, A. Al-Khalili, A. M. Derkatch, A. Le Padellec, A. Neau, S. Rosén, W. Shi, L. Vikor, H. Danared, M. af Ugglas, M. Larsson, and G. H. Dunn, Phys. Rev. A **62**, 042707 (2000).

8 A. Le Padellec, N. Djuric, A. Al-Khalili, H. Danared, A. M. Derkatch, A. Neau, D. B. Popovic, S. Rosén, J. Semaniak, R. Thomas, M. af Ugglas, W. Zong, and M. Larsson, Phys. Rev. A **64,** 012702 (2001).

9 N. Djuric, G. H. Dunn, A. Al-Khalili, A. M. Derkatch, A. Neau, S. Rosén, W. Shi, L. Vikor, M. Larsson, A. Le Padellec, H. Danared, and M. af Ugglas, Phys. Rev. A **64,** 022713 (2001).

10 A. Neau, A. Al-Khalili, S. Rosén, A. Le Padellec, A. M. Derkatch, W. Shi, L. Vikor, M. Larsson, J. Semaniak, R. Thomas, M. B. Någård, K. Andersson, H. Danared, and M. af Ugglas, J. Chem. Phys. **113**, 1762 (2000).

11 J. Semaniak, B. F. Minaev, A. M. Derkatch, F. Hellberg, A. Neau, S. Rosén, M. Larsson, H. Danared, A. Paál, and M. af Ugglas, Astron. Astrophys. Suppl. Ser. **135**, 275 (2001).

12 S. Datz, R. Thomas, S. Rosén, M. Larsson, A. M. Derkatch, F. Hellberg, and W. van der Zande, Phys. Rev. Lett. **85**, 5555 (2000).

13 R. A. Phaneuf, C. C. Havener, G. H. Dunn, and A. Müller, Rep. Progr. Phys. **62,** 1143 (1999).

14 H. Danared, G. Andler, L. Bagge, C. J. Herrlander, J. Hilke, J. Jeansson, A. Källberg, A. Nilsson, A. Paál, K.-G. Rensfelt, U. Rosengård, J. Starker, and M. af Ugglas, Phys. Rev. Lett. **72**, 3775 (1994).

15 A. Al-Khalili, H. Danared, M. Larsson, A. Le Padellec, R. Peveralll, S. Rosén, J. Semaniak, M. af Ugglas, L. Vikor, and W. J. van der Zande, Hyperfine Interact. **114,** 281 (1998).

16 T. Tanabe, I. Katayama, S. Ono, K. Chida, T. Watanabe, Y. Arakaki, Y. Haruyama, M. Saito, T. Odagiri, K. Hosno, K. Noda, T. Honma, and H. Takagi, J. Phys.B **31**, L297 (1998).

17 M. Larsson, Phil. Trans. R. Soc. Lond. A **358,** 2433 (2000).

18 A. E. Orel, I. F. Schneider, and A. Suzor-Weiner, Phil. Trans. R. Soc. Lond. A **358**, 2445 (2000).

19 M. J. Jensen, H. B. Pedersen, C. P. Safvan, K. Seirsen, X. Urbain, and L. H. Andersen, Phys. Rev. A **63,** 052701 (2001).

20 D. Strasser, L. Lammich, S. Krohn, M. Lange, H. Kreckel, J. Levin, D. Schwalm, Z. Vager, R. Wester, A. Wolf, and D. Zajfman, Phys. Rev. Lett. **86**, 779 (2001).

21 R. Thomas, S. Rosén, F. Hellberg, M. Larsson, R. Vane, and S. Datz, in *Abstracts of Contributed Papers, 22nd International Conference on Photonic, Electronic and Atomic Collisions*, Santa Fe, NM, 18-24 July, 2001, p. 348.

22 H. Kreckel, S. Krohn, L. Lammich, D. Schwalm, D. Strasser, R. Wester, A. Wolf, and D. Zajfman, *Abstracts of Contributed Papers, 22nd International Conference on Photonic, Electronic and Atomic Collisions*, Santa Fe, NM, 18-24 July, 2001, p. 342.

23 B. Peart, R. A. Forrest, and K. T. Dolder, J. Phys. B **12**, 3441 (1979).

24 F. B. Yousif, P. Van der Donk and J. B. A. Mitchell, J. Phys. B **26**, 4249 (1993).

25 H. H. Michels and R. H. Hobbs, Astrophys. J. **286**, L27 (1984).

Recent Studies of Three-Body Fragmentation Dynamics via Dissociative Recombination in CRYRING

R. Thomas,[1] S. Datz,[2] M. Larsson[1], W. J. van der Zande,[3] F. Hellberg,[1] A. Petrignani,[3] S. Rosén,[1] A.M. Derkatch,[1] A. Neau[1], and C. R. Vane[2]
[1]*Department of Physics, SCFAB, Stockholm University, SE-106 91 Stockholm, Sweden;* [2] Oak Ridge National Laboratory, Oak Ridge, TN 37831-6377, USA; [3] FOM Instituut AMOLF, PO Box 41883, 1009 DB Amsterdam, The Netherlands
rdt@physto.se

1. INTRODUCTION

Much recent work in the field of dissociative recombination (DR) has been related to the break-up of polyatomic ions since, in spite of the apparent simplicity of DR, developing a general theory to predict product branching ratios for even the simplest polyatomic ions, e.g. XH_2^+, has proven to be difficult. Early models suggested that such ions would predominantly fragment to H + XH. However, all recent storage ring studies for such ions show a propensity for three body breakup, i.e. X + H + H.[1-4] In the last year, the dynamics of some of these systems have been investigated, with results reported for H_2O^+ [5] and H_3^+,[6] and a brief review was also published.[7] Using trajectory calculations, we have also investigated the dynamics on some of the potential surfaces involved in the DR of H_2O^+.[8] The group at TSR looked at the dynamics in, and competition between, the two and three body fragmentation dynamics of H_3^+. Using a statistical model,[9] they reached good agreement with the experimental data previously obtained at CRYRING[1] as a function of reaction energy.

In this article, a selection of results from recent experimental work at CRYRING on H_2O^+, NH_2^+ and CH_2^+ is discussed. At 0 eV collision energy, the following channels, with their associated kinetic energy release (KER_n), are allowed:

Dissociative Recombination of Molecular Ions with Electrons
Edited by S. Guberman, Kluwer Academic/Plenum Publishers, 2003

$$
\begin{aligned}
H_2O^+ + e^- &\rightarrow O(^3P) + 2H(^2S) + KER_1 \quad (= 3.04\ eV) \\
&\rightarrow O(^1D) + 2H(^2S) + KER_2 \quad (= 1.07\ eV) \\
NH_2^+ + e^- &\rightarrow N(^4S) + 2H(^2S) + KER_3 \quad (= 3.67\ eV) \\
&\rightarrow N(^2D) + 2H(^2S) + KER_4 \quad (= 1.29\ eV) \\
&\rightarrow N(^2P) + 2H(^2S) + KER_5 \quad (= 0.09\ eV) \\
CH_2^+ + e^- &\rightarrow C(^3P) + 2H(^2S) + KER_6 \quad (= 2.25\ eV) \\
&\rightarrow C(^1D) + 2H(^2S) + KER_7 \quad (= 1.00\ eV)
\end{aligned}
$$

Using time and position sensitive detectors for the DR fragments in a triple coincidence experiment enables differentiation between the channels (1,2), (3-5) and (6,7). The detectors used in these experiments consist of a stack of multi-channel plates (MCPs), a phosphor screen and a multiple strip photomultiplier tube (PMT) (see Ref.'s 1, 10 and 11 for more detail).

As mentioned earlier, one of the techniques recently employed for studying these reactions is the use of a thin foil. The foil is a 5 μm Aluminum disc of 5 mm diameter and is held in front of, and parallel to, the first MCP, a few mm from its surface. With an ion beam energy of 250 keV/amu, the foil is thick enough such that any H atom arising from DR, or from other processes in the cooler, is stopped in the foil. The heavier atoms, N/O/C, have sufficient energy to pass through the foil and hit the MCP. Thus, of the three hits detected on the MCP, the hit corresponding to the heavier fragment is identified. Having identified the heavier fragment, the center of mass of the three fragments can be determined and from that, the kinetic energy released together with the intramolecular angle. It must be noted that the kinetic energy available to the fragments is fixed, i.e. KER_n, and that what we measure in the experiment are the projections of these 3D momentum vectors onto the 2D plane of the detector. The difference in arrival time of the fragments at the detector can be used to select specific events, as can the momentum vectors measured at the detector. Experimental data for H_2O^+ and NH_2^+ are given in Fig. 1 and data for CH_2^+ are plotted in Fig. 2.

The branching ratio between the competing channels was derived by fitting the data to results obtained from a Monte Carlo simulation, created to model the fragmentation.[8,12] We have determined that the DR of H_2O^+ favors the production of ground state, $O(^3P)$, over that of first excited state, $O(^1D)$ by a ratio of 3.5(0.5):1. The DR of NH_2^+ favors production of ground state, $N(^4S)$, and first excited state, $N(^2D)$, almost equally, with a ratio of ~1.3(0.3):1.

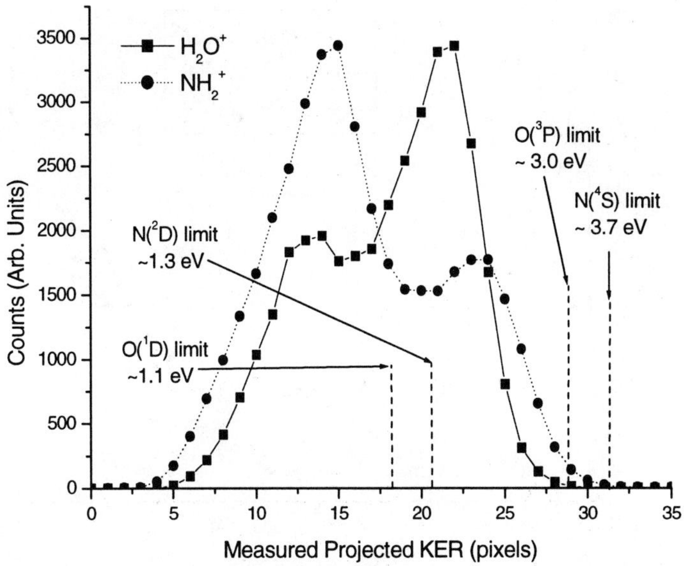

Figure 1. The measured projected kinetic energy release into the three fragments following the DR of H_2O^+ and NH_2^+.

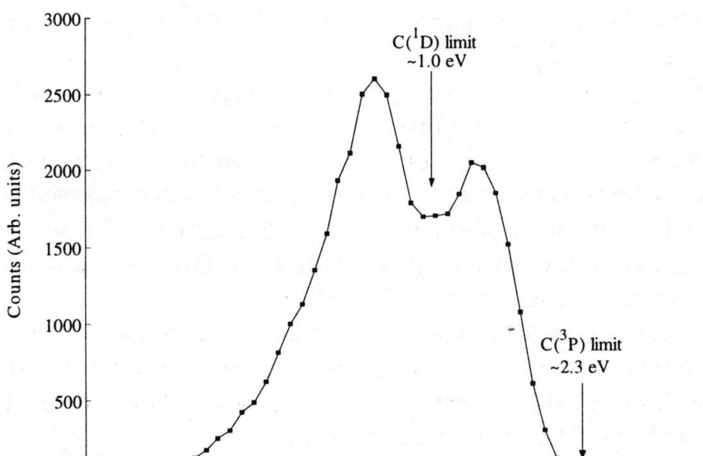

Figure 2. The measured projected kinetic energy release into the three fragments following the DR of CH_2^+.

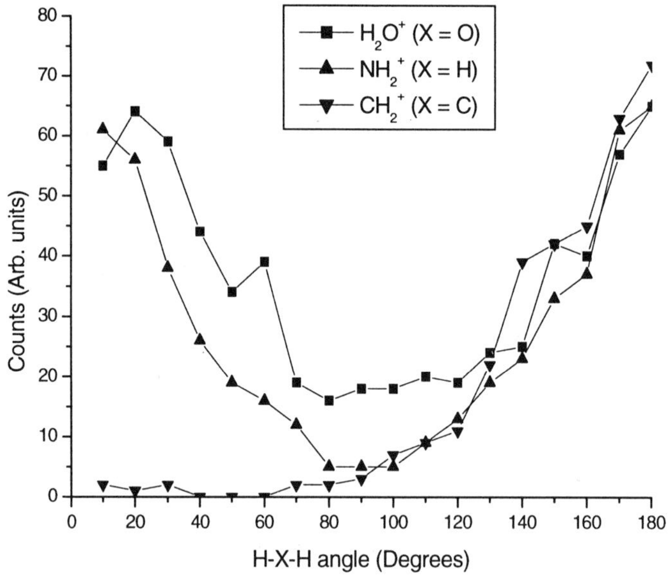

Figure 3: The measured H-X-H angle, where X = C, N and O, for those DR events which dissociated parallel to the detector.

Nitrogen atoms in the second excited state, $N(^2P)$, were not detected though they cannot be ruled out. Finally, the DR of CH_2^+ also favors the production of first excited $C(^1D)$ and ground state $C(^3P)$ carbon atoms almost equally with a ratio of 1.2(0.3):1.

The intramolecular angle of the fragmenting molecule was also investigated and we report results for the channel leading to the production of ground state atoms. It is possible to select from these experimental data those events having preferentially dissociated parallel to the detector. Since these events have the least projection onto the 2D detector, the angle measured in the plane of the detector is approximately the angle they had on dissociation. Fig. 3 shows these data for H_2O^+, NH_2^+ and CH_2^+.

For those events fulfilling this criterion, it is also possible to look at how the available reaction energy is distributed over the two hydrogen fragments. For H_2O^+, the data are compared to predicted results from the Monte Carlo simulation, both of which are plotted in Fig. 4.

Three cases were investigated for comparison: the energy is shared equally between both H atoms, one atom gets the majority of the energy and, lastly, the energy is shared randomly. These are denoted by $\rho = 1.0$, $\rho = 0.1$ and $\rho = c$ in Fig. 4, respectively. It can be seen that the best description of the experimental data is for a random distribution over the two H atoms.

Fig. 5 has a comparison of the H atom energy distributions for both H_2O^+ and NH_2^+.

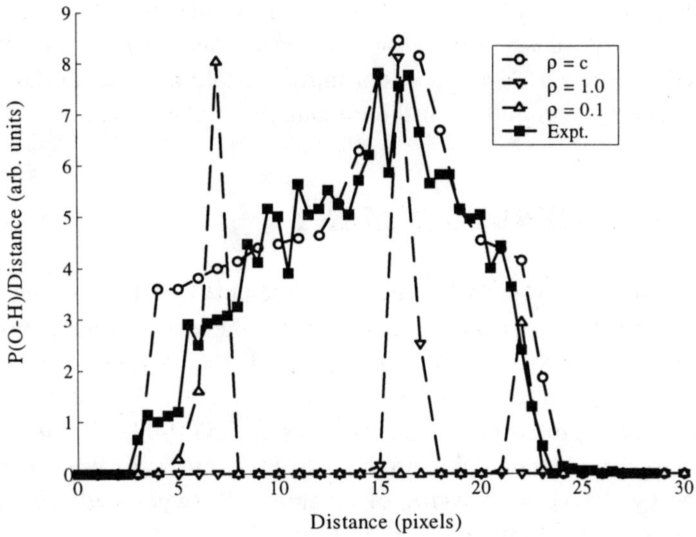

Figure 4. Energy distribution for the two hydrogen atoms for those events having preferentially dissociated parallel to the detector. Experimental data are compared to simulated data for ρ = random (c), ρ = 0.1 and ρ = 1.0.

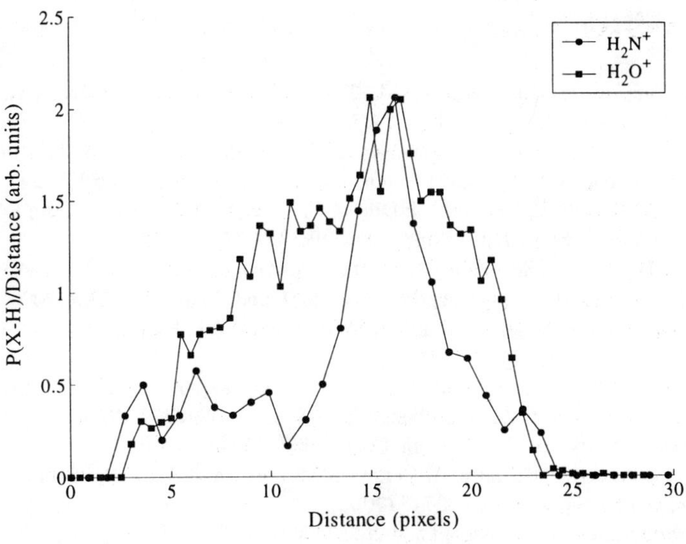

Figure 5. Comparison, between H_2O^+ and NH_2^+, of the energy distribution over the two hydrogen atoms for those events having preferentially dissociated parallel to the detector.

It can be seen that the two distributions are significantly different. Though none of the three cases simulated for comparison to H_2O^+ matches those experimental data for NH_2^+, it is not unreasonable to conclude that the case where the energy is equally distributed matches the data best.

ACKNOWLEDGEMENT

The authors are grateful to the staff members of the Manne Siegbahn Laboratory for their assistance and help in the experiment. This work is supported by: The Swedish Science Council, the Swedish Foundation for International Cooperation in Research and Higher Education, the Fifth Framework Programme of the EU (HPRN-CT-2000-00142), Stichting voor Wetenschappelijk Onderzoek, and the U.S. Department of Energy, Offices of Basic Energy Sciences, Division of Chemical Sciences under Contract No. DE-AC05-00OR22725 with UT-Battelle, LLC.

This contribution is dedicated to Sheldon Datz, who died suddenly on August 15th 2001, a few weeks before the ACS meeting. The areas of physics discussed here represent only a fraction of Sheldon's interest in physics, and science in general, and he will be greatly missed.

REFERENCES

1 S. Datz, G. Sundström, Ch. Biederman, L. Broström, H. Danared, S. Mannervik, J. R. Mowat and M. Larsson, Phys. Rev. Lett. **74**, 896 (1995).
2 Å. Larson, A. Le Padellec, J. Semaniak, C. Strömholm, M. Larsson, S. Rosén, R. Peverall, H. Danared, N. Djuric, G. H. Dunn, and S. Datz, Astrophys. J. **505**, 459 (1998).
3 L. Vikor, A. Al-Khalili, H. Danared, N. Djuric, G. H. Dunn, M. Larsson, A. Le Padellec, S. Rosén, and M. af Ugglas, Astron. and Astrophys. **344**, 531, (1999).
4 S. Rosén, A. Derkatch, J. Semaniak, A. Neau, A. Al-Khalili, A. Le Padellec, L. Vikor, R. Thomas, H. Danared, M. af Ugglas and M. Larsson, Faraday Disc. **115**, 295 (2000).
5 S. Datz, R. Thomas, S. Rosén, M. Larsson, A.M. Derkatch, F. Hellberg, and W. van der Zande, Phys. Rev. Lett. **85**, 5555 (2000).
6 D. Strasser, L. Lammich, S. Krohn, M. Lange, H. Kreckel, J. Levin, D. Schwalm, Z. Vager, R. Wester, A. Wolf and D. Zajfman, Phys. Rev. Lett. **86**, 779 (2001).
7 M. Larsson and R. Thomas, Phys. Chem. Chem. Phys. **3**, 4471, (2001).
8 R. Thomas, S. Rosén, F. Hellberg, A. Derkatch, M. Larsson, S. Datz, R. Dixon and W. J. van der Zande, Phys. Rev. A **66**, 032715 (2002).
9 See the contribution by D. Strasser et al. in this volume, p. 235.
10 Z. Amitay and D. Zajfman, Rev. Sci. Instrum. **68**, 1387 (1997).
11 J. R. Peterson, A. Le Padellec, H. Danared, G. H. Dunn, M. Larsson, Å Larson, R. Peverall, C. Strömholm, S. Rosén, M. af Ugglas and W. J. van der Zande, J. Chem. Phys. **108**, 1978 (1998).
12 U. Müller and P. C. Cosby, Phys. Rev. A **59**, 3632 (1999).

Dissociative Recombination of Vibrationally Excited Levels in Oxygen Molecular Ions

Philip C. Cosby, James R. Peterson, and David L. Huestis
Molecular Physics Lab, SRI International, Menlo Park, CA 94025
cosby@CRVAX.SRI.COM

1. INTRODUCTION

In the ionospheres of both Venus and Mars, O_2^+ is the most abundant molecular ion, followed at an order of magnitude lower abundance by CO_2^+ and NO^+.[1,2] This result was initially surprising, given that there is relatively little (<0.2%) O_2 in either atmosphere. At altitudes above 150 km, O_2^+ is formed by the fast exothermic reaction:[3]

$$O^+ + CO_2 \rightarrow O_2^+ + CO, \qquad (1)$$

where the O^+ reactant comes from dissociative ionization of the CO_2 atmosphere.

Reaction (1) has been studied in the laboratory[4,5] and is known to produce O_2^+ that is both rotationally hot and vibrationally excited, with about 40% of the reaction exothermicity distributed statistically as internal excitation of the O_2^+ product. The measured vibrational distribution[5] is 38% in v = 0, 30% in v = 1, 18% in v = 2, and 15% in v ≥ 3. On Earth, nascent O_2^+ is also vibrationally excited, but this excitation is relaxed by fast symmetric charge transfer collisions with O_2,[6]

$$O_2^+(v') + O_2(v{=}0) \rightarrow O_2(v{>}0) + O_2^+(v{<}v'), \qquad (2)$$

a process that is improbable in the oxygen-poor atmospheres of Mars and Venus. On those planets, quenching of the vibrational excitation can only occur slowly[7] by collisions with CO_2,

Dissociative Recombination of Molecular Ions with Electrons
Edited by S. Guberman, Kluwer Academic/Plenum Publishers, 2003

$$O_2^+(v') + CO_2 \rightarrow O_2^+(v<v') + CO_2, \qquad (3)$$

prior to the ions being neutralized by dissociative recombination (DR):

$$O_2^+(v) + e \rightarrow O(^3P, {}^1D, {}^1S) + O(^3P, {}^1D, {}^1S) + W, \qquad (4)$$

where W is the translational energy release. Fox[2] had earlier proposed that, above 150 km on Venus, DR of excited vibrational levels of O_2^+ must be considered, because vibrational relaxation in collisions with CO_2 would be relatively slow. That conclusion is supported by the subsequent laboratory experiments.

2. DISSOCIATIVE RECOMBINATION OF $O_2^+(v)$

The DR rates and product state distributions have recently been measured at CRYRING for O_2^+ in its ground vibrational level,[8] and at ASTRID for the ${}^{16,18}O_2^+$ isotopomer in both a nascent (uncharacterized) excited vibrational distribution and in a radiatively-relaxed vibrational distribution.[9] Each of these measurements made use of the translational energy released to the neutral O-atoms by the DR reaction (4), as shown schematically in Fig. 1. Unfortunately, without vibrationally resolved rates and branching ratios, these measurements cannot be extrapolated to the DR rates and distributions relevant to the vibrational distributions expected in the planetary atmospheres of Mars and Venus. The spatial extent of the vibrational wave functions varies from level to level, yielding different contributions in the final product states. Guberman[10] has recently shown that the DR yields of N_2^+ are extremely sensitive to the potential energy of the ion, with marked variations even for the small change in zero point vibrational energy between 14,14 and 14,15 isotopomers. Rather, the DR rates and product distributions of individual O_2^+ vibrational levels must be measured explicitly. In collaboration with Stockholm University and the FOM Institute for Atomic and Molecular Physics, we plan to make these measurements at CRYRING using a special ion source to control the ion beam vibrational distributions. Translational spectroscopy of O_2^+ + Cs dissociative charge-transfer products will be used to assay the ion beam populations.

3. HPEI ION SOURCE

The population distribution in a homonuclear ion beam is established by the method used to create the ions and by their subsequent collisions or reactions within the ion source. Two distinct ion sources have been used to

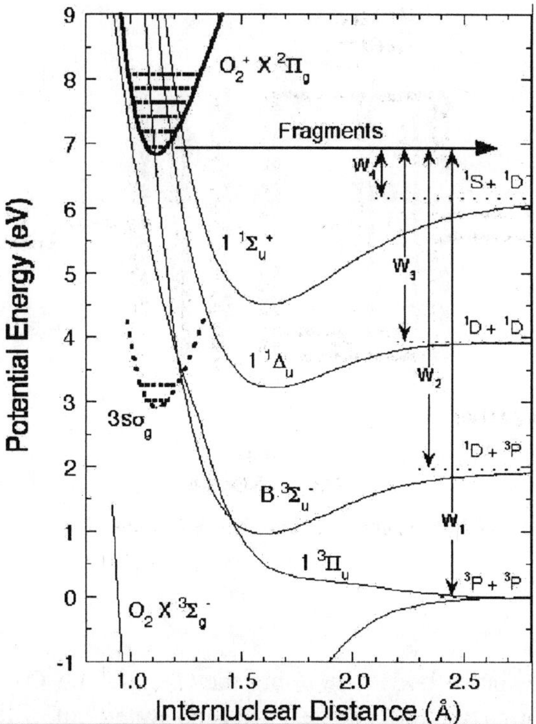

Figure 1. Potential energy curves of oxygen relevant to DR of O_2^+. Measurement of fragment translational energy W_1, W_2, W_3, or W_4 explicitly defines the electronic states of the DR products. The dashed curve shows the location of the $3s\sigma_g\ ^{1,3}\Pi_g$ Rydberg states that are populated by charge-transfer neutralization of O_2^+ in Cs vapor. Translational spectroscopy of the predissociation products of the $3s\sigma_g$ Rydberg states (the *gerade* continua are not shown in the Figure) is used to monitor the vibrational distribution in the ion beam.

date for preparing of molecular ion beams at CRYRING. The MINIS source uses a low pressure, hot filament discharge, which produces substantial rovibrational and electronic excitation in the extracted ion beam due to the high electron energies in the discharge (160 eV) and the high thermal temperature of the source (~900 K).[11] In contrast, the JIMIS source uses a cooled hollow cathode discharge at high pressure (1 Torr) to produce ion beams that are essentially in their ground electronic and vibrational states.[8] The proposed measurement of the DR rates and products for specific excited vibrational levels of O_2^+ requires that the ion beam population distribution be adjustable between these two extremes.

A high pressure, Nier-type electron-impact (HPEI) ion source will be used to achieve the required control of the O_2^+ internal energy distribution. The design, shown schematically in Fig. 2, is based on a design of T. O. Tiernan that has been used in our laboratory for many years.[5] A magnetically collimated electron beam emitted from an externally mounted filament is

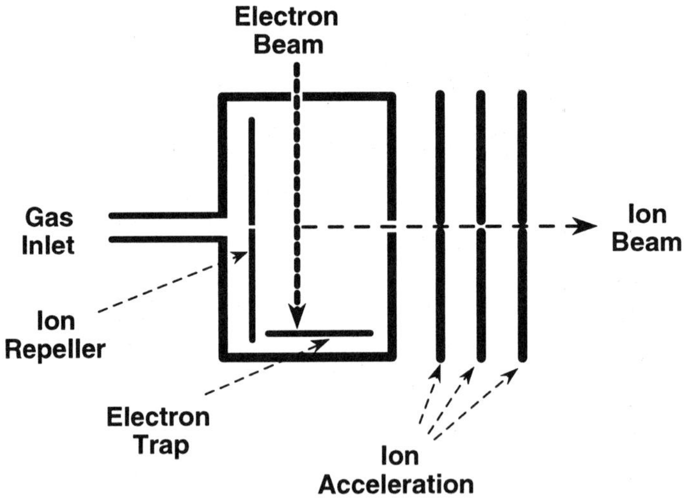

Figure 2. Schematic diagram of the HPEI ion source.

accelerated into the source body containing the O_2 gas. The O_2^+ ions created by electron-impact ionization of the gas, are accelerated out of the source by a weak electric field applied by the ion repeller electrodes. The broad vibrational distribution of the O_2^+, created by the ionization is modified by the collisions of these ions with the O_2 gas, Reaction (2), as they drift out of the source. The source gas pressure (0.1 – 50 mT) and the ion repeller voltage (0 – 10 V/cm), can be varied over wide ranges during normal source operation to continuously change the population distribution in the O_2^+ extracted from the source from the near-nascent populations in v = 0-8 created by the electron impact ionization[5] to a fully-relaxed O_2^+ (v=0) beam created by many near-resonant change-transfer collisions.[6]

4. POPULATION ASSAY BY DCT

The actual vibrational population in the O_2^+ beam will be determined using translational spectroscopy of its dissociative charge transfer (DCT) products in Cs vapor. Similarly to DR, DCT can be viewed as a two-step process:

$$O_2^+(v) + Cs \rightarrow O_2^*(v) + Cs^+ \tag{5}$$

and

$$O_2^*(v) \rightarrow O + O + W. \tag{6}$$

The cross section for Reaction (5) is large (>100 $Å^2$) and the reaction is near resonant for the production of O_2^* in the $3s\sigma_g$ Rydberg states,[12] shown by the dashed curve in Fig. 1. These properties allow the vibrational population produced in these bound Rydberg states to closely replicate the population in the parent ions. Further, the Rydberg states are predissociated by nearby valence states much more rapidly than they can radiate, hence the vibrational population appears in the kinetic energy release (W) spectrum of the dissociation fragments.

The experimental apparatus used for the DCT measurements is shown in Fig. 3. It incorporates the same principles used to determine the DR product distributions at CRYRING,[8] except that the electron-beam interaction region is replaced by a thin Cs-vapor charge-transfer cell inserted into the ion beam at L = 1.64 m upstream of the detector. The position sensitive detector for correlated fragments (PSD-C)[13] explicitly measures the radial, R, and temporal, Δt, separations of the two neutral fragments produced by a single O_2^+ collision in the cesium vapor to determine the translational energy release:

$$W = \frac{E_0}{4L^2}(R^2 + v_0^2 \Delta t^2) \tag{7}$$

where E_0 and v_0 are the energy and velocity, respectively, of the molecular ion beam.[14]

Both the versatility of DCT for monitoring the O_2^+ beam population distributions and the wide range of control over these populations that can be achieved with a high pressure electron impact ion source are illustrated in Fig. 4. The left panel shows the kinetic energy release spectra measured for the DCT in Cs vapor of a 5-keV O_2^+ beam created in the source using four different O_2 gas pressures and an ionization energy of 100 eV. At the very lowest pressure (0.01 mT), the upper spectrum reveals that the population in the O_2^+ beam is distributed over more than 9 vibrational levels in the $X^2\Pi_g$ state. Increasing the source gas pressure allows a quenching of this nascent excitation by low energy charge-transfer reactions with O_2 as the O_2^+ ions drift out of the source. At the highest pressure (28 mT), 90% of the O_2^+ beam population is in the ground vibrational level of this ground electronic state. Reducing the ionization energy from 100 eV to 20 eV eliminates all evidence of electronic excitation and increases the ground (electronic and vibrational) state population to 98% without serious reduction in the ion beam current. A small positive voltage applied to the ion repeller electrodes in the source (Fig. 2) decreases the residence time of the O_2^+ within the source and, thereby, the number of thermalizing collisions (Reaction 2). The electric field from this repeller thus provides a second variable with which to control the ion vibrational distribution. The effect on the ion vibrational distribution of increasing the ion repeller field is shown in the right panel of Fig. 4. We have

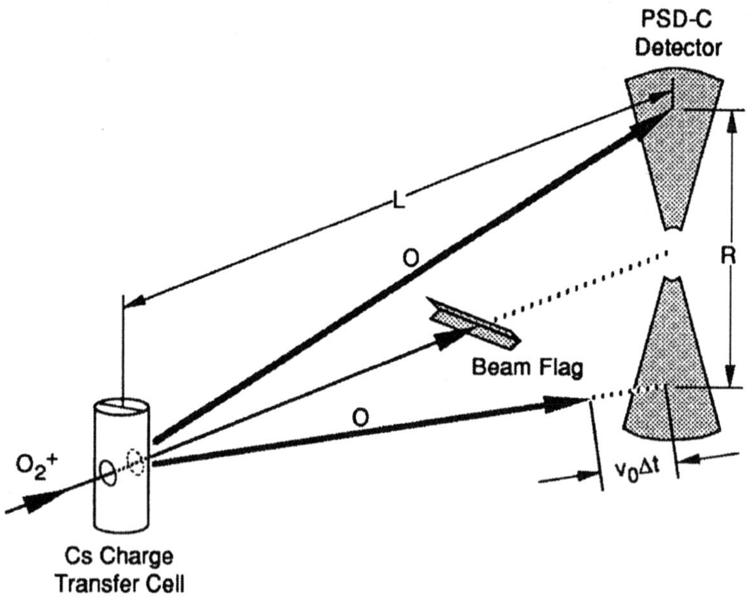

Figure 3. Schematic diagram of the DCT apparatus. Ions exiting the charge transfer cell are deflected out of the beam by a weak electric field (not shown) and collected.

found that the ion vibrational distributions created by a given combination of source gas pressure, ion repeller voltage, and electron energy are stable over long periods of time.

5. CONCLUSIONS

It is not practical to produce a molecular ion beam in a single excited vibrational level with sufficient intensity for DR measurements. However, O_2^+ beams with a fully characterized vibrational population *distribution* can be produced, and the product distribution and DR rate of this distribution accurately measured. By making a number of such measurements using beams with widely varying vibrational population distributions, we can determine the product distribution and rate of each vibrational level.

ACKNOWLEDGEMENTS

This work is partially funded by grant NAG5-11173 from the NASA Planetary Atmospheres program and is made in collaboration with Prof. Mats

Larsson (Stockholm University), Prof. Wim van der Zande (FOM Institute), and their colleagues.

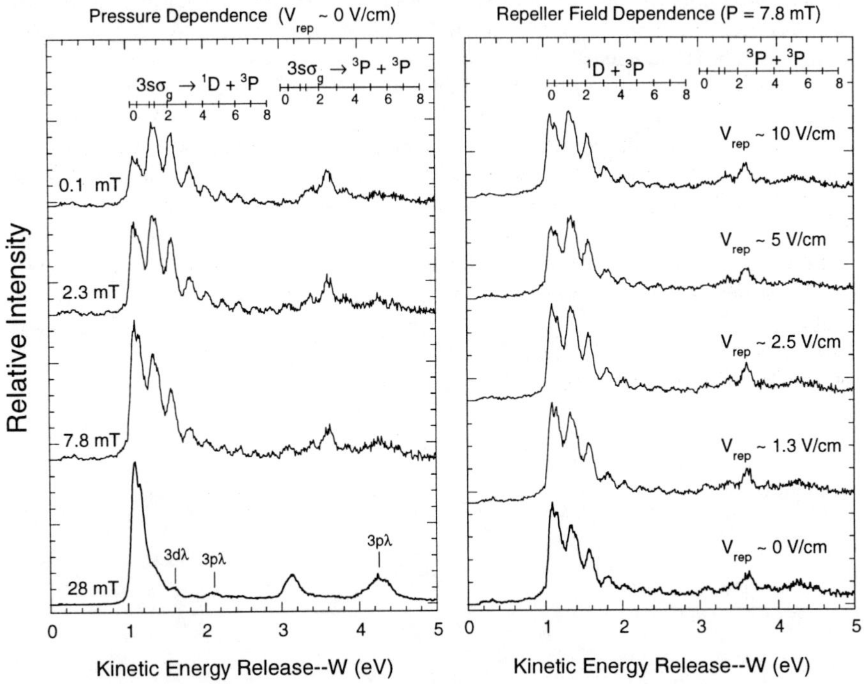

Figure 4. Dissociative charge-transfer spectra of 5 keV O_2^+ in Cs vapor observed for different pressures of O_2 gas (left) and ion repeller field strengths (right) using the high-pressure electron impact ion source operating at 100 eV electron impact energy. The tick marks label the energy releases expected for 3sσRydberg states on the $O_2^+ X^2\Pi_g$ (v) core dissociating to $O(^1D) + O(^3P)$ and $O(^3P) + O(^3P)$ products. The additional features labeled in the highest pressure (28 mT) spectrum correspond to non-resonant production of 3pσ and 3pπ Rydberg states in the charge transfer. These are prominent only because >90% of the O_2^+ population is in a single vibrational level (v = 0). When the spectra are corrected for the detector collection efficiency, the ion vibrational populations can be determined from the areas of the peaks.

REFERENCES

1 W. B. Hanson, S. Sanatani, and D. R. Zuccaro, J. Geophys. Res. **82**, 4351 (1977).
2 J. L. Fox, "Aeronomy," in *Atomic, Molecular, and Optical Physics Handbook*, ed. by G.W.F. Drake (AIP Press, Woodbury, NY, 1996).
3 J. L. Fox, Adv. Space Res. **5**, 165 (1985).
4 D. E. Hunton, A. A. Viggiano, R. A. Morris, and J. F. Paulson, J. Geophys. Res. **96**, 13881 (1991).
5 C. W. Walter, P. C. Cosby, and J. R. Peterson, J. Chem. Phys. **98**, 2860 (1993).

6 T. F. Moran, K. J. McCann, M. Cobb, R. F. Borkman, and M. R. Flannery, J. Chem. Phys. **74**, 2325 (1981).

7 H. Bohringer, M. Durup-Ferguson, D. W. Fahey, F. C. Fehsenfeld, and E. E. Ferguson, J. Chem. Phys. **79**, 4201 (1983).

8 R. Peverall, S. Rosen, J. R. Peterson, M. Larsson, A. Al-Khalili, L. Vikor, J. Semaniak, R. Bobenkamp, A. Le Padellec, A. N. Maurellis, and W. J. van der Zande, J. Chem. Phys. **114**, 6679 (2001) and references therein.

9 D. Kella, L. Vejby-Christensen, P. J. Johnson, H. B. Pederson, and L. H. Andersen, Science **276**, 1530 (1997); *ibid.* **277**, 167 (1997).

10 S. L. Guberman, "Isotope Effects in Dissociative Recombination" in *Dissociative Recombination: Theory, Experiment, and Applications IV*, ed. by M. Larsson, J. B. A. Mitchell, and I. F. Schneider, (World Scientific, Singapore, 1999), p. 111.

11 A. Le Padellec, J.B.A. Mitchell, A. Al-Khalili, H. Danared, A. Kallberg, A. Larson, S. Rosen, M. af Ugglas, L. Vikor, and M. Larsson, J. Chem. Phys. **110**, 890-901 (1999).

12 W. J. van der Zande, W. Koot, J. R. Peterson, and J. Los, Chem. Phys. Lett. **140**, 175 (1987).

13 H. Helm and P. C. Cosby, J. Chem. Phys. **86**, 6813 (1987).

14 D. P. deBruijn and J. Los, Rev. Sci. Instrum. **53**, 1020 (1982).

Studies of Electron Collisions with CN^+, CN^- and HCN^+/HNC^+

Arnaud Le Padellec
LCAR UMR 5589, Université Paul Sabatier-Toulouse III, 118 route de Narbonne, Bât. III R1B4, 31062 Toulouse Cedex 4, France
arnaud.lepadellec@irsamc.ups-tlse.fr

1. INTRODUCTION

We made use of two different merged beam setups, MEIBE and CRYRING to study electron collisions upon molecular ions. Absolute cross sections and branching fractions over the final product states were measured for the dissociative recombination (DR) of $CN^+(X^1\Sigma^+$ and $^3\Pi$, v=0). The same process was investigated for isomeric HCN^+/HNC^+ target ions. An isomeric mixture was first studied, and HNC^+ was then almost exclusively isolated. The two sets of measurements exhibit differences and the process is found to be overwhelmingly more efficient for HCN^+ than that for HNC^+. Electrons were also scattered on CN^- ($X^1\Sigma^+$, v=0, 1) target ions. Pure detachment is found to dominate over fragmentation channels. The cross section curve rises from threshold near 7 eV, up to a plateau.

2. MERGED BEAM SINGLE PASS APPARATUS: MEIBE

A sketch of the single pass merged beam (MEIBE) set-up is presented in Fig. 1. The ions are produced in a radio frequency ion source mounted in the terminal of a 400keV Van de Graaff accelerator. After mass analysis, the ion beam is injected into the collision chamber where a beam of electrons, produced by a barium oxide cathode, is merged with it using a combination

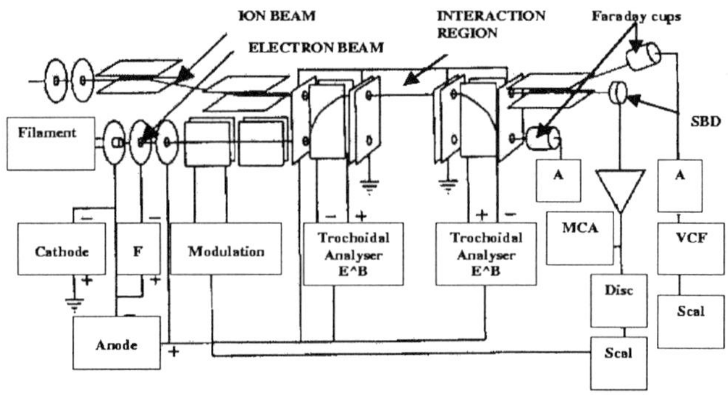

Figure 1. A sketch of the MEIBE apparatus at the University of Western Ontario. The ions come from the top left. Both trochoidal analyzers (entrance and exit of the 8.6cm interaction region) in which the electrons display jumps, are indicated.

of transverse electric and magnetic fields in a trochoidal analyzer. The beams interact with each other over a distance of 8.6 cm. After the interaction region, the electrons are de-merged, the electron beam collected and measured in a Faraday cup, which is connected to a Keithley electrometer. The ion beam is deflected electrostatically into another Faraday cup, also connected to an electrometer. Fast neutrals, formed in the interaction region continue undeflected on their original path, to be detected further downstream by an energy sensitive surface barrier detector. Recently, we performed a careful reanalysis of the quantum efficiency of our detector and this was indeed, found to be essentially unity.

2.1 Signal Analysis

The true recombination signal is extracted from the background arising from charge exchange collisions by modulating the electron beam and counting the signals in and out of phase with the modulation. Since the simultaneous arrival of two or three neutral particles is indistinguishable from that of a single particle having the same total mass, the DR signal appears in the "full energy" channel. This can be selected for counting using a single channel analyzer (SCA), with lower and upper gating levels set accordingly. The overlap of the two beams in the interaction region is monitored using moveable scanners and the effective collision area, F, determined.[1] The electron and ion currents, I_e and I_i, the electron and ion velocities in the laboratory frame, v_e and v_i, as well as the neutral signal count rate, C_n, are all measured.

The cross section $\sigma(v_{cm})$ is given by:

$$\sigma(v_{cm}) = \frac{C_n}{I_i}\frac{e^2}{I_e}\frac{F}{L}\left|\frac{v_i \cdot v_e}{v_i - v_e}\right| \quad (1)$$

where e is the elementary charge. The conventional relation gives the center of mass (CM) energy:

$$E_{cm} = \left(\sqrt{E_+} - \sqrt{E_e}\right)^2 \quad (2)$$

where $E_+ = (m_e/m_i)E_i$ is the reduced collision energy and E_e is the electron energy. Another important issue is the energy resolution of the MEIBE apparatus. Although we make use of Eq. 1 in order to determine the recombination cross section, one must take into account the finite energy resolution of the apparatus. The measurements yield an average of a velocity weighted cross section over a "flattened" anisotropic electron velocity distribution, f:

$$f(v_e) = \frac{m_e}{2\pi k T_{e\perp}}\left(\frac{m_e}{2\pi k T_{e//}}\right)^{1/2} \exp\left(-\frac{m_e v_{e\perp}^2}{2kT_{e\perp}} - \frac{m_e v_{e//}^2}{2kT_{e//}}\right) \quad (3)$$

(where $T_{e//}$ and $T_{e\perp}$ represent the longitudinal and transverse temperatures, respectively) according to:

$$\langle \sigma(v_{rel}) v_{rel} \rangle = \int_{-\infty}^{+\infty} v_{rel}\, \sigma(v_{rel}) f(v_e) d^3 v_e \quad (4)$$

which for small values of the energy resolution compared to the CM energy, can be simplified to:

$$\langle \sigma(v_{rel}) v_{rel} \rangle = \sigma(v_{cm}) v_{cm} \,. \quad (5)$$

In order to check the validity of our simplification, we extracted the rate coefficients $\langle \sigma(v_{rel}) v_{rel} \rangle$ from our measured cross sections at specific v_{cm} values for HCN^+,[2] and we then applied a Fourier transform based technique[3]:

$$\sigma(v_{cm}) = \frac{1}{v_{cm}} F^{-1}\left[\frac{F(\langle \sigma(v_{rel}) v_{rel} \rangle)}{F(f)}\right] \quad (6)$$

which is similar to that used in the storage ring community to extract the cross section from the rate coefficient $\langle \sigma(v_{rel}) v_{rel} \rangle$. F represents the Fourier Transform and f is the electron velocity distribution defined above. The assumption was made that the longitudinal electron temperature is negligible compared to the transverse one. This assumption has been shown to be valid in several circumstances.[4,5] The deconvoluted cross sections obtained are essentially those obtained directly from Eq.(1) and we therefore confirm the high energy resolution of our apparatus.

3. MERGED BEAM MULTI-PASS APPARATUS: HEAVY ION STORAGE RING CRYRING

The storage ring CRYRING is displayed in Fig. 2.

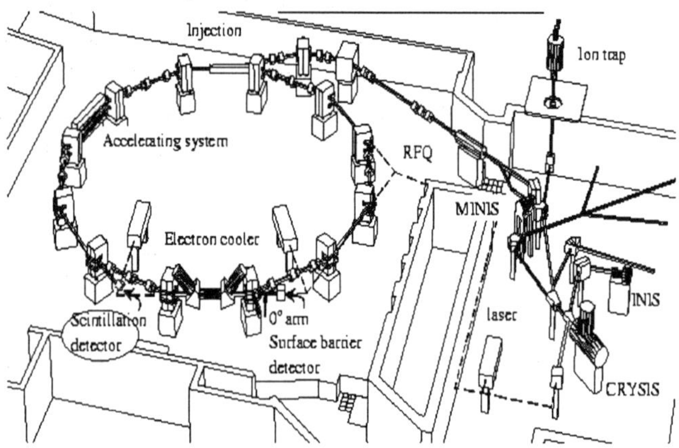

Figure 2. A view of the heavy ion storage ring CRYRING (Manne Siegbahn Institute - Stockholm University).

3.1 Rate Coefficient Determination

Each machine cycle consists of five phases: injection (following extraction (usually at 40 keV) from the ion source), acceleration, cooling, measurement and dumping of the beam. About one second is used for the two first phases. The relatively long storage time of the ions in the ring makes possible, in most cases, the relaxation of their internal degrees of freedom prior to measurement (the so-called "cooling" phase). By varying the electron velocity, collisions at different CM energies, E_{cm}, can be achieved, and this is given by an equation similar to Eq.(2). Various types of measurement protocols can be used. The CM energy can be varied either by jumping or ramping continuously the electron cooler cathode voltage. Three different types of detectors are used during the experiments: two are used for the monitoring of the ion beam during the recording of data and one is used to record the events from the process under study. A current transformer is used for an absolute ion current measurement. A drawback is its relatively low sensitivity and hence both a large ion beam and long integration times are required in order to get a statistically acceptable measurement. Another detector, a scintillator, monitors the ion current and is placed after one of the straight sections. It monitors the background associated with the neutralization of the ions in the beam, due to collisions with the rest gas. This detector has a large dynamical range but it cannot be used for an absolute cross section measurement since the signal depends on the pressure in the vacuum chamber. Energy sensitive surface barrier detectors (SBD) are placed in a suitable fashion in order to intercept the different particles. The signal from the SBD, proportional to the energy deposited by the impinging particles, is amplified and analyzed using pulse height discrimination. The output of the discriminator is fed into a multichannel scaler (MCS) that is triggered at the start of each machine cycle. The MCS records the signal as a function of time. By simply relating the time delay to energy, one can record the signal as a function of the CM energy. An experimental difficulty is that the detection of the events resulting from a given process and the ion current measurement require very different ion beam intensities. Indeed, a relatively low ion beam current is needed in order to avoid pile-up in the SBD, whereas the measurement of the ion beam current requires an intense ion beam to get a statistically acceptable recording. This problem is solved by relating the SBD and current transformer signals to the recording from the scintillator detector. The experimental rate coefficient at a given CM energy is given by

$$\langle v_{rel}\sigma\rangle = R_B \frac{C}{n_e\, l} \frac{N}{N_{Scint}} \tag{7}$$

In this expression, C is the circumference of the ring, l is the length of the interaction region and n_e is the electron density. N is the difference between the number of counts obtained with the collision energies set to E_{cool} (with velocity matched electrons) and E_{cm}. Simultaneously, another independent MCS spectrum, labeled N_{Scint}, is recorded with the scintillation detector. The last parameter in Eq. (7), R_B, is the destruction rate per ion. It can be derived from another MCS spectrum taken with the same scintillation detector, but acquired at the same time as the ion current measurement, I_i. This is given by

$$R_B = \frac{dN_{Scint}}{dt} \frac{1}{I_i} f\, e, \tag{8}$$

where *f* represents the orbital frequency of the ions in the ring.

The cross sections can be extracted from the corresponding rate coefficients using the Fourier transform technique mentioned above. Specific experimental difficulties that must be solved include: trapped ions in the cooler section, the spread in the collision energies and the contribution to the signal from the so-called "toroidal" regions of the electron cooler.[8]

3.2 Branching Fractions

3.2.1 Grid

Many neutral fragments can arise from the open dissociation channels. The counts detected by the SBD are recorded along with the energy of a "neutral event" (combination of neutrals) that occurs at a specific fraction of the full ion beam energy. Since many open channels may contribute to this same "neutral event", these channels are therefore, not distinguishable from each other. To remove this indetermination, one can make use of a grid that is placed in front of the energy-sensitive SBD. This grid is thick enough to stop all neutral fragments that do not pass through the grid holes. If the transmission probability for passage of a neutral fragment through a hole is T, then the absorption probability is (1-T). Since particles stopped by the grid are not recorded by the SBD, the counts originating from a specific "neutral event" (when the grid is not inserted) are redistributed among "neutral events" characterized by smaller fraction of the total ion beam energy. A set of linear equations can be constructed, which relate the number of dissociation events in the different open channels N_i, to the measured numbers of counts in the various "neutral events" N_{meas} that correspond to specific fractions of the full ion beam energy. The final result can be summarized by the following set of equations:

$$[N_{meas}] = [M][N_i], \tag{9}$$

with the transmission matrix, [M]. By solving this matrix equation, one can extract the branching ratios. The critical value of the grid transmission was recently carefully reanalyzed[7] and was found to be 0.297 (± 0.015).

3.2.2 Imaging detector

A position sensitive imaging detector can be used in order to record the distance separation between DR fragments.[5, 9] This detector is positioned 6.3 meters beyond the interaction region, in the zero degree section. DR fragments strike the first of a package of three multichannel plates (MCPs; 77 mm in diameter), the signal from which produces flashes on a phosphor screen. The light is detected by a CCD (Charge Coupled Device) camera that is used to image the fragment separations. The kinetic energy shared among the fragments can be determined and the dissociation limits can be identified. Moreover, a photo-multiplier tube is used as an event trigger in order to grab the image from the CCD camera and to reduce the high voltage (250 V/100 ns) over the MCP package by means of a fast switch. This is used to identify false coincidences in the camera frames. A background spectrum is measured separately. The data can be reduced using the following formula, which applies for an individual distance distribution and a given kinetic energy release (the finite length of the interaction region is also taken into account):

$$P(D) = \frac{1}{D_{L+l/2} - D_{L-l/2}} \left(\cos^{-1}\left(\min\left(1, \frac{D}{D_{L+l/2}} \right) \right) - \cos^{-1}\left(\min\left(1, \frac{D}{D_{L-l/2}} \right) \right) \right). \quad (10)$$

L is the distance to the center of the cooler (6.3 m) and l denotes the length of the interaction region (0.85 m). The maximum distance separations of the fragments when the dissociation occurs at the beginning of the cooler and at the end are $D_{L+l/2}$ and $D_{L-l/2}$, respectively.

4. CN^+

The cyanogen cation CN^+ can be found in the hot, high-density conditions, existing in flames.[3] One of its destruction pathways is the DR mechanism:[10-11]

$$\begin{aligned}
CN^+(X^1\Sigma^+, v=0) + e- &\rightarrow C(^3P) + N(^4S) + 6.3\text{ eV} && (11a)\\
&\rightarrow C(^1D) + N(^4S) + 5.0\text{ eV} && (11b)\\
&\rightarrow C(^3P) + N(^2D) + 3.9\text{ eV} && (11c)\\
&\rightarrow C(^1S) + N(^4S) + 3.6\text{ eV} && (11d)\\
&\rightarrow C(^3P) + N(^2P) + 2.7\text{ eV} && (11e)\\
&\rightarrow C(^1D) + N(^2D) + 2.6\text{ eV} && (11f)\\
&\rightarrow C(^5S) + N(^4S) + 2.1\text{ eV} && (11g)\\
&\rightarrow C(^1D) + N(^2P) + 1.4\text{ eV} && (11h)\\
&\rightarrow C(^1S) + N(^2D) + 1.2\text{ eV} && (11i)\\
&\rightarrow C(^1S) + N(^2P) + 0.0\text{ eV} && (11j)
\end{aligned}$$

The ions were produced in a conventional ion source, from a mixture of nitrogen and methane in a 9 to 1 ratio. The 12 s lower limit for the $a^3\Pi$ radiative lifetime, comparable to the 6s delay between injection of the beam and data acquisition, means that our measurement refers to a CN^+ beam which is a mixture of the $X^1\Sigma^+$ and $a^3\Pi$ states. An additional 0.08 eV would have to be added to all of the exothermicities mentioned above in order to get those for $CN^+(a^3\Pi, v = 0)$. For the $X^1\Sigma^+$ state, we obtain vibrational relaxation times of 38, 58 and 115 ms for the v = 3, v = 2 and v = 1 levels, respectively, and the equivalent quantities for the $a^3\Pi$ state are 101, 152 and 304 ms. Therefore, our measurement refers to vibrationally cold ions.

4.1 Experimental Findings

Fig. 3(a) displays the deconvoluted cross sections. We have found[11] a steeper cross section energy dependence over the range from 0.4 to 3 eV of:

$$\sigma(E_{cm}) = 1.49 \times 10^{-15} (0.4/E_{cm})^{1.41} \qquad (12)$$

than that expected from the so-called direct process, i.e. E^{-1}, which is observed below 0.1 eV :

$$\sigma(E_{cm}) = 8.00 \times 10^{-14} (0.01/E_{cm})^{1.05} \qquad (13)$$

The branching fractions were obtained after subtraction of the background contribution to the signal, by fitting each distance distribution spectrum with a linear combination of the individual distance distributions for a given Kinetic Energy Release, KER (see Eq. (10)). Moreover, we had to account for the presence of the $a^3\Pi$ state. Because of the small energy gap between the KER values corresponding to the $C(^3P)+N(^2D)/C(^1S)+N(^4S)$, $C(^3P)+N(^2P)/C(^1D)+N(^2D)$ and $C(^1D)+N(^2P)/C(^1S)+N(^2D)$ limits, it is not possible to give individual final state branching fractions for all of the products. This limitation is due to the spatial resolution of our detector. The distribution of distances measured at 0 eV CM energy is shown in Fig. 4. At this energy, the DR process yields only a small percentage of ground state atomic fragments (< 1.8 % $C(^3P)+N(^4S)$). The second most exothermic limit $C(^1D)+N(^4S)$ is not significantly populated either (3.8 %). The main contributing channels come in pairs, namely the $C(^3P)+N(^2D)/C(^1S)+N(^4S)$, $C(^3P)+N(^2P)/C(^1D)+N(^2D)$ and $C(^1D)+N(^2P)/C(^1S)+N(^2D)$ limits that account for 14.2, 56.1 and 25.5 %, respectively. The $C(^5S)+N(^4S)$ channel as well as the less exothermic $C(^1S)+N(^2P)$ channel were not found to be significantly populated (< 1 - 1.4 %, respectively).

4.2 Interpretation and Discussion

For the direct DR process to be efficient at low energies, there must be a favorable crossing of a repulsive neutral curve through the lower part of the ionic curve. Among the nine neutral curves[15] that are displayed in Fig. 5, not all are suitable to drive the DR via the direct process. The curves labeled $D^2\Pi i$, $E^2\Sigma^+$ or $^2\Pi(4)$ do not intersect the ionic curves near their minima. Two additional factors indicate that these states do not contribute significantly to the recombination of CN^+. The energy dependence at low energy is very close to the E_{cm}^{-1} dependence, which is the signature of the direct process and more persuasively, the dissociation limits for these states are not observed to be populated in our branching fraction measurements. For the ground state ion, we have found that the best Franck-Condon overlap is achieved with the state labeled $^2\Sigma^+(5)$. This state has the most favorable curve crossing with the ion, since it cuts the $X^1\Sigma^+$ state at the bottom of the potential well. For the $a^3\Pi$ state, we have found that the best overlap is achieved with the states labeled $^2\Sigma^+(6)$ and $^2\Pi(6)$. These findings are supported by the branching results, since it was shown that 81.6 % of the dissociating flux goes to the $C(^3P)+N(^2P)$, $C(^1D)+N(^2D)$, $C(^1D)+N(^2P)$ and $C(^1S)+N(^2D)$ limits. On the

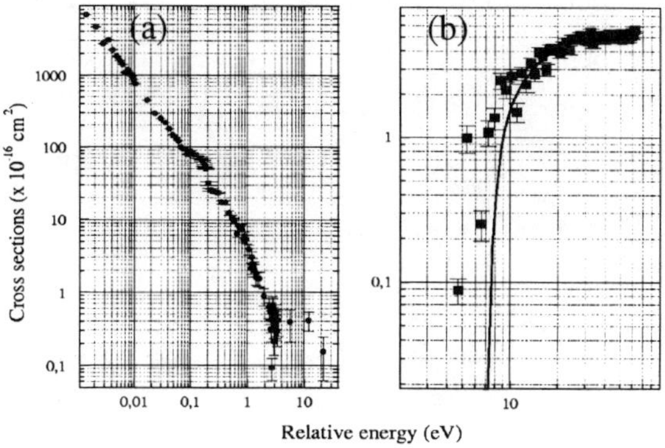

Figure 3. (a) DR cross sections as derived from the measured rate coefficients, using a deconvolution procedure.[3] (b) Electron impact detachment cross sections for CN^-.[12] The data points are the experimental data, and the solid line is a fit to the semi-classical model.[13, 14]

other hand, we have found that the "paired" limits $C(^3P)+N(^2P)/C(^1D)+N(^2D)$ are more populated than the $C(^1D)+N(^2P)/C(^1S)+N(^2D)$ limits. This certainly means that some of the other states from Fig. 5 contribute significantly to the DR process, especially those converging to the $C(^3P)+N(^2P)$ and $C(^1D)+N(^2D)$ limits (such as the $^2\Sigma^+(4)$ and $^2\Pi(5)$ states). The less exothermic $C(^1S)+N(^2P)$ channel is not found to be significantly populated. The exothermicity for this channel is indeed very small. A dissociative state of the CN molecule converging to this asymptotic limit and displaying a favorable curve crossing near the v = 0 of the ion, would have to be extremely shallow.

5. CN^-

Particular interest in the CN^- anion is associated with its relatively high degree of stability against detachment, with an electron affinity of 3.8 eV. This anion can readily form tightly bound complexes with several transition metal ions,[17] and it can also be used to dope alkali halide crystals in lattice defect studies.[18] CN^- ions were produced in a cesium sputter ion source[19] with a cathode made of boron nitride (BN) and graphite. An important issue is the internal excitation of the target anions. In particular, the $a^3\Pi$ excited state is expected to be long-lived since the $a^3\Pi \rightarrow X^1\Sigma^+$ transition is spin forbidden. The calculated lifetime, 21 ms, is small compared with the storage time of the CN^- ions in the storage ring, several seconds. Therefore, the internal state of the CN^- ions in our beam was exclusively the $^1\Sigma^+$ electronic ground state. It was also important to evaluate the different relaxation times for the vibrational levels of this state. We estimated lifetimes of 700, 930,

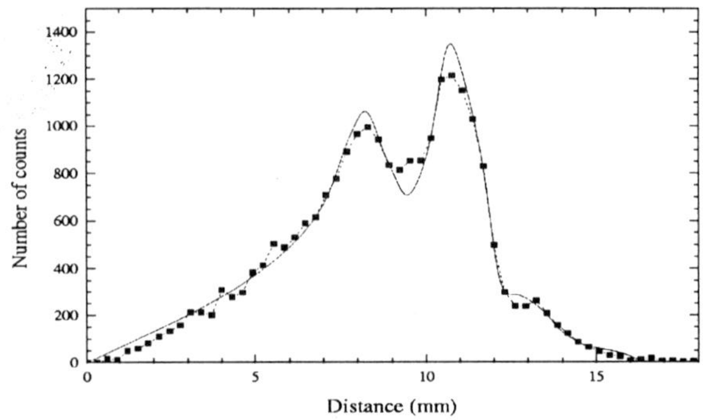

Figure 4. Distance spectrum for the branching fraction determination. The (--) symbol shows experimental data taken at 0 eV CM energy whereas the full drawn line shows a best fit to the experimental data. The scale is 0.31 mm per pixel.

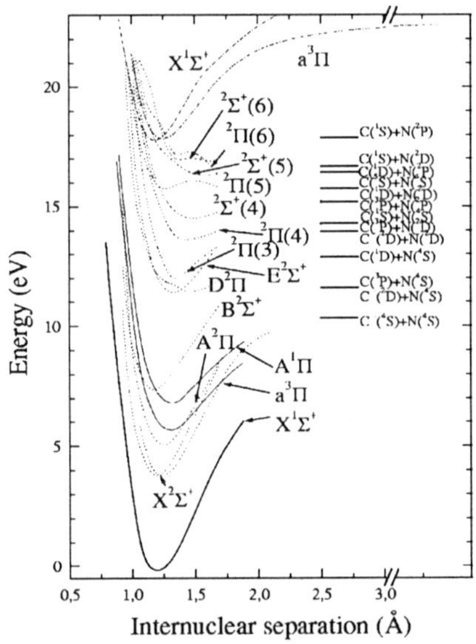

Figure 5. Potential energy curves relevant to the CN,[15] CN^+,[16] and CN^- [26] studies. The ten asymptotic limits are represented.

1390 and 2790 ms for the levels v = 4, 3, 2 and 1, respectively. We therefore conclude that our measurements refer to $CN^-(X^1\Sigma^+$, v=0,1) anions.

Over the energy range that was investigated, several channels had to be considered:

$$CN^- + e- \rightarrow CN + 2e \quad (14a)$$
$$\rightarrow C^- + N + e \quad (14b)$$
$$\rightarrow C + N + 2e \quad (14c)$$
$$\rightarrow C + N^+ + 3e \quad (14d)$$
$$\rightarrow C^+ + N + 3e \quad (14e)$$
$$\rightarrow CN^+ + 3e \quad (14f)$$
$$\rightarrow C^- + N^+ + 2e \quad (14g)$$
$$\rightarrow C^+ + N^+ + 4e \quad (14h)$$

An experimental complication arose from the fact that the energy sensitive SBD that was used to record the neutral fragments cannot distinguish between the channel 14a and 14c. To overcome this difficulty, we made use of a grid placed in front of the detector as described above.

5.1 Experimental Findings

A branching ratio measurement was made at a CM energy of 60 eV: 91(±4)% was found to branch into the pure detachment channel, (14a), and 8(±4)% into the dissociation channel C^- + N (14b). The flux into all other open channels represents less than 1%, but none were large enough separately to warrant quoting an accurate value. In Fig. 3(b), we present the measured cross section curve for the pure detachment channel, (14a), over the electron energy range 0-60 eV. The overall shape of the curve shows an onset at a threshold energy of about 7 eV. This threshold is determined by the binding energy of the anion (3.8 eV) plus a finite contribution due to the Coulomb repulsion. The cross section then increases monotonically up to a maximum around 25 eV followed by a plateau with at $\sim 5\text{x}10^{-16}$ cm^2.

5.2 Interpretation and Discussion

The $^3\Pi$ and $^1\Pi$ states, which are displayed in Fig. 5, correlate to $C^-(^2D)$ + $N(^4S)$ and $C^-(^2D)$ + $N(^2D)$ dissociation channels, respectively. Other curves (not shown here), corresponding to different molecular symmetries and multiplicities, also correlate to these two dissociation limits (34 in total). Moreover, three other Σ^+ states of different multiplicities correlate to the limit $C^-(^4S)$ + $N(^4S)$, in addition to the $^1\Sigma^+$ ground state.[26] Concerning the neutral states, it must be stressed that these curves, relevant to the DR process, are adapted from Lavendy et al.[15] who did not consider symmetries other than Σ^+ and Π. The omitted symmetries include those of Σ^-, Δ, Φ and Γ character that account for 13, 13, 5 and 1 of the molecular states, respectively (over a total of 73 neutral states and ten asymptotes). One might notice that of the thirteen calculated[15] Σ^+ and Π states, nine of them are the lowest bound states of the neutral CN radical. In fact, among the overall 73 states, a majority of them are expected to be associated with bound states. Many states can, in principle, be reached by electron collisions. For example, consider the neutral states produced in the electron detachment of molecular anions. The neutrals can further stabilize by radiative decay. The upper $^2\Sigma^+$ (5, 6) and $^2\Pi$ (6) states of CN, as displayed in Fig. 5, could very well contribute to the very weak observed dissociation channel 14c. As mentioned earlier, a total of 40 anionic potential curves correlate to the three lowest anionic limits. Electron impact excitation from CN^- ($X^1\Sigma^+$, v=0,1) can generate the $a^3\Pi$ and $A^1\Pi$ states. The population of these excited states would radiatively decay to the $^1\Sigma^+$ ground state with lifetimes of 21 ms and 269 ns, respectively. Of course, this process of pure excitation cannot be studied in a storage ring, since the excited CN^- anions will continue to circulate in the ring. This process might, however, significantly affect states other than the three bound anionic states displayed in Fig. 5. A purely repulsive anionic state might certainly contribute to the 8(±4)% of the flux that we measured for the dissociative channel C^- + N (14b).

We now turn to the second issue that we attempted to address in this work, namely the search for the possible existence of a doubly charged anion that should have appeared as a near-threshold resonance in the detachment curve. The full line in Fig. 3(b) represents a fit to our experimental data at medium energy, using a semi-classical formalism.[13,14] In a comparison of this fit and our data in the energy range 7-11 eV, one might notice an obvious discrepancy. This structure could imply the existence of a dianionic resonance in this region, but the statistical uncertainty in the experiment was, however, too large to either prove or disprove the existence of such a resonance. It is interesting to note that the CN^{-2} ion is unable to correlate at large internuclear separations to C^- + N^-, since N has a negative electron affinity. The resonance might be associated with a virtual state of CN^{-2}. Such a short lived state would perhaps explain why the observed structure is broad.

6. HCN^+/HNC^+

The MEIBE apparatus was used to examine the DR of [C-H-N^+] ions.[2] HCN^+ is an interesting ion because of its isomerisation chemistry. Ab initio calculations have indicated that the HCN^+ ion has a $^2\Pi$ ground state and is less stable by 0.98eV[20] than HNC^+, which has a $^2\Sigma^+$ ground state. Petrie et al.[21] have found that, after electron impact upon HCN, the relative concentration of HCN^+ and HNC^+ ground states are in the ratio of approximately 3 to 1. They showed that addition of carbon dioxide leads to a plasma that is completely dominated by HNC^+. It was also observed[22] that HCN^+ reacts rapidly with CH_4 but HNC^+ does not.

We carried out different sets of measurements, using different gas mixtures:

- *8.35% N_2 – 8.35% CH_4 – 83.3% CO_2*. The ion beam was comprised primarily of HNC^+ ions (96.2%). The HCN^+ population accounted for the remaining 3.8%.[23]
- *90% N_2 - 10% CH_4*. The HCN^+ population (12.5%) was more significant than with the above mixture but HNC^+ (87.5%)[23] remained the overwhelmingly populated isomer.

One has not only to question the possible presence of isomeric forms [C-H-N$^+$] in the target beam, but also their internal excitation. The $B^2\Sigma^+$ and $A^2\Sigma^+$ HCN^+ states lie 5.25 and 0.40eV above the $X^2\Pi$ state,[24] respectively. It is likely that the X and A states were populating our target beam since the HCN^+ A state radiative lifetime is 3ms. The $B^2\Sigma^+$ and $A^2\Pi$ HNC^+ states lie 10.63 and 2.13eV above $X^2\Sigma^+$ HNC^+,[25] respectively. Only the X state was likely to be populated in our target beam since HNC^+ A state has a much shorter radiative lifetime than that of its HCN^+ counterpart.

At 0 eV CM energy, three dissociation channels are open for ground state HNC^+ ions:

$$\begin{aligned} HNC^+ + e &\rightarrow CN + H + 7.2eV, && (15a)\\ &\rightarrow NH + C + 3.9eV, && (15b)\\ &\rightarrow CH + N + 2.9eV, && (15c) \end{aligned}$$

and four for HCN^+:

$$\begin{aligned} HCN^+ + e &\rightarrow CN + H + 8.2eV, && (16a)\\ &\rightarrow NH + C + 4.9eV, && (16b)\\ &\rightarrow CH + N + 3.9eV, && (16c)\\ &\rightarrow H + C + N + 0.4eV. && (16d) \end{aligned}$$

6.1 Experimental Findings

The cross sections are plotted in Fig. 6 over the range 1meV - 1eV.[2] The filled squares are for the target beam produced from the N_2/CH_4 source gas mixture. The open circles represent the data points taken under the $N_2/CH_4/CO_2$ mixture. The two sets of cross sections exhibit the same general trend, except in the 0.01-0.04eV and 0.1-0.3eV energy range. It appears that the HCN^+ DR process is considerably more efficient than that for HNC^+; the cross sections for HCN^+/HNC^+ being systematically larger than for HNC^+ "alone".

6.2 Interpretation and Discussion

We carried out a quantum chemical investigation of the DR of $HCN^+(X^2\Pi)$, $HNC^+(X^2\Sigma^+)$ and $HCN^+(A^2\Sigma^+)$.[23] Only the most thermodynamically favorable CN+H exit channels were considered. The relevant curves that drive the DR of HCN^+($X^2\Pi$ and $A^2\Sigma^+$) are displayed in Fig. 7(a) in the quasi-diabatic representation. There is a crossing between the lowest 1D$^1\Pi$ dissociative state and the lowest $^{1,3}\Pi$ Rydberg states at their minima. Similarly, the lowest 1D$^{1,3}\Sigma$ dissociative states intersect the lowest $^{1,3}\Sigma$ Rydberg states at the minima of the Rydberg curves. No curve crossing is observed between the ionic X$^2\Pi$, A$^2\Sigma^+$ and any of the dissociative $^{1,3}\Pi$ or lowest dissociative $^{1,3}\Sigma$ states. However, there is a higher repulsive state, 2D$^3\Sigma$ that crosses both HCN$^+$ X$^2\Pi$ and A$^2\Sigma^+$ at their minima.

The curves governing the DR of HNC$^+$(X$^2\Sigma^+$) can be seen in Fig. 7(b) in the quasi-diabatic representation. The lowest repulsive state 1D$^3\Sigma$ crosses the lowest Rydberg state of similar multiplicity at its minimum energy arguing for an efficient DR via the indirect mechanism. Similarly, the repulsive states 1D$^1\Pi$ and 1D$^3\Pi$, cross the lowest $^1\Pi$ and $^3\Pi$ Rydberg states at their minima. The second repulsive state, 2D$^3\Sigma$, crosses the ion curve. This crossing occurs at the third vibrational energy level of HNC$^+$.

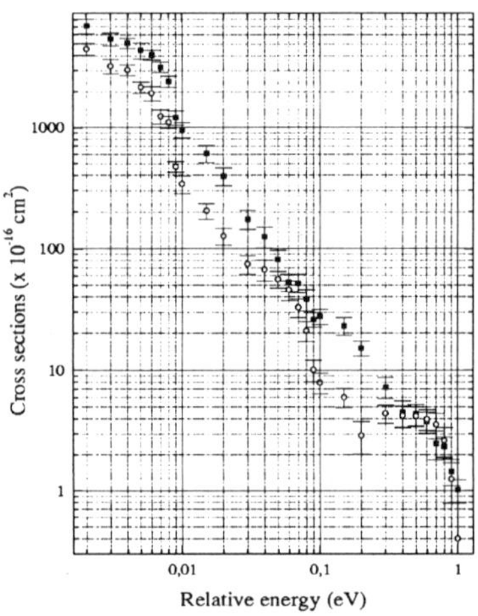

Figure 6. DR cross sections for the isomeric mixture HCN$^+$/HNC$^+$ (full squares) and for HNC$^+$ "alone" (open circles).

Fig. 7(a) suggests that the DR of HCN$^+$(X$^2\Pi$ and A$^2\Sigma^+$) could involve both the direct and indirect mechanisms. The direct process where the X$^2\Pi$ and A$^2\Sigma^+$ states of HCN$^+$ together with the neutral repulsive state HCN(2D$^3\Sigma$) will be involved and the indirect process where $^1\Pi$ Rydberg states together with the 1D$^1\Pi$ dissociative state and $^{1,3}\Sigma$ Rydberg states together with 1D$^{1,3}\Sigma$ will be involved. Fig. 7(b) suggests that the DR of HNC$^+$(X$^2\Sigma^+$, v=0) will involve an indirect mechanism, but that for HNC$^+$(X$^2\Sigma^+$,v=2) will proceed via the direct mechanism. Usually, the

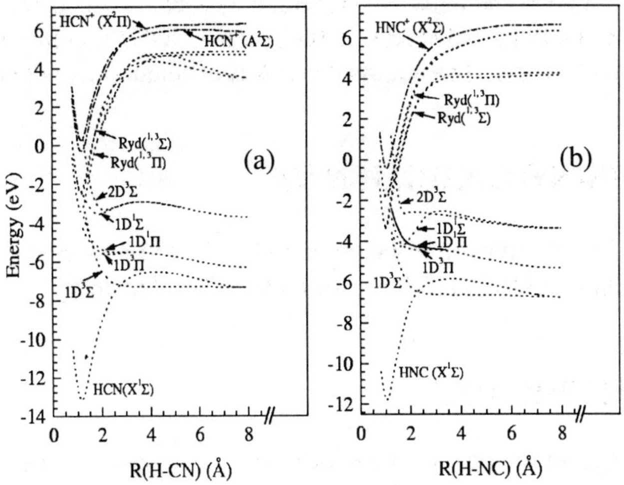

Figure 7. (a) Σ and Π HCN quasi-diabatic potential energy curves. (b) Σ and Π HNC quasi-diabatic curves. Energies are in atomic units and distances in Angstroms.

indirect DR process is less efficient than the direct one and our present results are in perfect agreement with this picture. Indeed, we extracted the DR thermal rates for pure HCN^+ and HNC^+ from the cross sections displayed in Fig. 6, and the HCN^+ rate is found to be about 28 times larger than that for HNC^+.

7. CONCLUSION

The DR of $CN^+(X^1\Sigma^+$ and $a^3\Pi$, v=0) is efficient, in agreement with the findings made for other diatomic ions. Cross sections of about 1.0×10^{-12} cm^2 at 1 meV and 1.0×10^{-14} cm^2 at 1 eV are found, with essentially a E_{cm}^{-1} dependence. The experimental results from the imaging detector indicate that at zero collision energy, the DR mechanism is dominated by a direct dissociation process which does not populate the ground state limit $C(^3P)+N(^4S)$. Concerning CN^-, we have measured cross sections for single detachment. The cross section rose from zero at threshold to a maximum of about 5×10^{-16} cm^2 just below 30 eV, after which it remained essentially constant up to about 60 eV. A slight enhancement of the cross section in the threshold region may be due to the presence of a resonance associated with the doubly charged negative ion CN^{-2}.

Two different source mixtures were used in order to study the DR of the isomeric HCN^+/HNC^+ ions: N_2-CH_4 (90-10%) to obtain HCN^+/HNC^+ ions in a 12.5/87.5% proportion and CO_2-N_2-CH_4 (85%-13.5%-1.5%) to obtain a beam containing predominantly HNC^+ with a 3.8/96.2% proportion. We have found that the DR rate coefficient of the isomer HCN^+ is much larger than that for the HNC^+ isomer, though the process is efficient for both cases.

ACKNOWLEDGMENTS

The author wants to express his gratitude to Mats Larsson and his group, to Brian Mitchell, to Dahbia Talbi as well as to Dag Hanstorp.

REFERENCES

1 C. J. Keyser, H. R. Froelich, J. B. A. Mitchell and W. J. McGowan, J. Phys. E: Sci. Instr. **12,** 316 (1979).
2 C. Sheehan, A. Le Padellec, W. N. Lennard, D. Talbi and J. B. A. Mitchell, J. Phys. B **32,** 3347 (1999).
3 J. R. Mowat, H. Danared, G. Sundström, M. Carlson, L. H. Andersen, L. Vejby-Christensen, M. af Ugglas and M. Larsson, Phys. Rev. Lett. **74,** 50 (1995).
4 A. Al-Khalili, H. Danared, M. Larsson, A. Le Padellec, R. Peverall, S. Rosén, J. Semaniak, M. af Ugglas, L. Vikor and W. van der Zande, Hyper. Int. **114,** 281 (1998).
5 S. Rosén, R. Peverall, M. Larsson, A. Le Padellec, J. Semaniak, Å. Larson, C. Strömholm, W. van der Zande, H. Danared and G. H. Dunn, Phys. Rev. A **57,** 4462 (1998).
6 S. Rosén, A. M. Derkatch, J. Semaniak, A. Neau, A. Al-Khalili, A. Le Padellec, L. Vikor, H. Danared, M. af Ugglas, R. Thomas and M. Larsson, Faraday Discuss. **115**, 295 (2000).
7 A. Neau, A. Al-Khalili, S. Rosén, J. Semaniak, A. Le Padellec, A. M. Derkatch, W. Shi, L. Vikor, M. Larsson, R. Thomas, M. Nagard, K. Andersson, H. Danared and M. af Ugglas, J. Chem. Phys. **113**, 1762 (2000).
8 A. Lampert, A. Wolf, D. Habs, J. Kettner, G. Kilgus, D. Schwalm, M. S. Pindzola and N. R. Badnell, Phys. Rev. A **53**, 1413 (1996).
9 D. Zajfman, Z. Amitay, C. Broude, P. Forck, B. Seidel, M. Grieser, D. Habs, D. Schwalm, and A. Wolf, Phys. Rev. Lett. **75**, 814 (1995).
10 A. Le Padellec, C. Sheehan and J.B.A. Mitchell, J. Phys. B **31**, 1725 (1998).
11 A. Le Padellec, J. B. A Mitchell, A. Al-Khalili, H. Danared, A. Källberg, Å. Larson, S. Rosén, M. af Ugglas, L. Vikor and M. Larsson, J. Chem. Phys. **110**, 890 (1999).
12 A. Le Padellec, K. Andersson, D. Hanstorp, F. Hellberg, M. Larsson, A. Neau, S. Rosén, H. T. Schmidt, R. Thomas, J. Semaniak, D. J. Pegg, F. Österdahl, H. Danared and A. Källberg, Phys. Script. **64**, 467 (2001).
13 L. H. Andersen, D. Mathur, H. T. Schmidt and L. Vejby-Christensen, Phys. Rev. Lett. **74**, 892 (1995).
14 L. Vejby-Christensen, D. Kella, D. Mathur, H. B. Pedersen, H. T. Schmidt and L. H. Andersen, Phys. Rev. A **53**, 2371 (1996).
15 H. Lavendy, G. Gandara and J. M. Robbe, J. Mol. Spectr. **106**, 395 (1984).
16 D. M. Hirst, Mol. Phys. **82**, 359 (1994).
17 F. A. Cotton and G. Wilkinson, "Advanced Inorganic Chemistry", (Wiley-Interscience, New York, 1972).

18 B. M. Chadwick and H. G. M. Edwards, "Molecular Spectroscopy", Vol. 1, (Chemical Society, London, 1972).
19 Peabody Scientific, Peabody, Massachusetts, USA.
20 K. A. Peterson, R. C. Mayrhofer and R. C. Woods, J. Chem. Phys. **93,** 4946 (1990).
21 S. Petrie, C. G. Freeman, M. Meot-Ner, M. J. McEwan and E. E. Ferguson , J. Am. Chem. Soc. **112,** 7121 (1990).
22 S. Petrie, C. G. Freeman, M. J. McEwan and E. E. Ferguson, Mon. Not. R. Astron. Soc. **248,** 272 (1991).
23 D. Talbi, A. Le Padellec and J. B. A. Mitchell, J. Phys. B **33**, 3631 (2000).
24 C. Fridh and J. Asbrink, J. Electron Spectrosc. Relat. Phenom. **7,** 119 (1975).
25 D. Forney, W. E. Thompson and M. E. Jacox, J. Chem. Phys. **97,** 1664 (1992).
26 T.K. Ha and G. Zumofen, Mol. Phys. **40**, 445 (1980).

Electron Induced Vibrational Deexcitation of the Molecular Ions H_2^+ and D_2^+

S. Krohn,[1,2] H. Kreckel,[1] L. Lammich[1], M. Lange,[1] J. Levin,[1] D. Schwalm,[1] D. Strasser,[2] R. Wester,[1] A. Wolf,[1] and D. Zajfman[2]
[1]*Max-Planck Institut fur Kernphysik, Heidelberg, Germany and*
[2]*Weizmann Institute of Science, Rehovot, Israel*
sven.krohn@mpi-hd.mpg.de

1. INTRODUCTION

The interaction of molecular ions with low kinetic energy electrons has been studied extensively at different heavy-ion storage rings during the last ten years. The usage of storage rings made it possible to store infrared-active molecular ion beams, such as HD^+, for a time that is sufficiently long so that these ions are completely relaxed to their vibrational ground state by spontaneous emission of radiation. The stored ion beams are then merged with an intense and cold electron beam by using the electron cooling device present in the ion storage ring as a target for electron-ion collision studies. Here we report on the extension of such experiments to the homonuclear hydrogen ions which do not cool radiatively and for which we hence can study electron-impact vibrational de-excitation.

2. INELASTIC COLLISIONS OF HYDROGEN MOLECULAR IONS WITH ELECTRONS

In the electron interaction with hydrogen molecular ions in a given vibrational state, v, two inelastic reactions can take place even at kinetic energies very close to zero: Firstly, the dissociative recombination (DR),

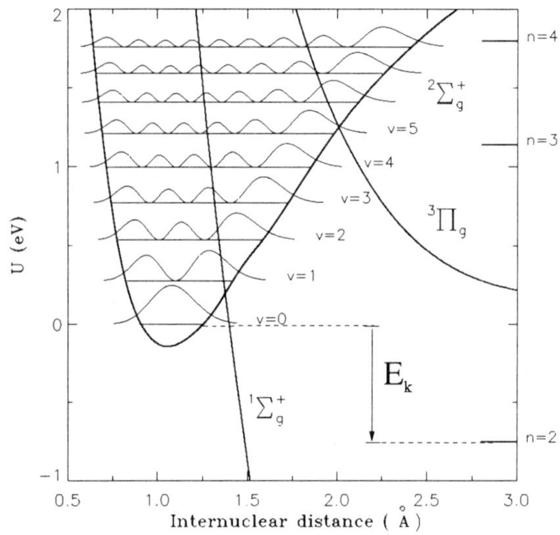

Figure 1. The ground electronic state, $^2\Sigma_g^+$, of H_2^+ and the doubly excited $^1\Sigma_g^+$ dissociative state[3,4] of H_2. Additionally shown are the next higher doubly excited state belonging to the Q_1 series and the nuclear wave functions squared $|\chi_v(R)|^2$ (J = 0). The final fragment state energies of the excited atomic hydrogen (n = 2-4) and the energy release E_k measured by the RFI method are also indicated.

which is described for H_2^+ by

$$H_2^+(v) + e^- \longrightarrow H_2^{**} \longrightarrow H(nl) + H(n'l') + E_k, \qquad (1)$$

where nl and $n'l'$ are the final fragment states of the hydrogen atoms and H_2^{**} is a doubly excited dissociative state (see Fig. 1) and secondly, the process of super-elastic collisions (SEC), where internal excitation energy is transferred to the outgoing electron:

$$H_2^+(v) + e^-(E) \longrightarrow H_2^{**} \longrightarrow H_2^+(v') + e^-(E + \Delta E), \qquad (2)$$
$$(v' < v, \Delta E > 0)$$

The two reactions (dissociation and autoionization) are competing with each other and can proceed via the same doubly excited state (see Eqs. (1) and (2)). As a further competing process, one has elastic resonant scattering via the intermediate state H_2^{**}, which corresponds to autoionization where v' = v (Eq. (2), for $\Delta E = 0$).

In addition to several detailed experiments on the heteronuclear isotopomer HD^+, the homonuclear species H_2^+ and D_2^+ have also been studied in storage rings. In particular, changes in the DR rate coefficient as a function of the storage time have been observed in experiments[1,2] with stored H_2^+ and D_2^+ ion beams and interpreted by variations of the vibrational distributions of the stored ions. Larsson et al.[2] proposed that the observed vibrational cooling might be due to collisions with the rest gas molecules and Van der Zande et al.[1] claimed that the larger DR rate coefficients for the high vibrational states cause the cooling. The first experimental hints that the SEC processes are important for vibrational cooling were found at the storage ring TARN II by Tanabe et al.[5] and Saito et al.[6] in recombination experiments with H_2^+ and D_2^+, respectively. Recent studies[7] at the TSR facility of cooling in a stored H_2^+ ion beam used vibrational population diagnostics based on the Coulomb explosion imaging technique (see Sec. 2.1). It was unambiguously shown that the observed vibrational cooling is dominated by SECs and that the SEC rate coefficients are unexpectedly high.

3. EXPERIMENTAL METHOD

We here discuss new experiments on H_2^+ and D_2^+ carried out at the heavy-ion storage ring TSR located at the Max-Planck Institut für Kernphysik, Heidelberg. Both ions were produced by direct electron impact ionization of the neutral molecule. The H_2^+ ions were accelerated to about 1 MeV by a Van-de-Graaf accelerator and injected into the TSR,[7] whereas the D_2^+ ions were accelerated to about 2 MeV using two radio frequency quadrupole (RFQ) structures. Approximately 10^7 ions were typically stored after each injection. The circulating ion beams were merged with an intense electron beam over a length of ≈1.5 m. The electron velocities were set in each experiment close to those of the injected ion beams. Hence, the electron-ion collision energies were approximately zero. The electron densities ranged up to 1.6 x 10^7 cm^{-3} and the longitudinal and transverse electron temperatures were $kT_{\parallel} \approx 0.1$ meV and $kT_{\perp} \approx 12$ meV, respectively. The diameter of the electron beam was 2 - 3 cm (see Refs. 7 and 8). A Coulomb explosion imaging (CEI) setup was used as a tool to monitor the vibrational population of the stored molecular ions.[9,10] The relative contributions of different initial vibrational states to DR were probed by a recombination fragment imaging (RFI) setup mounted downstream from the electron cooler.[11]

3.1 Coulomb Explosion Imaging

The CEI technique provides information about the vibrational population distribution in stored molecular ion beams as a function of the storage time.

Using a slow extraction scheme, a part of the stored ion beam is extracted and steered to the CEI setup. A detailed discussion of the CEI principle and the experimental setup has been given elsewhere[9,12] and therefore only a brief overview is given here.

In a CEI measurement, fast molecular ions collide with a thin target foil ($\leq$70 Å thick), where all binding electrons are stripped on a time scale of $\approx 10^{-17}$ s. The dwell time in the foil ($\approx 10^{-16}$ s) is short compared to vibrational (10^{-14} s) and rotational times (10^{-12} s). After passing the target foil, the charged fragments repel each other due to the strong Coulomb force and the initial internuclear distance is converted into several centimeters after a free flight of a few meters. Using a three dimensional imaging technique, the asymptotic kinetic energy, E_k, released in the center-of-mass frame of reference after the Coulomb explosion can be determined. E_k is derived from the observed distance d of the fragments as measured by three dimensional imaging. For details about the detector see Ref. 9.

The measured kinetic energy release, E_k, carries information about the vibrational state of the extracted molecule, as it is directly related to the initial internuclear distance R at the time the binding electrons were stripped. For the two singly charged fragments produced in the Coulomb explosion of hydrogen molecular ions, the relation is

$$E_k = \frac{e^2}{R}, \tag{3}$$

where e is the electron charge. These measurements were performed for one molecular ion at a time and a distribution of released kinetic energies $P^{E_k}(E_k;t)$ is obtained after the accumulation of many events. Due to the fast stripping process inside the target foil, the distribution of internuclear distances $P^R(R;t)$ of the stored molecular ions at a storage time t is transformed to a distribution of macroscopic fragment distances on the CEI detector corresponding to a distribution of kinetic energy releases $P^{E_k}(E_k;t)$. For an ensemble of molecules in a vibrational state v these distributions are related to each other by:

$$P_v^{E_k}(E_k) = P_v^R(R)\left|\frac{dE_k}{dR}\right|^{-1} \tag{4}$$

$$\text{with} \qquad P_v^R(R) = \Psi_v^2(R),$$

where Ψ_v is the nuclear wave function of the vibrational state. Thus, by fitting a linear combination of normalized distributions, $P_v^{E_k}(E_k)$, to the normalized experimental kinetic energy release spectrum, the relative population $p_v(t)$ of the vibrational states can be extracted:

$$P^{E_k}(E_k;t) = \sum_v p_v(t) P_v^{E_k}(E_k). \tag{5}$$

This simplified description does not take into account several processes inside the target foil as well as the initial momentum of the exploding fragments and the reader is referred to Ref. 13 for more details on the data analysis. In the following, the distributions, $P^d(d,t)$, of the measured three-dimensional fragment distances, d, are plotted instead of the corresponding kinetic energy release distributions $P^{E_k}(E_k;t)$. These are similar in shape.

3.2 Recombination Fragment Imaging

This technique measures the kinetic energy release, E_k, of recombination events in the electron cooler, reflecting the different initial vibrational states v of the molecular ions and the final atomic states n, n' of the atomic hydrogen fragments (see Fig. 1). In the present experiments the electrons had near-zero kinetic energy. Thus, the final quantum state of one of the hydrogen fragments is always the ground state (i.e. $n'=1$). The distribution of projected (two-dimensional) distances, D, of the DR fragments was measured at a distance of about 6.5 m downstream from the center of the electron cooler using a RFI detector.[11] As has been shown by Amitay et al.,[11] the distribution of projected fragment distances, D, of each DR channel (i.e. initial vibrational state v and final atomic fragment states n and $n' = 1$) can be described by an analytic function $P_{vn}(D)$ ($P_{vn1}(D) \equiv P_{vn}(D)$ in the notation of Ref. 11). Thus, relative contributions, q_{vn}, to a DR channel at a storage time t were obtained by a fit of the linear combination of the individual functions $P_{vn}(D)$ to the measured and normalized distributions $P(D;t)$,

$$P(D;t) = \sum_{vn} q_{vn}(t) P_{vn}(D) \tag{6}$$

The relative contribution, q_{vn}, is proportional to the DR rate coefficient, α_{vn}, and the population p_v of the vibrational state v,

$$q_{vn} \propto \alpha_{vn} p_v \tag{7}$$

The Boltzmann distributed rotational excitation of the stored ions (T =600K)[13] as well as the different degeneracies of rotational states with odd and even J, due to nuclear statistics, have been taken into account in modified functions, $P_{vn}^J(D)$.

4. RESULTS

Using the CEI and RFI techniques, we extensively studied DR and SEC of stored H_2^+ and D_2^+ beams. For both processes, absolute rate coefficients are presented below.

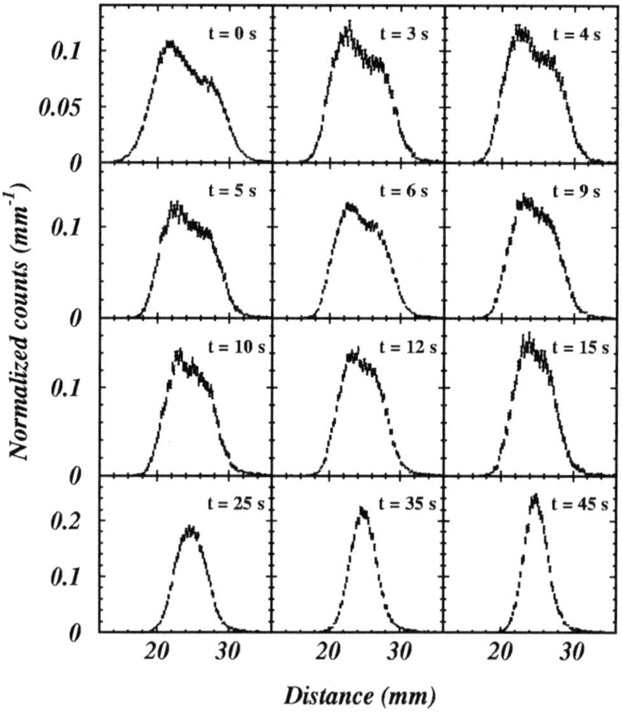

Figure 2. Spectra of the three-dimensional fragment distance for D_2^+ Coulomb explosion measured after different interaction times (t = 0-45 s) with near-zero relative energy electrons (electron density 5.6 x 10^{-6} cm^3/s). The spectra are normalized to unity area.

4.1 Observation Of Electron-Induced Vibrational Cooling

As discussed above, the homonuclear species of the hydrogen molecular ions are not infrared active, i.e. they do not cool radiatively. Thus, we studied the influence of electron-ion collisions on the vibrational population of stored H_2^+ and D_2^+ ion beams by merging the ions with an intense electron beam in the electron cooler. The collision energy E was approximately zero in all experiments. The ion beam was injected and stored for a time t. After this, the beam was continuously extracted towards the CEI setup. Before each new injection the remainder of the stored beam was kicked out. Fragment distance spectra for Coulomb explosion of D_2^+ measured after different storage times with electron interactions are shown in Fig. 2. Significant changes in the vibrational distribution are evident. The decrease of the width of the

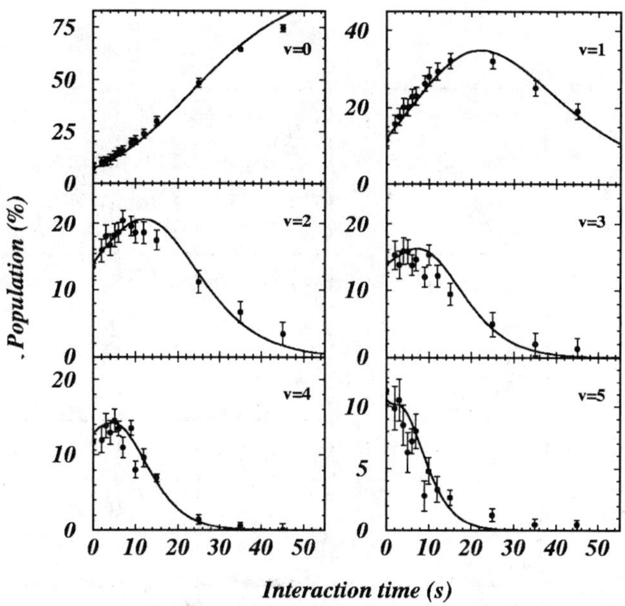

Figure 3. Time evolution of the vibrational population of the first six vibrational states of D_2^+ as a function of the interaction time with near-zero relative energy electrons. The full lines represent the modeled time behavior as obtained from a simultaneous fit to all shown spectra. The fits were performed using the SEC rate coefficients as free parameters and keeping the DR rate coefficients fixed to the experimental values.

distribution is a sign (c.f. Sec. 3.1) of vibrational de-excitation. Fig. 3 shows the time evolution of the vibrational population for the first six vibrational states of D_2^+ obtained following the fitting procedure discussed in Sec. 3.1.

Additionally plotted are modeled time evolutions that will be discussed in Sec. 4.4. After a storage time of about 40 s at an electron density of 5.6 x 10^6 cm^3/s more than 98% of the ions were found in the two lowest vibrational states indicating a high efficiency of vibrational de-excitations by electron-ion collisions. Similar results for H_2^+ can be found in Ref. 7.

In order to prove that the observed changes in the vibrational population are due to interactions with the electron beam, the H_2^+ and D_2^+ ion beams were stored over 12 s and 20 s, respectively, while the electron cooler was off. CEI measurement were performed over the whole storage time. The acquired events were divided in several time slices and analyzed. In contrast to Fig. 3, no changes of the CEI spectra and hence of the vibrational population distributions could be observed for both molecular ions.[7,8] This demonstrates that the vibrational cooling is only due to electron interaction and that the contribution of rest gas collisions is negligible.

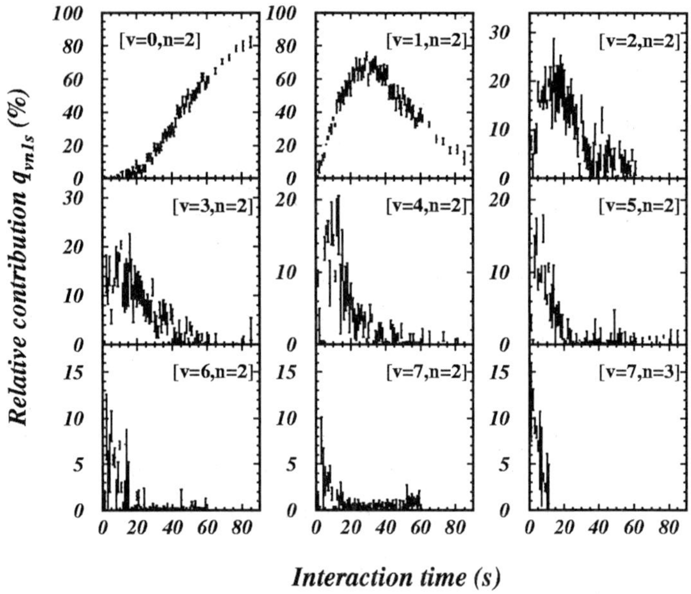

Figure 4. Time evolution of the relative contributions, q_{vn}, to the DR of D_2^+, where v describes the initial vibrational state and n the final state of one atomic fragment (i.e. with D(1s)+D(n)). The data were obtained at different interaction times (t = 0-85 s) and a collision energy $E \approx 0$ eV.

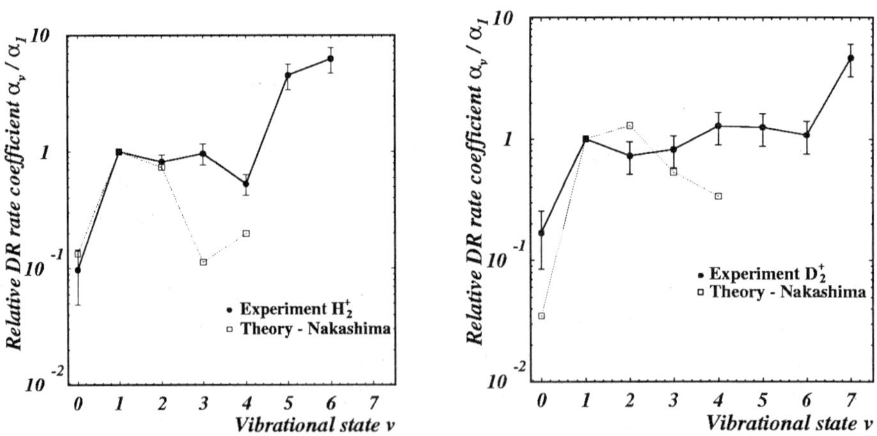

Figure 5. Experimental results for the relative DR rate coefficients of H_2^+ (left) and D_2^+ (right) at E ≈ 0 eV as a function of the initial vibrational quantum number (solid circles). Also shown are the theoretical calculations by Nakashima et al.[14] (open boxes). Lines are drawn to guide the eye through the plotted values.

4.2 Vibrational State Specific Relative DR Rate Coefficients

At electron beam parameters identical to the CEI measurement, two dimensional DR fragment distance distributions (see Sec. 3.2) were measured over storage times of t = 0-14 s and t = 0-85 s for H_2^+ and D_2^+, respectively. The time evolution of the relative contributions, q_{vn}, of the different DR channels to the total DR rate of D_2^+ were extracted following the fitting procedure discussed in Sec. 3.2. The fitting results for the measurement with a D_2^+ beam are shown in Fig. 4. The DR channels are v, n with $n = 2$ for $v \leq 6$ and additionally $n = 3$ for $v = 7$ as expected for the vibrational level positions of D_2^+. A clear vibrational cooling is observable and after about 40 s of storage only the two lowest vibrational states still contribute to the DR. Similar results for H_2^+ are presented in Ref. 8.

Knowing the relative contribution $q_v(t) = \Sigma_n q_{vn}$ of a vibrational state to the DR and the vibrational population $p_v(t)$ of that state it is possible to derive relative DR rate coefficients (see Eq. 7) for both molecular ions, H_2^+ and D_2^+. We choose to normalize these relative rate coefficients to $v = 1$, which has a much higher DR cross section than $v = 0$. The experimental results are shown in Fig. 5. Also shown in Fig. 5 are results from calculations based on the multichannel quantum defect theory (MQDT).[14] In these calculations only the $^1\Sigma_g^+$ dissociative state (see Fig. 1) of the neutral molecule has been considered and rotational excitation has been neglected. In the experiment, a clear increase of the DR rate coefficient is observed between $v = 0$ and $v = 1$ and for $v > 4$ in the case of H_2^+ ($v > 6$ for D_2^+); the initial rise can probably be attributed to a better Franck-Condon-overlap with the $^1\Sigma_g^+$ dissociative curve, while the rise at higher v seems to be due to the accessibility of additional higher lying dissociative states and the opening of the n = 3 final fragment channel. Similar results have been obtained for the vibrational states v = 0-7 of HD^+ by Amitay et al.[10,13] The exact reasons for the observed differences between theory and experiment are not yet understood.

4.3 Absolute DR Rate Coefficients

As demonstrated in the previous sections, H_2^+ and D_2^+ can be vibrationally cooled via collisions with electrons. This has been used to provide vibrationally cold ion beams for absolute DR rate coefficient measurements. In these experiments, the vibrational population was monitored using the RFI technique together with the measured relative DR rate coefficients. The electron density was increased for these particular measurements by almost a factor of 3 to 1.6 x 10^7 cm^{-3} and the transverse and

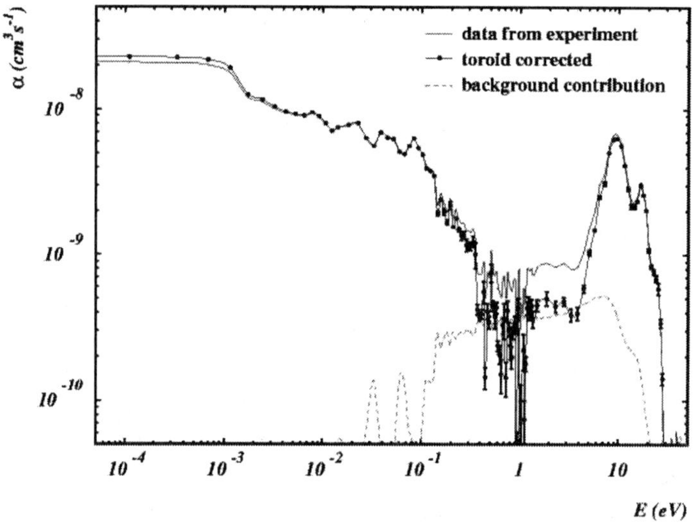

Figure 6. Measured rate coefficient for DR of H_2^+ with electrons as a function of their relative energy *E*, shown by the thick solid line with dots and error bars. The spectrum was taken after a storage time of 40s (see text). The overall systematic error on the plotted rate coefficient is 8.4 %. The background contribution from electron collisions in the toroid region of the electron cooling device (dashed line) was subtracted from the experimental data (solid line).

longitudinal temperatures were $kT_\perp$ = 11.8 meV and $kT_\parallel$ = 0.1 meV, respectively. For vibrational cooling before the measurement, the electron cooler was adjusted to zero collision energy during 40 s. Due to the higher electron density, as compared to the experiments discussed in the previous sections, more than 99% of the molecular ions were found to be in the vibrational ground state after the 40 s long cooling period. Thus, the DR rate coefficient measurement were performed for an essentially $v = 0$ beam. For our merged-beam conditions, we measured in the limit of zero energy detuning between the electron and ion beam a DR rate coefficient for H_2^+ of

$$\alpha_{mb}(v=0) \quad = \quad (2.1 \pm 0.3) \times 10^{-8}\,\mathrm{cm}^3/\mathrm{s} \tag{8}$$

and for D_2^+ of

$$\alpha_{mb}(v=0) \quad = \quad (1.2 \pm 0.2) \times 10^{-8}\,\mathrm{cm}^3/\mathrm{s}. \tag{9}$$

The error limits are preliminary and include the uncertainties due to the ion current measurement and the remaining $v \geq 1$ fractions of < 1% in the stored beam. Using these results, the relative DR rate coefficients of Fig. 5 can be converted to absolute values for the vibrational states v = 0-6 and v = 0-7 for H_2^+ and D_2^+, respectively. We have also obtained the energy dependence of the DR cross section for the essentially $v = 0$ H_2^+ and D_2^+ beams after collisional cooling. The result for H_2^+ is shown in Fig. 6.

4.4 Absolute SEC Rate Coefficients

As has been shown above the observed vibrational de-excitation of D_2^+ and H_2^+ is caused by electron-ion collisions (DR and SEC) in the electron cooler. Thus, the observed cooling due to vibrational state specific DR and SEC rate coefficients can be modeled by solving a set of coupled rate equations.[7] Several calculations of SEC rate coefficients, based on MQDT[14] and R-matrix theory,[15] showed the dominance of vibrational transitions with $\Delta v = 1$. The model is therefore restricted to such SEC transitions. The absolute v-state specific DR rate coefficients are known from our measurements (see Secs. 4.2 and 4.3). Thus, the time evolutions of the first six vibrational states can be simultaneously fitted to the measured vibrational populations using the SEC rate coefficients as free parameters. The measured and fitted time evolutions of the vibrational populations of D_2^+ are shown in Fig. 3. The absolute SEC rate coefficients, β_v, obtained from the fits are (in units of 1×10^{-6} cm^3/s followed by the uncertainties in parentheses): β_1 = 0.36(7), β_2 = 0.89(18), β_3 = 1.25(25), β_4 = 1.50(30), β_5 = 2.20(44) for D_2^+ and β_1 = 0.39(8), β_2 = 0.76(16), β_3 = 1.21(26), β_4 = 1.46(30), β_5 = 2.10(42) for H_2^+. These values are up to an order of magnitude larger than values predicted by various theoretical calculations[14,15] and show that the SEC processes are primarily responsible for the fast vibrational de-excitation. More detailed information can be found in Refs.7 and 8.

5. SUMMARY

In the present overview, the vibrational de-excitation of the homonuclear hydrogen molecular ions H_2^+ and D_2^+ in collisions with low kinetic energy electrons has been discussed. For both SEC and DR, absolute rate coefficients were derived for the lower vibrational states. Furthermore, it has been shown that the observed cooling is primarily caused by SEC. The comparison of the absolute SEC and DR rate coefficients shows that the doubly excited intermediate neutral state formed in the electron-ion collisions autoionizes much more strongly than it dissociates.

ACKNOWLEDGMENTS

This work has been funded by the German Federal Minister for Education, Science, Research and Technology (BMBF) within the framework of the German-Israeli Project Cooperation in Future-Oriented Topics (DIP) and by the European Community within the Research-Training Network "Electron Transfer Reactions".

REFERENCES

1 W.J. van der Zande, J. Semaniak, V. Zengin, G. Sundström. S. Rosen, C. Stromhölm, S. Datz, H. Danared, and M. Larsson, Phys. Rev. A **54**, 5010 (1996).

2 M. Larsson, M. Carlson, H. Danared, L. Broström, S. Mannervik and G. Sundström, J. Phys. B: At. Mol. Opt. Phys. **27**, 1397 (1994).

3 T.E. Sharp, Atomic Data and Nuclear Data Tables **2**, 119 (1971).

4 S.L. Guberman, J. Chem. Phys. **78**, 1404 (1983).

5 T. Tanabe, H. Takagi, I. Katayama, K. Chida, T. Watanabe, Y. Arakaki, Y. Haruyama, M. Saito, I. Nomura, T. Honma, K. Noda, and K. Hosono, Phys. Rev. Lett. **83**, 2163 (1999).

6 M. Saito, Y. Haruyama, T. Tanabe, I. Katayama, K. Chida, T. Watanabe, Y. Arakaki, I. Nomura, T. Honma, K. Noda, and K. Hosono, Phys. Rev. A **61**, 062707 (2000).

7 S. Krohn, Z. Amitay, A. Baer, D. Zajfman, M. Lange, L. Knoll, J. Levin, D. Schwalm, R. Wester, and A. Wolf, Phys. Rev. A **62**, 032713 (2000).

8 S. Krohn et al., in preparation.

9 R. Wester et al., Nucl. Instrum. Methods A **413**, 379 (1998).

10 Z. Amitay, A. Baer, M. Dahan, L. Knoll, M. Lange, J. Levin, I. F. Schneider, D. Schwalm, A. Suzor-Weiner, Z. Vager, R. Wester, A. Wolf, D. Zajfman, Science **281**, 75 (1998).

11 Z. Amitay, D. Zajfman, P. Forck, U. Hechtfischer, B. Seidel, M. Grieser, D. Habs, R. Repnow, D. Schwalm, and A. Wolf, Phys. Rev. A **54**, 4032 (1996).

12 Z. Vager, R. Naaman, and E.P. Kanter, Science **244**, 244 (1989).

13 Z. Amitay, A. Baer, M. Dahan, J. Levin, Z. Vager, D. Zajfman, L. Knoll, M. Lange, D. Schwalm, R. Wester, A. Wolf. I. F. Schneider and A. Suzor-Weiner, Phys. Rev. A **60**, 3769 (1999).

14 K. Nakashima, H. Takagi and H. Nakamura, J. Chem. Phys. **86**, 726 (1987).

15 B. K. Sarpal and J. Tennyson, Mon. Not. R. Astron. Soc. **263**, 909 (1993).

Wave Packet Studies of Dissociative Recombination

A. E. Orel[1] and Å. Larson[2]
[1]*Department of Applied Science, University of California, Davis, CA 95616;* [2]*Physics Department, Royal Institute of Technology (KTH), S-100 44 Stockholm, Sweden*
aeorel@ucdavis.edu

1. INTRODUCTION

Wave packet methods, that is, numerical solutions of the time-dependent Schrödinger equation, are standard techniques in the study of the dynamics of chemical reactions. We have used these techniques with great success in the calculation of cross sections for dissociative recombination (DR) and dissociative excitation of molecular ions by electron impact. In this paper, we will outline the basic method and highlight two areas where the wave packet method has been shown to be robust. The first is in the determination of branching ratios into final states. This will be illustrated by the examples of ion-pair formation in HD^+ and the branching ratios for final states produced in the dissociative recombination of HeH^+ with low energy electrons. The second is the applicability of the method to polyatomic systems. This will be illustrated by the example of the dissociative recombination of the water ion.

2. WAVE PACKET METHOD

The wave packet method proceeds by the direct integration of the time-dependent Schrödinger equation:

$$i\frac{\partial}{\partial t}\Psi(\mathbf{r},t) = H(\mathbf{r},t)\Psi(\mathbf{r},t) \tag{1}$$

where, as throughout this paper, atomic units are used. As written, the method is exact. The approximations come in the definition of the Hamiltonian, and the initial conditions for the propagation.

In the cases described below DR proceeds by capture of the electron into a resonant dissociative state. The molecule begins to fragment, moving on the excited state (resonant) potential energy surface. During this process, the molecule can re-emit the electron (autoionize). If no autoionization occurs, the molecule fragments into neutral products. For the systems studied here, we have used the so-called 'boomerang' model for nuclear motion[1]:

$$\left(-\frac{1}{2\mu}\frac{\partial^2}{\partial R^2}+E_{res}(R)-i\frac{\Gamma(R)}{2}-E\right)\Psi(R)=-V_E(R)\chi_{v_i}(R) \quad (2)$$

In this case, the Hamiltonian in Eq. (1) becomes:[2]

$$H=-\frac{1}{2\mu}\frac{\partial^2}{\partial R^2}+E_{res}(R)-i\frac{\Gamma(R)}{2} \quad (3)$$

where χ_{v_i} is the initial vibrational state, μ is the reduced mass of the diatomic, Ψ(*R*) is the wave function on the dissociating resonant state, and V_E is the electronic coupling between the electronic scattering continuum and the resonant state. V_E is related to the autoionization width, Γ(*R*), by:

$$\Gamma(R)=2\pi|V_E(R)|^2 \quad (4)$$

The initial wave packet is given by

$$\Psi(R,t=0)=V_E(R)\chi_{v_i}(R) \quad (5)$$

For long times, when the wave packet, Ψ(*R*), has reached the asymptotic region of the potential and the autoionization loss has gone to zero, it is projected onto asymptotic final states ϕ_k to obtain the T-matrix[3]:

$$T(E)=\lim_{t\to\infty}\left(-i\int_0^\infty \phi_k\Psi(R)dR\right) \quad (6)$$

where k is the wave number, and the cross section is obtained:

$$\sigma(E)=\frac{2\pi^3}{E}|T(E)|^2 \quad (7)$$

The generalization of these equations to polyatomic systems is obvious.

3. BRANCHING RATIOS

There have been a limited number of studies of final state branching ratios produced by dissociative recombination. The wave packet is easily used to study this problem. We will illustrate this with two examples, ion-pair formation in HD^+ and branching ratios in HeH^+.

3.1 Dissociative Recombination of HD^+

Recent experiments in the CRYRING ion-storage ring[4] on low energy electron scattering by HD^+ have shown oscillations as a function of electron energy in the ion-pair formation cross section, that is, the reaction:

$$HD^+ + e^- \rightarrow H^- + D^+. \tag{8}$$

Zong et al.[4] proposed that the source of these oscillations was an interference effect, since multiple paths exit to the ion-pair state arising from crossing of the resonant state with the Rydberg manifold of states. Preliminary semiclassical calculations showed these oscillations and were in qualitative agreement with the experiment.[5] We used the time dependent wave packet method to accurately determine the ion-pair formation cross section.

We solved Eq. 1 with T(R) as the diagonal kinetic energy operator and V(R) as a complex potential energy matrix:

$$\mathbf{V}(R) = \begin{pmatrix} V_{res} & V_{r1} & V_{r2} & \cdots & V_{rn} \\ V_{1r} & V_1 & V_{12} & \cdots & V_{1n} \\ \cdots & \cdots & \cdots & \cdots & \cdots \\ V_{nr} & V_{n1} & V_{n2} & \cdots & V_n \end{pmatrix}$$

where

$$V_{res}(R) = V_0(R) + i\frac{\Gamma_{tot}(R)}{2}. \tag{9}$$

$V_0(R)$ is the real part of the resonance energy surface and $\Gamma_{tot}(R)$ is the total autoionization width of the resonance. Here, only the lowest $^1\Sigma_g^+ (2p\sigma_u)^2$ resonant state of the HD molecule is included. The total width for this state is given by a sum, where only the partial widths with angular momentum $l = 0$ and $l = 2$ (s and d waves) of the external electron contribute:[6]

$$\Gamma_{tot}(R) = \Gamma_s(R) + \Gamma_d(R). \tag{10}$$

V_i are the Rydberg potential energy curves. V_{ri} is the coupling between the Rydberg and the resonance, (note that $V_{ri} = V_{ir}$) and V_{ij} is the coupling between the Rydbergs that is neglected in the present calculation. The potential energy curve for the ion and the quantum defects as a function of internuclear separation are taken from Schneider et al.[7] The potential energy curves for the $^1\Sigma_g^+$ Rydberg states are given by:

$$V_n(R) = V_{ion}(R) - \frac{0.5}{n_{eff}(R)^2} \tag{11}$$

where $V_{ion}(R)$ is the ionic ground state potential ($^2\Sigma_g^+(2p\sigma_u)$) and $n_{eff}(R)$ is the effective quantum number given by:

$$n_{eff}(R) = n - \mu_l(R) \qquad (12)$$

Here, μ_l is the quantum defect for Rydberg states with the configurations ($1s\sigma_g\, ns\sigma_g$) and ($1s\sigma_g\,(n + 1)d\sigma_g$), corresponding to an angular momentum of $l = 0$ and $l = 2$ for the excited electron. At large R, where $V_{res}(R)$ correlates diabatically to the ion-pair curve, this curve goes asymptotically as:

$$V_{res}(R) \to V_{final} - \frac{1}{R} - \frac{\alpha}{R^4}, \qquad (13)$$

where the polarizability,[8] α, is 230 a.u. The resonance curve at small R given by Schneider et a1.[7] was connected smoothly to this asymptotic form.

The autoionization width $\Gamma(R)$ was taken from Schneider et a1,[7] with autoionization width set equal to zero after the resonance crosses the ion curve. At small R, the coupling between the higher ($n \geq 3$) Rydberg states with configurations ($1s\sigma_g\, nl\sigma_g$) and the resonance is the scaled autoionization width, that is,

$$V_{ri}(R) = \sqrt{\frac{\Gamma_l(R)}{2\pi}}(n_{eff}(R))^{\frac{3}{2}}. \qquad (14)$$

This scaling does not hold for the lowest n = 2 Rydberg states.[9,10] For these states the diabatic couplings are estimated using a two-state approximation. At large internuclear separations, it is necessary to include the coupling between the ion-pair state and the Rydberg states at the second crossing. These were obtained from the coefficients of the ion-pair and Rydberg terms in configuration-interaction (CI) wave functions. The infinite number of Rydberg states converging up to the ion were included by using two 'effective' Rydberg states, one with $l = 0$ and the other with $l = 2$ of the excited electron.

The ion-pair cross section was obtained from Eq. (6) using the energy normalized eigenfunctions with eigenvalues above the dissociation limit in the projection.[12] These results are compared to experiment in Fig. la. The same calculation that produces the ion-pair cross section can, of course, produce the total dissociative recombination cross sections as well as the branching ratios into the H(1s) + D(n) or the H(n) + D(1s) channels. These results are compared to experiment[13] in Fig. 1b. Note that as each new channel opens, the branching ratios into the other channels drop. These results are also in good agreement with those calculated by Zajfman et al.[13] who used a Landau-Zener treatment for the curve crossings.

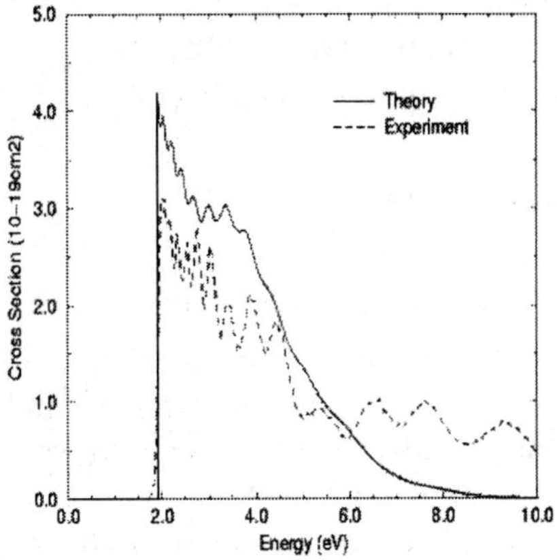

Figure 1a. **Cross section for formation of ion-pair in the collision of low-energy electrons with HD^+ Theory - solid line Experiment - dashed line.**

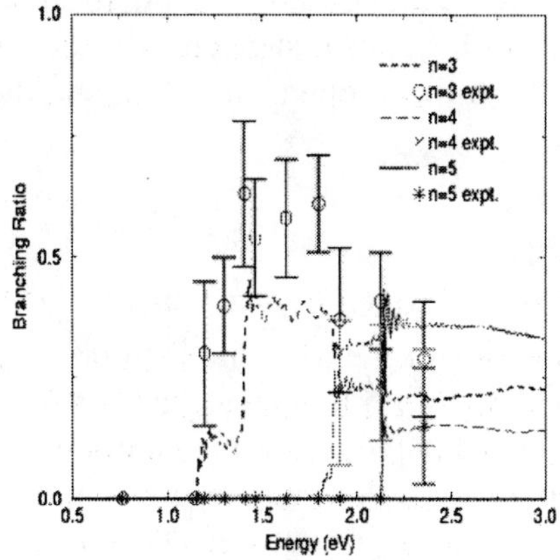

Figure 1b. **Branching ratios for final states of H produced after dissociative recombination of low energy electrons with HD^+**

3.2 Dissociative Recombination of HeH$^+$

In the case of HD$^+$, the initial ion curve was crossed by a resonant state. In the case of HeH$^+$, a significant cross section for dissociative recombination has been obtained at low energy where no such crossing exists. It was shown by Guberman[14] using the MQDT method and Sarpal[15] using the R-matrix method that the DR proceeds via nonadiabatic coupling to the Rydbergs and then to a dissociative state. There have been later, more extensive calculations[16] that confirmed this. Although these calculations agree well with each other and experiment as to the magnitude of the DR cross section, they predict different final state fragment distributions. Experiments have shown that at zero electron energy only the H(n=2) + He final channels are significantly populated. In addition, as the electron energy is raised to above the H(n=3) + He threshold, the flux switches over a very narrow energy range from zero (below threshold) to about 80% in H(n=3).[17]

If we label by Ψ_i and Ψ_j as two electronic states, the non-adiabatic coupling between them is given by:[18]

$$F^1_{i,j} = -\frac{1}{\mu}\langle\Psi_i|\frac{\partial}{\partial R}|\Psi_j\rangle\frac{\partial}{\partial R} \tag{15}$$

where the term:

$$F^2_{i,j} = -\frac{1}{2\mu}\langle\Psi_i|\frac{\partial^2}{\partial R^2}|\Psi_j\rangle \tag{16}$$

has been neglected[14,18] and the matrix elements represent integration over electronic coordinates only. In the above formulas, μ is the reduced mass.

In a time-dependent treatment, the initial wave function is given by:[19]

$$\Psi(R; t = 0) = F^1_{i,j}(R)\chi_{v_i}(R) \tag{17}$$

where $F^1_{i,j}(R)$ is given in Eq. (15) and χ_{v_i} is the initial vibrational wave function of the ion. One then propagates this initial wave packet using Eq. (1) with the Hamiltonian as just T + V where T is the kinetic energy operator and V is the potential for the neutral potential energy surface. The cross section is obtained as before by projection onto the asymptotic states. When more than one surface is involved, the initial wave packet becomes a vector, and the Hamiltonian becomes a matrix, with the couplings between states appearing in the off-diagonal elements.

In order to calculate the non-adiabatic coupling between an (ion + electron) state and a neutral state, we used analytic gradient techniques[20] to calculate derivative couplings between a series of Rydberg states of HeH. These states were represented by large-scale multi-reference configuration wave functions in a basis sufficient to describe the n=4 Rydbergs.

Since the shape of the neutral curves in the Franck-Condon region parallels the ion, and their energies are well represented as a Rydberg series, the coupling from the ground state to the members of a Rydberg series can be scaled to produce the coupling from the ground state to the ion, that is,

$$\langle\Psi_{continuum}|\frac{\partial}{\partial r}|\Psi_{gd}\rangle = \langle\Psi_{Rydberg}|\frac{\partial}{\partial r}|\Psi_{gd}\rangle n_{eff}^{1.5} \tag{18}$$

where n_{eff} is the effective quantum number given by Eq. (12).

The potential curves for HeH are shown in Fig. 2. Using these curves and couplings we have begun calculations on the DR in this system over a range of energies up to the n=4 threshold. The wave packet method allows us to obtain a snapshot of the dynamics as they proceed, allowing us to isolate the various physical processes that determine the cross section and final state distributions. By manipulating the couplings we can determine which aspect of the problem determines the cross section and final state distribution: the initial ion-Rydberg coupling, subsequent Rydberg-Rydberg coupling, coupling to the ground state of the neutral, etc. Our preliminary results which only include the coupling between the ion + electron continuum and the Rydbergs, but not the Rydberg-Rydberg coupling are shown in Figure 3. At zero energy, essentially 100% of the flux goes into the n=2 channel. As the energy is increased such that the n=3 channels are energetically open, most of the flux (about 70%) goes into those channels. There is another sharp jump as the n=4 channels become open.

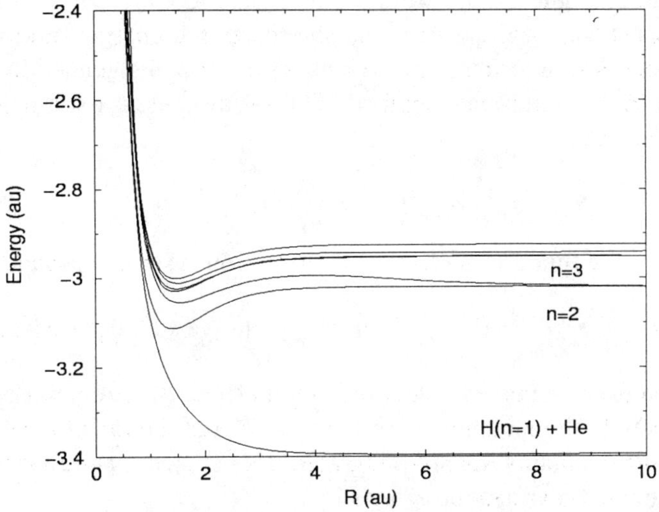

Figure 2. Potential Energy Curves for HeH.

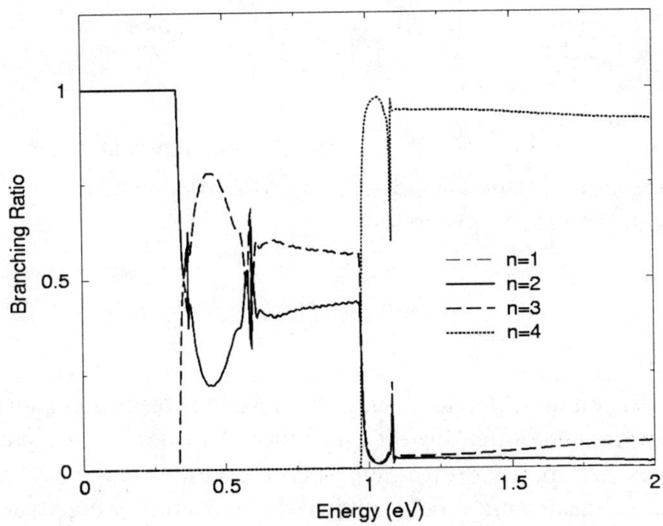

Figure 3. Branching ratios for HeH^+ dissociative recombination. The branching ratios into the various Rydberg states of the H atom as a function incident electron energy.

4. DISSOCIATIVE RECOMBINATION OF THE WATER ION

As an example of DR from a polyatomic ion, we will begin with some preliminary data on the water ion. The resonance energies and widths come from electron-scattering calculations from the molecular ion using the complex Kohn variational method. This method uses a stationary principle for the T-matrix:

$$T_{stat}^{\Gamma\Gamma_o} = T_{trial}^{\Gamma\Gamma_o} - 2\int \Psi_\Gamma (H-E)\Psi_{\Gamma_o}. \tag{19}$$

The trial wave function for the (N + 1)-electron system is expanded as

$$\Psi_{\Gamma_o} = \sum_\Gamma A[\Phi_\Gamma(\vec{r}_1..\vec{r}_N)F_{\Gamma\Gamma_o}(\vec{r}_{N+1})] + \sum_\mu d_\mu^{\Gamma_o}\Theta_\mu(\vec{r}_1..\vec{r}_{N+1}) \tag{20}$$

where the first sum is over electronic target states, Φ_Γ, A antisymmetrizes the coordinates of the incident electron with those of the target electrons and the second sum contains square-integrable, (N+1)-electron terms that describe correlation and polarization effects.

In the Kohn method, the scattering functions are further expanded in a combined basis of Gaussian (ϕ) and continuum (regular Coulomb wave, F_l, and outgoing Coulomb wave, G_l^+) basis functions:

$$F_{\Gamma\Gamma_o}(\vec{r}) = \sum_i c_i^{\Gamma\Gamma_o}\phi_i(\vec{r}) + \sum_{lm}[F_l(k_\Gamma r)\delta_{ll_o}\delta_{mm_o} + T_{ll_omm_o}^{\Gamma\Gamma_o}G_l^+(k_\Gamma r)]Y_{lm}(\hat{r})/r. \tag{21}$$

The T-matrix elements, $T_{ll_0mm_0}^{\Gamma\Gamma_0}$, are the fundamental dynamical quantities from which all fixed-nuclei cross sections are derived.

The ground state of water is:

$$(1a_1)^2(2a_1)^2(1b_2)^2(3a_1)^2(1b_1)^2 \tag{22}$$

The water ion has three low-lying excited states resulting from removing one electron from the three lowest lying orbitals. In order to obtain good target functions, we first performed an SCF calculation on the neutral system. We then froze the lowest orbital, and performed an all singles and doubles CI calculation to obtain the natural orbitals. In order to produce a balanced treatment we took natural orbitals from averaging the density matrices for the lowest three states of the ion. It was necessary to include three natural orbitals ($4a_1$, $2b_1$, $2b_2$) besides the occupied orbitals. The basis set was expanded to include additional diffuse orbitals. The target states were determined from a full CI in the natural orbital space (of eight orbitals).We will concentrate on scattering in A_1 symmetry. We first looked at the eigenphase sum as a function of initial electron energy for several bond angles and O-H bond distances. This shows the shift of the resonances with the two normal coordinates of bend and symmetric stretch. We found four resonances in this symmetry. The eigenphase sum was fit to a Breit-Wigner form and the

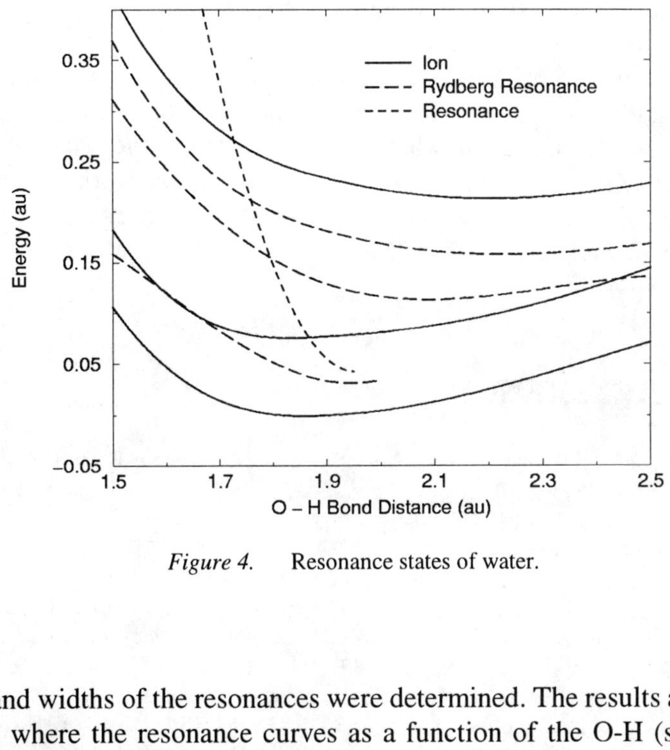

Figure 4. Resonance states of water.

position and widths of the resonances were determined. The results are shown in Fig. 4, where the resonance curves as a function of the O-H (symmetric stretch, at fixed angle) are plotted and compared to the potential energy curves for the three ion states.

Three of the four A_1 resonances parallel the highest B_2 ion curve. They represent three members of a Rydberg series converging to this ionic state. Note that as they lie above the ground state of the ion, they appear as scattering resonances, the so-called 'core-excited' resonances, seen in OH^+ and CH^+ DR. In addition, there is another low-lying dissociative resonance.

These results, though interesting, are quite preliminary. Calculations are continuing to determine the three-dimensional potential energy surfaces for the resonances. These will be used as input into a full-dimensional treatment of the DR process in this system.

5. SUMMARY

There are only a few wave packet calculations for dissociative recombination. However, the method shows great promise for the study of final state branching ratios and dissociative recombination of polyatomics.

ACKNOWLEDGEMENTS

The authors would like to thank Mats Larsson, Chris Greene, Stig Stenholm, Ioan Schneider, Annick Suzor-Weiner, Markku Jaaskelainen and Xavier Urbain for fruitful discussions. We are also grateful for the experimental data obtained from the groups of Mats Larsson and Daniel Zajfman. A. E. O. acknowledges support from the National Science Foundation, Grant No. PHY-99-87877. Part of this work was performed under the auspices of the U.S. Department of Energy by University of California Lawrence Livermore National Laboratory under contract number W-7405-Eng-48.

REFERENCES

1 A. Herzenberg, J. Phys. B: At. Mol.Phys. **1**, 548 (1968); D. T. Birtwistle and A. Herzenberg, J. Phys. B: At. Mol.Phys. **4**, 53 (1971); L. Dube and A. Herzenberg, Phys. Rev. A **11**, 1314 (1975).

2 C. W. McCurdy and J. L. Turner, J. Chem. Phys. **78**, 6773 (1983).

3 K. C. Kulander and E. J. Heller, J. Chem. Phys. **69**, 2439 (1978).

4 W. Zong, G. H. Dunn, N. Duric, M. Larsson, C. H. Greene, A. Al-Khalili, A. Neau, A. M. Derkatch, L. Vikor, W. Shi, A. Le Padellec, S. Rosen, H. Danared, and M. af Ugglas, Phys. Rev. Lett. **83**, 951 (1999).

5 A. Larson, N. Djuric, W. Zong, C. H. Greene, A. E. Orel, A. Al-Khalili, A. M. Derkatch, A. Le Padellec, A. Neau, S. Rosen, W. Shi, L. Vikor, H. Danared, M. af Ugglas, M. Larsson, and G. H. Dunn, Phys. Rev. A **62**, 042707 (2000).

6 X. Urbain, A. Giusti-Suzor, D. Fussen and C. Kubach, J. Phys. B: At. Mol. Phys. **19**, L273 (1986).

7 I. F. Schneider and A. Giusti-Suzor, J. Phys. B: At. Mol. Phys. **24**, L289 (1991).

8 D. R. Bates and J. T. Lewis, Proc. Phys. Soc. (London) A **68**, 173 (1955).

9 I. Shimamura, C. J. Noble and P. G. Burke, Phys. Rev. A **41**, 3445 (1990).

10 J. Tennyson, At. Data Nucl. Data Tables **64**, 253 (1996).

11 I. F. Schneider, O. Dulieu and A. Giusti-Suzor, J. Phys. B: At. Mol.Phys. **24**, L289 (1991).

12 A. Larson and A. E. Orel, Phys. Rev. A. **59**, 3601 (1999).

13 D. Zajfman, Z. Amitay, M. Lange, U. Hechtfischer, L. Knoll, D. Schwalm, A. Wolf and X. Urbain, Phys. Rev. Lett. **79**, 1829 (1997).

14 S. L. Guberman, Phys. Rev. A **49**, R4277 (1994).

15 B. K. Sarpal, J. Tennyson and L. A. Morgan, J. Phys. B: At. Mol.Phys. **27**, 5943 (1994).

16 T. Tanabe, I. Katayamai, S. Ono, K. Chida, T. Watanabe, Y. Arakaki, Y. Harnyama, M. Saito, T. Odagiri, K. Noda, T. Houma and H. Takagi, J. Phys. B: At. Mol.Phys. **31**, L297 (1998).

17 J. Semaniak, S. Rosen, G. Sundstrom, C. Stromhold, S. Datz, H. Danared, M. af Ugglas, M. Larsson, W. J. van der Zande, A. Amitay, U. Hechtfischer, M. Grieser, R. Repnow, M. Schmidt, D. Schwalm, R. Wester, A. Wolf, and D. Zajfman, Phys. Rev. A **54**, R4627 (1996).

18 H. Lefebvre-Brion and R. W. Field, *Perturbations in the Spectra of Diatomic Molecules,* (Academic Press Inc., Orlando, Florida 1986), p. 80.

19 J. L. Krause, A. E. Orel, B. H Lengsfield III, and K. C. Kulander, *"Wave Packet Studies of the Predissociation of H_3"* in *Time Dependent Quantum Molecular Dynamics: Experiments and Theory,* ed. by J. Broeckhove and L. Lathouwers (Plenum Press, New York, 1992), p. 131.

20 B. H. Lengsfield III and D. R. Yarkony, Adv. Chem. Phys. **82**, 1 (1992).

Stark Mixing in Rydberg Systems by Ultralow Energy Collisions with Ions

M. R. Flannery and D. Vrinceanu
School of Physics, Georgia Institute of Technology, Atlanta, Georgia 30332
ray.flannery@physics.gatech.edu

1. INTRODUCTION

The ion-Rydberg atom/molecule collisional process

$$A^{Z+} + Ry(n, \ell) \rightarrow Ry(n, \ell') + A^{Z+} \tag{1}$$

is called collisional Stark mixing since the ℓ-changing transitions ℓ arrow ℓ' occur within the same energy shell n of the Rydberg species and are induced by the time-dependent electric field generated by the passing ion A^{z+}. The process is significant over a broad range of interest and applications, which include:

- The formation[1] of anti-hydrogen by three-body recombination $e^+ + \overline{\mathrm{p}} + e^+ \rightarrow \overline{\mathrm{H}} + e^+$ at ultralow energies, where the sequence is collisional capture into high ℓ → Stark Mixing → Radiative Relaxation. Since the n-changing collisions are relatively unimportant at ultralow energies, the ℓ-mixing collisions are essential in producing the low angular momentum states required to radiatively decay at a relatively high rate to low n-levels, thereby stabilizing the recombination.
- Experiments[2] on single ion collisions with alkali-metal Rydberg atoms have measured large ℓ-mixing cross sections for slow projectiles, including dipole-forbidden transitions.

- ZEKE (Zero Kinetic Energy) Spectroscopy[3] where high ℓ states are produced from low ℓ states by electric fields.
- Dissociative Recombination, as discussed in References 4 and 5.

The process is also interesting from a theoretical point of view and has remained largely unsolved for four decades. Notable theoretical progress has been achieved by the following treatments:

- modified first-order impact parameter treatment.[6]
- classical diffusion theory.[7]
- hydrogen atom in weak E-B fields.[8]
- truncated closely-coupled channel system of equations.[9]
- Monte-Carlo simulations.[10]
- $0 \rightarrow \ell'$ transitions.[11-14]

Exact solutions of Stark Mixing in atomic hydrogen induced by the time--dependent (dipole) electric field generated by (adiabatic) collision with a slow ion were recently presented in both classical[15,16] and quantal[16,17,18] formulations. The exceptionally rich dynamical symmetry of the hydrogen atom provides the key foundation that enables both the classical and quantal solutions to be constructed in a unified way, by using group representation theory.

2. DISSOCIATIVE RECOMBINATION

The indirect mechanism of dissociative recombination,

$$e_s^-(\varepsilon) + AB^+(\mathrm{v}\ J) \rightleftarrows [\ AB^+\ (v'\ J') - e^-\ ]_{n\ell} \tag{2}$$

$$\rightleftarrows (AB^{**})_r \rightarrow A + B^*,$$

involves the rotational/vibrational excitation of the ion with the subsequent capture of the colliding electron into a Rydberg state $n\ell$. The Rydberg electron is bound to the rotationally/vibrationally excited ion $AB^+(\mathrm{v}'\ J')$ in a state $n\ell$ with high n and low. This angular momentum ℓ can be redistributed to intermediate and high ℓ by the very efficient process of angular momentum (Stark) mixing. The much larger lifetimes of these high ℓ-states towards autoionization, predissociation, radiative relaxation and other internal processes can therefore increase the probability that deactivation/rearrangement of the Rydbergs by collision with electrons and neutrals can occur.

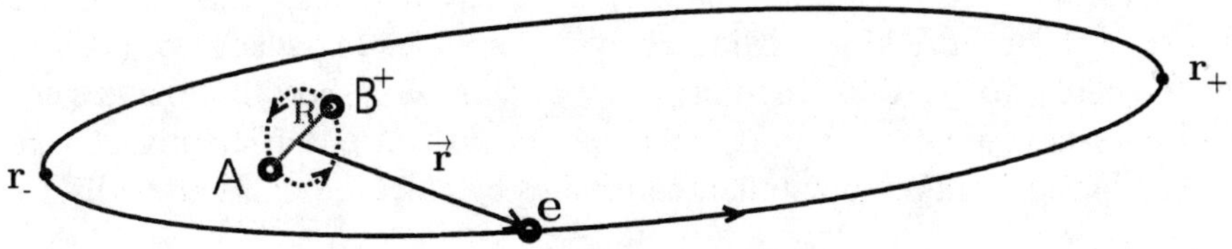

Figure 1. Rydberg Molecule.

In addition to their "direct" formation in the intermediate step (2) of "indirect" DR , molecular Rydbergs are also produced "indirectly" in the sequence

$$e_s^-(\varepsilon) + AB^+(v\ J) \rightleftarrows (AB^{**})_r \rightarrow A + B^* \tag{3}$$

for "direct" recombination, via the coupling with the intermediate electronically repulsive states $(AB^{**})_r$, as in

$$(AB^{**})_r \rightleftarrows [AB^+(v'\,J') - e^-]_{n\ell}$$

3. INTERNAL ELECTRON-MOLECULAR ION INTERACTION

Strong interaction between core-penetrating-electrons and AB^+ results in internal processes within the Rydberg molecule e.g., rapid autoionization at frequency ν_a and predissociation at frequency ν_{pd}, internal conversion, and fragmentation, as in (2) and (3) and Jahn-Teller effects.[19] These interactions can be treated by Multichannel Quantum Defect Theory.[20,21] Also, short-range coupling of $e^-_{n\ell}$ to AB^+ (anisotropic interactions) provides exchange of angular momentum with the core which results in enhancement[22] of Rydberg lifetimes.

Core-penetrating electrons (cpe's)[23] with $\ell = 0, 1, 2$, responsible for these internal processes, can be distinguished from non-penetrating core electrons (npe) with $\ell \geq 3$. For example, the atomic Cs quantum defects are 4.1, 3.6 and 2.5 for $\ell = 0$, 1 and 2 but drop abruptly to 0.03, 0.01 and 0.002 for $\ell = 3$, 4 and 5. The transit time of cpe interaction with the core of radius r_c, is $2\sqrt{2}\,[r_c - \ell(\ell+1)/2][r_c + \ell(\ell+1)]/3$ which is independent of n. Since the orbital time $\tau_n = 2\pi/\omega_n$ is $2\pi n^3$, the probability of interaction with the core i.e., for the internal processes varies as n^{-3}. External processes induced by electric fields (applied, stray or ion-produced) result in enhancement of the lifetimes of molecular Rydbergs.[24,3,25] In ZEKE measurements, analysis of the $e_s^-(\varepsilon)$spectrum in field ionization of AB^{**} provides knowledge of (v', J') of the ion core (AB^+). A distribution of low ℓ-states with short lifetime (towards autoionization, predissociation, radiative decay) can be efficiently changed into a high-ℓ distribution with large lifetimes by electric fields and collisional Stark mixing.

The long range interaction between the Rydberg electron and its parent ion can be decomposed as

$$V(\mathbf{R},\mathbf{r}) = -\frac{e^2}{r} + \frac{eD(R)}{r^2}\, P_1(\hat{r}\cdot\hat{R}) - \frac{eQ(R)}{r^3}\, P_2(\hat{r}\cdot\hat{R}) - \frac{\alpha_S(R)e^2}{2r^4} - \frac{\alpha_T(R)e^2}{3r^4}\, P_2(\hat{r}\cdot\hat{R}) + \dots$$

where D and Q are the permanent dipole and quadrupole moment of the ion with averaged dipole polarizability α_S and anisotropic polarizability α_T. The averaged values

$$\langle\frac{1}{r}\rangle = \frac{1}{n^2}; \qquad \langle\frac{1}{r^2}\rangle = \frac{2}{n^3(2\ell+1)} \tag{4}$$

$$\langle\frac{1}{r^3}\rangle = \frac{1}{n^3\ell(\ell+1)(2\ell+1)} \tag{5}$$

$$\langle\frac{1}{r^4}\rangle = \frac{4[3n^2-\ell(\ell+1)]}{n^5\ell(\ell+1)(2\ell+3)(2\ell+1)(2\ell-1)} \tag{6}$$

in a.u. indicate that the Coulomb interaction is dominant for $n \geq 5$, particularly for core non-penetrating orbits. Competition between the dynamics of (a) internal processes with low ℓ and n-independent interaction time, τ_C, and (b) external processes with large el and long lifetime presents a problem that is both interesting and challenging to both theory and experiment.

Adopting $V(R, r) = -e^2/r$ for the e - AB^+ interaction will allow exploitation of the Group Dynamic Symmetry of the Coulomb Potential which in turn will lead to the development of exact results in both classical and quantal formulations.

4. GROUP SYMMETRY

In addition to the unperturbed Hamiltonian,

$$H_0 = \frac{p^2}{2m_e} - \frac{e^2}{r} ,$$

the angular momentum L = r x p and the classical Runge-Lenz vector

$$\mathbf{A} = p_n^{-1} \left[\mathbf{p} \times \mathbf{L} - m_e e^2 \frac{\mathbf{r}}{r} \right]$$

directed toward the pericenter and normalized to angular momentum units, are all conserved. The operator A has the symmetrized (Pauli-Lenz) quantal form

$$\mathbf{A} = \left[\frac{1}{2} (\mathbf{p} \times \mathbf{L} - \mathbf{L} \times \mathbf{p}) - m_e e^2 \hat{r} \right] / p_n$$

and the following properties:

$$\begin{aligned} [A_j, H] &= 0 && \text{It is a conserved quantity,} \\ [L_j, A_k] &= i\hbar \, \epsilon_{jkn} A_n && \text{It is also a system vector,} \\ [A_j, A_k] &= i\hbar \, \epsilon_{jkn} L_n && \text{Its components do not commute,} \\ \mathbf{A} \cdot \mathbf{L} &= \mathbf{L} \cdot \mathbf{A} = 0 && \text{The vector } \mathbf{A} \text{ is orthogonal on } \mathbf{L}, \\ A^2 + L^2 &= (n^2 - 1)\hbar^2 && \text{Is constant for intrashell transitions,} \end{aligned}$$

where ε_{kjn} is the Levi-Civita antisymmetric symbol for any *k, j* = 1, 2, 3. These commutation relations define the SO(4) dynamic symmetry group for the restricted motion of the orbital electron to the energy shell. The SO(4) operators can be disentangled by introducing **M** = (**L**+**A**)/2 and **N** = (**L** - **A**)/2. Each **M** and **N** operator separately generates an SO(3) subalgebra, such that SO(4) ≡ SO(3) ⊕ SO(3). The classical analogy is that the set of coupled equations for **L** and **A** are decomposed into an uncoupled set for **M** and **N**, which then evolve independently with time. In classical mechanics, the operators L and A are replaced by their classical vectors and the commutators $(i\hbar)^{-1}[...]$ by the corresponding Poisson brackets.

5. COLLISION INTERACTION

Fig. 2 illustrates the collision dynamics. The frequencies of the Rydberg electron and the collision are $\omega_n = v_n/a_n$ and $\omega_R = \dot{\Phi}$ respectively. Here

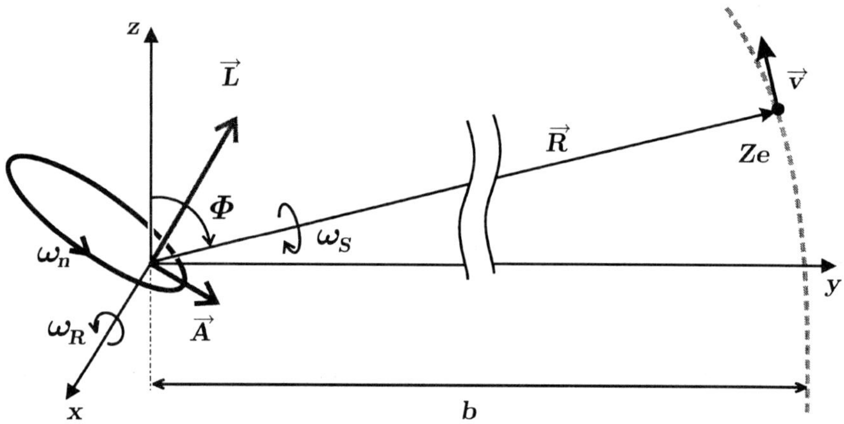

Figure 2. Ion-Rydberg Collision dynamics and the characteristic frequencies ω_n (orbital), ω_R (collisional) and ω_S (Stark).

$a_n = n^2 a_o$ and $v_n = v_o/n$ are the averaged orbital radius and velocity. The vectors **L** and **A** precess about the direction of the electric field $\mathcal{E}$ with the frequency $\omega_S = (3/2)a_n v_n(\mathcal{E}/e)$. Simulation of the collision in Fig.2 shows that the frequencies are ordered as $\omega_S < \omega_n < \omega_R$. It also shows that although L and A change little during one orbit, significant change is accomplished during the collision over the course of many (-50) electron orbits. It also indicates that the ℓ-changing collision at ultralow impact energies satisfies the following three approximations:

- **Weak field**: The collision occurs at large impact parameters b where the ion-Rydberg atom interaction potential is the ion-dipole potential,

$$V(\mathbf{r},\mathbf{R}) = (e\mathrm{r})\cdot\vec{\mathcal{E}} = -Ze^2\frac{\hat{\mathbf{R}}\cdot\mathbf{r}}{R^2}$$

- **Classical path**: The angular momentum of relative motion, $-\mu R^2\dot{\Phi}$, of the massive projectile is »ℓħ, the angular momentum of the Rydberg electron. It is therefore conserved with value μbv so that the interaction potential is

$$V(\mathbf{r},\Phi) = -\frac{Ze^2}{vb}\frac{d\Phi}{dt}\hat{\mathbf{R}}\cdot\mathbf{r} = \frac{Ze^2}{vb}\frac{d\Phi}{dt}(y\sin\Phi + z\cos\Phi)$$

- **Orbital adiabatic**: The collision frequency, $\omega_R = \dot{\Phi} << \omega_n$, the orbital frequency of the Rydberg electron. In classical treatments, the electron position vector **r** can then be replaced by its average $\langle r\rangle$. In quantal treatments, then

$$\langle n\ell' m'|\mathbf{r}|n\ell m\rangle = -\frac{3}{2p_n}\langle n\ell' m'|\mathbf{A}|n\ell m\rangle \tag{7}$$

for matrix elements within the same energy shell. This is Pauli's replacement rule.[26] Then

$$V(\mathbf{r},\alpha) = -\alpha\frac{d\Phi}{dt}(A_2 \sin\Phi + A_3 \cos\Phi) \tag{8}$$

where the dimensionless Stark parameter is

$$\alpha = \frac{3Za_n v_n}{2bv} \equiv \frac{3Z}{2}(\tilde{b}\tilde{v})^{-1}$$

where $\tilde{b} = b/a_n$ and $\tilde{v} = v/v_n$, are the impact parameter and speed scaled to the averaged Rydberg values. Equation (8) has the advantage that it is expressed only in terms of the generators of the SO(4) group. Moreover, the components $\{L_1, A_2, A_3\}$ generate a subgroup of the original symmetry group. Under the above three approximations, the resulting set of quantal and classical equations governing the problem of collisional Stark Mixing at ultralow energies can be solved exactly.

6. EXACT SOLUTIONS

6.1 Quantal Transition Probability

The quantal probability[16,17] for transition $\ell \to \ell'$ between any states with given angular momentum can be expressed[18] in the compact form

$$P_{\ell'\ell}^{(n)}(\chi) = \frac{2\ell'+1}{n}\sum_{L=|\ell'-\ell|}^{n-1}(2L+1)\begin{Bmatrix} \ell' & \ell & L \\ j & j & j \end{Bmatrix}^2 H_{jL}^2(\chi) \tag{9}$$

which involves only one summation. Here {...} is the 6-j symbol for coupling of three angular momenta and $H_{jL}(\chi)$ is a special matrix element of the irreducible representation of the group O(4). This function (also called the generalized character χ_L^j, associated with the irreducible representation of the rotation group) is well studied (see, for example, the books by Talman[27] and Varshalovich et al[28]). It can be written in terms of ultraspherical polynomials $C_n^{(\alpha)}$ by

$$H_{jL}(\chi) = (2L)!!\sqrt{\frac{(2j+1)(2j-L)!}{(2j+L+1)!}}\,(\sin\chi)^L\, C_{2j-L}^{(1+L)}\,(\cos\chi)$$

The angle χ is determined by

$$\cos\chi = \frac{1+\alpha^2\cos\Delta\Phi\sqrt{1+\alpha^2}}{1+\alpha^2} \tag{10}$$

and is called a *collision* parameter defined by the Stark parameter $\alpha = 3Z/2\tilde{b}\tilde{v}$ and by the polar angle $\Delta\Phi$ swept out by the internuclear vector during the collision time. The formula given by Eq. (9) is capable of providing exact numerical results even for large quantum numbers *n*, in contrast to previous expressions[16,17] where the number of terms to be summed increases dramatically with *n*. Note that the capacity of the projectile to produce angular momentum changes is governed solely by the parameter χ, the argument of H_{jL} in Eq. (9). Since $\Delta\Phi$ varies with time t, expression (9) furnishes $P_{\ell'\ell}^{(n)}$ as a function of t and Stark parameter α, all absorbed within *collision* parameter χ. It can be show that both detailed balance $(2\ell+1)P_{\ell'\ell}^{(n)} = (2\ell'+1)P_{\ell\ell'}^{(n)}$ and probability conservation $\sum_{\ell'} P_{\ell'\ell}^{(n)} = 1$, are satisfied.

It is important to note that the above solution (9) for $P_{\ell\ell'}^{(n)}$ is equivalent to the numerical solution that would, in principle, be obtained from the very large (n^2 x n^2) set of coupled differential equations given by the closely-coupled channels formulation of the Stark mixing problem.[9]

6.2 Classical Transition Probability

It can be shown that the classical transition probability provided in Ref.s 15 and 16 can be expressed as the analytic expression

$$P_{\ell'\ell}^{(n)}(\chi) = \frac{2(\ell'/n)}{\pi\hbar|\sin\chi|} \tag{11}$$

$$\times \begin{cases} 0 & \text{, if} \quad |\sin\chi| < |\sin(\eta_1 - \eta_2)| \\ \dfrac{K\left[\dfrac{\sin^2(\eta_1+\eta_2) - \sin^2(\eta_1-\eta_2)}{\sin^2\chi - \sin^2(\eta_1-\eta_2)}\right]}{\sqrt{\sin^2\chi - \sin^2(\eta_1-\eta_2)}} & \text{, if} \quad |\sin\chi| > |\sin(\eta_1+\eta_2)| \\ \dfrac{K\left[\dfrac{\sin^2\chi - \sin^2(\eta_1-\eta_2)}{\sin^2(\eta_1+\eta_2) - \sin^2(\eta_1-\eta_2)}\right]}{\sqrt{\sin^2(\eta_1+\eta_2) - \sin^2(\eta_1-\eta_2)}} & \text{, if} \quad |\sin\chi| < |\sin(\eta_1+\eta_2)|\,. \end{cases}$$

It is a function of the same collisional parameter χ given by (10), which is also basic to the quantal result (9), and of the angles η_1 and η_2 determined by

$$\cos\eta_1 \equiv \frac{\ell}{n} = \epsilon \quad \text{and} \quad \cos\eta_2 \equiv \frac{\ell'}{n} = \epsilon', \tag{12}$$

where the initial and final state ratios are $\varepsilon = \ell/\mathrm{n}$ and $\varepsilon' = \ell'/\mathrm{n}$. The eccentricities $\tilde{\varepsilon}_{1,2}$ of the initial and final classical Kepler orbits are $\sin\eta_{1,2} = (1-\varepsilon_{1,2}^2)^{1/2}$. The collision parameter χ increases monotonically with the Stark parameter $\alpha = 3\mathrm{Z}/2\tilde{\mathrm{b}}\tilde{\mathrm{v}}$ as

$$\chi \approx 2\alpha\sin(\Delta\Phi/2) + \mathcal{O}(\alpha^3)$$

The transition probability $\mathrm{P}_{\ell\ell'}^{(\mathrm{n})}$ can be also interpreted as the distribution of final states with angular momentum ℓ', which originate from one initial state of angular momentum ℓ and result from collision with a charged projectile. The probability (11) satisfies detailed balance $2\ell\mathrm{P}_{\ell'\ell} = 2\ell'\mathrm{P}_{\ell\ell'}$, where 2ℓ is the classical weight of the state $\mathrm{n}\ell$. Moreover, the classical limit of the full quantal probability can be directly and easily obtained from (9) to provide a result identical with (11).

7. RESULTS

7.1 Quantal-Classical Convergence

Note that the classical function $n\mathrm{P}^{(n)}_{\varepsilon' n,\varepsilon n}(\chi)$ is independent of n. Such universal functions are an advantage of classical treatments, in general and this one will be exploited as in Fig. 3 to demonstrate the rapid convergence of the quantal results onto the classical frame as n is increased. The structure apparent in the universal classical frame of Fig. 3 can be explained in terms of two distinct classical accessible and two classical inaccessible regions. It remains obscured within the quantal results.

7.2 Structure

Representative probabilities as a function of ℓ' are presented in Fig. 4 for the n = 28, $\ell = 18 \rightarrow \ell' = 28$ transition in atomic hydrogen. The increasing values of $\chi \sim 1/bv$ correspond to a series of collisions either at fixed impact parameter b and decreasing v, or vice-versa. Increasing χ therefore corresponds to lengthening the duration of the collision. Since the dipole interaction couples $\ell \rightarrow \ell \pm 1$ transitions, the sequence of transitions that occur during the time, τ_C, of collision is then $\ell \rightleftarrows (\ell+1) \rightleftarrows (\ell+2) \rightleftarrows \ldots\rightleftarrows \ell'_{max}$ and $\ell \rightleftarrows (\ell-1) \rightleftarrows (\ell-2) \rightleftarrows \ldots\rightleftarrows \ell'_{min}$. For the shorter collision times, τ_C, (at small χ), ℓ'_{max}, and ℓ'_{min} can remain $< n-1$ and > 0, respectively; there is not enough time to sequentially access the higher or lower values of ℓ' during the collision. Steps then appear in the classical structure. Within the classical inaccessible regions, the quantal results exhibit the characteristic exponentially decreasing and increasing variation. When the extreme limit $\ell' = 0$ and $\ell' = n - 1$, can be accessed during the collision, then these end values act in turn as additional sources which then proceed to populate $\ell' = 0, 1, 2$ and $\ell = n - 2, n - 3 \ldots$ for as long as the collision lasts. The classical probabilities then exhibit two distinct cusps arising from these secondary sources. These cusps merge into one as the collision duration time further increases with increasing values of the collision parameter χ.

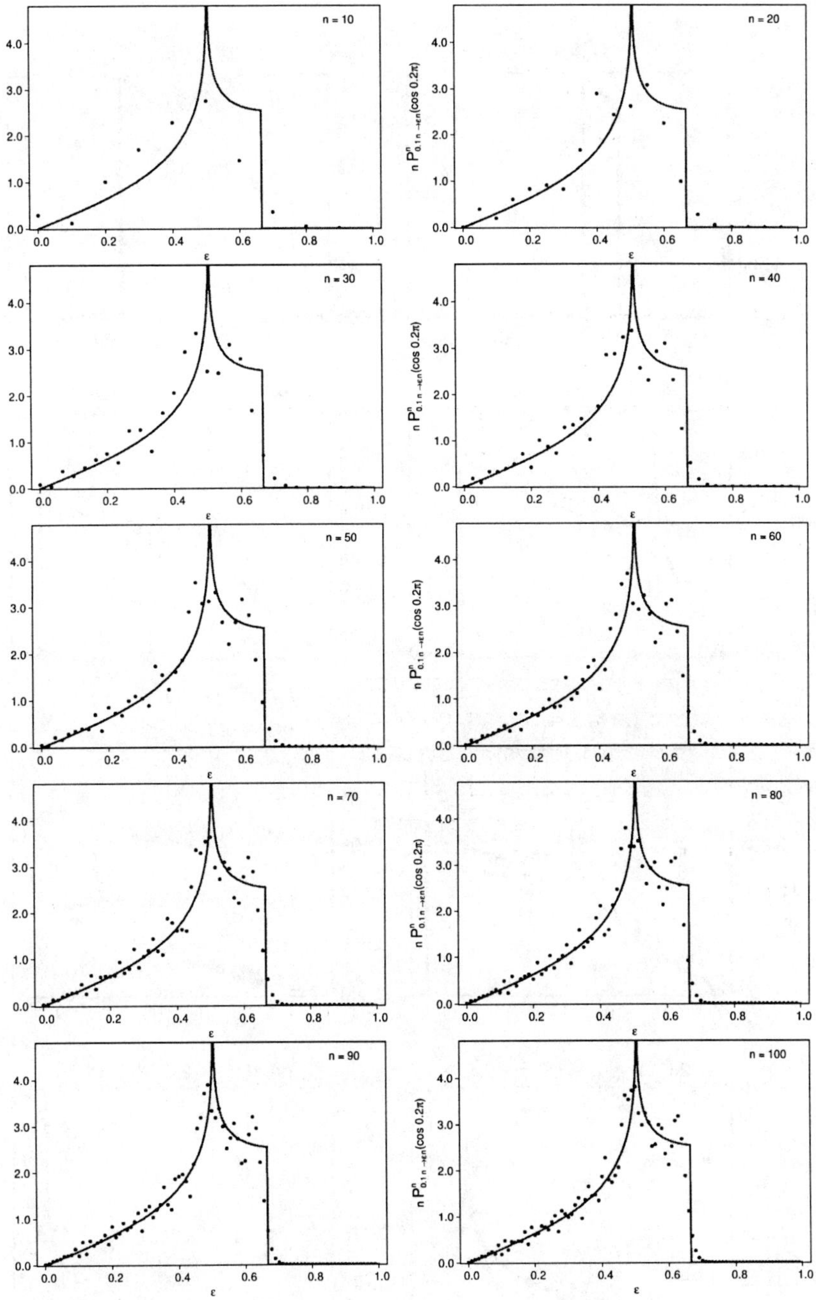

Figure 3. Plots of the scaled transition probability $nP^{(n)}_{\ell\to\varepsilon n}(\chi)$ as a function of the ratio $\varepsilon = \ell'/n$ illustrating, as n increases, the convergence of the exact quantal results (dots) onto the fixed classical framework (solid line), for constant values of the ratios $\ell/n = 0.1$ and $\chi = 0.2\pi$.

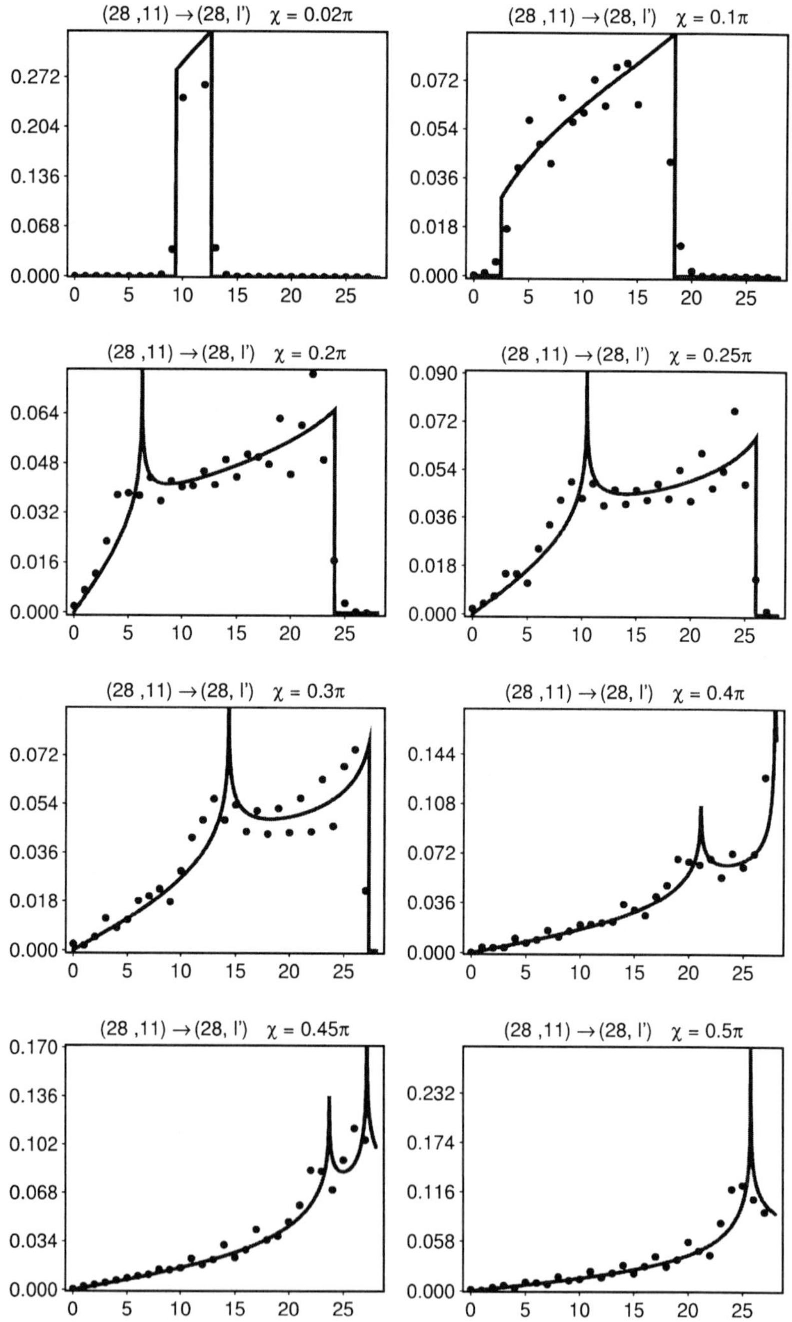

Figure 4. Quantal and Classical transition probabilities $P_{\ell'\ell}^{(n)}(\chi)$ for Stark Mixing from initial $\ell=1$ to final ℓ' states within the n = 28 energy shell for various values of collision parameter χ.

8. RYDBERG STATES WITH QUANTUM DEFECTS

Rydberg atoms and molecules have quantum defects only for states with core penetrating electron orbits i.e., those with $\ell = 0\text{-}2$. These low ℓ Rydberg wave functions, which are assumed known, can be expanded in terms of a hydrogenic basis set $|n\,\ell\,m>$ as,

$$|RyN\ell m\rangle = \sum_n \int w_n^{(N\ell)} |n\ell m\rangle$$

and the amplitudes $w_n^{(N\ell)}$ determined. There is a small energy defect $\varepsilon_{\text{fi}} = \hbar\omega_{\text{fi}}$ involved in transitions from low ℓ.

In the limit of weak coupling (first order perturbation), the Born approximation

$$P^B_{\ell\to\ell'} = \frac{1}{\hbar^2}\sum_{mm'}\left|\int_{-\infty}^{\infty} V_{if}[R(t)]e^{i\omega_{if}t}\,dt\right|^2$$

for the probability for $\ell \to \ell' = \ell \pm 1$ transitions that involve an energy defect $\varepsilon_{\text{fi}} = \hbar\omega_{\text{fi}}$ yields

$$P^B_{\ell\to\ell'} = \frac{\ell_>}{3(2\ell+1)}\left(\frac{2Ze^2\omega_{fi}a_0}{\hbar v^2}\right)^2 (R_{fi}/a_0)^2\left[K_0^2(\omega_{fi}b/v) + K_1^2(\omega_{fi}b/v)\right] \tag{13}$$

where the dipole radial matrix element, R_{fi}, involving the radial wave functions, $R_{n\ell}$, is $\int_0^\infty R_{n\ell}R_{n'\ell'}r^3 dr$ and where $K_{0,1}$ are modified Bessel functions.

In order to account approximately for the probability of small energy transfer, it can be shown that the Stark mixing quantal transition probability (9) becomes modified to

$$P_{N\ell\to N'\ell'} = (2\ell'+1)\sum_L (2L+1) \tag{14}$$

$$\times\left[\sum_n\int\frac{(-1)^n}{\sqrt{n}} w_n^{(N'\ell')} w_n^{(N\ell)} \begin{Bmatrix} \ell' & \ell & L \\ j & j & j \end{Bmatrix} H_{jL}(\chi)\right]^2$$

$$\times\left[K_0^2(\omega_{fi}b/v) + K_1^2(\omega_{fi}b/v)\right]$$

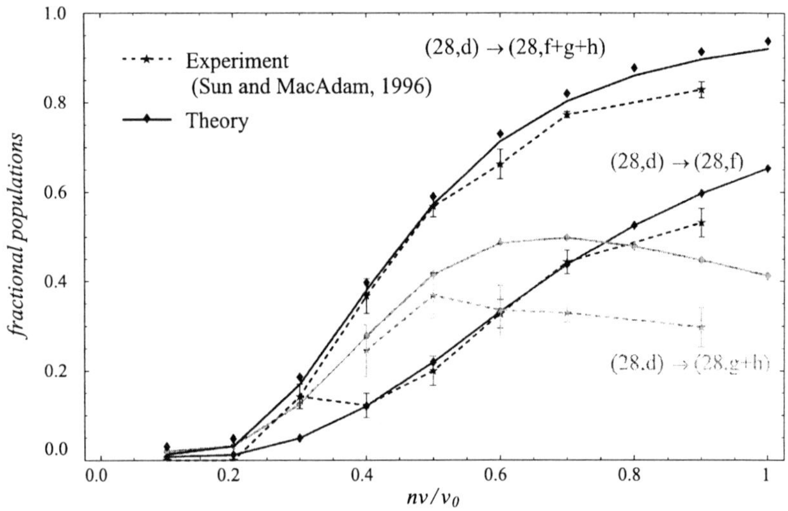

Figure 5. Fractional populations for *Na (28d)* ⟶ *Na (28f)*, *Na (28g + 28h)*, *Na (28f + 28g + 28h)* transitions.

The weak coupling limit of (14) is (13). This expression (14) was used to provide the agreement in Fig. 5 of the present theoretical results with the normalized measured fractional populations for *Na(28d)* ⟶ *Na(28f)*, *Na(28g + 28h)*, *Na(28f + 28g + 28h)* transitions.

9. INTERNAL VERSUS EXTERNAL PROCESSES

The rate ($cm^3\,s^{-1}$) of Stark mixing varies as

$$v\sigma_{\ell'\ell}^{(S)} = \frac{n^3}{\tilde{v}}\left(\frac{9I_{\ell'\ell}}{4}\right)(\pi a_0^2 v_0)$$

where the integral

$$I_{\ell'\ell} = \int_{\alpha_{min}}^{1} P_{\ell'\ell}^{(n)}(\alpha)\,\frac{d\alpha}{\alpha^3}$$

does not depend on the projectile properties (velocity, impact parameter or charge) but depends only on the initial and final states of the target. Due to the $1/\alpha^3$ singularity, the cross sections for elastic ($\ell = \ell'$) and near elastic transitions are very much enhanced. Various cut-off (α_{min}) procedures can be deduced[16] from physical considerations, specific to the actual problem under consideration. The ratio of the Stark frequency (s^{-1})

$$\nu_S = v\sigma_{\ell'\ell}^{(S)} N^+ \sim \frac{n^3}{\tilde{v}}$$

to the radiative frequency (s^{-1})

$$A_{n\ell} = 1.071\ 10^{11} n^{-3}(\ell + 1/2)^{-2},$$

taken as characteristic of internal processes involving core-penetrating electrons, is

$$\frac{\nu_S}{A_{n\ell}} = \left(\frac{n^6}{17}\right)\left(\frac{(\ell+1/2)^2}{\tilde{v}}\right)\left(\frac{I_{\ell'\ell}^{(n)}}{10^3}\right)\left(\frac{N^+}{10^8}\right)$$

which exhibits a strong dependence on n. For example, the collisional Stark mixing frequency for Na(28d) by ions with concentration $N^+ = 10^8$ cm^{-3} and mean speed $\tilde{v} = 1$ is $\nu^S_{28d} \approx 10^8\ s^{-1}$ which is much faster than the frequency $A_{28d} \approx 7.8\ 10^5\ s^{-1}$ of radiative decay. The two frequencies are equal for $n \approx 17$.

10. SUMMARY AND CONCLUSIONS

- Case study of collisional Stark mixing for the full transition array $n\ell \rightarrow n\ell'$ has been presented.
- The inherent Group Dynamic Symmetry permits development of exact solution in both quantal and classical formulations.
- Collisional Stark mixing at ultralow energies is probably one of the last remaining problems in collision physics capable of an exact analytical solution.
- The solutions are presented in compact form reflecting the mathematical beauty of the problem as well as pragmatic value.
- Classical results complement quantal results.
- Quantal-classical correspondence and convergence is exhibited.
- Modified theory can account for quantum defects (low ℓ) in non-hydrogenic systems.
- Essential agreement obtained with MacAdam's measurements.
- Depending on the n-levels populated in dissociative recombination, Stark mixing can be important since its frequency can exceed the frequency of characteristic internal processes.

ACKNOWLEDGMENTS

This research has been supported by AFOSR Grant No. 49620-99-10277 and NSF Grant No. 01-00890.

REFERENCES

1 M. R. Flannery and D. Vrinceanu, in *Atomic processes in Plasmas: 11th APS Topical Conference,* ed. by E. Oks and M. S. Pindzola (AIP Press, New York, 1998), p. 317.
2 X. Sun and K. B. MacAdam, Phys. Rev. A **47**, 3913 (1993).
3 F. Merkt and R. N. Zare, J. Phys. Chem. **101**, 3495 (1994).
4 S. R. Lundeen, in *Dissociative Recombination: Theory, Experiment, and Applications,* ed. by J. B. A. Mitchell and S. L. Guberman (World Scientific, Singapore, 1989), pp. 182-192.
5 T. Gougousi, R, Johnsen and M. F. Golde, Int. J. Mass Spectrom. Ion Processes **149/150**, 131 (1995).
6 R. M. Pengelly and M. J. Seaton, Monthly Notices Roy. Astronom. Soc. **127**, 165 (1964).
7 I. C. Percival and D. R,ichards, J. Phys. B **12**, 2051 (1979).
8 Y. N. Demkov, B. S. Monozon, and V. N. Ostrovskiĭ, Sov. Phys. JETP **30**, 775 (1970).
9 I. L. Beigman and V. S. Lebedev, Phys. Rep. **250**, 95 (1995).
10 P. Bellomo, D. Farrelly, and T. Uzer, J. Chem. Phys **107**, 2499 (1995).
11 K. Kazanski and V. N. Ostrovski, Phys. Rev. A **53**, 11,1811 (1995).
12 K. Kazansky and V. N. Ostrovsky, J. Phys. B **29**, 3651 (1996).
13 K. Kazansky and V. N. Ostrovsky, Phys. Rev. Lett **77**, 3094 (1996).
14 K. Kazansky and V. N. Ostrovsky, Sov. Phys. JETP **83**, 1095 (1996).
15 D. Vrinceanu and M. R. Flannery, Phys. Rev. Lett **85**, 4880 (2000).
16 D. Vrinceanu and M. R. Flannery, Phys. Rev. A **63**, 032701 (2001).
17 D. Vrinceanu and M. R. Flannery, J. Phys. B **33**, L721 (2000).
18 D. Vrinceanu and M. R. Flannery, J. Phys. B **34**, L1 (2001).
19 C. H. Greene, V. Kokoouline and B. D. Esry, in *Dissociative Recombination of Molecular Ions with Electrons,* ed. by S. L. Guberman (Kluwer Academic/Plenum Press, New York, 2002).
20 M. J. Seaton, Rep. Prog. Phys. **46**, 167 (1983).
21 C. H. Greene and C. Jungen, Adv. At. Mol. Phys. **21**, 51 (1985).
22 F. Remacle and R. D. Levine, J. Chem. Phys. **105**, 4649 (1996).
23 M. S. Child, in *Molecular Rydberg dynamics,* ed. by M. S. Child (Imperial College Press, London, 1997), p. 1.
24 W. A. Chupka, J. Chem. Phys. ***98***, 4520 (1993).
25 M. J. J. Vrakking, in *Molecular Rydberg dynamics,* ed. by M. S. Child (Imperial College Press, London, 1997), p. 229.
26 W. Pauli, Z. Phys. **36**, 336 (1926).
27 J. D. Talman, *Special Functions: A Group Theoretic Approach* (W. A. Benjamin, New York, 1968), p 176.
28 D. A. Varshalovich, A. N. Moskalev, and V. K. Khersonskii, *Quantum Theory of Angular Momentum* (World Scientific, Singapore, 1988).

Analytical Treatment of the K Matrix Integral Equation in the Dynamics of Superexcited Molecules

Lukas Pichl,*[§] Miyabi Hiyama*[†], and Hiroki Nakamura*[‡]
**Department of Functional Molecular Science, The Graduate University for Advanced Studies, [§]present address: Atomic Physics Lab. The Institute of Physical and Chemical Research RIKEN, Hirosawa 2-1, Wake, Saitama 351-0198, Japan; †present address: Department of Pure and Applied Sciences, Graduate School of Arts and Sciences, The University of Tokyo, Komaba, Meguro-ku, Tokyo 153, Japan; [‡]Department of Theoretical Studies, Institute for Molecular Science, Myodaiji, Okazaki 444-8585, Japan*
nakamura@ims.ac.jp

1. INTRODUCTION

Superexcited states (SES) constitute one peculiar class of electronically highly excited states of molecules whose internal energies exceed the corresponding lowest ionization potentials.[1-3] They play important roles in various fields of chemistry and physics as intermediate states of dynamic processes. Fig. 1 shows a schematic view of this feature. It is crucial to understand the characteristics of the central SES M*, in order to systematically comprehend the various dynamic processes. Dissociative recombination represents one of several good examples. According to the difference in the autoionization mechanisms, the SES are classified into the following two kinds: (i) multiply or inner-shell excited states, which are called "SES of the first kind", and (ii) rovibrationally excited Rydberg states, which is called "SES of the second kind." The dynamics of SES are governed by the two basic interactions: the electronic coupling, $V(R,\varepsilon)$, between the first kind of SES and the electronic continuum, and the quantum defect function $\mu(R)$, where R is the internuclear distance and ε is the electron energy. If these two quantities are available together with the potential curves $E_d(R)$ of the first kind of SES, then the MQDT (multi-channel quantum defect theory) presents a very powerful tool to investigate the various

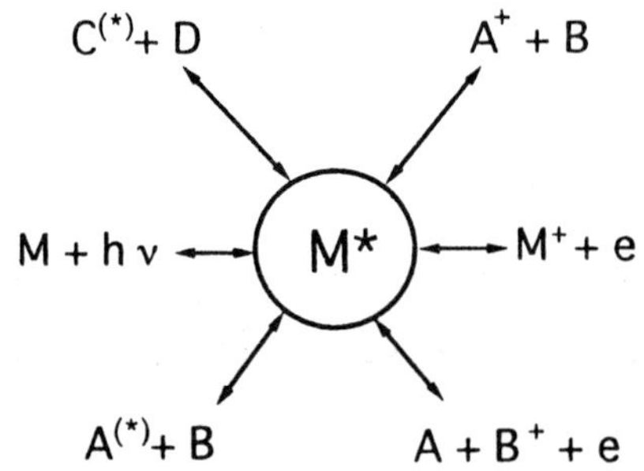

Figure 1. A Schematic unified view of the various dynamical processes and spectroscopies as well as scattering. The most important issue is to understand the central part M*.

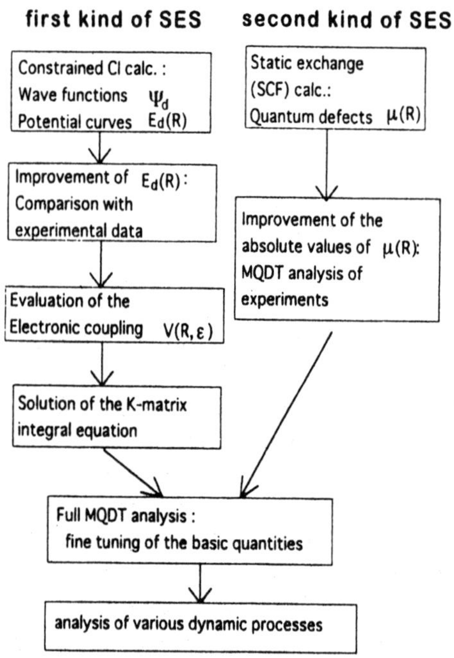

Figure 2. A general theoretical procedure to reveal the characteristics and dynamics of superexcited states of diatomic molecules.

dynamics. These quantities should be basically evaluated by quantum chemical electronic structure theory, but it is not easy to obtain accurate absolute values, especially in the case of $V(R,\varepsilon)$. It is crucial to use spectroscopic experiments to improve the information. In this sense the interplay among quantum chemistry, dynamics theory, and experiment is very important. The general procedure presented here is pictorially depicted in Fig. 2. There is still one theoretical obstacle in this procedure, the solution of the K-matrix singular integral equation. The first order perturbation theory or the grid method is usually employed; but they are not accurate enough or not efficient enough. Here we present a new method, which enables us to deal with the equation even analytically. The first part, i.e., the general theoretical procedure, has been presented,[4] and the second part, i.e., the analytical solution of the K-matrix, has been proposed.[5] Finally, when nonadiabatic transitions at potential curve crossings are to be analyzed, for instance in order to find out the final state branching in dissociative recombination, the compact and accurate Zhu-Nakamura theory can be used[6] instead of the Landau-Zener formula.

2. GENERAL THEORETICAL PROCEDURE

2.1 Quantum Chemical Calculations

The difficulty of the first kind of SES consists in the fact that they are embedded in the electronic continuum and are not eigenstates of the electronic Hamiltonian H_{el} and we have to deal with the electronic continuum wave function. In order to avoid the first problem we can use the projection operator formalism and from the eigenvalue problem in the bound states subspace Q we can obtain the wave function ψ_d and the potential curve $E_d(R)$. Since the electronic correlation is very important in the first kind of SES, this eigenvalue problem should be solved within the CI (configuration interaction) scheme. Another care to be taken is that diffuse functions should not be included in the Q-space, since the states we want should diabatically cross the Rydberg manifold and enter into the continuum.

The quantum defect function, $\mu(R)$, characterizes Rydberg states and its R-dependence determines the vibrational autoionization. The absolute value of $\mu(R)$ is not easy to estimate correctly, but its R-dependence can be well estimated by the SCF method. From the excited orbital energy $\varepsilon(R)$ the quantum defect $\mu(R)$ can be easily evaluated. The absolute value can be corrected by analyzing spectroscopic experiments with use of MQDT.

2.2 MQDT Analysis Of Spectroscopic Experiments

The ab initio quantum chemical calculations, i.e. the constrained CI calculation of $E_d(R)$ and the SCF method for $\mu(R)$, are not accurate enough to determine the relative position of $E_d(R)$ with respect to the ionic potential curve and the absolute values of $\mu(R)$. It is very important to use spectroscopic experiments to improve these quantities. The relative position of $E_d(R)$ can be corrected by using the experimental data on the relative predissociation and autoionization widths of vibrationally excited Rydberg states. This is because the relative width is primarily determined by the Franck-Condon factor between the vibrational wave function and the dissociative continuum wave function of the dissociative state. It is, of course, desirable to have the data available for a wide range of vibrational states. The predissociation and autoionization peak widths can be used to obtain a first rough estimate of the average electronic coupling strength $V_0 \simeq < V(R,\varepsilon) >$. The absolute value of $\mu(R)$ can be determined by evaluating the term values of Rydberg states with use of MQDT. If necessary, in this analysis, effects of the dissociative SES may be incorporated approximately by using the perturbative solution of the K-matrix equation and the average electronic coupling strength V_0. It is important to point out that in order to make the above analysis effective and precise, the spectroscopic experiments should clarify the following things:(1) Symmetries of the electronic states involved in autoionization and predissociation. (2) Level assignment of the Rydberg spectrum. (3) Wide coverage of the principal quantum number n and vibrational states v of the Rydberg manifold. (4) Estimates of the predissociation and autoionization widths.

2.3 Evaluation of Electronic Coupling

The electronic coupling, which characterizes the electronic autoionization of the first kind of SES, is defined by

$$V(R,\varepsilon) = < \Psi_\varepsilon \mid H_{el} \mid \psi_d > , \tag{1}$$

where ψ_d is an electronic wave function of the SES and Ψ_ε represents the ionization continuum (energy normalized continuum wave function). The wave function, ψ_d, can be taken from the constrained CI calculations. The multi-electron continuum wave function, Ψ_ε, can be constructed with use of an ionic wave function combined with the two-center Coulomb function ψ_ε.

The wave function of the ionizing electron is replaced by the two-center Coulomb function in the field of two positive charges corresponding to the residual diatomic ion. It is a rough approximation to use the pure Coulomb function, but it is known from the detailed studies of the doubly excited states of H_2 that the autoionization rate can be well reproduced by this approximation, if the bound state wave function, ψ_d, is an accurate one.[7, 8] Although the two-center Coulomb functions can be produced numerically,[9] these are not convenient for the evaluation of the electronic coupling $V(R,\varepsilon)$. If we expand the functions in terms of Gaussians, then the evaluation of $V(R,\varepsilon)$ can be done straightforwardly with use of a conventional CI code with a little modification. Since the long range part of the Coulomb function is not required to evaluate $V(R,\varepsilon)$, the Gaussian expansion is useful.[10]

2.4 The *K* Matrix Integral Equation

The effects of the electronic coupling $V(R,\varepsilon)$ on the dynamics can be incorporated by solving the following Lippman-Schwinger type singular integral equation for the reactance (K)-matrix:

$$K_{mn}(E,E') = V_{mn}(E,E') + \sum_p \mathrm{P}\int \frac{V_{mp}(E,E'')}{E-E''} K_{pn}(E'',E')dE'', \quad (2)$$

where E is the total energy, m, n, and p represent channels of the nuclear motion (vibrational or dissociative), and $V_{mn}(E, E')$ is defined by

$$V_{mn}(E,E') =< E,m|V(R,\epsilon)|E',n >_R . \quad (3)$$

Here $|E,n>$ represents the nuclear motion. P indicates the Cauchy's principal value. What we need finally are the eigenvalues and diagonalization matrix of $K_{mn}(E, E)$. The energy dependence of $V(R,\varepsilon)$ is crucial to obtain the non-perturbative solution of Eq.(2), because that makes the second term nonvanishing. In order to solve Eq.(2), the first or second order perturbation theory or the grid method has been usually utilized. Unfortunately, however, the perturbation theory is not accurate enough and the grid method is very inefficient. Here we propose a much more efficient method to deal with this equation.

2.5 Studies of Dynamic Processes

The various dynamics displayed in Fig. 1 can be nicely treated by the MQDT formalism. The most basic quantity is the χ matrix defined by $\chi = (E^+)(E^-)^{-1}$ with

$$\mathrm{E}^{\pm} = \begin{cases} \sum_{v'} e^{\pm i\eta_\alpha} < v|e^{i\pi\mu(R)}|v' > U_{v\alpha} & \text{for ionization channel,} \\ U_{d\alpha} e^{\pm i\eta_\alpha} & \text{for dissociation channel,} \end{cases} \quad (4)$$

where η and $\mathbf{U}$ are the eigenvalues and diagonalizing matrix of the K matrix, respectively. Once χ is constructed, the S matrix for scattering processes and the D matrix for photo-impact processes can be easily evaluated. As for the dissociative recombination, this formalism gives the total dissociation probability. If the branching ratios to various dissociation products are required, valence type dissociation channels can be incorporated with appropriate couplings or we can use the theory of nonadiabatic transition due to curve crossing.[6]

3. SOLUTION OF THE *K* MATRIX EQUATION

For simplicity we consider here the case of only one dissociative channel, *d*. Then, because of the property of the coupling matrix element, $V_{nm}(E, E')$ =0 for $(n, m) = (d, d)$ and (v, v'), the K matrix equation decouples into the following two sets of equations:

$$K_{dd'}(E) = \sum_{v'} \mathrm{P} \int \frac{V_{dv'}(E, E')}{z - E'} K_{v'd'}(E') dE' \qquad (5)$$

$$K_{vd}(E) = Vvd(E, E') + \sum_{d'} \mathrm{P} \int \frac{V_{vd'}(E, E')}{z - E'} K_{d'd}(E') dE' \qquad (6)$$

and

$$K_{uv}(E) = \sum_{d'} \mathrm{P} \int \frac{V_{ud'}(E, E')}{z - E'} K_{d'v}(E') dE' \text{ and} \qquad (7)$$

$$K_{dv}(E) = V_{dv}(E, E') + \sum_{v'} \mathrm{P} \int \frac{V_{dv'}(E, E')}{z - E'} K_{v'v}(E') dE' \qquad (8)$$

Unfortunately, the shape and magnitude of the electronic coupling, $V(R, \varepsilon)$, is not well known. However, calculations on the dissociative attachment process,[11-13] show that the coupling element is, in many cases, nearly separable with respect to R and ε as

$$V(R, \epsilon) \simeq V_R(R) V_\epsilon(\epsilon), \qquad (9)$$

where V_R and V_ε are certain functions.

The most thoroughly studied system is H_2 for which the electronic coupling of the doubly excited state, $^1\Sigma_g(2p\sigma_u)^2$, has been calculated by Hara and Sato in the static exchange approximation.[7] Their values are not precisely separable, but the separable approximation works well as is shown in Fig. 3. This separability may be qualitatively understood by the fact that the integration in Eq.(1) is limited to a relatively short range, $r < r_0$, and the ε-dependence comes mainly from the energy normalization factor of the outer wave function and the matching at $r = r_0$.[5]

It should be noted that the separable approximation, Eq. (9), is different from the ordinary one, but it can simplify the solution of the *K* matrix equation. Eq. (9) gives the energy kernel, defined by Eq. (3), as

$$V_{dv}(E, E') \simeq V_\epsilon(E' - E_v) V_v(E), \qquad (10)$$

where

$$V_v(E) = \int_0^\infty \Psi_{diss}(R : E) V_R(R) \Phi_{vib}(R : E_v) dR. \qquad (11)$$

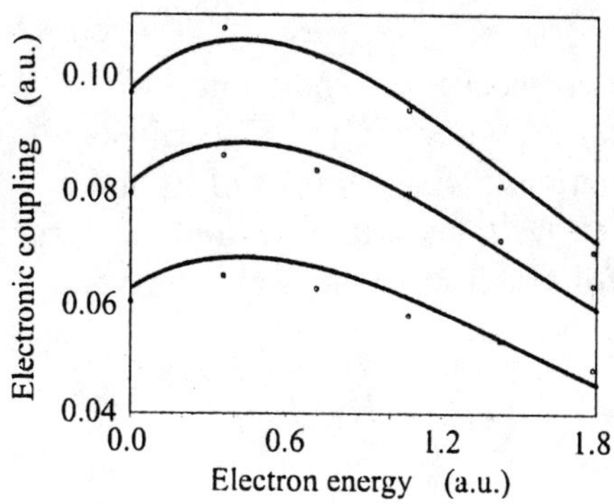

Figure 3. Separable approximation of the electronic coupling of $H_2(^1\Sigma_g(2p\sigma_u)^2)$. The abscissa, ε, is the energy of the continuous electron. The ordinate measures the electronic coupling. The lowest, middle and highest curves correspond to R = 1.4, 2.0, and 2.6 respectively. Circles show the data by Hara and Sato[7] and the full lines indicate the separable approximation.

The solution of Eq.(2) with the kernel in Eq.(10) can be obtained analytically as shown below for fixed E' and z only as a function of E. For the final determination of the phase shift, we finally set $E = E' = z$.[5]

From Eqs. (5) - (8) we have

$$K_{dd}(E) = \sum_v \alpha_v V_v(E), \quad K_{vd}(E) = \beta_v V_d(E - E_v), \tag{12}$$

and

$$K_{v_1 v_2}(E) = \gamma_{v_1 v_2} V_d(E - E_{v1}), \quad K_{dv}(E) = \sum_{v'} \delta_{vv'} V_{v'}(E). \tag{13}$$

If we introduce the principal value integrals,

$$\pi_v(E) \equiv \mathrm{P} \int_{E_v + \epsilon_n}^{\infty} \frac{V_d^2(E' - E_v)}{E - E'} dE' \tag{14}$$

and

$$\pi_{vv'} \equiv \mathrm{P} \int_{E_0}^{\infty} \frac{V_v(E') V_{v'}(E')}{E - E'} dE', \tag{15}$$

then the coefficients in Eqs. (12) and (13) satisfy the following linear equations:

$$\beta_v = V_v(E) + \sum_{v'} \pi_{vv'} \pi_{v'} \beta_{v'}, \quad \alpha_v \equiv \beta_v \pi_v, \tag{16}$$

$$\gamma_{v_1 v_2} = V_d(E - E_{v2}) \pi_{v_1 v_2} + \sum_{v'} \pi_{v_1 v'} \pi_{v'} \gamma_{v' v_2}, \tag{17}$$

and

$$\delta_{vv'} = V_d(E - E_v)\delta_{vv'}^{Kron} + \pi_{v'}\gamma_{v'v}, \tag{18}$$

where $\delta_{vv'}^{Kron}$ is the Kronecker δ. Thus, once we can evaluate the singular integrals π_v and $\pi_{vv'}$, the solution of the K-matrix can be easily obtained from the above linear equations. The estimates of π_v and $\pi_{vv'}$ can be carried out as explained below. First we transform the infinite integration range to (-1, 1) as exemplified by the following example. The integral,

$$I = \mathrm{P}\int_{E_0}^{\infty} \frac{f(E')}{E - E'} dE', \tag{19}$$

is transformed into

$$I = \frac{2C}{(E - E_0) + C}\mathrm{P}\int_{-1}^{1} \frac{f(E'(x))}{(x - y)(x - 1)} dx \tag{20}$$

by using the variable transformations $E'(x) = C\,(1 + x)/(1 - x)$ with $y = (E - E_0 - C)/(E - E_0 + C)$. The singularity at x = 1 has no problem at all as far as the original integral converges at $E \to \infty$. Since $V_\varepsilon(\varepsilon)$ can be nicely approximated by a regular function as

$$V_\epsilon(\epsilon) = \epsilon^a e^{-b\epsilon}, \tag{21}$$

where a and b are certain constants,[14] the quantity π_v can be easily evaluated by using the formulas (A.8) and (A.11) of Ref. 5. In the case of $\pi_{vv'}$, since $V_v(E)$ has a singularity as $V_v(E) \propto (E - E_0)^{-1/4}$, Eq.(15) can be reduced to the form of Eq.(A.5) or (A.2) of Ref. 5. In this way, under the separable approximation, all the calculations can be carried out easily.

In a general non-separable case, or in the case where we want to check the validity of the separable approximation, the coupling matrix elements and K matrix elements are expanded in terms of Chebyshev polynomials; and the K-matrix equation can be reduced to coupled linear equations with respect to the expansion coefficients of the K-matrix.

4. NUMERICAL EXAMPLES

First, we show a comparison between the present method and the grid method for the integral,

$$I(E) = \mathrm{P}\int_{1/18}^{\infty} \frac{E'e^{-E'}}{E - E'} dE'. \tag{22}$$

Table 1 shows the results for $I(E)$ at $E = 1/18 + 0.05$. It can be easily seen that the present method gives the converged result very quickly. It should be noted that there is a plateau around N = 200 ~ 300 in the grid method. The

Table 1. Principal value integrals by Chebyshev and grid methods. E = 1/18+0.05 is the total energy in Eq.(22), and *N* is the highest Chebyshev polynomial index in the expansion as well as the number of mesh-points.

N	Chebyshev expansion	grid method	N	grid method
1	-0.13721765 ×10[1]	-0.08059284	100	-0.11055386 ×10[1]
5	-0.12194187 ×10[1]	-0.28652549	200	-0.11118244 ×10[1]
10	-0.11701910 ×10[1]	-0.49621710	300	-0.11118664 ×10[1]
15	-0.11709514 ×10[1]	-0.64225435	1000	-0.11578406 ×10[1]
20	-0.11708547 ×10[1]	-0.75076966	10000	-0.11694942 ×10[1]
25	-0.11708652 ×10[1]	-0.83310492	50000	-0.11705546 ×10[1]
30	-0.11708646 ×10[1]	-0.89616934	100000	-0.11706870 ×10[1]

calculation is seemingly converged, but actually is not.

Fig. 4 shows the dissociative recombination cross sections for $H_2^+(\upsilon = 0) + e \rightarrow H + H$ as a function of electron energy in comparison with the first order perturbation approximation. The separable approximation gives an indistinguishable result from the exact one, while the perturbation approximation is noticeably different from them.

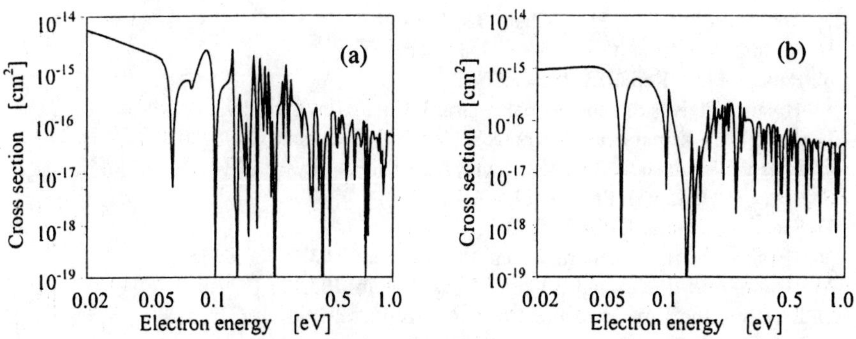

Figure 4. The DR, cross section for H_2^+ (v = 0) + e $\rightarrow$ H + H as a function of electron energy (in eV). The incident electron energy is displayed on a logarithmic scale, -(log ε - log(0.02))/log(0.02), and the cross section (in cm^2) is also displayed on a logarithmic scale, $\log_{10}(\sigma_{\upsilon=0\text{-}d})$. (a) The general non-separable case and the separable approximation (indistinguishable). (b) First order perturbation theory.

5. CONCLUDING REMARKS

A general theoretical procedure is reviewed to reveal and investigate the characteristics and dynamics of superexcited states (SES) of diatomic molecules. Since SES are targets like elephants touched by blind men, interplay among quantum chemistry, dynamics theory, and experiments is definitely needed. One obstacle in the theoretical treatment, namely the solution of the K-matrix singular integral equation, was overcome by using the separable approximation to the electronic coupling matrix element and the Chebyshev polynomial expansion. Even if the approximation cannot be applied, the singular equation can be reduced to ordinary coupled linear equations. The method was proved to be very effective and promising for the study of various dynamic processes involving SES such as dissociative recombination.

ACKNOWLEDGMENTS

The present research was partially supported by a Grant-in-Aid for Scientific Research on Priority Area "Molecular Physical Chemistry" and Grant No.13440182 from The Ministry of Education, Science, Culture, and Sports of Japan.

REFERENCES

1 H. Nakamura, Int. Rev. Phys. Chem. **10**, 123 (1991).
2 H. Nakamura, J. Chin. Chem. Soc. **42**, 359 (1995).
3 Y. Hatano, Phys. Rep. **313**, 109 (1999).
4 M. Hiyama, N. Kosugi, and H. Nakamura, J. Chem. Phys. **107**, 9370(1997).
5 L. Pichl, H. Nakamura, and J. Horacek, Comp.Phys.Comm. **124**,1 (2000).
6 C. Zhu, Y. Teranishi, and H. Nakamura, Adv. Chem. Phys. **117**,127 (2001).
7 S. Hara and H. Sato, J. Phys. B **17**, 4301 (1984).
8 H. Sato and S. Hara, J. Phys. **19**, 2611 (1986).
9 M. Hiyama and H. Nakamura, Comp. Phys. Comm. **103**, 209 (1997).
10 M. Hiyama and H. Nakamura, Comp. Phys. Comm. **103**,197 (1997).
11 D. E. Atems and J. W. Wadehra, Phys. Rev. A **42**, 5201 (1990).
12 A. P. Hickman, Phys. Rev. A **43**, 3495 (1991).
13 P. L. Gertitschke and W. Domcke, Phys. Rev. A **47**, 1031 (1993).
14 H. Takagi, in *"Dissociative Recombination: Theory, Experiment and Applications"*, ed. by B. R. Rowe, J. B. A. Mitchell, and A. Canosa (Plenum Press, New York, 1993), p.75.

Extension of the Quantum Defect Theory and Its Application to Electron and Molecular Ion Collisions

Hidekazu Takagi
Physics Laboratory, School of Medicine, Kitasato University, Kitasato 1-15-1, Sagamihara, Kanagawa, 228, Japan
takagi@kitasato-u.ac.jp

1. INTRODUCTION

Seaton has formulated the quantum defect theory for multichannel problems.[1] For an electronic state of the electron-ion system, he gave a formula showing the relation between the Coulombic wave functions defined near the force center (inner region) and those of the asymptotic zone (outer region). This formulation allows not only a uniform description of both scattering and Rydberg states, but also makes use of an almost energy independent quantity called the quantum defect. The idea of dividing physical space into the inner and outer regions is a powerful approach for describing the Rydberg and scattering states. The eigenstates of the inner region contain all the information about the interaction. The boundary conditions in the outer region provide the results that are relevant to physical observations.

Fano and his school showed that the eigenstates can be obtained by connecting coordinate systems.[2-4] The interactions contained in the K matrix are given by frame transformations between different systems of coordinates. For example, rotational transitions are described by a frame transformation between body and lab fixed coordinates.[2] The Stark effect frame transformation is between the parabolic and spherical coordinates.[3] The eigenstates have been further extended to suitable function spaces. Jungen and Atabek showed that the vibrational and rotational transitions can be described by a frame transformation between the Born-Oppenheimer bases and those of the close coupling method.[5]

The main theme of this paper is the extension of the multichannel quantum defect theory (MQDT) to dissociative processes. This has been achieved by hybridizing the method of Jungen and Atabek (JA) with another method. The most popular method is the so-called two-step method introduced by Giusti[6] in order to investigate dissociative recombination (DR). Since the dissociative states have different electronic configurations from those of the incident electron and ionic ground state, we first determine the eigenstates that diagonalize the configuration interaction (CI) matrix. Next, we apply the JA method using these eigenstates as the basis. In order to represent the dissociative valence (non-Rydberg) doubly excited states, the two-step method is effective and has been applied to the DR of various molecules.

In the first calculations, the perturbation theory, especially the first order approximation, was often employed in the first step of solving the CI scattering problem. On the other hand, the algebraic method of solving the Lippmann-Schwinger equation has proved to be effective when used with the Chebychev quadrature.[7-9] This method enables the estimation of the off-the-energy-shell contribution completely although it is usually difficult to obtain this contribution except for a few simple molecules.

Ross and Jungen proposed a hybrid method using R matrix theory.[10] They successfully represented the photoabsorption spectra of H_2 including the dissociative states. However, there seems to be no application to dynamic problems like DR.

The author has proposed a method that uses discretized dissociative states in the JA method.[11] In the following section, I introduce this method and confirm its validity. As an application for a crude model of H_3^+ DR, a sensitive dependence on the initial rotational state is shown. I discuss the DR of vibrationally excited ions and rotational transitions in superelastic collisions.

2. MQDT WITH DISCRETIZATION

2.1 Method

In order to extend the MQDT to the dissociative processes, the simplest way is to include the dissociative states in addition to the bound vibrational states as the basis functions of the JA method. Since this method has been described previously,[8,12] I review here the point of extension. In the inner region, a modification from the JA method is only to add the discretized dissociative states as basis functions. In order to adopt the usual MQDT, where the electronic basis function is normalized by energy, the dissociative wave function, $F_E(R)$, of dissociation energy E should be normalized by state (i), that is for the i-th discretized energy E_i, $\left\langle F_{E_j}(R) \middle| F_{E_i}(R) \right\rangle_R = \delta_{ji}$, where δ_{ji} is Kronecker's δ. In the outer region, $F_{E_i}(R)$ must vanish if the E_i does not conserve the total energy defined by the initial state of the target ion and the incident energy of the electron. We call such a state 'closed dissociative channel' (CDC). The CDC could appear when the incident electron is captured in a certain discrete electronic level. For the E_i satisfying the energy conservation law, $F_E(R)$ constitutes an open channel and should be normalized by energy, i.e. $\left\langle F_{E'}(R) \middle| F_E(R) \right\rangle_R = \delta(E - E')$, where $\delta(E - E')$ is Dirac's δ function.

For a dissociative channel with a continuum electron, we must choose which particle to observe. In the storage ring experiment of dissociative excitation (DE), the dissociated ion is observed. For such a situation, the dissociative state is normalized by energy, and the electronic state is averaged with its energy within an interval of discretized energy, i.e. the electronic state is normalized by state. It is noteworthy that the MQDT treats a single continuum, *i.e.* all channels must consist of one energy normalization function and the other state normalized function(s). Although the MQDT analytically changes the normalization of the open and closed wave functions, we need to change it numerically for the dissociative channels.

The frame transformation similar to the JA method and Seaton's MQDT enable us to obtain the DR and DE cross sections.[12,13] The present approach is valid only for the dissociative Rydberg state. For the valence state, we must take the other approaches mentioned in the previous section.

2.2 DR of HeH^+

For several isotopes of HeH^+, the above mentioned method reproduces the energy dependence[14] and the absolute values[8] of the observed DR cross section.

An important feature of HeH^+ DR is the absence of potential curve crossings between the ionic ground state and neutral dissociating molecular states. Nevertheless, the cross section is almost as large as for cases having favorable crossings. The reason is the large change of adiabatic quantum defect with the internuclear distance. That of the C state changes by almost one unit. We call such large change a 'quantum defect jump.' In order to reproduce the experimental result, the CDC also plays an important role.

2.3 DR of H_3^+

There are large differences among the experimental DR cross sections of H_3^+.[15] The measured rate coefficients at 300K range from < 10^{-11} cm^3/s to > 10^{-7} cm^3/s. The recent storage ring measurements support the large value,[16] which is not supported by the observations of diffuse interstellar clouds.[17] At the united atom limit of two hydrogen atoms, the H_3^* is isoelectronic to HeH*. Guberman has pointed out this similarity and its connection with DR.[18]

Here, I apply my method, which has successfully described HeH^+ DR. As the first step, I used a crude model, regarding the H_3^+ as HeH^+ except for its electronic states and mass. The electronic states are taken from the result of Michels and Hobbs,[19] where a nuclear configuration of the C_{2v} symmetry is assumed. The $^2B_2$ state calculated by them gives a quantum defect jump. That state belongs to the pσ Rydberg series and the jump is seen only in this series. In order to dissociate from $^2B_2$, a transition to $^2B_1$ is necessary. That transition is induced by the three-atom effect and non-adiabatic interaction. I assumed that the $^2B_2$ state correlates to the $^2B_1$ state in the asymptotic region. Hence, we only discuss the upper limit of the rate coefficient by the present crude model.

Fig. 1 shows the calculated DR rate coefficient of the vibrational ground state of H_3^+. The rate coefficient is here defined by the velocity of the incident electron times the cross section. A strong dependence on the initial rotational state is seen. For the rotational states lower than N=4, the rate is smaller than 10^{-11} cm^3/s below the collision energy of 5 meV. This energy region covers the temperature of a diffuse interstellar cloud, where H_3^+ is considered to be in the rotational ground state.

On the other hand, a large resonance structure is seen for the N=4 state around the collision energy of 1meV. Over the whole energy region shown in the Figure, the rate coefficient becomes as large as 10^{-8} cm^3/s for the higher rotational states. The present result indicates that the disagreement among the measurements and astronomical observation might come from the rotational dependence of the rate. The rotational temperature of the storage ring experiment is about several thousand K if it is estimated from the temperature of the ion source.

The strong rotational dependence comes from the asymmetry of the electronic interaction, i.e., only the pσ symmetry has a quantum defect jump. That fact also provides a reason for adopting the present diatomic model. The frame transformation gives the transformation matrix of the angular part for a symmetric top molecule as[20]

$$\langle NK|\Lambda n\rangle_{\ell}^{JM} = \sqrt{\frac{2N+1}{2J+1}}\, C(\ell, N, J; \Lambda, n, K), \qquad (1)$$

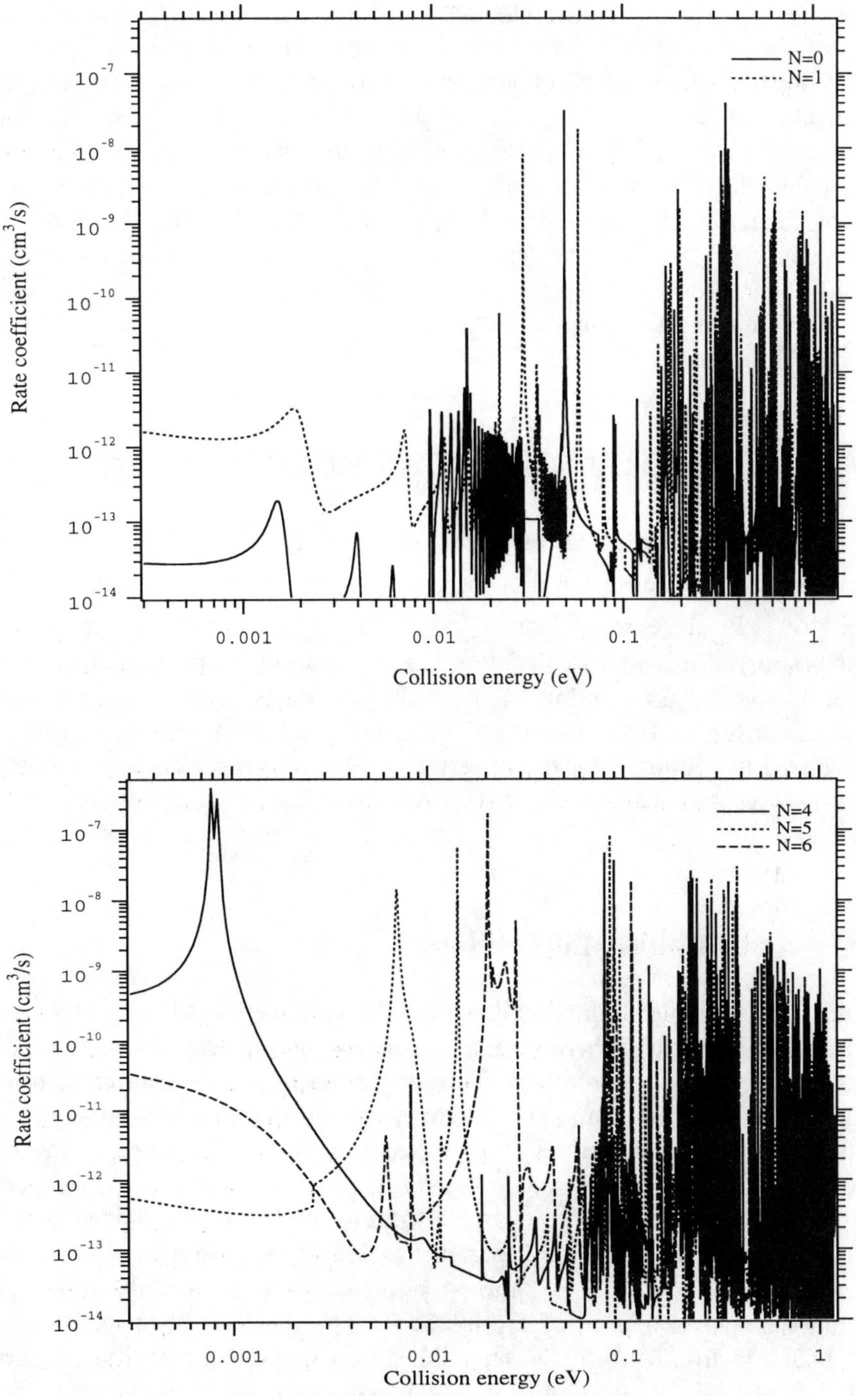

Figure 1. The DR rate coefficient of H_3^+ for vibrational state v=0 and rotational state N.

where N and K are rotational quantum numbers of the molecular axis and the figure axis respectively, ℓ is the angular momentum of the incident electron, Λ is the projection of ℓ on the molecular axis, J is the total angular momentum, and $n = K - \Lambda$, the nuclear part of K. The Clebsch-Gordan coefficient C was taken after Rose.[21] The difference between the diatomic molecule and the symmetric top rotor is only whether n is restricted to zero or not. Since the present rotational transition is induced by the peculiar ℓ, Λ dependence of the quantum defect, n does not make the transition of K. Besides that, for $\Lambda = 0$, the rotational energy levels are not sensitive to n,[20] i.e. they are almost degenerate. Thus, the effect of K is only to make the spectra of the rate coefficient diffuse.

3. DR OF THE HYDROGEN MOLECULAR ION

Among the excited states of H_2 converging to $H_2^+(2p\sigma_u)$, only the lowest, $^1\Sigma_g\ (2p\sigma_u)^2$, has valence (non-Rydberg) character. The magnitude of the CI times $\nu^{1.5}$ is close to the same value for the full Rydberg series ($^1\Sigma_g$ $(2p\sigma_u)(np\sigma_u)$). Tennyson's calculation[21] shows that this quantity is almost the same for the second and third $^1\Sigma_g\ (2p\sigma_u)(np\sigma_u)$ states. However, the lowest state is different from the others. Thus we adopt a two-step method introduced in Section 1 for the lowest state. The other core excited Rydberg states are represented by the discretized dissociative states introduced in Section 2.

3.1 Separable Approximation

In the separable approximation, the CI strength, V(R,E), is given as V(R,E) = F(R)· G(E), where the R and E are the internuclear distance and the electronic energy, respectively. This approximation is often adopted in dissociative attachment studies.[23] There is a report that this approximation is quite good for the DR of H_2^+.[9] Fig. 2 shows calculated values of the CI strength due to Hara and Sato,[11] the result of a nonseparable least square fitting, and the fitting by the separable form where G(E) is optimized at R = 2.0 au. The separable fitting does not represent the behavior of the low energy CI strength. Fig. 3 shows the dissociation probability from the vibrational state v=0 and 1 of H_2^+ by the CI. Those probabilities are affected by about a factor of two by the separable approximation at collision energies below 0.5eV (about –0.58 au of the total energy) for the algebraic solutions. The separable approach is of questionable accuracy in this case.

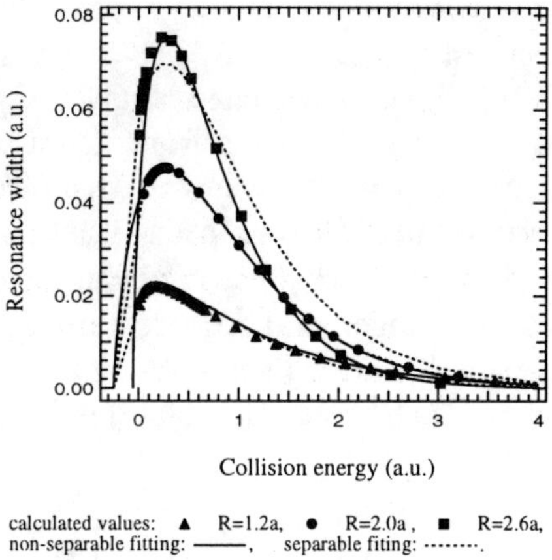

Figure 2. Fitting of the CI strength with separable and nonseparable functions.

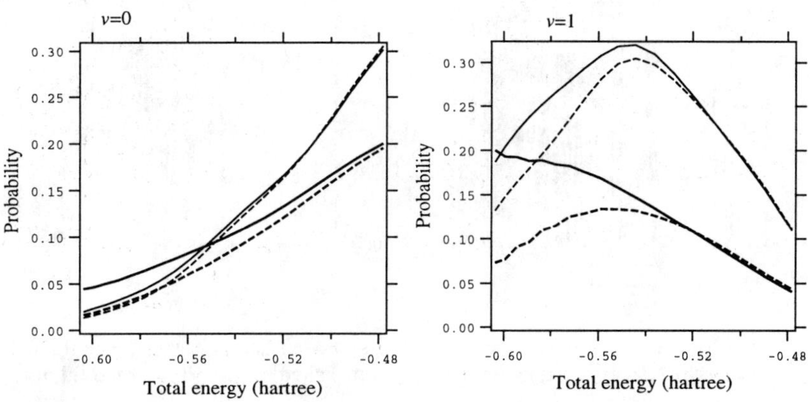

Figure 3. Probability of dissociation from the vibrational state ν=0 and 1. The lines have the following meanings: solid: non-separable fitting, broken: separable approximation, bold: the algebraic solution, thin: first order perturbation.

3.2 Higher Two Electron Excited States

The two electron excited states other than the lowest can be regarded as the Rydberg states converging to the first excited state of HD^+. Thus we introduce two types of discretized dissociative states, of which nuclear motions are driven on the potential energy curves of the ground and the first excited electronic states of HD^+. The CDC improves the agreement between the calculation and the experiment for the DR and DE of HD^+ [8,24] as well.

Let us here investigate the DR of the vibrationally excited HD^+, paying attention to the effect of the two electron excited states other than the lowest.

The considered two-electron excited Rydberg manifolds are the same as previous studies[8,12]: states of the $^1\Sigma_g$, $^1\Sigma_u$, $^1\Pi_g$, $^3\Pi_g$, and $^3\Pi_u$ symmetries. The CI strength adopted is the same as the latest study,[8] of which more details will appear separately.[24] The CI scattering problem was solved numerically without separable approximation for the lowest two electron excited state. For the others, the first order perturbation theory was adopted. In the second step applying JA method, we took the adiabatic quantum defect of pσ and pπ states from Fig. 2 (n=2) of Jungen and Atabek,[5] and for the dσ states, the values in our previous paper[13] were adopted. As to the triplet scattering, we assumed the quantum defect to be zero. The rotational motion was neglected.

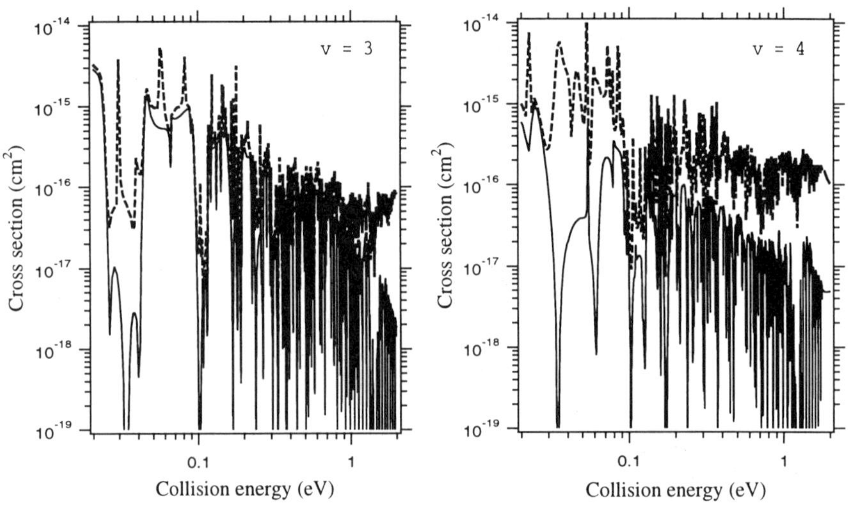

Figure 4. DR cross section of HD^+ of v=3 and 4. The broken line indicates the total DR cross section. The contribution from the lowest two electron excited state is shown by solid line.

For $v < 2$, almost the entire DR cross section is due to the lowest two electron excited state at energies lower than 1 eV. The contribution from the states above the lowest can be seen for $v \geq 2$. Fig. 4 shows the DR of v=3 and 4. For $v \geq 4$, the major contribution comes from states other than the lowest. The higher two electron excited states are indispensable except for a few low vibrational states at the low collision energies. The MQDT with discretized dissociative states allows for the calculation of the DR cross section for various vibrational states for which data are needed in fusion plasma studies.[24]

3.3 Rotational Transitions

An accurate treatment for the CI scattering is possible in the two step method of MQDT applied to H_2^+. Thus, we can perform a reliable calculation of rotational transitions induced by electron impact, since the frame transformation of the angular part is known even for the system including the two electron excited states.[25] Fig. 5 shows an example of rotational transitions of H_2^+, where N denotes the rotational state. In the present treatment, the rotational transitions of H_2^+ are essentially caused by the anisotropy that arises from the dominant dσ partial wave. Overall, the cross sections of various excitation and deexcitation processes are the same order of magnitude at the same collision energies.

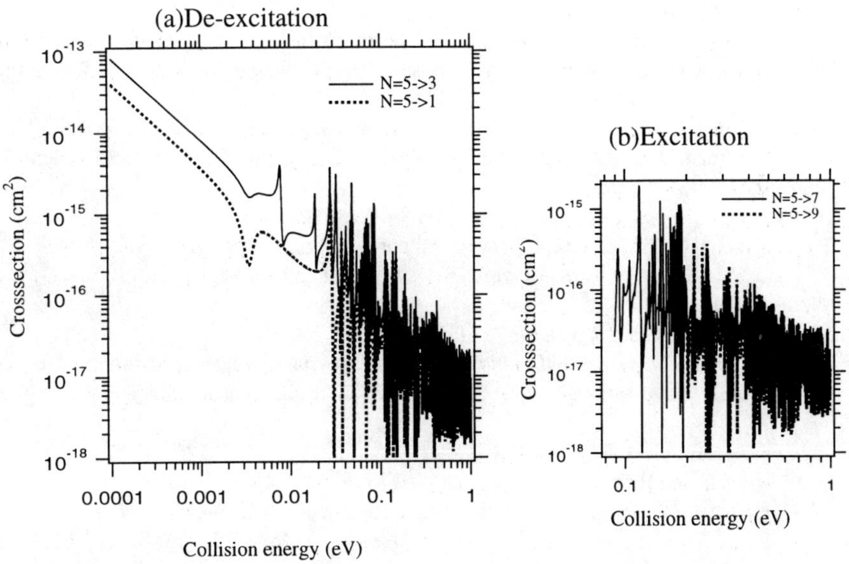

Figure 5. The cross section of the rotational transition of $H_2^+(v=0, N=5)$ by electron impact.

4. SUMMARY

I have discussed the inclusion of discretized dissociative states in the MQDT. As the result of its application to the crude model of H_3^+ DR, I propose that the rotational dependency is the cause of the enigma in the DR of H_3^+. It also found out that the higher two electron excited states are indispensable for the DR of vibrationally excited HD^+. With the accurate estimation of the off-the-energy-shell effect, we discussed the rotational transitions of H_2^+ by electron impact.

REFERENCES

1 M. J. Seaton, Proc. Phys. Soc., 801; ibid **88**, 815 (1966).
2 U. Fano, Phys. Rev. A **2**, 353 (1970).
3 U. Fano, Phys. Rev. A **24**, 619 (1981).
4 K. T. Lu, Phys. Rev. A **4**, 579 (1971).
5 Ch. Jungen and O. Atabek, J. Chem. Phys. **66**, 5584 (1977).
6 A. Giusti, J. Phys. B **12**, 3867 (1980).
7 H. Takagi, Atomic Collision Research in Japan **24**, 3 (1998).
8 H. Takagi in *Dissociative Recombination, Theory, Experiment, and Applications IV*, ed. by M. Larsson, J. B. A. Mitchell and I. F. Schneider, (World Scientific, Singapore, 2000), p. 180.
9 L. Pichl, H. Nakamura and J. Horacek, Comp. Phys. Commun. **124**, 1 (2000).
10 S. Ross and Ch. Jungen, Phys. Rev. Lett. **59**, 1297 (1987).
11 H. Takagi in *Dissociative Recombination, Theory, Experiment, and Applications*, ed. by B. R. Rowe, J. B. A. Mitchell and A. Canosa, (Plenum, New York, 1993), p75, (*see* Fig. 6).
12 H. Takagi in *Dissociative Recombination, Theory, Experiment, and Applications III*, ed. by D. Zajfman, J. B. A. Mitchell, D. Schwalm and B. R. Rowe (World Scientific, Singapore, 1996), p.174.
13 H. Takagi and H. Nakamura, J. Chem. Phys. **84**, 2431 (1986).
14 T. Tanabe et al., J. Phys. B **31**, L297 (1998); corrigendum J. Phys. B **32**, 5221 (1999).
15 A. Canosa, J. C. Gomet, B. R. Rowe, J. B. A. Mitchell and J. L. Queffelec, J. Chem. Phys. **97**, 1028 (1992).
16 M. Larsson et al., Phys. Rev. Lett. **70**, 430 (1993).
17 T. Oka in *Dissociative Recombination, Theory, Experiment, and Applications IV*, ed. by M. Larsson, J. B. A. Mitchell and I. F. Schneider, (World Scientific, Singapore, 2000), p. 13.
18 S. L. Guberman, Phys. Rev. A **49**, R4277 (1994).
19 H. H. Michels and R. H. Hobbs, Ap. J. **286**, L27 (1984).
20 W. G. Harter, C. W. Patterson, F. J. da Paixao, Rev. Mod. Phys. **50**, 37 (1978).
21 M. E. Rose, *Elementary Theory of Angular Momentum* (John Wiley & Sons, New York, 1957).
22 J. Tennyson, At. Data Nucl. Data Tables **64**, 253 (1996).
23 W. Domcke and C. Mündel, J. Phys. B **18**, 4491 (1985).
24 H. Takagi, Physica Scripta **T96**, 52 (2002).
25 H. Takagi, J. Phys. B **26**, 4815 (1993).

The Dissociative Recombination of N_2^+

Rate Coefficients and Isotope Effects

Steven L. Guberman
Institute for Scientific Research, 22 Bonad Road, Winchester, MA 01890
slg@sci.org

1. INTRODUCTION

The dissociative recombination (DR) of N_2^+,

$$N_2^+ + e^- \rightarrow N + N, \qquad (1)$$

plays a central role in determining the chemical and physical properties of planetary ionospheres. At 300 km above the surface of the Earth, DR is a well known loss process[1] for N_2^+ and an important source of excited N. Reaction (1) yields as much $N(^2D)$ near 300 km as all other sources combined.[2] The dayglow emissions at 3466Å and 10,400Å due to $N(^2P)$ arise in part from the DR of N_2^+. At Mars, DR of $^{14}N_2^+$ imparts sufficient kinetic energy to the product atoms to allow for atmospheric escape. However, in DR of $^{15}N^{14}N^+$ from v=0, although ^{14}N can escape the atmosphere, ^{15}N receives insufficient energy for escape.[3] This mechanism may explain why the Viking missions to Mars found $^{15}N/^{14}N$ to be 1.6 times greater than the value at Earth.[4] (Escape of N by DR of N_2^+ is not energetically possible at Earth.) DR of N_2^+ may also contribute to $^{15}N/^{14}N$ isotope enrichment at Titan.[5]

Here, I describe theoretical quantum chemical calculations of the DR cross sections and rate coefficients for the lowest three ion vibrational levels. The calculation of the potential curves needed to describe the DR of these three levels is described in the next section. In Section 3, the calculation of the electronic widths is described. A description of the calculation of the rate coefficients and a comparison to experimental values are in Section 4. Isotope effects are discussed in Section 5. The results are summarized in Section 6.

Dissociative Recombination of Molecular Ions with Electrons
Edited by S. Guberman, Kluwer Academic/Plenum Publishers, 2003

2. POTENTIAL CURVES AND MECHANISMS

The calculation of rates coefficients for Reaction (1) requires an accurate description of the potentials that describe the motion of the ion nuclei and the dissociating neutral atoms. For DR from the first excited ion vibrational level at low electron energies (e.g. < 1 eV), 5 atomic asymptotes are energetically accessible: $^4S + {}^4S$, $^4S + {}^2D$, $^4S + {}^2P$, $^2D + {}^2D$, and $^2D + {}^2P$. A total of 90 molecular states arise from these asymptotes. At first glance the problem seems somewhat daunting since for DR it is necessary to determine which of these states cross the ion near the turning points for the v=0,1 vibrational levels. However, molecular symmetry considerations allow for the elimination of many states. For example, since the ion is of $^2\Sigma_g^+$ symmetry, only singlet and triplet spin states can be formed after electron capture. Therefore, all states of septet and quintet symmetry can be eliminated. States of negative reflection symmetry also cannot be formed by electron capture. In order to identify the important states, small scale Complete Active Space Self Consistent Field (CASSCF)[6] calculations were done. These calculations and the larger scale calculations were done with nuclear centered contracted Gaussian basis sets. For these calculations, I have used the [5s, 4p, 3d, 2f, 1g] basis of Dunning.[7] Here the letters represent the angular momenta of the basis functions and the numbers are the number of contracted Gaussian functions. Each contracted Gaussian is a superposition of individual Gaussians with optimized contraction coefficients and exponents.

The description of DR used here involves potential curves that describe diabatic states. The states are diabatic because they intentionally exclude Rydberg character from the basis set. The exclusion is needed in order to allow the repulsive states to cross the ion. The electrostatic widths described in the next Section are consistent with this use of diabatic states. Alternatively, fully optimized states including Rydberg character could have been used. In this case, the widths are not electrostatic but instead are Born-Oppenheimer coupling matrix elements. None of the repulsive curves cross through the ion in this case. However, for the N_2 potential curves, the diabatic approach is the most convenient.

The CASSCF calculations identified 25 states that were candidates for large scale calculations. The large scale calculations used the CASSCF orbitals in Contracted Multireference Configuration Interaction (CMRCI)[8] calculations for the potential curves. The detailed results of these calculations will be reported separately. For seven states of N_2 for which there are experimental values for comparison,[9,10] the average differences between experimental and calculated excitation energies, force constants and equilibrium internuclear separations are only 0.04 eV, 22 cm^{-1}, and 0.019 Bohr respectively.

For v=0, the CMRCI potential curves which intersect the ion within or

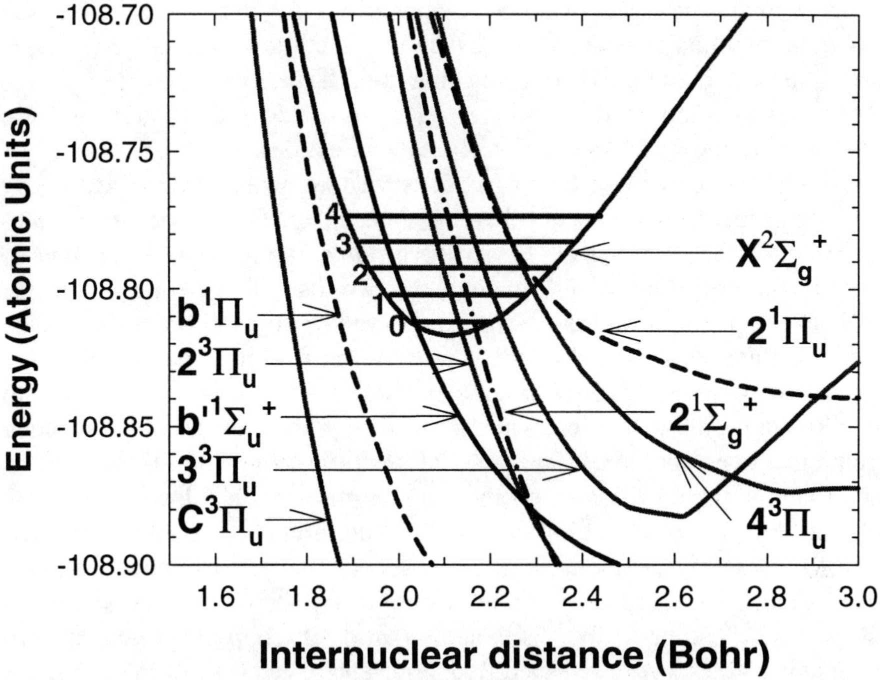

Figure 1. Ion and dissociative potential curves for N_2^+ DR of v=0 and 1.

near the endpoints of the vibrational motion are shown in Fig. 1 and are $2^3\Pi_u$, $3^3\Pi_u$, $2^1\Sigma_g^+$, and $b'^1\Sigma_u^+$. For the v=1 level, the important states are those for v=0 plus $4^3\Pi_u$ and $2^1\Pi_u$. Several additional states with favorable intersections are not shown in Fig. 1 because they have very small electron capture widths (see next Section). The $^3\Pi_u$ curves are the same as those reported earlier[11,12] and were calculated in a [4s, 3p, 2d, 1f] basis. These curves have been shifted in energy in order to plot them on the same Figure with curves calculated in the larger basis set.

The $b'^1\Sigma_u^+$ curve crosses the ion curve at the inner turning point of v=0 and leads to $N(^2P) + N(^2D)$ atoms which are not energetically accessible from v=0 until the electron energy is above 0.137 eV. Therefore, this state will not affect room temperature rate constants. Below 0.137 eV electron energy, DR along this route is energetically possible for the excited vibrational levels.

Dissociative routes that do not cross within the ion vibrational turning points may also be important for DR via the indirect mechanism. Capture into a highly excited vibrational level of a Rydberg state can lead to predissociation along a repulsive curve that is not near the ion. While this mechanism does occur in N_2^+ DR, its magnitude is not large enough to affect the values for the rate constants reported below.

Additional potential curves included in the rate coefficient calculations (see below) are the Rydberg series having the same symmetry as each of the important dissociative states identified above. In addition, the rate coefficient calculations include the $^3\Pi_u$ states having the $A^2\Pi_u$ excited ion core.[12] These curves were determined from the quantum defects that were in turn determined in the width calculations (see next Section).

Because there is more than one $^3\Pi_u$ state that takes part in DR, a very interesting mechanism occurs here.[13] Since the $^3\Pi_u$ dissociative states predissociate the same set of $^3\Pi_u$ Rydberg states, the Rydberg states can be viewed as intermediates that "connect" the two dissociative states. In a time independent picture, the dissociative states mix with each other due to the Rydberg states. In a time dependent picture, the flux proceeding along one dissociative state can transfer to another state of the same symmetry prior to dissociation or autoionization. This mechanism allows for dissociation along a route that does not have a favorable intersection with the ion if capture can occur into a repulsive state of the same symmetry that has a favorable crossing. Fig. 1 shows that initial electron capture into $2^3\Pi_u$ can be followed in part by dissociation along $3^3\Pi_u$ and $4^3\Pi_u$. A similar mechanism can also connect repulsive states of different electronic symmetry. An example is the DR of O_2^+ leading to $O(^1S)$ where spin-orbit coupling connects two dissociative states[14] via connected Rydberg states. The effect of the spin-orbit coupling is quite dramatic and amounts to over an order of magnitude increase in the DR rate coefficient. However, the spin-orbit coupling mechanism requires that the ion have both nonzero angular momentum and spin projection on the internuclear axis. Therefore, this mechanism does not occur in the DR of N_2^+ $X\,^2\Sigma_g^+$ but can occur in the DR of N_2^+ excited states.

3. ELECTRONIC WIDTHS

The probability of electron capture and autoionization is governed by an electronic width, Γ(R). The direct DR cross section is approximately directly proportional to Γ(R), which is given by the Golden Rule expression,

$$\Gamma_{r,d}(R)=2\pi\rho<\{\Phi_{ion}(x,R)\phi_r(x,R)\}|H|\Phi_d(x,R)>^2. \quad \textbf{(2)}$$

In (2), H is the electronic Hamiltonian, R is the internuclear distance, the integration is over the electronic coordinates represented by x, and the electronic wave functions from left to right are for the ion core, the Rydberg electron, and the dissociative neutral state. The antisymmetrized product of the ion core and Rydberg electron wave functions is denoted by { } and the density of states is given by ρ.

The calculation of the width has been described previously[15-17] and involves the placement of a large basis set of diffuse Gaussians[18] at the molecular midpoint coupled with the use of the Improved Virtual Orbital (IVO) method for calculating the Rydberg orbitals.[19] These calculations allow for the accurate description of a Rydberg series having principal quantum numbers as high as n=10. The width is calculated as in (2) with high Rydberg orbitals used to represent the free electron. The matrix element is then multiplied by a density of states factor in order to convert the normalization of the Rydberg orbital to that for a continuum coulomb-like orbital. The wave functions for the ion and dissociative states in (2) are full valence space CI wave functions. The full Rydberg state wave function is obtained by coupling the Rydberg orbitals determined in the IVO procedure with the ion multiconfiguration wave function. In addition, the multiconfiguration wave function for the ion plus a Rydberg electron is allowed to mix in higher nonphysical correlating roots from the space of configurations describing the dissociating states. This procedure allows important additional correlation to mix into the electron-ion states[15,20]

From the energy difference, $\Delta E(R)$, between the multiconfiguration Rydberg and ion wave functions, the quantum defect, $\mu(R)$, is calculated. The quantum defect is given by $\mu(R) = n - (1/(2\,\Delta E(R)))^{1/2}$ where n is the principal quantum number. The quantum defect is an important input parameter for the calculation of the cross sections and rate coefficients.

An example of the calculated widths (from Eq.(2)) for a full range of effective principal quantum numbers, n* (where $n^*=n-\mu$), is given in Fig. 2 for the C and $2\,^3\Pi_u$ states. In both cases, the widths must be determined up to $n^* \approx 9$ in order to assure convergence to two significant figures.

4. RATE COEFFICIENTS

For the calculation of DR cross sections, we use the Multichannel Quantum Defect Theory (MQDT) approach.[17,21] This approach allows one to treat direct and indirect DR simultaneously in order to accurately describe the important interference between these processes. In this approach, an electronic matrix is constructed which takes into account the interactions between all channels. Using the width matrix elements in Eq.(2), the matrix elements of the electronic interaction matrix or K matrix can be calculated. The first order matrix elements are[21]

$$K_{ion,d} = (1/2\pi)^{1/2} < X_{ion} \,|\, \Gamma_{r,d}^{1/2}(R) \,|\, X_d > , \qquad (3)$$

where X_{ion} and X_d are vibrational wave functions for the ion and dissociative states respectively, Γ is from Eq. (2), and the integration in (3) is over the internuclear distance, R. Higher order contributions to this matrix element are also included.[17] These contributions include the second order electronic

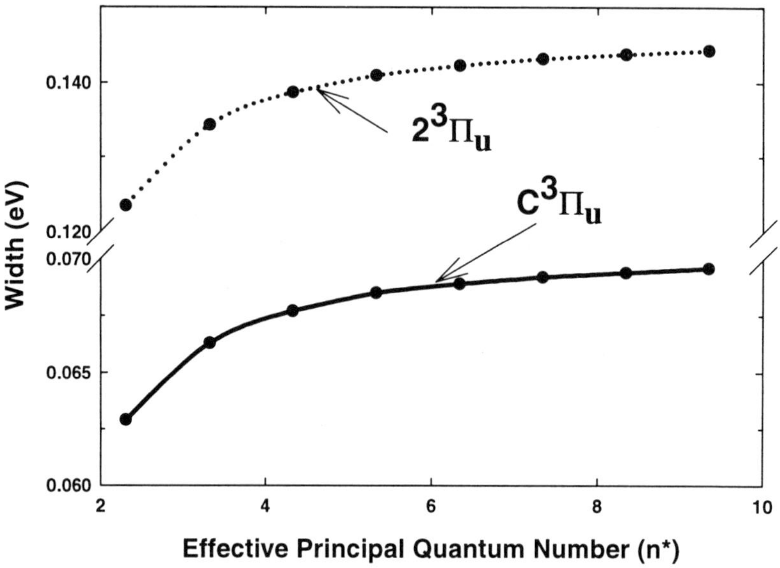

Figure 2. Electronic widths from Eq.(2) for two states of N_2.

coupling, which has already been described in the first chapter of this volume. In addition, the excitation of vibrationally excited intermediate neutral Rydberg states (indirect DR) by Born-Oppenheimer breakdown is included as is the interference between all these processes. The eigenvalues and eigenvectors obtained from the diagonalization of the K matrix are used to construct frame transformation matrix elements from which the scattering matrix can be determined by a method due to Seaton.[22] The cross sections and rate coefficients are calculated from the energy dependent scattering matrix. For the $^3\Pi_u$ cross sections, the K matrix includes 18 vibrational levels in the ion ground state and 18 in the $A^2\Pi_u$ excited core state. With the dissociative states, the K matrix is 40x40 for $^3\Pi_u$. For each of the remaining states shown in Fig. 1, excited core states are not included. For $^1\Pi_u$, the K matrix is 20x20. For the remaining states, the K matrices are 19x19. The $\ell=1$ π partial waves for the 'free' electron are included.

The calculated rate coefficients for the lowest three vibrational levels of N_2^+ are shown in Table 1. The rate coefficients are obtained by Maxwellian averaging of the calculated cross sections. The coefficient for v=0 at 300 K is $2.1 \times 10^{-7} \times (T_e/300)^{-0.20}$ cm^3/sec. The previously reported calculated coefficient, ($1.6 \times 10^{-7} \times (T_e/300)^{-0.37}$cm^3/sec),[11] is for a single dissociative route, $2^3\Pi_u$. The 300K rate coefficient is in excellent agreement with experimental results,[23-25] which range from $1.70\pm0.09 \times 10^{-7} \times (T_e/300)^{-0.30\pm0.09}$cm^3/sec[23] to 2.6×10^{-7}cm^3/sec.[24] This agreement gives us confidence that the calculated results for v=1 and 2 can also be expected to be

Table 1. Total Calculated DR rate coefficients (x 10^7) for N_2^+

	Electron Temperature (K)	
v	300	800
0	2.1 x $(T_e/300)^{-0.20}$	1.7 x $(T_e/800)^{-0.35}$
1	2.9 x $(T_e/300)^{-0.69}$	1.5 x $(T_e/800)^{-0.63}$
2	1.1 x $(T_e/300)^{-0.40}$	0.84 x $(T_e/800)^{-0.26}$

quite accurate. For excited vibrational levels, there are no experimental measurements or prior theoretical calculations. However, the most recent storage ring results indicate that the rate coefficients are not "strongly dependent" upon vibrational level.[23] The calculations reported here indicate otherwise. At 300 K, the calculated rate coefficient for v=1 is about 38% higher than that for v=0. However, at 800K (ionospheric temperature), the v=0 and 1 rate coefficients differ by only 12%. For v=2, the calculated rate coefficients are about ½ those for v=0 at 300 and 800K. For each vibrational level, the DR rate coefficient decreases with increasing electron temperature. One can show that if the total effective (over several dissociative states) Franck-Condon factor (between the dissociative and bound vibrational levels) remains constant as the electron energy increases, the rate coefficient will have a temperature dependence of $T_e^{-0.50}$. The values of the exponents for the temperature dependencies of the v=0 and 2 rate coefficients are more positive than -0.50 indicating that the effective Franck-Condon factors increase with electron energy. However, for v=1 the exponents indicate that the Franck-Condon factor decreases with increasing electron energy.

An analysis of Atmosphere Explorer data,[26] led to the proposal that the DR rate coefficient from excited vibrational levels had to be high, i.e. about 1.7 x 10^{-6} for v=1 at electron temperatures between 1400 and 2400K, in order to explain why the measured N_2^+ density was lower than expected. At these temperatures, my calculations give a rate coefficient of 0.92 x 10^{-7} x $(T_e/1900)^{-0.50}$ cm^3/sec for v=1 and 1.1 x 10^{-7} x $(T_e/1900)^{-0.50}$ cm^3/sec for v=0. The proposal[26] that the rate constant for v=1 may be 1.7 x 10^{-6} cm^3/sec at these electron temperatures, is not supported by these results. However, the final word rests with calculations currently underway, in which all of the core excited Rydberg states are included. Nevertheless, it is unlikely that the calculated rate coefficients will increase by an order of magnitude. Instead, other aspects of the model that predicts the N_2^+ density will have to be reexamined in order to understand the disagreement between the model results and the measurements. Objections, supported by these calculations, have been raised concerning the possibility of very high DR rate coefficients from excited vibrational levels.[27]

5. ISOTOPIC SUBSTITUTION

Isotopic substitution is an important issue because recent storage ring experiments[28] have used $^{15}N^{14}N^+$ instead of $^{14}N_2{}^+$ to measure DR rate coefficients and quantum yields. The rationale for using $^{15}N^{14}N^+$ is that due to its small dipole moment, it may vibrationally relax in a storage ring prior to DR and thereby guarantee that DR from only v=0 is being measured. However, $^{14}N_2{}^+$ has no dipole moment and will not radiatively relax within the duration of the experiment. Unfortunately, the measurements[28] show that even $^{15}N^{14}N^+$ is not relaxed prior to recombination. Are the storage ring measurements of rate coefficients for $^{15}N^{14}N^+$ relevant to the $^{14}N_2{}^+$ found in ionospheres?

Using the same approach described above, for the DR of $^{15}N^{14}N^+$ I find that the total DR rate coefficient at 300K from v=0 is 26% higher than the corresponding rate coefficient for $^{14}N_2^+$ (2.1 x 10^{-7} cm^3/sec). At an ionospheric temperature of 1000K, the $^{15}N^{14}N^+$ DR rate coefficient is 11% higher than the $^{14}N_2^+$ value (1.43 x 10^{-7} cm^3/sec). However, the storage ring DR experiment[28] on $^{15}N^{14}N^+$ has reported quantum yields at only $T_e = 0$ K where the rate coefficient is expected to differ by more than 26% from the $^{14}N_2^+$ value. Future work is needed to determine if there may be a fortuitous cancellation of errors in calculating the quantum yields for $^{14}N_2^+$ from the data for $^{15}N^{14}N^+$. In the mean time, the experimental $^{15}N^{14}N^+$ yields should be used with caution when modeling the $^{14}N_2^+$ ionospheric chemistry.

6. SUMMARY

Quantum theoretical calculations have been completed for the DR of N_2^+ from the lowest three ion vibrational levels. The calculated rate coefficient from v=0 agrees well with experimentally determined values. However, at room temperature, the calculated v=1 rate coefficient is 38% higher than that for v=0. The v=2 rate coefficient is about ½ of the v=0 value. These results contrast with recent storage ring experiments[23] where different ion vibrational distributions generated in different ion sources were studied. The results indicated that the rate coefficients would have little variation with v. Calculations on $^{15}N^{14}N^+$, show that at room temperature, the rate coefficient for v=0 DR is 26% higher than that for $^{14}N_2^+$, with the difference decreasing as the electron temperature increases. While the DR rate coefficients are considerably different for the two isotopomers at room temperature and below, it is not clear if this difference also holds for the quantum yields. Quantum yield calculations, in progress in this laboratory, will answer this question.

ACKNOWLEDGEMENTS

This material is based upon work supported by the National Science Foundation under Grants ATM-9812034 and ATM-0225256. The research was also supported by NASA Grants NAG5-12220, NAG5-11827, and NAG5-11428.

REFERENCES

1 M. R. Torr and D. G. Torr, Geophys. Res. Lett. **6**, 775 (1979).
2 J. E. Frederick and D. W. Rusch, J. Geophys. Res. **82**, 3509 (1977).
3 J. L. Fox and A. Hać, J. Geophys. Res. **102**, 9191 (1997).
4 A. O. Nier, M. B. McElroy and Y. L. Yung, Science **194**, 68 (1976); M. B. McElroy, Y. L. Yung and A. O. Nier, Science **194**, 70 (1976).
5 H. Lammer, W. Stumptner, G. J. Molina-Cuberos, S. J. Bauer, T. Owen, Planet. Space Sci. **48**, 529 (2000).
6 P. J. Knowles, and H.- J. Werner, Chem. Phys. Lett. **115**, 259 (1985); H.-J Werner and P. J. Knowles, J. Chem. Phys. **82**, 5053 (1985).
7 T. H. Dunning Jr., J. Chem. Phys. **90**, 1007 (1989).
8 H.- J. Werner and P. J. Knowles, J. Chem. Phys. **89**, 5803 (1988); P. J. Knowles and H.-J. Werner, Chem. Phys. Lett. **145**, 514 (1988). All calculations in the [54321] basis were done with the MOLPRO programs. MOLPRO is a package of ab initio programs written by H.- J. Werner and P. J. Knowles, with contributions from J. Almlöf, R. D. Amos, M. J. O. Deegan, S. T. Elbert, C. Hampel, W. Meyer, K. Peterson, R. Pitzer, A. J. Stone, P. R. Taylor, and R. Lindh.
9 A. Lofthus and P. H. Krupenie, J. Phys. Chem. Ref. Data **6**, 113 (1977).
10 K. P. Huber and G. Herzberg, *Molecular Spectra and Molecular Structure. IV Constants of Diatomic Molecules* (Van Nostrand Reinhold, New York, 1979).
11 S. L. Guberman, Geophys. Res. Lett. **18**, 1051 (1991).
12 S. L. Guberman, New Mechanisms for Dissociative Recombination, in *The Physics of Electronic and Atomic Collisions, XIX International Conference*, Book of Invited Papers, ed. by L. J. Dube et al. (American Institute of Physics Press, New York, 1995), p.307.
13 S. L. Guberman, Electron-Ion Continuum-Continuum Mixing in Dissociative Recombination, in *Dissociative Recombination: Theory, Experiment, and Applications*, ed. by B. R. Rowe, J. B. A. Mitchell, and A. Canosa (Plenum, New York, 1993), p.47.
14 S. L. Guberman, Science **278**, 1276 (1997).
15 S. L. Guberman, Ab Initio Studies of Dissociative Recombination, in *Dissociative Recombination: Theory, Experiment, and Applications*, ed. by J. B. A. Mitchell and S. L. Guberman (World Scientific, Singapore, 1989), p. 45. An extrapolation of the width to the continuum is no longer used.
16 S. L. Guberman, J. Chem. Phys. **102**, 1699 (1995).
17 S. L. Guberman and A. Giusti-Suzor, J. Chem. Phys. **95**, 2602 (1991).
18 K. Kaufmann, W. Baumeister and M. Jungen, J Phys. B **22**, 2223 (1989).
19 W. J. Hunt and W. A. Goddard III, Chem. Phys. Lett. **3**, 414 (1969).
20 A. U. Hazi, Molecular Resonance Phenomena, in *Electron-Atom and Electron-Molecule Collisions*, ed. by J. Hinze (Plenum Press, New York, 1983), p.103.
21 A. Giusti, J. Phys. B **13**, 3867 (1980).
22 M. J. Seaton, Rep. Prog. Phys. **46**, 167 (1983).

23 J. R. Peterson, A. Le Padellec, H. Danared, G. H. Dunn, M. Larsson, Å. Larson, R. Peverall, C. Strömholm, S. Rosén, M. af Ugglas and W. J. van der Zande, J. Chem. Phys. **108**, 1978 (1998).
24 A. Canosa, J. C. Gomet, B. R. Rowe and J. L. Queffelec, J. Chem. Phys. **94**, 7159 (1991).
25 F. J. Mehr and M. A. Biondi, Phys. Rev. **181**, 264 (1969); E. C. Zipf,, Geophys. Res. Lett. **7**, 645 (1980); M. Geoghegan, N. G. Adams and D. Smith, J. Phys. B **24**, 2589 (1991).
26 N. Orsini, D. G. Torr, H. D. Brinton, L. H. Brace, W. B. Hanson, J. H. Hoffman and A. O. Nier, Geophys. Res. Lett. **4**, 431 (1977).
27 M. A. Biondi, Geophys. Res. Lett. **5**, 661 (1978).
28 D. Kella, P. J. Johnson, H. B. Pedersen, L. Vejby-Christensen and L. H. Andersen, Phys. Rev. Lett. **77**, 2432 (1996).

Dissociative Recombination of Slow Electrons and Molecular Oxygen Ions in a Strong Laser Field

M. G. Golubkov, G. V. Golubkov, and A. N. Romanov
N. N. Semenov Institute of Chemical Physics, Russian Academy of Sciences, Kosygin str 4, Moscow 117977, Russia
golubkov@chph.ras.ru

1. INTRODUCTION

The reaction of low-temperature dissociative recombination

$$e^-(l\Lambda) + O_2^+(^2\Pi_g, v=0) \xrightarrow{k\omega_f} \left[O_2^{**}(n_k p\pi_u(B^3\Sigma_u^-)) \Leftrightarrow O_2^*(B^3\Sigma_u^-)\right] \rightarrow O(^3P) + O(^1D) \quad (1)$$

was investigated. It takes place due to direct and resonance radiationless transitions (k=0) into the Schumann-Runge continuum ($B^3\Sigma_u^-$) and due to free-bounded radiation transitions (k=1) into predissociative Rydberg $n_1 p\pi_u(B^3\Sigma_u^-)$ states by the action of monochromatic laser field with frequency ω_f in the visible region. Index k denotes the change in the number of photons in the system, n_k is the principal quantum number at the prescribed k value, v is the vibrational quantum number, and l and Λ are the electron orbital momentum and its projection on the molecular axis, respectively. The electron energy is assumed to be small and lies in the interval $0 < E_e < \omega$, where ω is the ion vibrational frequency. Under the condition ($\hbar = m_e = e = 1$)

$$pf\omega_f^{-2} << 1, \quad p = \sqrt{2E_e}, \quad (2)$$

where p is the impulse of incident electron and f is the amplitude of electromagnetic field intensity. The electron oscillation amplitude is many times smaller than its wavelength. The external field in that case has no influence on the electron motion, i.e. the electron has the definite energy, E_e. The field mostly affects the states of the intermediate complex XY^{**} if it does not lead to dipole allowed transitions in the isolated ion XY^+. In this case the electron motion is multichannel. The second restriction on the external field strength is

$$fD << 1, \tag{3}$$

where D is the dipole moment of the transition. With this restriction, it is only necessary to consider two main types of transitions: radiationless (k=0) and stimulated (k=1) with one photon radiation. Therefore we neglect the bound-bound dipole allowed transitions from resonance intermediate Rydberg $n_0 l\Lambda_g$ states into predissociative $n_1 p\pi_u(^3\Sigma_u^-)$ states since the contribution relative to the leading $n_0 p\pi_u$ series turns out to be small.[1] We exclude also direct radiation transitions from $n_0 l\Lambda_g$ states to Schumann-Runge continuum and low-lying $3p\pi_u(^3\Sigma_u^-)$ Rydberg state. These transitions occur in a region where the typical wavelengths are several times greater than the maximum wavelength of visible light (≈ 750 nm). In this paper, we are concerned only with the visible region. Moreover, we neglect radiationless transitions into the dissociative $2^3\Pi_g$ state, which correlates at infinite nuclear separation with the products of Reaction (1). Its potential curve crosses the ion potential curve in the vicinity of the right-hand classical turning point of third excited vibrational level[2] (v=3) and configuration coupling is small.

2. METHOD

The analysis was carried out in the framework of the multichannel quantum defect theory (MQDT) using the stationary formalism of the radiative collision matrix.[1] The theory is based upon the information about adiabatic potential curves of the Rydberg complex created at an intermediate stage of the process. We have investigated the rotationally adiabatic spectrum region, i.e. $Bn^3 << 1$, where B is the ion rotational constant. The molecular axis is assumed to be fixed during the collision and the calculations are carried out in the molecular frame. The unperturbed Hamiltonian, H_0, is chosen in a Coulomb basis set in such a way that all the interactions in the isolated (diabatic) configuration $X + Y^*$ are taken into account exactly. The complete Hamiltonian of the system is $\mathrm{H} = \mathrm{H}_0 + \mathrm{V}$. The operator **V** can be represented as the sum

$$\mathbf{V} = \mathbf{V}^{\mathrm{e}} + \mathbf{V}^{\mathrm{f}}, \tag{4}$$

where the operator of the electrostatic interaction,

$$\mathbf{V^e} = \mathbf{V_{nc}} + \mathbf{V^{CI}}, \tag{5}$$

involves the non-Coulomb part of the interaction with the ion core, $\mathbf{V_{nc}}$, and the configuration interaction, $\mathbf{V^{CI}}$. The latter operator couples the Rydberg and dissociative configurations denoted by indices q and β respectively. Matrix elements $\langle q_k|\mathbf{V^e}|q'_k\rangle$ are responsible for vibronic transitions in the system $e^- + XY^+$ and are diagonal in *k*. The quantities $\langle q_k|\mathbf{V^e}|\beta_k\rangle$ describe the interaction between $e^- + XY^+$ and $X + Y^*$ configurations. The elements of the field operator $\mathbf{V^f}$ couple the states with *k* and $k' = \pm 1$.

MQDT leads to the following fundamental equation for the **T**-matrix of radiative multichannel scattering[1]

$$\mathbf{T} = \mathbf{t} + \mathbf{t}\sum_{v,k}|q_k\rangle\langle q_k|\, ctg(\pi\nu_{vk})\mathbf{T} - i\mathbf{t}\sum_{\beta,k}|\beta_k\rangle\langle\beta_k|\mathbf{T}, \tag{6}$$

where $\nu_{vk} = (-2\varepsilon_{vk})^{-1/2}$ is the effective principal quantum number which corresponds to the energy $\varepsilon_{vk} = E_e - v\omega_v - k\omega_f$. Operator **t** describes interaction of an electron with the ion core in the presence of an electromagnetic field.

Eq. (6) allows for the use of an algebraic method for construction of the unitary scattering **S**-matrix (related to the **T**-operator by the expression $\mathbf{S} = \mathbf{I} - 2i\mathbf{T}$) without regard to the number of channels of motion taken into account. The elements $T_{q_m q'_{m'}}$ and $T_{q_m \beta_{m'}}$ with open-channel indicies (where $E > v\omega_v + k\omega_f$ and the functions $ctg(\pi\nu_{vk}) = -i$) characterize the amplitudes of inelastic scattering and dissociative recombination, respectively. The reaction matrix in Eq. (6) represents the sum of two operators characterizing the electrostatic and field interaction

$$\mathbf{t} = \mathbf{t^e} + \mathbf{t^f}. \tag{7}$$

The total cross section of dissociative recombination for the transition to the dissociative β-channel is

$$\sigma_{0\beta}(\theta_e) = \frac{8\pi^2}{E_e} g \sum_{k=0,1}\left\langle \left| \sum_{l\Lambda} i^l e^{i\delta_l} Y^*_{l\Lambda}(\mathrm{n})\, T_{0_0\beta_k}(l\Lambda \to l_\beta\Lambda_\beta) \right|^2 \right\rangle, \tag{8}$$

where $l\Lambda$ are Rydberg states allowed by the selection rules for radiative transitions into the final $l_\beta\Lambda_\beta$ state of the system, δ_l is the Coulomb phase, vector **n** specifies the direction of incident electron beam with respect to the axis of the XY molecule, θ_e is the angle between vectors **n** and **f**, *g* is a spin factor and angle brackets <...> mean an average over all possible orientations of the molecules.

The potential curves of the diabatic valence configuration, $U_\beta(R)$, and the oxygen molecular ion, $U_i(R)$,were calculated using the GAMESS program[3] by the Multi Reference Double Configuration Interaction method

(MRD-CI)[4] with reference configurations obtained by the CASSCF method[5] over the full valence space ($2\sigma_g$, $2\sigma_u$, $3\sigma_g$, $1\pi_u$, $1\pi_g$, $3\sigma_u$) in the basis [6s,3p,2d,1f].[6] Our calculated potential curves are in good agreement with data.[7] The energy of the direct transition, $\varepsilon_0 = U_\beta(R_e^+) - U_i(R_e^+)$, is 0.713 eV. For the electron configuration coupling, $V_\beta(R)$, we have used data from Ref. 7 . The calculation of the Reaction (1) cross section was carried out in a seven channel approximation (five open $v=0$ channels (p_π for k=0 and s_σ, d_σ, d_π, and d_δ for k=1), one closed channel ($v=1$) and one open dissociative β – channel) using Formula (8). Since in the energy region under consideration, $E_e \le 0.6$ eV, the vibronic resonances with $v \ge 2$ are absent and this approximation is well justified. The information about corresponding quantum defects and their derivatives was taken from Ref.'s 8 and 9.

3. RESULTS

The cross section dependence upon the incident electron energy, E_e, for different values of field strength, f, is shown in Fig.1. These results were obtained for $\omega_f = 24040$ cм$^{-1}$ and $\theta_e = \pi/2$. The typical Fano-Feshbach resonant structure can be observed at $f=0$. It is due to a radiationless transition with the Rydberg $n_0 p\pi_u(^3\Sigma_u^-, v_0 = 1)$ series, which dissociate into Schumann-Runge continuum (^1D+^3P). The corresponding resonance lines are denoted by parenthesis (n_0, v_0) and arrows. The first resonance (with wide minimum in the vicinity of $E_e = 0.03$ eV) relates to the principal quantum number $n_0 = 9$. The frequency ω_f is chosen in such a way that the field induced $n_1 p\pi_u(^3\Sigma_u^-, v_1 = 0)$ – resonances from the initial s_σ, d_σ, d_π, and d_δ – channels of the electron motion are situated close to this minimum.

The dependence of the Reaction (1) cross section on the frequency of the external field is bell-shaped. The position of the maximum, $\omega_f^{\max}$, is a linear function of the incident electron energy, E_e. The cross section rises approximately two orders of magnitude in the case of $\omega_f = \omega_f^{\max}(E_e)$. For an electron energy, $E_e = 0.03$ eV, the field influence is the most effective for $\omega_f^{\max} = 24040$ cm^{-1}. We have found that for $0 < E_e \le 0.6$ eV, the frequency, $\omega_f^{\max}(E_e)$, should be placed in the interval $\Delta\omega_f = 23000 - 25000$ cm^{-1}, which is in the visible spectrum region.

The dependence on field strength, *f*, is presented in Fig. 2 for $E_e = 0.03$ eV, $\omega_f^{\max} = 24040$ cm^{-1} and $\theta_e = \pi/2$. The maximum cross section is at $f^{\max} = 1.15 \cdot 10^8$ V/cm, where it is eight orders of magnitude above the cross section for zero field strength.

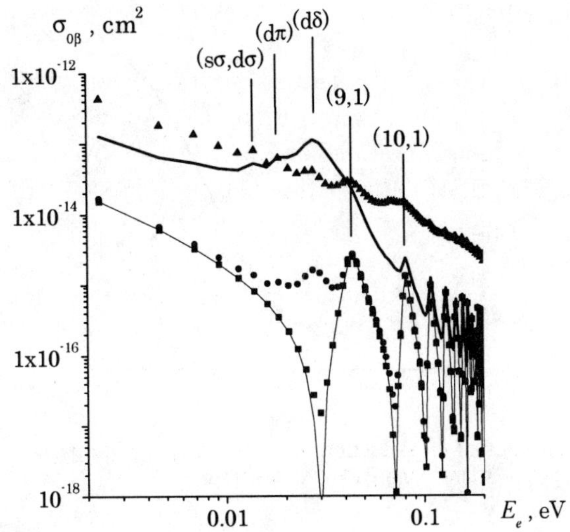

Figure 1. The dependence of reaction (1) cross section on the electron energy for different values of field strength: thin line - $f=0$, squares - $f=5\cdot 10^{5}$ V/cm, circles - $f=5\cdot 10^{6}$ V/cm, thick line $f=5\cdot 10^{7}$ V/cm, triangles - $f=5\cdot 10^{8}$ V/cm.

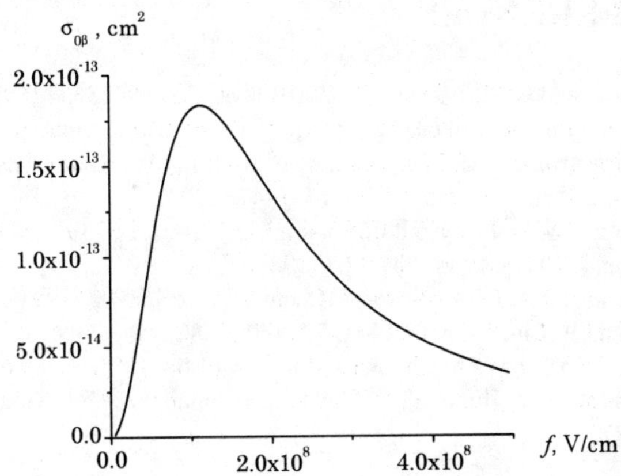

Figure 2. The dependence of reaction (1) cross section on the field strength *f*.

The angular dependence of the cross section ratio, $\sigma_{0\beta}(\theta_e)/\sigma_{0\beta}(\pi/2)$, for $E_e=0.03$ eV and $\omega_f=24040\ \mathrm{cm}^{-1}$ is given in Fig.3. The maximum is reached at $\theta_e=\pi/2$ and doesn't depend upon *f*.

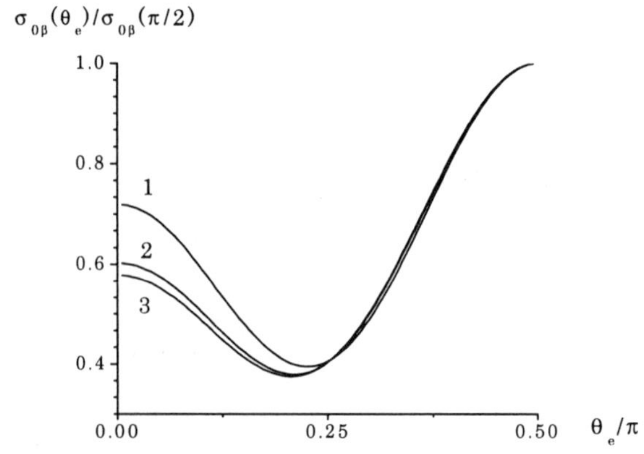

Figure 3. The dependence of cross section relation $\sigma_{0\beta}(\theta_e)/\sigma_{0\beta}(\theta_e = \pi/2)$ on θ_e: (1) $f = 5 \cdot 10^8$ V/cm; (2) $f = 5 \cdot 10^7$ V/cm; (3) $f = 5 \cdot 10^6$ V/cm.

ACKNOWLEDGEMENT

This work was carried out with financial support of INTAS (project #99-00039).

REFERENCES

1 G. V. Golubkov and G. K. Ivanov, Zh. Eksp. Teor. Fiz. **104,** 3334 (1993).

2 S. L. Guberman, Potential Energy Curves for Dissociative Recombination, in *Physics of Ion-Ion and Electron-Ion Collisions* ed. by F. Brouillard (Plenum, New York, 1983).

3 M. W. Schmidt et. al., J. Comp. Chem. **14**, 1347 (1993).

4 R. J. Buenker and S. D. Peyerimhoff, Theor. Chim. Acta. **35**, 33 (1974).

5 T. H. Dunning, J. Chem. Phys. **90**, 1007 (1989).

6 S. L. Guberman and A. Giusti-Suzor, J. Chem. Phys. **95**, 2602 (1991).

7 R. P. Saxon and B. Liu, J. Chem. Phys. **73**, 870 (1980).

8 A. Sur, C. V. Ramana, W. A. Chupka and S. D. Colson, J. Chem. Phys. **84**, 69 (1986).

9 D. C. Cartwright, W. J. Hunt, W. Williams, S. Trajmar and W. A. Goddard III, Phys. Rev. **A8**, 2436 (1973).

Dissociative Recombination of c-$C_3H_3^+$

Potential Energy Curves

Dahbia Talbi
Laboratoire d'étude théorique des milieux extrêmes
Université de Nice Sophia-Antipolis, Parc VALROSE 06108 NICE cedex 2
talbi@letmex.unice.fr

1. INTRODUCTION

The three-membered ring molecule, cyclopropenylidene (c-C_3H_2), discovered in space in 1985,[1] has gained attention very rapidly because of its ubiquity in space. It is not a common molecule on Earth and it is not used in the laboratory for any synthesis. However, it is a stable interstellar species surviving even in UV exposed regions. The formation is due to the dissociative recombination (DR) of c-$C_3H_3^+$:

$$c\text{-}C_3H_3^+ + e^- \rightarrow c\text{-}C_3H_2 + H$$

In 1993, Abouelaziz et al.[2] attempted an experimental determination of the rate coefficient for the above reaction. At 300K, they measured this coefficient to be $7 \pm 2 \times 10^{-7}$ cm^3s^{-1}. This value, indicating an efficient DR process, is in agreement with the 1×10^{-7} cm^3s^{-1} determined earlier at 2000K.[3] However, there are two uncertainties related to the Abouelaziz et al. experiment. These are the state of excitation and the isomeric form of the recombining ion including whether or not $C_3H_3^+$ was in its cyclic form.

Because it is important to understand the formation of c-C_3H_2 in the galaxy as well as to know with more certainty the value of the DR reaction rate constant, a quantum chemical study of the DR of *c*-$C_3H_3^+$ has been undertaken.

Quantum chemical calculations have proven to be extremely powerful for understanding these recombination processes. Such processes may occur by direct and/or indirect mechanisms. While the former involves a direct transition from the ionic state to the final neutral dissociative state, the second one is a two-step process involving the formation of a rovibrationally excited state of the neutral molecule. For this last case, either autoionization takes place or the state formed after binding the electron, predissociates. Understanding such processes requires knowledge of the potential energy surfaces of ionic, Rydberg and dissociative states. This is not a trivial task because, very often, the nature of the states changes along the reaction coordinate. Carefully designed wave functions are needed to accurately follow those transformations. But the difficulty is also in the even handed treatment required for all the states involved in the mechanism.

2. METHODOLOGY

In terms of localized orbitals, the ground state electronic configuration of $c\text{-}C_3H_3^+$ can be written as:

$$\text{(Core)}\ \ 3\ \text{x}\ (\sigma CC)^2\ 3\ \text{x}\ (\sigma CH)^2\ (\pi CC)^2 \qquad (1)$$

DR requires capture of an electron that will lead to the formation of a doubly or singly excited state of the neutral ground state. If singly excited, the additional electron, in direct DR, occupies a valence virtual orbital. If the virtual orbital is one of antibonding character, CH*, the resulting electronic configuration is

$$\text{(Core)}\ \ 3\ \text{x}\ (\sigma CC)^2\ 2\ \text{x}\ (\sigma CH)^2\ (\pi CC)^2(\sigma CH)^2(\sigma CH^*)^1 \qquad (2)$$

This state, for which the antibonding CH* orbital is singly occupied, is the lowest one of those governing the DR of $c\text{-}C_3H_3^+$. It dissociates to H + $c\text{-}C_3H_2$, both in the ground electronic states. The additional electron can also occupy a diffuse (i.e. Rydberg) orbital, R, far from the $c\text{-}C_3H_3^+$ core with the electronic configuration:

$$\text{(Core)}\ \ 3\ \text{x}\ (\sigma CC)^2\ 2\ \text{x}\ (\sigma CH)^2\ (\pi CC)^2(\sigma CH)^2(R)^1. \qquad (3)$$

These Rydberg states allow for the indirect DR mechanism.[4,5] For a quantitative study of the DR of $c\text{-}C_3H_3^+$, the even-handed treatment of both the ionic, Rydberg and dissociative states of $c\text{-}C_3H_3$ is required. Correlation energy must be treated through the same CI (Configuration Interaction) procedure so as not to favor one state over another. In order to fulfill this requirement, we have adopted a methodology, already defined,[4,5] that is based on the use of orbitals of strong localized character to build the CI wave functions. It is then possible to separate, valence bonding from antibonding orbitals and antibonding from Rydberg orbitals, each of them having a well-defined spatial expansion and a clear physicochemical meaning. The advantage of such a method is that the CI n-particle space is defined in terms of orbitals whose chemical significance is clearly identified. Configurations can be selected according to chemical evidence and the type of arbitrariness associated with energy related selection processes can be avoided. More important, the CI expansion is directly obtained in a configuration space that contains the approximate diabatic representation of the problem.[6] It is therefore possible to extract an approximate diabatic representation of the states (here after quasi-diabatic) according to the weights of their relevant configurations in the CI wave function.

In order to obtain orbitals of localized character, several procedures can be used. We have, after a series of test calculations, chosen to use natural orbitals from carefully designed MCSCF calculations. A 6-311G(d,p) atomic basis set has been used to construct the MCSCF molecular basis set. For the Rydberg components, the valence space obtained from the MCSCF procedure has been augmented by two pairs of atomic diffuse functions: s (0.026, 0.016) and p (0.023, 0.015). The exponents were determined by averaging the diffuse exponents of Dunning and Hay.[7] The resulting set of MOs have been Schmidt-orthogonalized prior to the CI calculations. This method of building a complete set of orbitals has been shown to reliably position the Rydberg states.

The CI n-particle space, shown in Table 1, has been designed in the following manner: the valence $^2A_1$ and $^2B_2$ orbitals as well as their corresponding virtuals, of secondary importance for the dissociation process, are distributed respectively in subsets 1 and 3. In these subsets, electrons are distributed to insure the geometry relaxation of the molecule when the CH bond breaks apart. In subset 2 are included all orbitals essential to describe the possibly important excited states (Rydberg and lowest valence) including the dissociative one. In addition, semi-internal effects are considered with one electron in the external space.

3. THE POTENTIAL ENERGY CURVES

The adiabatic representation: The potential energy curves for the two lowest $^2B_1$, two lowest $^2A_2$, and three lowest $^2A_1$ states of c-C_3H_3 as well as that of the $^1A_1$ ion, calculated along one of the C-H reaction coordinates, are shown in Fig.1. While $^2B_1$ and $^2A_2$ surfaces smoothly tend to the dissociation limits, a strongly avoided crossing can be observed on the lowest surface of $^2A_1$ symmetry, indicating the change in character of this state. An analysis of the $^2A_1$ CI natural orbitals reveals, that this state has Rydberg character at short CH internuclear distances (configuration (3)), changing to a state of

Table 1: CI n-particle space designed for the c-$C_3H_3^+$ + e^- study. The three 1s core orbitals are kept frozen. The number of configurations is 566,293 CSF's (Configuration State Functions) for the $^2A_1$ states, 97,834 CSF's for the $^2A_2$ states and 385,250 CSF's for the $^2B_1$ states. Electron distributions of the positive ion are given in parentheses.

	Electronic		Configurations	
Sym	Set 1	Set 2	Set 3	External set
A_1	σCCσCC σCH	σCH(lpc) σCH*(H) R R	σCC*σCH*	22 MOs
B_1		Π Π*R		10 MOs
B_2	σCCσCH	R	σCH*σCC* σCC*	17 MOs
A_2		Π*		6 MOs
	10 (10)	5 (4)	0 (0)	0 (0)
	9 (9)	6 (5)	0 (0)	0 (0)
	8 (8)	7 (6)	0 (0)	0 (0)
	9 (9)	5 (4)	1 (1)	0 (0)
	10 (10)	4 (3)	1 (1)	0 (0)
	8 (8)	5 (4)	2 (2)	0 (0)
	9 (9)	4 (3)	2 (2)	0 (0)
	10 (10)	3 (2)	2 (2)	0 (0)
	9 (9)	5 (4)	0 (0)	1 (1)
	10 (10)	4 (3)	0 (0)	1 (1)
	8 (8)	5 (4)	1 (1)	1 (1)
	10 (10)	3 (2)	1 (1)	1 (1)
	9 (9)	4 (3)	1 (1)	1 (1)

repulsive character (configuration (2)) for CH distances above 1.25 Å. This change in character indicates the existence of a curve crossing, in a quasi-diabatic representation, between a bound $^2A_1$ Rydberg state and the lowest $^2A_1$ repulsive state. Fig. 1 also shows that there is a $^2A_2$ repulsive state crossing the curve of the $^1A_1$ ion. However, the intersection is at much too high an energy to be interesting for this study.

The quasi-diabatic representation: Quasi – diabatic potential energy surfaces for the three lowest $^2A_1$ states of c-C_3H_3 as well as that of the ion are reported as a function of the CH dissociative reaction coordinate in Fig. 2. As expected from the adiabatic representation, there is a curve crossing between the lowest $^2A_1$ repulsive state and the two lowest Rydberg states. The crossing points are located close to the minima of the Rydberg curves, below the first vibrational levels. The lowest $^2A_1$ repulsive state also intersects the $^1A_1$ curve of the ion near its minimum energy.

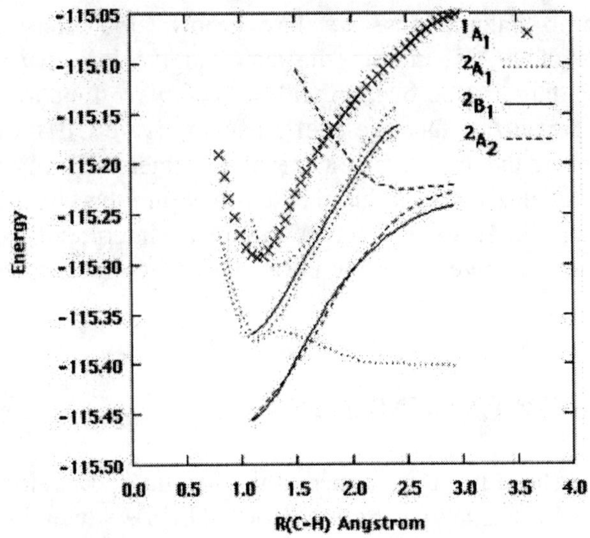

Figure 1. Adiabatic potential energy surfaces for c-$C_3H_3^+$ + e^-. The $^1A_1$ potential energy curve of the ion is the line with crosses.

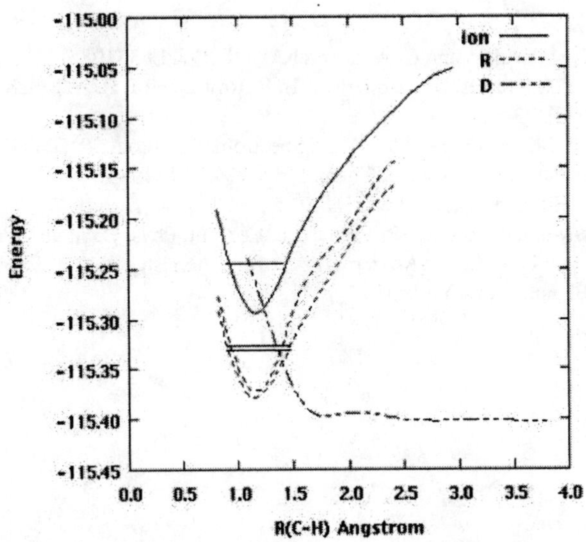

Figure 2. Quasi-diabatic $^2A_1$ potential energy curves for c-$C_3H_3^+$ + e^-. The $^1A_1$ potential energy curve of the ion is shown as a full line.

4. CONCLUSIONS

Using ab initio calculations, we have shown that the lowest $^2A_1$ dissociative state of c-C_3H_3 crosses the curve of the ion at its minimum energy, thus providing a good overlap with its v=0 wave function. It is then reasonable to expect a fast DR rate coefficient to give c-C_3H_2, through the direct mechanism at the low temperature of the interstellar medium. The assumption of Abouelaziz et al.,[2] that they did measure the DR of the cyclic isomer, is supported. However, for a complete interpretation of their experiment, we need to investigate the DR of the $C_3H_3^+$ linear isomer, a task that has started.

ACKNOWLEDGMENTS

All calculations have been performed at the Institut de Développement et des Ressources en Informatique Scientifique (IDRIS), which is gratefully acknowledged.

REFERENCES

1 P. Thaddeus, J. M. Vrtilek, and C. A. Gottlieb, Ap. J. **299**, L63 (1985).
2 H. Abouelaziz, J. C. Gomet, D. Pasquerault, B. R. Rowe, and J. B. A. Mitchell, J. Chem. Phys. **98**, 1 (1993).
3 S. M. Graham, J. M. Goodings, Int. J. Mass Spectrom. Ion Proc. **56**, 205 (1984).
4 D. Talbi, F. Pauzat, and Y. Ellinger, Chem. Phys. **126**, 291 (1988).
5 D. Talbi and Y. Ellinger, Chem. Phys. **288**, 155 (1998).
6 M. C. Bacchus-Montabonnel and P. Vermeulin, Comput. Phys. Commun. **163**, 30 (1983).
7 T. H. Dunning and P. J. Hay, in Modern Theoretical Chemistry, Vol. 3, ed by H. F. Schaeffer III (Plenum, New York, 1977) p. 1.

Help!!! Theory for H_3^+ Recombination Badly Needed

Takeshi Oka

Department of Chemistry, and Department of Astronomy and Astrophysics, the Enrico Fermi Institute, The University of Chicago, 5735 South Ellis Ave., Chicago, IL 60637

t-oka@uchicago.edu

1. INTRODUCTION

A spectrum is haunting the field of dissociative recombination—the spectrum of H_3^+ in the diffuse interstellar medium. This spectrum, discussed in detail in a separate paper of this volume,[1] has revealed the surprisingly high abundance of H_3^+ in the diffuse interstellar medium. The intensities of the infrared absorption lines observed toward some 10 reddened visible stars correspond to H_3^+ column densities of $1–4\times10^{14}cm^{-2}$, which is one to two orders of magnitude higher than the value estimated from canonical laboratory and observational astrophysical parameters. The understanding of dissociative recombination is crucial in this estimate since it is the major destruction process of H_3^+ in the diffuse interstellar medium.

The abundance of hydrogen in space and the simplicity of hydrogen plasma chemistry allow us to narrow down possible sources causing the problem to the following three parameters: (1) the cosmic-ray ionization rate[2] of $\zeta{\sim}10^{-17}$ s^{-1}, (2) the ratio of electron number density n(e) to hydrogen number density n(H), x_e = n(e)/n(H) which is taken to be comparable to the relative carbon abundance $x_c{\sim}10^{-4}$ (because all carbon atoms are ionized), and (3) the rate constant of the H_3^+ dissociative recombination $k_e{\sim}10^{-7}cm^3s^{-1}$. The first two parameters, ζ and x_e are important astrophysical quantities, and the observed H_3^+ column densities should provide crucial information about them. Before we can do this, however, we need to understand the laboratory parameter k_e, and this is the subject of this paper. This is an important and general problem because the ubiquity of H_3^+ is being observed in cosmic plasmas, and it is an urgent one since H_3^+ is being observed rapidly in many astronomical sources.

In this paper I briefly sketch the plasma chemistry of H_3^+ in the diffuse interstellar medium and critically review the past and present studies of the H_3^+ dissociative recombination. Clearly it is presumptuous of a non-specialist to do the latter; this is an act of a man desperately needing help. (The number of exclamation marks in the title of this paper has increased with the successive announcements of the ACS meeting, and it is in logarithmic scale!) My cry of SOS is directed to theorists since k_e has been measured by many groups using several entirely different techniques.[3] They seem to give $k_e \sim 10^{-7} cm^3 s^{-1}$ at room temperature except for a few cases as we discuss in Section 3. On the other hand, theoretical studies give much lower values and seem to be in a more uncertain state of affairs.[4] Since laboratory experiments cannot be conducted in the extremely low density, field free condition of interstellar space, the following statement of Bates et al.[5] still applies. "Although several independent experiments provide strong evidence that the dissociative recombination coefficient of H_3^+ at 300K is around $1.5 \times 10^{-7} cm^3 s^{-1}$, an element of unease must remain as long as a theory seems to exclude this possibility."

2. CHEMISTRY OF H_3^+ IN THE DIFFUSE INTERSTELLAR MEDIUM

2.1 General remarks

Four special characteristics of interstellar chemistry are stressed here.

(i) Hydrogen and helium constitute 99.9% of interstellar atoms. Since He is not chemically active, interstellar chemistry is hydrogen dominated chemistry.

(ii) Ions are produced by ubiquitous cosmic-ray particles and stellar radiation and destroyed by electron recombination. Interstellar gas is in the state of weakly ionized plasmas.

H_3^+ has been observed in dense and diffuse clouds.[6]

(iii) Dense (molecular) clouds have number density of $10^5 \sim 10^3 cm^{-3}$; they are gravitationally bound and are on their way to star formation. Stellar radiation does not penetrate the clouds. Stars are invisible and gas is in molecular form; hydrogen as H_2 and carbon as CO.

(iv) Diffuse clouds have a number density of $10^3 \sim 10^1 cm^{-3}$. Stars are visible through the clouds and hydrogen is a mixture of H and H_2. Carbon stays atomic and is ionized by star radiation since carbon has the lowest ionization potential among abundant species, i.e., H, H_2, He, C, N and O. This causes the high relative abundance of electrons in the clouds.

2.2 Ion Chemistry of H_3^+

H_3^+ is produced by cosmic ray ionization of H_2

$$H_2 \xrightarrow{\text{C.R.}} H_2^+ + e \tag{1}$$

followed by the ion-neutral reaction

$$H_2^+ + H_2 \rightarrow H_3^+ + H \tag{2}$$

with the typical Langevin rate constant[7] of $k_L = 2\times10^9 cm^3s^{-1}$. Since the rate of reaction (1), $\zeta \sim 10^{-17}s^{-1}$ is very much lower than the rate of reaction (2), $k_L n$ (H_2) for H_2 number density of clouds, (1) is the rate determining process and the production rate of H_3^+ (in cm^3s^{-1}) is given by ζ n (H_2).

The major destruction process is different in dense and diffuse clouds. In dense clouds H_3^+ is destroyed by proton hop reaction to CO,

$$H_3^+ + CO \rightarrow HCO^+ + H_2$$

with the rate of $k_L n(H_3^+) n(CO)$. In diffuse clouds H_3^+ is destroyed by electron recombination

$$H_3^+ + e \begin{cases} \nearrow H + H + H \\ \searrow H_2 + H \end{cases} \tag{3}$$

with the rate of $k_e n(H_3^+) n(e)$.

Equating the production and destruction rates, we obtain the steady state H_3^+ number density

$$n(H_3^+)_{dense} = \frac{\zeta}{k_L} \bullet \frac{n(H_2)}{n(CO)} \tag{4}$$

for dense clouds, and

$$n(H_3^+)_{diffuse} = \frac{\zeta}{k_e} \bullet \frac{n(H_2)}{n(e)} \tag{5}$$

for diffuse clouds. Therefore, if we assume a simple model that all hydrogen atoms are in the form of H_2 and all carbon atoms in the form of CO in dense clouds (which is valid), and hydrogen H_2 and carbon C^+ in diffuse clouds (which is not entirely valid and discussed later), $n(H_2)/n(CO) = n(H_2)/n(C^+) = n(H_2)/n(e)$ since they are simply half of the abundance ratio of hydrogen and carbon. We therefore have

$$\frac{n(H_3^+)_{diffuse}}{n(H_3^+)_{dense}} \sim \frac{k_L}{k_e} \sim 4\times10^{-3}$$

where the recombination rate constants recommended by Sundström et al,[8]

$$k_e = 4.6\times10^{-6}/(T/K)^{0.65}\ cm^3s^{-1} \tag{6}$$

for a temperature, T, of 30K and $k_L = 2\times10^{-9}cm^3s^{-1}$ (which is temperature independent) are used. Various uncertainties and assumptions in the simplified model lead to

$$n(H_3^+)_{diffuse} = (10^{-2} \sim 10^{-3})n(H_3^+)_{dense} \tag{7}$$

2.3 Observed H_3^+ column densities and the mystery

Examples of H_3^+ column densities observed in dense and diffuse clouds are listed in Table 1.

Table 1. Observed column densities (in $10^{14}cm^{-2}$) of H_3^+ in dense and diffuse clouds.

Dense clouds[a]		Diffuse clouds[b]	
AFGL 2136	3.8	Cyg OB2 12	3.8
W33A	5.2	Cyg OB2 5	2.6
MonR2 IRS 3	1.4	HD183143	2.3
AFGL 961E	1.7	HD20041	3.5
AFGL 490	1.1	WR121	2.2

[a] from McCall et al.[9]
[b] from McCall et al.[10]

Clearly the column densities in dense clouds and diffuse clouds are comparable and on the order of $N(H_3^+) \sim 10^{14}cm^{-2}$.

Now Eqs. (4) and (5) demonstrate a special characteristic of H_3^+ as an astronomical probe; $n(H_3^+)$ does not depend on cloud density since $n(H_2)/n(CO)$ and $n(H_2)/n(e)$ are constant as long as we are discussing typical dense or diffuse clouds, as shown in Fig. 1. This special property makes H_3^+ a powerful probe to measure cloud dimensions since the optical pathlength of a cloud L is simply given as a ratio of observed H_3^+ column density to

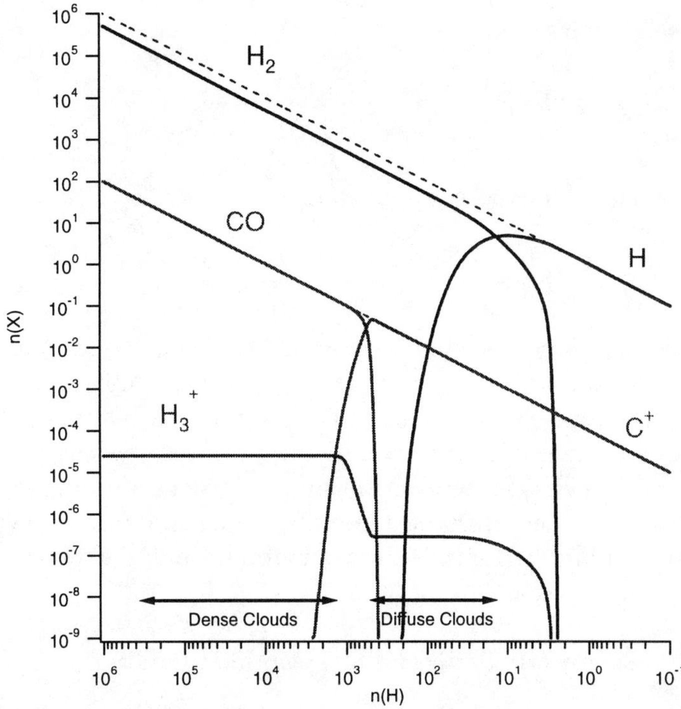

Figure 1. Schematic diagram indicating number densities n(X) of key chemical species in dense and diffuse clouds relative to total hydrogen density n(H) (drawn by B. J. McCall).

estimated number density,

$$L = \frac{N(H_3^+)}{n(H_3^+)}. \tag{8}$$

Thus using (4) we obtain a dimension of dense clouds L_{dense} ~ lpc (3×10^{18}cm) which is in accord with astrophysical estimates.

The problem in diffuse clouds, however, is that since its H_3^+ column density is comparable to dense clouds, Eqs. (7) and (8) give $L_{diffuse}$ = (10^2~10^3) L_{dense} = 100pc~1000 pc. This is at least one order of magnitude higher than the canonical dimensions of diffuse clouds. If this is true, clouds occupy a sizeable fraction of the total distance from us to stars. Since clouds are not in general pencil shaped, our galaxy must be filled with H_3^+! Although this is a dream of H_3^+ aficionados like me, it runs into trouble with other observational evidence. For one thing such a huge low density, low temperature cloud cannot be in pressure equilibrium with surrounding gas.

2.4 More Details and What Ifs

Since hydrogen atoms are not all in the form of H_2 and carbon C^+, Eq. (5)

can be elaborated as

$$n(H_3^+)_{diffuse} = \frac{\varsigma}{k_e} \bullet \frac{f}{2} \bullet \frac{1}{1-\alpha} \bullet \frac{1}{z_C} \tag{9}$$

where f is a fraction of hydrogen atom in H_2,

$$f \equiv \frac{2n(H_2)}{2n(H_2)+n(H)} = \frac{2n(H_2)}{n(\Sigma H)},$$

α is a fraction of carbon not in the form of C^+, and $z_C \equiv n(\Sigma C)/n(\Sigma H)$ is a fraction of total carbon in the gas phase. In our discussion in 2.2 and 2.3 we assumed $f = 1$ and $\alpha = 0$. By definition f is lower than 1 but cannot be much lower since H_2 is needed to produce H_3^+. Also a lower value of f simply makes the problem worse. There are electrons released from atoms other than carbon. This effect is not included in Eq. (9) but again it makes the problem worse. Solution of this mystery has to come from one of the four parameters in Eq.(9), i.e., k_e, ζ, α or z_C.

2.4.1 Rate constant of dissociative recombination; k_e

Clearly the simplest way out of this puzzle is to use the value of k_e that is lower by one to two orders of magnitude. However, in view of the good agreement of the laboratory values of $k_e \sim 10^{-7}\text{cm}^3\text{s}^{-1}$ at T=300, obtained by several different techniques, I do not want to use this escape route easily. The values of k_e are discussed in the next section. The value of k_e is smaller if the temperature of electrons is very high in interstellar space. This is highly unlikely, however, because of the large cross section of electron-neutral collisions, which thermalize electrons rapidly.

2.4.2 Ionization rate; ζ

The puzzle can be explained if the ionization rate is one to two orders of magnitude higher than the canonical value of 10^{-17}s^{-1}. Since our argument is based on a comparison of dense and diffuse clouds, this difference has to be between the two categories of clouds. There is a possibility that a soft component of cosmic ray is shielded in dense clouds but penetrates the diffuse clouds, making the ionization rate higher in diffuse clouds than in dense clouds. This is a very interesting astrophysical problem. There are chemical arguments based on observational abundance of HD by O'Donnell and Watson[11] and OH by Hartquist et al.[12] that constrain the value of ζ to $\sim 10^{-17}\text{s}^{-1}$ although they may not be conclusive.

When H_3^+ was discovered in Cygnus OB2 No. 12 by McCall et al.,[13] Black[14] proposed that photo-ionization by intense X-ray from the extraordinarily bright star may make the ionization rate higher by two orders of magnitude. Gredel, Black and Yan[15] showed that the greatly enhanced ionization rate of $\zeta = (0.6–3)\times10^{-15}s^{-1}$ consistently explains the observations of C_2 and CN as well as our observation of H_3^+. However, this will not be a general solution since other stars that show large column densities of H_3^+ such as HD183143 and HD20041 are more ordinary stars.

2.4.3 States of carbon; α and z_C

The puzzle can be explained if z_C is much less than its canonical value due to depletion of carbon onto grains and large carbon molecules. This is not very likely in view of similar C/H ratio of 1.4×10^{-4} observed in diffuse clouds[16] and dense clouds.[17] The puzzle can be explained also if $\alpha \approx 1$, that is, if carbon atoms are mostly in neutral species. We have already confirmed observationally that a very small fraction of carbon atoms are in the form of CO in diffuse clouds;[13,10] therefore, carbon has to be in the form of a neutral atom. A model calculation by van Dishoeck and Black[18] shows it is unlikely to have carbon as a neutral atom without much CO or C^+.

Both of these possibilities can be checked by observing spectra of C^+ and C directly. I am just writing a proposal to the Hubble Space Telescope to do these observations.

3. DISSOCIATIVE RECOMBINATION OF H_3^+

Since the discovery of the high efficiency of dissociative recombination by Biondi and Brown[19] in plasmas of H_2, N_2 and O_2, and Bates's[20] theoretical estimate of dissociative recombination rate constants to be $10^{-7}cm^3s^{-1}$, a great many experiments have been conducted using many different experimental techniques and many theoretical works have been published. I summarize below experimental results published on H_3^+ recombination.

3.1 Experimental Values of H_3^+ Recombination Rate Constant

Dissociative recombination rate constants, k_e, at T=300K are listed in Table 2.

The period of 1947–1955 was that of discovery and establishment of the fast rate of dissociative recombination using microwave afterglow (MA). The measurements were done under high pressure of 1 to 50 Torr where some H_3^+ are clustered in the form of cluster cations $H_3^+(H_2)_n$. New experimental techniques of inclined beam (IB) and merging beam (MB), ion-trap (IT), and flowing afterglow (FA) were introduced from 1973–1984 which all gave k_e

on the order of $10^{-7}cm^3s^{-1}$. The period of 1984–1990 saw a sudden plunge of k_e by four orders of magnitude to $10^{-11}cm^3s^{-1}$ reported by Smith and Adams[31] using the FA/Langmuir probe (FALP) method. This extremely low value, now retracted by the authors, seriously misled a generation of astrophysical model calculations. In 1988 Amano[33] published a paper titled "Is the dissociative recombination of H_3^+ really slow?" describing his measurement using spectroscopic techniques and restoring k_e to ~$10^{-7}cm^3s^{-1}$. It took a few years, however, before his value was accepted by the community. Since 1992, FALP experimental values have returned to $10^{-7}cm^3s^{-1}$ except the value by Smith which was revised to $10^{-8}cm^3s^{-1}$.[38] The year 1993 saw the advent of the novel and clean technique of merging beam using an ion storage ring by Larsson et al.[39] The value of k_e given in Eq. (6) has been obtained by this method and is now being used in model calculations. More in-depth review has been given by Larsson.[3] Very recently Glosík et al.[43] reported a low value of $<1.3\times10^{-8}cm^3s^{-1}$ and the value is getting even lower[44] to $<1.3\times10^{-9}cm^3s^{-1}$. This plunge of the upper limit of k_e is reminiscent of the period of 1984–1990. Does history repeat itself?

3.2 Relation to the Mystery

The mystery of abundant H_3^+ in the diffuse interstellar medium will be immediately lessened if we adopt the value of Glosík et al.[43] However, I cannot accept their value easily since I do not understand the drastic dependence of k_e on number density of H_2 reported by them. They note that k_e changes from 1.3×10^{-8} to $1.5\times10^{-7}cm^3s^{-1}$ when H_2 number density is changed from 1×10^{11} to $2\times10^{12}cm^{-3}$. Such a drastic pressure dependence has not been reported in previous works except those in the earliest papers using microwave afterglow (MA) under higher pressure.[19,21-23] The values of k_e listed in Table 2 have been measured over a wide range of pressure, from the ultrahigh vacuum of a storage ring[39] to near 1 Torr pressure of a hollow cathode.[33] The approximate constancy of the k_e value over this wide range of number density (about 10 orders of magnitude) is clearly in conflict with the drastic dependence of Glosík et al.[43] The dependence is hard to explain theoretically also since three-body collisions with H_2 are not likely to be effective at the density of $10^{12}cm^{-3}$. Glosík et al.[43] stress "the excellent agreement" of their result and that of Smith and Španel,[38] but the latter experiment was conducted at an H_2 density of $2\times10^{14}cm^{-3}$.

Merging beam experiments of Mitchell's group[32,36] conducted during the "depression" period gave ~$10^{-8}cm^3s^{-1}$ but the value has been revised to $(1.2\pm0.2)\times10^{-7}cm^3s^{-1}$ (quoted in Laubé et al.[42]).

At this stage $10^{-7}cm^3s^{-1}$ at 300K and the value given in Eq.(5) are the experimental results to be adopted in astrophysical calculations. There remains uneasiness, however, since the laboratory experiments cannot be

Table 2. Experimental rate constants of dissociative recombination of H_3^+

k_e (in $10^{-7} cm^3 s^{-1}$)	Method[a]	Authors	Year
25[b]	MA	Biondi, Brown[19]	1949
20[b]	MA	Richardson, Holt[21]	1951
3.4[b]	MA	Varnerin[22]	1951
<0.3[b]	MA	Persson, Brown[23]	1955
2.3	MA/MS	Leu, Biondi, Johnsen[24]	1973
2.5	IB	Peart, Dolder[25]	1974
2.1	MB	Auerbach et al.[26]	1977
1.5	IT	Mathur, Khan, Hasted[27]	1978
4.2	MB	McGowan et al.[28]	1979
1.6	FA	MacDonald, Biondi, Johnsen[29]	1984
<0.2	FALP	Adams, Smith, Alge[30]	1984
≤0.0001	FALP[c]	Smith, Adams[31]	1987
0.2	MB	Hus et al.[32]	1988
1.8	IR	Amano[33]	1988
≤0.0001	FALP	Adams, Smith[34]	1989
≤0.001	FALP	Smith, Adams, Ferguson[35]	1990
0.2	MB	Yousif et al.[36]	1991
1.5	FALP/MS	Canosa et al.[37]	1992
0.1~0.2	FALP	Smith, Ŝpanel[38]	1993
1.15	SR[d]	Larsson et al.[39]	1993
<2	IR/MS	Féher, Rohrbacher, Maier[40]	1994
	FALP	Gougousi, Johnsen, Golde[41]	1995
0.78	FALP/MS	Laubé et al.[42]	1998
<0.13	ISA	Glosík et al.[43]	2000
<0.03	ISA	Glosík[44]	2001

[a] MA: Microwave afterglow
IB: Inclined beam
IT: Ion trap
LP: Langmuir probe
SR: Storage ring
MS: Mass Spectroscopy
MB: Merged beam
FA: Flowing afterglow
IR: Infrared spectrum
ISA: Integrated stationary afterglow

[b] These are measurements under fairly high pressures of 1 to 50 Torr. Under high pressure H_3^+ is clustered with H_2 in the form of $H_3^+(H_2)_n$.

[c] Adams and Smith advocated the low value also in refs. 45 and 46.

[d] There are other storage ring results that agree with this.[47]

conducted in the low density, field free condition of interstellar space—the limitation laboratory astrophysicists always experience in trying to mimic nature. Even the cleanest ion storage ring experiment is conducted with an electron number density of $10^{-7}cm^{-3}$ and in a magnetic field of 300 Gauss.[48,49] The effect of stray Coulomb fields from other electrons and relativistic electric fields (although the magnetic field is parallel to ion velocity to a good approximation.) may not be completely dismissed. Hence the need of theoretical confirmation and the remark of Bates et al.[5] quoted earlier in Section 1.

4. THEORISTS, HELP!

Since the prescient paper by Bates and Massey[50] on the basic reactions in the upper atmosphere, the paper by Bates[51] ascribing the rapid neutralization of He to dissociative recombination to He_2^+, and the paper by Bates[20] on the estimate of k_e, a great many theoretical papers have been published. Theoretical results on H_3^+, however, seem to be in an uncertain state. Michels and Hobbs[52] presented theory that supported Adams and Smith and this influenced the acceptance of the extremely low k_e value by the community. The most recent results,[4,53] are still more than two orders of magnitude lower than the laboratory value of $10^{-7}cm^3s^{-1}$. Like Michels and Hobbs, they are based on the assumption of C_{2v} symmetry of H_3 during the dissociative process which is not in accord with the recent experimental results of Datz et al.[54] and Strasser et al.[55]

Dissociative recombination of H_3^+ in interstellar space is a clean, well defined process. The enigma is a fundamental problem with wide impact on various areas of astrophysics. In its generality this enigma of H_3^+ is reminiscent of the problem of its isoelectronic sibling H^- and the opacity of the Sun; observation and theory by Massey and Bates[56] were orders of magnitude off till it was finally clarified by Chandrasekhar and Breen.[57] This clarified the very general problem related to opacity of all main sequence stars. Likewise a conclusive theory of H_3^+ recombination will clarify a major uncertainty of interstellar chemistry. Extensive theoretical calculations are eagerly awaited.

Finally a word of caution. This astronomical mystery should not be taken to constrain the value of k_e in any definitive way. The assumed canonical astrophysical parameters may be wrong. It is even possible that clouds containing H_3^+ are indeed very large. *What we need is the TRUE value of k_e and not a value, which explains the problem quickly.* To this end scientists in different disciplines need to work together. Physicists, chemists and astronomers of the world, unite!

ACKNOWLEDGEMENT

I wish to thank B. J. McCall, M. Larsson, and J. Glosík for reading this paper. Research in this paper has been supported by NSF Grant PHY-0099442 and NASA grant NAG5-9583.

REFERENCES

1 B. J. McCall and T. Oka, Enigma of H_3^+ in Diffuse Interstellar Clouds, this volume, p.365.
2 L. Spitzer, Jr. and M. G. Tomasko, Astrophys. J. **152**, 971 (1968).
3 M. Larsson, Phil. Trans. R. Soc. Lond. A **358**, 2433 (2000)
4 A. E. Orel, I. F. Schneider, and A. Suzor-Weiner, Phil. Trans. R. Soc. Lond. A **358**, 2445 (2000).
5 D. R. Bates, M. F. Guest, and R. A. Kendall, Planet. Space Sci. **41**, 9 (1993).
6 T. R. Geballe, Phil. Trans. R. Soc. Lond. A **358**, 2503 (2000).
7 V. G. Anicich and W. T. Huntress, Jr., Astrophys. J. **62**, 553 (1986).
8 G. Sundström, J. R. Mowat, H. Danared, S. Datz, L. Broström, A. Filevich, A. Källberg, S. Mannervik, K. G. Rensfelt, P. Sigray, M. af Ugglas and M. Larsson, Science **263**, 785 (1994).
9 B. J. McCall, T. R. Geballe, K. H. Hinkle, and T. Oka, Astrophys. J. **522**, 338 (1999).
10 B. J. McCall, K. H. Hinkle, T. R. Geballe, G. H. Moriarty-Schieven, N. J. Evans, II, K. Kawaguchi, S. Takano, V. V. Smith, and T. Oka, Astrophys. J. submitted.
11 E. J. O'Donnell and W. D. Watson, Astrophys. J. **191**, 89 (1974).
12 T. W. Hartquist, J. H. Black, and A. Dalgarno, Mon. Not. R. Astr. Soc. **185**, 643 (1978).
13 B. J. McCall, T. R. Geballe, K. H. Hinkle, and T. Oka,, Science **279**, 1910 (1998).
14 J. H. Black, Phil. Trans. R. Soc. Lond. A **358**, 2515 (2000).
15 R. Gredel, J. H. Black, and M. Yan, Astron. & Astrophys. **375**, 553 (2001).
16 J. A. Cardelli, D. M. Meyer, M. Jura, and B. D. Savage, Astrophys. J. **467**, 334 (1996).
17 J. H. Lacy, R. Knacke, T. R. Geballe, and A. Tokunaga, Astrophys. J. **428**, L69 (1994).
18 E. F. van Dishoeck and J. H. Black, Astrophys. J. **334**, 771 (1988).
19 M. A. Biondi and S. C. Brown, Phys. Rev. **76**, 1697 (1949).
20 D. R. Bates, Phys. Rev. **78**, 492 (1950).
21 J. M. Richardson, and R. B. Holt, Phys. Rev. **81**, 153 (1951).
22 L. J. Varnerin Jr., Phys. Rev. **84**, 563 (1951).
23 K-B. Persson and S. C. Brown, Phys. Rev. **100**, 729 (1955).
24 M. T. Leu, M. A. Biondi, and R. Johnsen, Phys. Rev. **A8**, 413 (1973).
25 B. Peart and K. T. Dolder, J. Phys. B **7**, 1948 (1974).
26 D. Auerbach, R. Cąčak, R. Caudano, T. D. Gaily, C. J. Keyser, J. Wm. McGowan, J. B. A. Mitchell, and S. F. J. Wilk, J. Phys. B **10**, 3797 (1977).
27 D. Mathur, S. U. Khan, and J. B. Hasted, J. Phys. B **11**, 3615 (1978).
28 J. Wm. McGowan, P. M. Mul, V. A. D'Angelo, J. B. A. Mitchell, P. Defrance, and H. R. Froelich, Phys. Rev. Lett. **42**, 373 (1979).
29 J. A. MacDonald, M. A. Biondi, and R. Johnsen, Planet. Space Sci. **32**, 651 (1984).
30 N. G. Adams, D. Smith, and E. J. Alge, Chem. Phys. **81**, 1778 (1984).
31 D. Smith and N. G. Adams, J. Chem. Soc. Faraday Trans. **II83**, 149 (1987).
32 H. Hus, F. Yousif, A. Sen, and J. B. A. Mitchell, Phys. Rev. A **38**, 658 (1988).
33 T. Amano, Astrophys. J. **329,** L121 (1988).

34 N. G. Adams and D. Smith, in *Dissociative Recombination: Theory, Experiment and Applications*, ed. by J. B. A. Mitchell and S. L. Guberman, (World Scientific, Singapore, 1989), pp. 124–140.
35 D. Smith, N. G. Adams, and E. E. Ferguson, in *Molecular Astrophysics*, ed. by T. W. Hartquist (Cambridge University Press, Cambridge, 1990), pp. 181–210.
36 F. B. Yousif, P. J. T. van den Donk, M. Orakzai, and J. B. A. Mitchell, Phys. Rev. A **44**, 5653 (1991).
37 A. Canosa, J. C. Gomet, B. R. Rowe, J. B. A. Mitchell, and J. L. Queffelec, J. Chem. Phys. **97**, 1028 (1992).
38 D. Smith and P. Španel, Chem. Phys. Lett. **211**, 454 (1993).
39 M. Larsson, H. Danared, J. R. Mowat, P. Sigray, G. Sundström, L. Broström, A. Filevich, A. Källberg, S. Mannervik, K. G. Rensfelt, and S. Datz, Phys. Rev. Lett. **70**, 430 (1993).
40 M. Féher, A. Rohrbacher, and J. P. Maier, Chem. Phys. **185**, 357 (1994).
41 T. Gougousi, R. Johnsen, and M. F. Golde, Int. J. Mass. Spectrom. Ion Proc. **149/150**, 131 (1995).
42 S. Laubé, A. Le Padellec, O. Sidko, C. Rebrion-Rowe, J.B.A. Mitchell, and B. R. Rowe, J. Phys. B **31**, 2111 (1998).
43 J. Glosík, R. Plašil, V. Poterya, P. Kudrna, and M. Tichý, Chem. Phys. Lett. **331**, 209 (2000).
44 J. Glosík, this volume.
45 N. G. Adams and D. Smith, in *Astrochemistry IAP Symp.120,* ed. by M. S. Vardya and S. P. Trafadar (Reidel, Dordrecht, 1987), pp 1-18.
46 N. G. Adams and D. Smith, in *Rate Coefficients in Astrochemistry,* ed. by T. J. Millar and D. A. Williams (Kluwer, Dordrecht, 1988), pp 173-192.
47 T. Tanabe, et al., in *Dissociative Recombination: Theory, Experiment and Applications IV,* ed. by M. Larsson, J. B. A. Mitchell, and I. F. Schneider (World Scientific, Singapore, 2000), pp. 170–179.
48 M. Larsson, H. Danared, Å. Larson, A. Le Padellec, J. R. Peterson, S. Rosén, J. Semaniak, and C. Strömholm, Phys. Rev. Lett. **79**, 395 (1997).
49 A. Le Padellec, M. Larsson, H. Danared, Å. Larson, J. R. Peterson, S. Rosén, J. Semaniak, and C. Strömholm, Phys. Scripta **57,** 215 (1998).
50 D. R. Bates and H. S. W. Massey, Proc. Roy. Soc. A **187**, 261 (1946).
51 D. R. Bates, Phys. Rev. **77**, 718 (1950)
52 H. H. Michels and R. H. Hobbs, Astrophys. J. **286**, L27 (1984).
53 I. F. Schneider, A. E. Orel, and A. Suzor-Weiner, Phys. Rev. Lett. **85**, 3785 (2000).
54 S. Datz, G. Sundström, Ch. Biederman, L. Brodström, H. Danared, S. Mannervik, J. R. Mowat, and M. Larsson, Phys. Rev. Lett. **74,** 896 (1995).
55 D. Strasser, L. Lammich, S. Krohn, M. Langer, H. Kreckel, J. Levin, D. Schwalm, Z. Vager, R. Wester, A. Wolf, and D. Zajfman, Phys. Rev. Lett. **86,** 779 (2001).
56 H. S. W. Massey and D. R. Bates, Astrophys. J. **91**, 202 (1940).
57 S. Chandrasekhar and F. H. Breen, Astrophys. J. **104**, 430 (1946).

Importance of Jahn-Teller Coupling in the Dissociative Recombination of H_3^+ by Low Energy Electrons

Hyperspherical Approach

C. H. Greene, V. Kokoouline, B. D. Esry[†]
Department of Physics and JILA, University of Colorado, Boulder, Colorado 80309-0440, USA
†Department of Physics, Cardwell Hall, Kansas State University, Manhattan,Kansas 66506,USA
chg@jilacg.Colorado.edu

1. INTRODUCTION

Dissociative recombination (DR) of the H_3^+ ion, namely the process

$$H_3^+ + e^- \rightarrow H_2 + H \text{ or } H + H + H, \qquad (1)$$

has attracted extensive attention since the discovery of H_3^+ in diffuse interstellar clouds.[1] But despite significant experimental[2,3] and theoretical[4,5] progress, the mechanism of the Reaction (1) with slow electrons is still poorly understood. For one thing, there is no theoretical model compatible with the large rate constant of Reaction (1). For another, there is no consistent explanation of the fact that different experiments yield such different DR rates.

From a theoretical point of view the analysis of the process is greatly complicated by two factors. First, the analysis should take into account at least one ionic potential surface and the infinite number of Rydberg states attached to this ionic surface. Second, all these ionic and Rydberg states are three-dimensional. Therefore the dissociative dynamics should be treated in three dimensions, something that has almost never been attempted for a target ion with three atoms or more.

Several years ago, Stephens and Greene[6,7] successfully interpreted the experimental photoabsorption spectrum of a metastable state of H_3.[8] The broadest resonances in the spectrum were found to decay via Jahn-Teller coupling (JTC). Scattering of an electron from one rovibrational state into another in that treatment was described by quantum defect theory, combined with a clamped-nuclei electronic reaction matrix whose structure parallels the structure of the Longuet-Higgins Jahn-Teller Hamiltonian. The corresponding scattering S (or reaction K) matrix elements then emerge after applying a frame transformation.[9,10]

The success of the earlier work[7] encouraged us to include JTC into a nonperturbative three dimensional description of DR in H_3^+. Accordingly, in this work JTC is represented in the same way using a quantum-defect reaction matrix. The process of recombination and subsequent dissociation is then described in two steps, which map onto methods as familiar in the description of DR for diatomic molecules. First, a complete set of effective *one-dimensional* potential curves of H_3 and H_3^+ is calculated in an adiabatic hyperspherical description of the full three-dimensional dynamics. The resulting adiabatic hyperspherical potential curves for H_3 have autoionization widths that are usually nonzero. The only difference (but an important one) compared to the diatomic case is the fact that one-dimensional curves of H_3^+ are obtained from the initial three-dimensional H_3^+ surface by solving the clamped-hyperradius Schrödinger equation in the space of two hyperangles. This solution must be carried out for many different values of hyperradius. In short, the hyperradius in the case of H_3^+ or other polyatomics can be viewed as a generalization of the internuclear distance in diatomics. H_3 potential curves are then obtained from H_3^+ curves using multichannel quantum defect techniques. Once the potential curves of H_3^+ and H_3 are known, we apply the O'Malley method[11] to calculate the DR cross-section. This method has been used for DR in diatomic molecules. We present in this paper a test of the new method, in which we recalculate the spectrum of H_3 that has good experimental and theoretical results for comparison, e.g. Ref.'s 7 and 12.

2. JAHN-TELLER EFFECT IN H_3

Jahn-Teller coupling is a generic feature of highly symmetric polyatomic molecules. Some insight into its nature emerges by considering the H_3^+ molecule.

When the nuclei of H_3^+ are clamped in their most probable equilateral triangle (C_{3v} symmetry) configuration, the electronic Hamiltonian of an H_3 Rydberg electron belongs to the symmetry group D_{3h}. Among the set of D_{3h} symmetry electronic energy eigenstates, there exist states that belong to a two-dimensional irreducible representation, labeled E. This means that for the C_{3v} nuclear geometry there are two states with the same electronic energy,

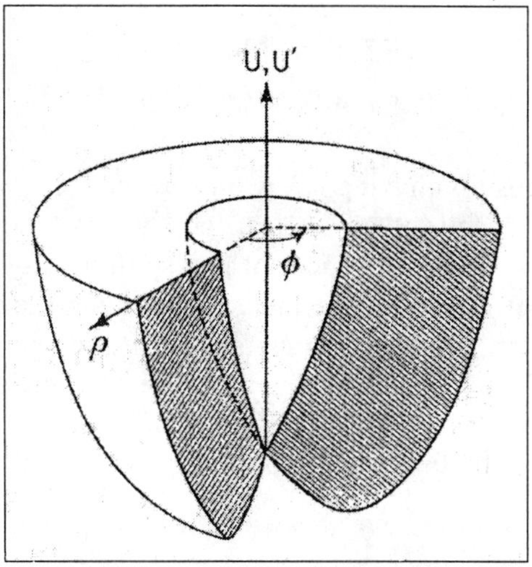

Figure 1. Two E electronic states coupled by the Jahn-Teller interaction. The figure is taken from Ref. 13.

and any linear combination of these states still remains an eigenstate of the Hamiltonian. However, any distortion of the nuclei away from that equilateral triangle C_{3v} nuclear symmetry breaks this exact D_{3h} symmetry of the electronic eigenstates. In particular, electronic states of E symmetry are no longer exact eigenstates of the Born-Oppenheimer Hamiltonian. Nevertheless, it is convenient to retain the E-state labels for small distortions away from equilibrium, provided the coupling of the two states is incorporated explicitly. Since the lifting of the D_{3h} symmetry causes a coupling between the two E symmetry basis states, the two fixed-nuclei eigenenergies for *almost* E eigenstates of the Hamiltonian are different. For the Born-Oppenheimer potential surfaces, this effect shows up as an energy splitting between the two "E" states at distorted nuclear geometries. We retain the notation E for such "almost E" states, under the assumption that distortions from D_{3h} symmetry are small in the region where DR capture takes place.

Potential surfaces of the E states are schematically shown on Fig. 1. Here, different nuclear configurations are described by normal coordinates, (Q_1, Q_a, Q_b), of the equilateral symmetry. Q_1, the symmetric stretch, is kept constant. A change of Q_1 retains the D_{3h} (C_{3v}) symmetry, and thus produces no Jahn-Teller coupling. The graph is made for one particular value of Q_1. Two other coordinates are varied. They both break the D_{3h} (C_{3v}) symmetry. It is more convenient to use not the pair (Q_a, Q_b), but the polar phase, φ, and radius, ρ, determined by them $(Q_a, Q_b) \rightarrow (\rho, \varphi)$:

$$\rho = \sqrt{Q_a^2 + Q_b^2}$$
$$\phi = \arctan(Q_b / Q_a) \qquad (2)$$

For the lone position having C_{3v} geometry, namely ρ=0, the two E states are degenerate. For all other nuclear geometries, they have an energy splitting. The physical consequences of this splitting effect are known as the static Jahn-Teller effect. When the molecule vibrates, the coupling influences the dynamics, which is referred to as the dynamical Jahn-Teller effect. One can see that JTC is a linear effect near the C_{3v} nuclear geometry. It is important to mention that the C_{3v} configuration is not stable, since the lower of potential surfaces on Fig. 1 has a local maximum at that point.

3. THEORETICAL APPROACH

3.1 Hyperspherical Method

The hyperspherical approach we employed in the present study was described in detail in our previous paper.[14] We recall here important points of the hyperspherical step of our treatment.

Our theoretical description begins by treating the ionic vibrational dynamics on the Jaquet H_3^+ electronic potential surface[15] within the adiabatic hyperspherical approximation.[16] The hyperspherical radius, R, is a useful collective coordinate for any few-body system: for three atoms of equal mass, the squared hyperradius is defined by $R^2 = \sqrt{3}\left(r_1^2 + r_2^2 + r_3^2\right)$, where r_1, r_2, r_3 are the distances of each nucleus from the center of mass. The adiabatic hyperspherical potential curves, $U_i^+(R)$, for the H_3^+ ion are energy eigenvalues of the fixed-R ionic nuclear motion Schrödinger equation,

$$H_{R=\text{const}}\,\Phi_i(\theta,\phi;R) = U_i^+(R)\Phi_i(\theta,\phi;R) \qquad (3)$$

Here (R,θ,ϕ) are the nuclear coordinates, rewritten in terms of a hyperradius R and two hyperangles θ,ϕ. The potential curves $U_i^+(R)$ are shown on Fig. 1 of Ref. 14.

The adiabatic potential curves for the neutral H_3 molecule can be calculated by the same way. Fig. 2 demonstrates two sets of adiabatic potential H_3 curves corresponding to the two lowest potential surfaces of the neutral molecule H_3. Also one (the lowest) hyperspherical curve of H_3^+ is shown together with the lowest vibrational level.

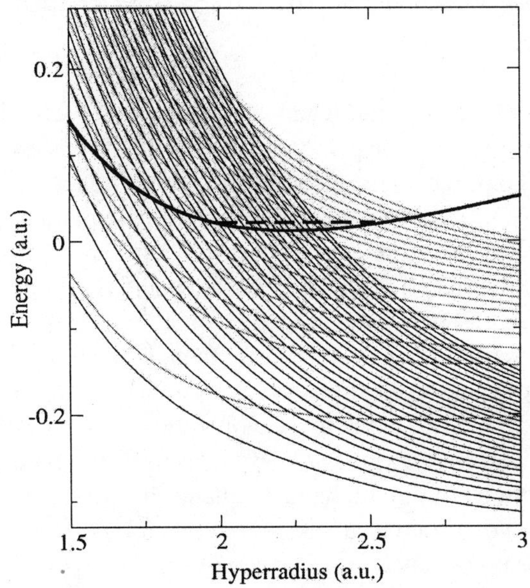

Figure 2. Adiabatic hyper-spherical potential curves of the two lowest surfaces of the H_3 molecule (thin lines) and the lowest hyperspherical curve of the H_3^+ ion (thick full line). Because of the high density of H_3 curves, only one per ten curves is shown. A thick dashed line marks the energy of the ground vibrational level of H_3^+. Two sets of (thin) H_3 potential curves can be distinguished, some of which dissociate to H+H+H, while the others dissociate to H_2+H.

Fig. 2 demonstrates how the process of DR takes place in hyperspherical coordinates. In the three-dimensional space of three nuclear coordinates, the lowest potentials of H_3 do not cross the H_3^+ ground vibrational level. But a crossing is possible in hyperspherical coordinates. This simply corresponds to the fact that a transition probability can be not negligible for a transition between the ground H_3^+ vibrational level and an excited H_3 vibrational level. One can see from Fig. 2 that low H_3 curves do not cross the ground H_3^+ vibrational level, but many excited hyperspherical curves do cross it. One should also remember that Fig. 2 shows only the two lowest electronic states of the H_3 molecule, but the DR process involves an infinite number of Rydberg states, associated with the ion. In order to treat all the Rydberg complexity we use quantum defect theory, where all neutral (Rydberg) states are represented by one ion potential surface and by a set of quantum defect parameters.

3.2 JTC Represented By Multichannel Quantum Defect Theory

A starting point of the multichannel quantum defect theory (MQDT) analysis is a set of ionic hyperspherical potential curves calculated as described above. To describe the electron-ion interaction, a set of *ab initio* H_3 potential surfaces and their mutual couplings are first parameterized at fixed values of the nuclear positions using MQDT[10,12] by several quantum defect parameters. Such a description includes also the Jahn-Teller coupling for all principal quantum numbers and the electronic continuum. The adiabatic hyperspherical potential curves of H_3 are then extracted by searching for autoionizing resonances of the e-H_3^+ system. Each potential curve $U_\beta(R)$, generally exhibits a nonzero autoionization width $\Gamma_\beta(R)$.

Our treatment can be viewed as an application of the familiar vibrational frame transformation of multichannel quantum defect theory that has successfully treated diatomic Rydberg state interactions.[9,10] However, the role of the internuclear distance in diatomic systems is replaced by the hyperradius *R*. The quantum mechanical scattering amplitude for a Jahn-Teller type symmetry changing *p* wave collision, with the nuclei clamped, is written as $S_{\Lambda\Lambda'}(R,\theta,\phi)$. The fixed nuclei scattering matrix, $S_{\Lambda\Lambda'}(R,\theta,\phi)$, represents an amplitude for the reaction:

$$e(1\Lambda') + H_3^+(R,\theta,\phi) \rightarrow e(1\Lambda) + H_3^+(R,\theta,\phi). \quad (4)$$

Here, the electron scatters from one irreducible representation (irrep), Λ',of the D_{3h} symmetry group into another, Λ, while the nuclei do not have time to move.

For a *p* wave electron, the $p\sigma$ component corresponds to the lone D_{3h} irrep A_1, while the $p\pi$ components, $\Lambda = \pm 1$, are associated with the Jahn-Teller active modes of irrep *E*. The Jahn-Teller-induced coupling between the $p\pi$ components can be represented by a 2×2 reaction matrix:[7,17]

$$K = \begin{pmatrix} \delta\rho^2 & \lambda\rho\exp(i\phi) \\ \lambda\rho\exp(-i\phi) & \delta\rho^2 \end{pmatrix}. \quad (5)$$

Here, (λ,δ) are the Jahn-Teller coupling constants.[7,12] In Eq. (5), the constant, equilibrium geometry quantum defect $\mu^0 = \mu_{\Lambda=\pm1}(Q=0)$ is not shown. It is included into the treatment in the way adopted in Ref. 7. The Jahn-Teller and MQDT parameters $(\lambda,\delta,\mu_\Lambda)$ are taken to be independent of the nuclear configuration, which tests show is a good approximation over the range of the vibrational ground state of H_3^+.

An element of the short-range reaction matrix for this process is now calculated by a generalized vibrational frame transformation[9,10] as an integral:

$$\overline{K}_{i\Lambda,i'\Lambda'}(R) = \left\langle\!\left\langle \Phi_i \middle| K_{\Lambda\Lambda'}(R,\theta,\phi) \middle| \Phi_{i'} \right\rangle\!\right\rangle_{(\theta,\phi)} \tag{6}$$

In Eq. (6), the double brackets and the symbol (θ,φ) denote an integral over the hyperangles only, at a fixed R. The dimension of this $\overline{K}$ matrix is $3N \times 3N$, where N is the number of adiabatic eigenstates of H_3^+ that are retained.

4. ANALYSIS OF AN H_3 SPECTROSCOPY EXPERIMENT

JTC[7] together with the quantum defect theory[7] for the description of the interaction between ionic and Rydberg states has allowed for a successful interpretation of the photoabsorption spectrum[8] of H_3 near the lowest ionization thresholds. The rotational and vibrational couplings were described by the frame-transformation formalism.[9] Our approach can be tested by comparing with both this work and a more recent work[12] in another energy region of the H_3 spectrum.

4.1 Theoretical Procedure

In the reaction matrix of Eq. (6) has the electron scattering amplitudes for a transition of the ion from one hyperspherical channel to another, at a fixed hyperradius *R*. When used in conjunction with the one-dimensional adiabatic hyperspherical potential curves, it becomes possible to apply the frame-transformation approach used for diatomic molecules.[9] The full procedure that we use is described below.

After the potential curves $U_i^+(R)$ are determined, we calculate vibrational hyperradial wave functions, $\Psi_{i,v}$ and corresponding energy levels. Each $U_i^+(R)$ potential curve is classified as belonging to a particular nuclear group symmetry (A_1, A_2 and the degenerate E pair). Consequently each $\Psi_{v,i}$ energy eigenstate is assigned a conventional set of vibrational quantum numbers, $\{v_1,v_2^1\}$. The agreement with accurate calculations[15] for vibrational energies calculated in our approach is of order of 10-100 cm^{-1} for the first 20 vibrational levels.

A hyperradial vibrational frame transformation is used to describe the scattering from one vibrational state, characterized by (i,v,Λ), to another state (i',v',Λ'). The resulting reaction matrix has the form:

$$\widetilde{K}_{iv\Lambda,i'v'\Lambda'}(R) = \left\langle\left\langle \Psi_{iv}(R) \middle| \overline{K}_{i\Lambda,i'\Lambda'}(R) \middle| \Psi_{i'v'}(R) \right\rangle\right\rangle_{(R)} . \quad (7)$$

The next two steps are carried out as in the work by Stephens and Greene.[7] First, the equilibrium quantum defect for states with $\Lambda=1$ is introduced into the reaction matrix (Eq. 11).[7] Second, the rotational frame transformation is applied (Eqs. 3, 5, 12-14),[7] which accounts for the phenomenon of l-uncoupling. Owing to lack of space we do not repeat here the formulas used for these stages of treatment, but they can be found in Ref. 7. The resulting full reaction matrix, $K_{iv\Lambda,i'v'\Lambda'}$, for a p-wave electron collision with the triatomic ion then has the structure:

$$\left\langle v, i, N^+, K^+ \middle| K \middle| v', i', N^{+'}, K^{+'} \right\rangle^{(N_f, l_f)} \quad (8)$$

where N^+ and K^+ are the total ion angular momentum (exclusive of spin) and its projection on the molecular symmetry axis, respectively; N_f is the total angular momentum of the ion electron system; l_f is the orbital angular momentum of the excited (Rydberg or continuum) electron. The K matrix is exactly diagonal in N_f and approximately diagonal in the l_f quantum number. Note, that the matrix has the same structure as the K matrix of Ref. 7 except for the labeling of vibrational levels of H_3^+.

Having obtained the K matrix, the physical reaction matrix, K^{phys}, is constructed in the usual way,[9] which amounts to an "elimination" (or enforcement of asymptotic boundary conditions) in the closed channels whose ionic vibrational energy level exceeds the total energy E of interest:

$$K^{phys} = K^{oo} - K^{oc}\left[K^{cc} + \tan\beta(E)\right]^{-1} K^{co} \quad (9)$$

Where the matrix $\tan\beta(E)$ is diagonal with $\beta(E) = \pi/\sqrt{2(E_I - E)}$ and E_I is the energy of the ionic state (ivN^+K^+).

4.2 Results

Fig. 3 compares experimental[12] and calculated spectra for photoionization of an excited p-state of H_3. This Figure is an analogue of the Fig. 5 of Ref. 12 where the experimental spectrum is compared with the calculated one, obtained by the "previous" approach.[6,7] The lower part of our figure is the calculated time-delay spectrum. The Wigner time-delay is determined as a derivative, $d\delta/dE$, where δ is the eigenphase sum of the physical reaction matrix K^{phys}.[9] The three theoretical curves in the lower graph correspond to three different ($N_f = 0, 1, 2$) total angular momenta of the H_3 molecule. In experiment their contributions are summed up.

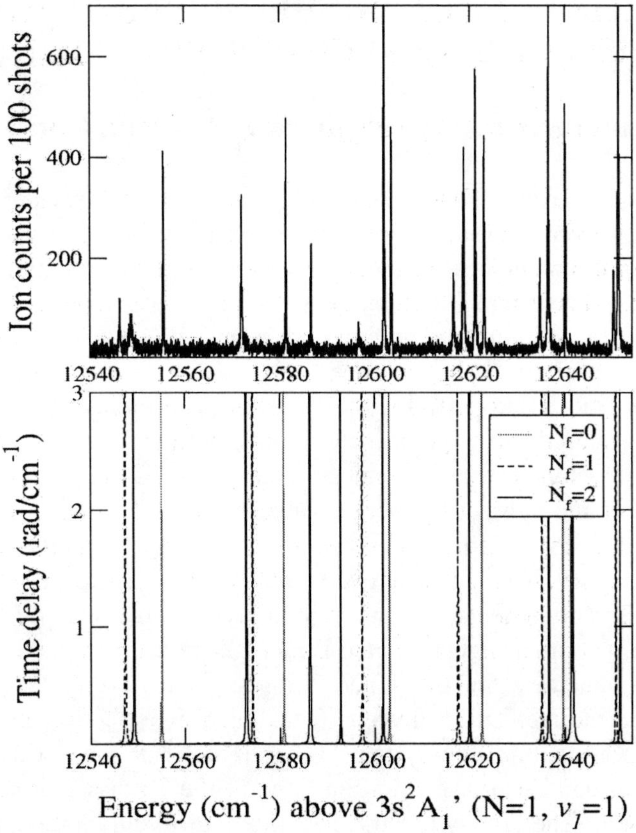

Figure 3. Comparison of the calculated (lower panel) and autoionization experimental[12] spectra of H_3 molecule.

The overall agreement with the experiment is generally quite good in Fig. 3. The agreement is similar over the much wider range studied experimentally, which is not shown here. However, we note that for some experimental levels the agreement is poorer. For several levels, the positions disagree by up to several dozen cm^{-1}. While a more careful analysis of these discrepancies remains desirable, we believe that the present results provide reason to be encouraged about the applicability of the hyperspherical vibrational frame transformation method that has been developed here to describe dissociative recombination of H_3^+.

5. CALCULATION OF THE CROSS-SECTION FOR DISSOCIATIVE RECOMBINATION OF H_3^+

5.1 Autoionizing Hyperspherical Potential Curves of H_3

The reaction matrix, $\overline{K}(R)$ (see Eq. 6), describes the scattering of a p-wave electron from the ionic core at some given value of the hyperradius, *R*. From this point on, our treatment can be viewed as being essentially identical to the familiar theoretical DR techniques that have been applied to diatomics. Specifically, these elements include: a standard MQDT closed-channel "elimination", followed by a time-delay analysis that determines the resonance positions and partial widths at each hyperradius. In principle, our formulation should also permit the application of more sophisticated treatments developed for diatomics,[18] that incorporate the competition between direct and indirect Rydberg state pathways.

In general, each H_3^+ adiabatic potential curve supports an infinite number of autoionizing Rydberg states, at every value of R. The direct contributors to DR are the 2p components, which are the only ones that lead diabatically to dissociation. All other higher excited Rydberg states are energetically closed dissociation channels in this energy range, but nevertheless they can contribute as indirect DR pathways. Once such a Rydberg state is excited during electron capture, the system can oscillate in R many times and can undergo a Landau-Zener transition into one of the 2p dissociative pathways that are open. The 2p symmetry has three different components (with $\Lambda = 0, \pm 1$). The coupled $2p_1$ and $2p_{-1}$ components form two diabatic potential surfaces of H_3 that correlate asymptotically either to H_2+H or to H+H+H. The $2p_0$ state does not lead to a dissociation channel that is open in this sub-eV energy range. Yet an initially captured np_0 electron can be scattered into the $np_{\pm 1}$ states by Coriolis coupling, which gives another indirect pathway to DR although it is a minor contributor to the final rate.

5.2 Cross Section And DR Rate Constant

O'Malley's expression[11] is used to calculate the DR cross section:

$$\sigma = \sum_{\beta} \frac{\pi^2}{E_{el}} \frac{\Gamma_\beta(R_\beta)}{|U'_\beta(R_\beta)|} |\Psi(R_\beta)|^2 \tag{10}$$

The sum in Eq. (10) is over all resonant curves crossing the ground state hyperspherical potential curve for H_3^+. $\Gamma_\beta(R_\beta)$ is the partial autoionization

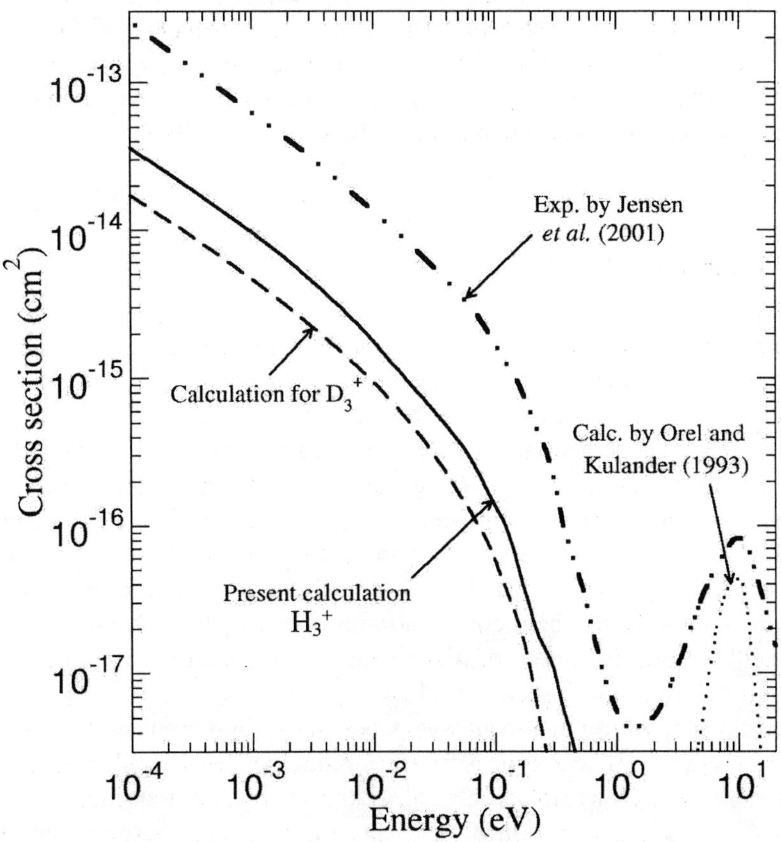

Figure 4. Calculated and experimental DR cross-sections. Full line shows the present calculations. Dash-dotted line is the cross-section obtained in the storage-ring experiment[3]. The doted line presents another theoretical calculations for higher energies. In this graph we give also the DR calculations for the D_3^+ ion (dashed line).

width in the incident channel. E_{el} is the energy of the captured electron in a.u. and $\Psi(R_\beta)$ is the normalized hyperradial wave function of the lowest H_3^+ vibrational level. R_β is the Condon point, i.e. the hyperradius at which the βth curve lies at an energy, E_{el}, above the H_3^+ ground state energy. $U'_\beta(R_\beta)$ is the derivative of the excited state H_3 hyperspherical potential curve into which capture occurs. More details about how the cross-section is calculated are given in Ref. 14.

For a thermal distribution of incident electron energies, the DR rate coefficient $\alpha = \langle v\sigma \rangle$ is defined as

$$\alpha(kT) = \frac{8\pi}{(2\pi kT)^{3/2}} \int_0^\infty \sigma(E) e^{-E/kT} EdE \qquad (11)$$

The calculated rate constant, which includes all indirect paths is 1.2×10^{-8} cm^3s^{-1} at 300 K electron temperature. Experimental values for $\alpha(300K)$ are $<1.32\times10^{-8}$ cm^3s^{-1} for a flowing-afterglow experiment[19] and $1\text{-}1.15\times10^{-7}$ cm^3s^{-1} for storage ring experiments.[2,3] A more complete comparison with earlier experiments can be found in Refs. 20 and 21.

6. CONCLUSION

In the present paper we show that the Jahn-Teller distortion effect plays an important role in the dissociative recombination of the H_3^+ ion. A new approach is proposed for the treatment of this process.

The approach can be viewed as a two-stage treatment. First, we calculate one-dimensional hyperspherical potential surfaces of H_3^+ using the full three-dimensional H_3^+ potential. One-dimensional potentials of H_3 are calculated using the H_3^+ curves and familiar quantum defect theory. At this stage the Jahn-Teller coupling is included. H_3 curves calculated in this way already contain information about the recombination process. At the second stage of the treatment we use the theory of dissociative recombination developed for diatomics to describe the dissociation of H_3.

The second step of the treatment was widely used for diatomics and it was proven to be quite efficient. The first step, reducing the three-dimensional dynamics to the one-dimensional one, is less known. For this reason, as a test, we analyze an H_3^+ autoionization experiment.[12] This experiment is successfully interpreted using the full three dimensional analysis. Good agreement is found with the experiment and with the full treatment.[12] Thus, we show that the hyperspherical approach gives reliable results and can be used in other cases.

The rate and cross section of the dissociative recombination of H_3^+ are found to be in the range of the microwave afterglow results and about an order of magnitude smaller than the storage ring experiments.

ACKNOWLEDGEMENTS

This work was supported by the National Science Foundation. Computational work was carried out at the National Energy Research Supercomputer Centre at Lawrence Berkeley National Laboratory using the Bulk Synchronous Parallel (BSP) library.

REFERENCES

1 T. Oka, Phil. Trans. R. Soc. Lond. A **358**, (2000).
2 G. Sundstrom, J.R. Mowat, H. Danared, S. Datz, L. Brostrom, A. Filevich, A. Kallberg, S. Mannervik, K.G. Rensfelt, P. Sigray, M. Afugglas, and M. Larsson, Science **263**, 785 (1994).
3 M.J. Jensen, H.B. Pedersen, C.P. Safvan, K. Seiersen, X. Urbain, and L.H. Andersen, Phys. Rev. A **63**, 052701 (2001).
4 A.E. Orel and K.C. Kulander, Phys. Rev. Lett. **71,** 4315 (1993).
5 A.E. Orel, I.F. Schneider, and A. Suzor-Weiner, Phil. Trans. R. Soc. Lond. A **358**, 3293 (2000).
6 J.A. Stephens and C.H. Greene, Phys. Rev. Lett. **72**, 1624 (1994).
7 J.A. Stephens and C.H. Greene, J. Chem. Phys. **102**, 1579 (1995).
8 M.C. Bordas, L.J. Lembo, and H. Helm, Phys. Rev. A **44**, 1817 (1991).
9 C.H. Greene and Ch. Jungen, Adv. At. Mol. Phys. **21**, 51 (1985).
10 Ch. Jungen, *Molecular Applications of Quantum Defect Theory*, Institute of Physics (Bristol, 1996).
11 T.F. O'Malley, Phys. Rev. **150**, 14 (1966).
12 I. Mistrík, R. Reichle, U. Muller, H. Helm, M. Jungen, and J.A. Stephens, Phys. Rev. A **61**, 033410 (2000).
13 H.C. Longuet-Higgins, in Advances in Spectroscopy (Interscience, New York, 1961), Vol. II, p.429.
14 V. Kokoouline, C.H. Greene, and B.D. Esry, Nature **412**, 891 (2001).
15 R. Jaquet, W. Cencek, W. Kutzelnigg, and J. Rychlewski, Chem. Phys. **108**, 2837 (1998).
16 C.D. Lin, Phys. Rep. **257**, 2 (1995).
17 A. Staib and W. Domcke, Z. Phys. D **16**, 275 (1990).
18 A. Giusti, J. Phys. B **13**, 3867 (1980).
19 J. Glosík, R. Plasil, V. Poterya, P. Kurdna, and M. Tichý, Chem. Phys. Lett. **331**, 209 (2000).
20 M. Larsson, Phil. Trans. R. Soc. Lond. A **358**, 2433 (2000).
21 S. Laubé, A. Le Padellec, O. Sidko, C. Rebrion-Rowe, J.B.A. Mitchell, and B.R. Rowe, J. Phys. B: At. Mol. Opt. Phys. **31**, 2111 (1998).

A Model for Calculating Branching Ratios in H_3^+ Dissociative Recombination

D. Strasser,[1] J. Levin,[2] H. B. Pedersen,[1] O. Heber,[1] A. Wolf,[2] D. Schwalm,[2] and D. Zajfman[1,2]

[1]*Department of Particle Physics, Weizmann Institute of Science, Rehovot, 76100, Israel*
[2] *Max-Planck Institut für Kernphysik, D-69019 Heidelberg, Germany.*
fndaniel@wicc.weizmann.ac.il

1. INTRODUCTION

The dissociative recombination (DR) of molecular ions is a key process for the formation of complex molecules in dense interstellar clouds.[1,2] Among all species, H_3^+ attracts particular interest owing to its key role in astrochemistry.[3] Moreover, the low energy DR rate for H_3^+ has been a controversial issue for many years, and there are still unexplained differences reaching orders of magnitude among the experimental and theoretical results.[3]

For H_3^+, the DR process with low kinetic energy electrons can be depicted as (cf. insert in Fig. 1)

$$H_3^+ + e^-(E) \rightarrow \begin{cases} H(1s) + H(1s) + H(1s) & (\alpha) \\ H_2(X^1\Sigma_g)(v) + H(1s) & (\beta) \end{cases} \tag{1}$$

with total product kinetic energies of $E_\alpha = 4.76$ eV and $E_\beta = 9.23$ eV (the latter being given for the vibrational ground state (v = 0) of the H_2 fragment). The first complete branching ratio (BR) measurement for the DR of vibrationally cold H_3^+, as a function of electron kinetic energy *E,* was carried out by Datz et al.[4] The three-body product channel α was found to be dominant for low E with 75% of the total reaction rate and only 25% remaining for the molecular channel β. So far, no explanations for this ratio or for the structure of the BR observed at higher electron energies

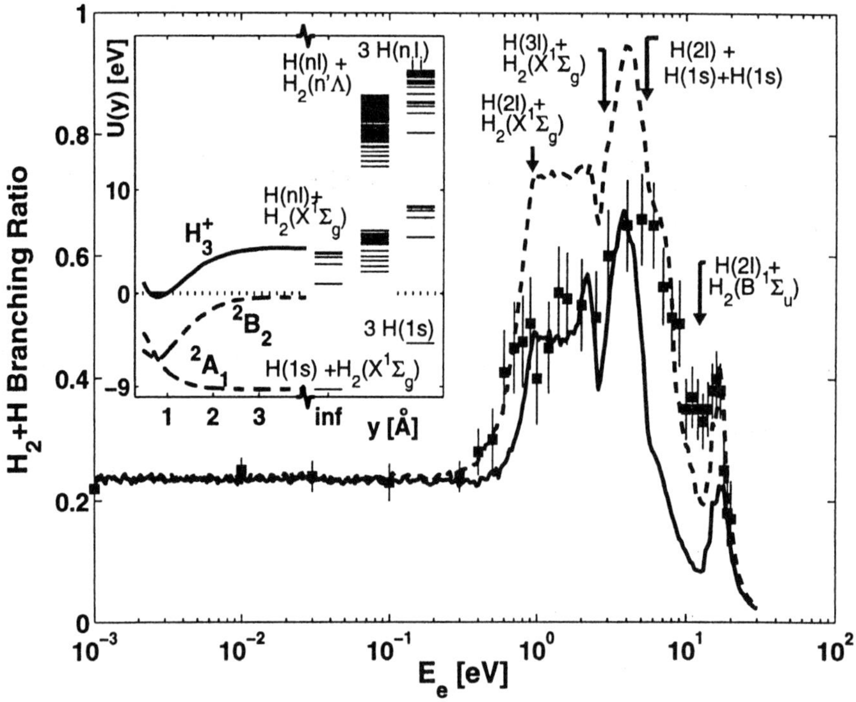

Figure 1. Branching ratio for the two-body channel as a function of electron energy as obtained from the model including all final states (dashed curve) and final states with $\ell=0$ only (full curve). Some important final state threshold energies are also indicated. The model includes rotational excitation of the H_3^+ ions and the experimental electron energy distribution. The inset shows the molecular potentials and final state energies relevant for the DR of H_3^+. Indicated are the H_3^+ ground state (solid curve) and final energies for various electronic excitations (ro-vibrational ground state for all H_2 fragment energies). Also shown are the H_3 potential curves (dashed) correlating to the lowest exit channels, reached at near-zero initial energy.

(~0.4-20 eV) have been given. Detailed experimental studies of the dissociation dynamics following the DR of H_3^+ have been performed recently,[5] yielding also the vibrational level populations of the H_2 fragment produced in channel β. For more complex molecular ions (such as H_2O^+, CH_2^+, H_3O^+, and others[6-11]) DR measurements also showed a strongly dominant many body fragmentation.

It has been seen as an appealing goal for more than two decades to provide a theoretical model capable of explaining these experimental results on a unified basis. The first attempt to predict the DR product branching ratios was performed by Herbst and co-workers,[12,13] applying the statistical phase-space model of chemical reactions. In essence, the BR was obtained by counting all energetically allowed electronic and ro-vibrational states of the neutral products. Galloway and Herbst[14] later modified the phase-space theory, considering the effects of Franck-Condon factors and curve crossings as also emphasized by Bates.[15] However, after comparisons with experimental results for several tetraatomic systems the authors of Ref. 14 reached the conclusion that although phase-space theory can describe some of the DR branching ratios, it is lacking "general predictive power, free from parametrization."

Here we present a model for calculating the product branching ratios for the DR of polyatomic molecular ions, based on simple statistical assumptions about the redistribution of energy among the dissociating nuclei and about the population of electronic potential surfaces. The development of this model was initially motivated by our recent experimental studies[5] of the dissociation process following electron capture on H_3^+ at the heavy-ion storage ring TSR which are described in the contribution by D. Zajfman.

2. DESCRIPTION OF THE MODEL AND RESULTS

The DR of a molecular ion can in its main features be described as a two step process, a basic picture that we shall adopt also here. In the first step an electron is captured into an electronic state of the neutral characterized by a particular potential energy surface (PES). In the second step, this capture is followed by a rapid dissociation of the system on the PES into an exit channel. Of all electronic states, some (referred to as repulsive) yield only atomic products while others (denoted as bound) may also support molecular fragments. Let us denote by E_i the asymptotic *electronic* energy after fragmentation associated with an energy surface i. Then, at an initial energy E, a total energy $E_{k,i} = E - E_i$ is available for the relative translational motion between the fragments plus all rovibrational motion within those fragments which are molecules. For low-energy recombination of H_3^+, the exit channel of three ground-state H atoms correlates to the $^2B_2$ surface, which represents a repulsive state, while the exit channel $H + H_2$ can be reached via the $^2A_1$ PES of H_3, which represents a bound state. Whereas a repulsive state always yields fully dissociated fragments, a bound state may contribute not only to product channels with bound molecular fragments but, if sufficient energy $E_{k,i}$ is available, with a certain probability also to fully dissociated exit channels. Thus, for any product channel γ we can write its formation probability, P_γ, as a sum over contributions from various electronic pathways i,

$$P_\gamma = A \sum_i \lambda_i P_\gamma^i, \qquad (2)$$

where λ_i denotes the relative contribution of an electronic state i and P_γ^i is the probability that fragmentation on the associated PES leads to the products

γ; A is a normalization factor so that $\sum_\gamma P_\gamma = 1$. To estimate the probability, P^i_γ, we assume that the energy release $E_{k,i}$ is randomized among all the nuclei which evolve on the multidimensional PES, taking into account the standard restrictions of energy and momentum conservation. If the state i is of the bound type and γ denotes a product channel where a subset of the nuclei can form a molecule in an electronic state *j*, we derive the relative kinetic energy distribution for the specific nuclei in this subset and then assume a bound molecule to be formed in all cases where the relative kinetic energy remains below the molecular binding energy $E_{b,j}$. Otherwise atomic fragments are produced and the electronic state contributes to the fully dissociated product channel. For the cases where a molecule is formed, the relative energy distribution for the subset predicts, in addition, the degree of rovibrational excitation for this product.

The corresponding relative kinetic energy distribution for complete phase-space randomization can be obtained in a straightforward way for a dissociating three-body system with a total kinetic energy E_k by using the Dalitz coordinates[16] $\eta_1 = (2E_3 - E_2 - E_1)/3E_k$ and $\eta_2 = (E_2 - E_1)/\sqrt{3}E_k$, where E_1, E_2 and E_3 are the kinetic energies of each of the three fragment atoms and $E_k = E_1 + E_2 + E_3$ is the total kinetic energy. The relative kinetic energy in the center-of-mass system of two selected particles can be expressed in terms of the Dalitz coordinates. For example, for three equal masses one obtains

$$E_{1,2}/E_k = ½ - 3/2\eta_1 \tag{3}$$

The Dalitz coordinates are defined[16] such that a random phase-space distribution for the three particles corresponds to a uniform distribution in the (η_1,η_2) plane, confined within a region given by energy and momentum conservation. For three equal masses this region is a circle of radius 1/3, and using Eq. 3, the normalized relative kinetic energy distribution $P(E_{12})$ is obtained by integration over η_2 as

$$\begin{aligned} P(E_{12})dE_{12} &= \frac{9}{\pi}\int d\eta_2 \left|\frac{d\eta_1}{dE_{12}}\right| dE_{12} = \\ &= \frac{4}{\pi E_k}\sqrt{1-\left(1-\frac{2E_{12}}{E_k}\right)^2}\, dE_{12}. \end{aligned} \tag{4}$$

This distribution has the shape of an arc of an ellipse spanning the range between $E_{12} = 0$ and $E_{12} = E_k$.

For the case of H_3^+ recombination with low-energy electrons towards H + H_2 we have $E_k = E_{k,^2A_1} = 9.23$ eV and $E_{b,j} = E_{b,H_2(X\Sigma_g^+)} = 4.48$ eV. As a first step, we have derived the vibrational distribution $P_\beta(v)$ of the H_2 (X $^1\Sigma$) products for this channel, using a simple quantization of the continuous energy distribution of Eq. (4) according to

$$P_\beta(v) = \int_{(E_{v-1}+E_v)/2}^{(E_v+E_{v+1})/2} P(E_{12})dE_{12} \tag{5}$$

where the lower limit is 0 and the upper limit is $E_{b,j}$ for $v = 0$ and $v = v_{max}$, respectively. The predicted relative vibrational populations are in very good agreement with our recent experimental data[5] (see Fig. 2).

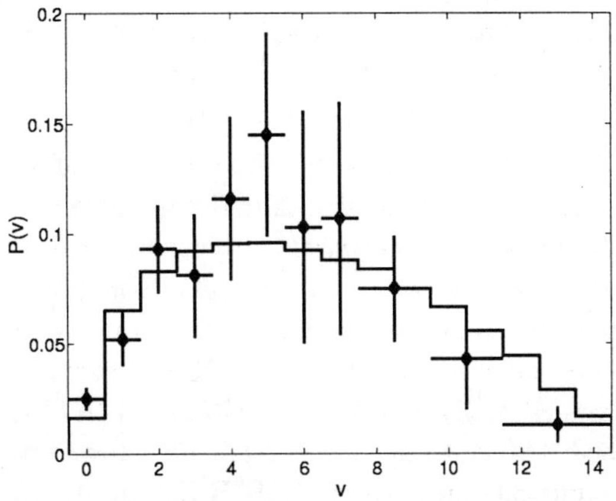

Figure 2. Relative vibrational state populations P(v) of H_2 fragments from the DR of H_3^+ at near-zero initial energy as obtained from the model (full curve), compared to the experimental results[5] (circles).

By the same reasoning, the molecular product fraction at near zero electron energies is obtained by

$$P_\beta^{^2A_1} = \int_0^{E_{b,j}} P(E_{12})dE_{12}. \tag{6}$$

where $E_{b,j}$ and E_k are defined as above, the two-body product fraction is $P_\beta^{^2A_1} = 0.48$, while the dissociation fraction is given by the complement ($P_\alpha^{^2A_1} = 0.52$). However, the overall branching ratio between the three and two-body product channels as represented by P_α and P_β (see Eq. (2)) also depends on the relative probabilities λ_i at which electronic states are populated in the electron capture step. We now assume that *each PES that is correlated to an energetically open asymptotic exit channel contributes to the reaction rate according to the statistical weight of the corresponding electronic state.* This seems to be a very rough assumption as it disregards the issue of potential-surface crossings and favorable Franck-Condon overlap factors which are usually assumed to determine the electron capture probability; however, as opposed to diatomic systems for which the latter considerations have been found important, we would expect a simple statistical assumption to become increasingly useful for polyatomic systems where often many pathways in a multidimensional coordinate space can contribute to a given reaction. For low-energy recombination of H_3^+, the statistical assumption implies $\lambda_{2_{B_2}} = \lambda_{2_{A_1}} = 2$, i.e., it is determined by the spin multiplicity, $\lambda_s = 2$. Remarkably, the normalized product branching ratios of $P_\alpha = 0.76$ and $P_\beta = 0.24$, following from the model of random fragment energy distribution *and* a statistical population of the relevant electronic energy surfaces, are in perfect agreement with the experimental result of $P_\alpha/P_\beta = 3$ for the BR in the low-energy DR of H_3^+.[4]

We have tested the more general predictive power of our model based on these two statistical assumptions by calculating also the electron energy dependence of the two- and three-body product branching ratio for H_3^+ for which experimental data by Datz et al.[4] are available. Bound type states are added with increasing final-state energy, correlating to the following exit channels (see inset in Fig. 1): $H(nl)+H_2(X^1\Sigma_g)$ (atomic Rydberg states, starting with $H(2l)+H_2(X^1\Sigma_g)$ at 0.97 eV), $H(1s)+H_2(n'\lambda)$ (molecular excited states, starting with $H(1s)+H_2(B^1\Sigma_u)$ at 1.95 eV) and combinations of both ($H(nl)+H_2(n'\lambda)$, starting with $H(2l)+H_2(B^1\Sigma_u)$ at 12.15 eV). We have calculated the two body BR, by summing the contributions of all the two body products that were derived with the described model. The energy dependence of the theoretical results has been convoluted with the rotational temperature as estimated from the experimental results of Strasser et al.[5] and the electron temperature present in the CRYRING storage ring experiment.[4]

Fig. 1 shows the comparison of the model calculation with the experimental results.[4] The dashed curve includes all the electronic final states as mentioned above. It reproduces the overall shape of the BR as a function of the electron energy, although the calculated values between 0.7 and 7 eV overestimate the measured ones. The comparison between the model and the experimental data provides a rather detailed interpretation of the observed structures. In particular, the rise of the two-body product fraction at ~0.6 eV is due to the opening of the $H(2l)+H_2(X\,^1\Sigma_g)$ electronic channel. The fact that the branching ratio is observed to be approximately independent of the incident electron energy below the $H(2l)+H_2(X\,^1\Sigma_g)$ threshold (0-0.7 eV) and again below $H(3l)+H_2(X\,^1\Sigma_g)$ threshold (~1-2 eV) clearly demonstrates that, once a channel is open electronically, the growing number of ro-vibrational states energetically accessible for increasing electron energy does not enhance the reaction rate into this channel. This supports the approach of combining separate statistical assumptions for the electronic and the nuclear degrees of freedom, as opposed to considering the total weight of electronic states plus internal ro-vibrational states.[12] Due to the opening of further channels $H(nl)+H_2(X^1\Sigma_g)$ (starting at ~2.86 eV for n = 3) the two-body fraction keeps increasing until ~4 eV. At ~5.5 eV, the opening of additional repulsive electronic channel leads to a strong drop in the two-body fraction. Starting from ~12 eV, the excited bound states $H(nl)+H_2(n'\lambda)$ open, producing an additional peak at ~15 eV, and the statistical model gives the correct size of this peak once the predissociation and decay of the excited molecular hydrogen states are taken into account.

The assumption that all possible l values are actually populated in the dissociative process should be regarded as an upper limit as electron capture into states with high electronic angular momentum is expected to be suppressed. In addition, a considerable fraction of the electronic energy surfaces correlated to high-l states of the fragments may be raised up in energy in the molecular complex as a consequence of molecular orbital promotion, so that their Franck-Condon overlap with the initial state of the DR reaction becomes very small. To obtain the lower limit for the branching ratio in our model we therefore repeated our calculation counting only final electronic states with $l = 0$. The result is shown by the full line in Fig. 1; the two limiting branching ratio predictions are seen to enclose the measured values over most of the energy range.

The model proposed here is general and does not require a detailed knowledge of the PESs involved in the DR process. It yields a straightforward explanation of the finding from many experiments[6-11] that a large degree of fragmentation occurs during the DR of polyatomic molecular ions as, by the energy randomization among the nuclei, exit channels where molecules can be produced also contribute to those with higher fragmentation. This will enhance fragmentation if bound and repulsive states among themselves are populated statistically (λ_i in Eq. (2)). First attempts to apply the model to H_2O^+ yield results close to measured data.[10]

ACKNOWLEDGMENTS

This work has been funded in part by the German Federal Minister of Education, Science, Research and Technology (BMBF) within the framework of the German-Israeli Project Cooperation in Future Oriented Topics (DIP), and by the European Community within the Research Training Network "Electron Transfer Reactions."

REFERENCES

1 E. Herbst, in *Dissociative Recombination: Theory, Experiment and Applications,* ed. by J. B. A. Mitchell and S. L. Guberman (World Scientific, Singapore, 1989), p. 303.

2 T. J. Millar, D. J. DeFrees, A. D. Mclean, and E. Herbst, Astron. Astrophys. **194**, 250 (1977).

3 B. J. McCall and T. Oka, Science **287**, 1941 (2000).

4 S. Datz et al., Phys. Rev. Lett. **74**, 896 (1995); Phys. Rev. A **52**, 2901(1995).

5 D. Strasser et al., Phys. Rev. Lett. **86**, 779 (2001).

6 C. R. Herd, N. G. Adams and D. Smith, Astrophys. J. **349**, 388 (1990).

7 N. G. Adams et al., J. Chem. Phys. **94**, 4852 (1991).

8 T.L. Williams, N.G. Adams, L.M. Babcock and M. Geoghegan, Mon. Not. R. Astron. Soc. **282**, 4I3 (1996).

9 L. Vejby-Christensen et al., Astrophys. J. **483**, 531 (1997).

10 M. J. Jensen et al., Phys. Rev. A. **60**, 2970 (1999).

11 S. Rosen et al., Faraday Discuss. **115**, 295 (2000).

12 E. Herbst, Astrophys. J. **222**, 508 (1977).

13 S. Green and E. Herbst, Astrophys. J. **229**, 121 (1979).

14 E. T. Galloway and E. Herbst, Astrophys. J. **376**, 531 (1991).

15 D. R. Bates, J. Phys. B. **24**, 3267 (1991), and references therein.

16 R. H. Dalitz, Phil. Mag. **44**, 1068 (1953).

Quantum Dynamical Study of H_3^+ Recombination

Motomichi Tashiro[1] and Shigeki Kato[2]
[1]Department of Physics, Kyoto University, Kyoto 606-8502, Japan; [2]Department of Chemistry, Graduate School of Science, Kyoto University, Kyoto 606-8502, Japan
tashiro@tap.scphys.kyoto-u.ac.jp; shigeki@kuchem.kyoto-u.ac.jp

1. INTRODUCTION

Discrepancies between experiment and theory have long been problems in the study of H_3^+ dissociative recombination. (See Ref.'s 1 and 2 for recent theoretical works). The discovery of H_3^+ in the diffuse interstellar medium has brought a renewed interest to this problem, because it suggests a lower rate constant than the experimentally derived values.[3] In order to shed light on this complicated situation, it is necessary to carry out more detailed studies about the reaction mechanism. Our aim is to provide theoretical information about the dynamics of this process.

Because the H_3 2p dissociative surfaces do not cross the H_3^+ surface, the recombination of H_3^+ is considered to proceed via transitions between vibrationally excited Rydberg states, followed by transition to the 2p dissociative states. In this picture, the essential points are (1) the relaxation dynamics of the Rydberg states, and (2) the predissociation of vibrationally excited Rydberg states. Our approach is to construct the effective Hamiltonian for both the Rydberg electron and the H_3^+ core vibration, including the effect of predissociation. We carried out *ab initio* calculations for the non-adiabatic coupling elements as well as the Rydberg state potential energy surfaces and performed wave packet propagations for describing the relaxation dynamics of Rydberg states. We considered vibronic coupling only, so that A_1' and E' electronic states were included in the Hamiltonian. For simplicity, the rotation of the molecule was not considered.

2. PREDISSOCIATION WIDTHS OF VIBRATIONALLY EXCITED H_3 2S STATES

We considered the predissociation of the 2s state, because it is the closest state to the dissociative states. The time-dependent version of Fermi's golden rule was used to obtain predissociation widths as follows,

$$\Gamma_i = \frac{1}{\hbar}\int_{-\infty}^{+\infty} dt \exp\left(\frac{iE_i t}{\hbar}\right)\langle\phi(t)|\phi(0)\rangle \tag{1}$$

$$|\phi(t)\rangle = \exp\left(-\frac{i\hat{H}t}{\hbar}\right)\hat{V}|\chi_i\rangle, \tag{2}$$

where χ_i is the i'th vibrational state on the H_3 2s adiabatic surface and E_i is its energy. The Hamiltonian $\hat{H}$ has the form[4]

$$\hat{H} = -\frac{\hbar^2}{2\mu}\left\{\frac{\partial^2}{\partial\rho^2} + \frac{16}{\rho^2}\left[\frac{1}{\sin\theta}\frac{\partial}{\partial\theta}\sin\theta\frac{\partial}{\partial\theta} + \frac{1}{4\sin^2(\theta/2)}\frac{\partial^2}{\partial\phi^2}\right]\right\} + \frac{15\hbar^2}{8\mu\rho^2} + W(\rho,\theta,\phi) \tag{3}$$

where μ is the reduced mass and W is the potential function of the dissociative states. To represent the nuclear geometry of H_3, hyperspherical coordinates ρ, θ, and ϕ were used.

The operator, $\hat{V}$, in Eq. (2) represents non-adiabatic coupling:

$$\hat{V} = -\frac{\hbar^2}{2\mu}\left[\, 2\langle\psi_d|\frac{\partial}{\partial\rho}|\psi_{2s}\rangle\frac{\partial}{\partial\rho} + \frac{32}{\rho^2}\langle\psi_d|\frac{\partial}{\partial\theta}|\psi_{2s}\rangle\frac{\partial}{\partial\theta} + \frac{16\cos\theta}{\rho^2\sin\theta}\langle\psi_d|\frac{\partial}{\partial\theta}|\psi_{2s}\rangle + \frac{8}{\rho^2\sin^2(\theta/2)}\langle\psi_d|\frac{\partial}{\partial\phi}|\psi_{2s}\rangle\frac{\partial}{\partial\phi}\,\right]. \tag{4}$$

Here $\langle\psi_d|\,\partial/\partial\rho\,|\psi_{2s}\rangle$, $\langle\psi_d|\,\partial/\partial\theta\,|\psi_{2s}\rangle$, and $\langle\psi_d|\,\partial/\partial\phi\,|\psi_{2s}\rangle$ are the non-adiabatic coupling matrix elements between dissociative state ψ_d and 2s state ψ_{2s}. These coupling elements were calculated using the MOLPRO98 package[5].

Initial vibrational wave functions, χ_i, were prepared by diagonalization of Eq. (3), where the potential term W was replaced by the H_3 2s surface. The DVR basis was used for the hyperradius, ρ, and hyperspherical harmonics[6] were used for the hyperangles, θ and ϕ, respectively. The same basis set was used for the wave packet propagation.

The wave packet, $\hat{V}|\chi_i\rangle$, was propagated on the surface of the upper/lower dissociative state to obtain the predissociation width. The Chebychev polynomial method was used for the propagation.

The predissociation widths of 600 vibrationally excited 2s states were calculated. The results are shown in Fig. 1. For the ground vibrational level,

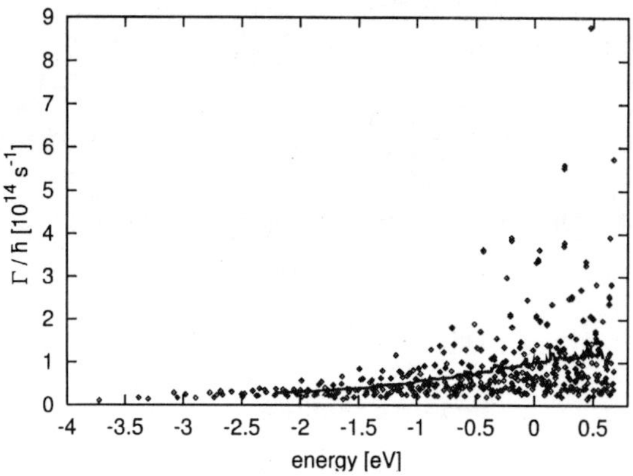

Figure 1. Predissociation widths, Γ_i , of vibrationally excited H_3 2s states. The energy of each state is given with respect to the ground vibrational level of H_3^+. The solid line represents the average width.

the calculated rate is 1.0×10^{13} s^{-1} which is close to the experimental value[7] of about 6×10^{12} s^{-1}. The widths were averaged over 70 states at each energy window and the result is also shown in Fig. 1. The average width increases with increasing the energy, and becomes about 1.0×10^{14} s^{-1} at the zero energy, which corresponds to the energy of H_3^+ vibrational ground state. The scatter around the average widths is large, however.

3. RELAXATION DYNAMICS OF H_3 RYDBERG STATES

3.1 Theoretical Model

We treat the dynamics of H_3 as the combined motion of the Rydberg electron and the H_3^+ core vibration. The evolution of the wave packet was investigated using the time-dependent Schrödinger equation,

$$i\hbar\frac{\partial}{\partial t}\Psi(\mathbf{r},\mathbf{R},t)=\hat{H}_{eff}\Psi(\mathbf{r},\mathbf{R},t)\,. \tag{5}$$

The effective Hamiltonian $\hat{H}_{eff}$ can be written as

$$\hat{H}_{eff}=\hat{H}_N(\mathbf{R})+\hat{H}_e(\mathbf{r},\mathbf{R})-\frac{i}{2}\hat{\Gamma}(\mathbf{r},\mathbf{R}) \tag{6}$$

where $\hat{H}_N$ is the Hamiltonian for the nuclear motion of H_3^+ and $\hat{H}_e$ is the one-electron Hamiltonian for Rydberg electron. We introduce the effect of predissociation as the operator $\hat{\Gamma}$. The coordinates of the nuclei and the

Rydberg electron are denoted as **R** and **r**, respectively.

The wave function $\Psi(\mathbf{r}, \mathbf{R}, t)$ can be expanded as

$$\Psi(\mathbf{r},\mathbf{R},t)=\sum_{n,m} C_{nm}(t)\chi_n(\mathbf{R})\psi_m(\mathbf{r},\mathbf{R}_0) \tag{7}$$

where $\mathbf{R}_0$ stands for the equilibrium nuclear geometry of H_3^+. The basis function, χ_n, is the eigen function of $\hat{H}_N$ as follows

$$\hat{H}_N\chi_n(\mathbf{R})=E_n^{(N)}\chi_n(\mathbf{R}) . \tag{8}$$

The basis function, ψ_m, is the eigen function of $\hat{H}_e$ as

$$\hat{H}_e\psi_m(\mathbf{r},\mathbf{R}_0)=E_m^{(e)}(\mathbf{R}_0)\psi_m(\mathbf{r},\mathbf{R}_0) . \tag{9}$$

The one-electron Hamiltonian, $\hat{H}_e$, has the form

$$\hat{H}_e=\hat{T}_e+\hat{V}_{cl}+\hat{V}_{ex} \tag{10}$$

where $\hat{V}_{cl}$ is the Coulomb potential of the H_3^+ core, and $\hat{V}_{ex}$ is the local exchange potential. For $\hat{V}_{cl}$, we placed a partial charge on each proton of H_3^+. The shielding effect of H_3^+ electrons was accounted for. We used the analogue of Hara's exchange potential,[8] where the multiplication factor and ionization potential were treated as free parameters. These parameters were chosen to reproduce the electronic energy levels at the equilibrium nuclear geometry.

The Hamiltonian, $\hat{H}_{eff}$, is represented in terms of the basis set, $|\psi_m\chi_n\rangle$, as

$$\begin{aligned}\langle\chi_{n'}\psi_{m'}|\hat{H}_{eff}|\chi_n\psi_m\rangle &= \left(E_n^{(N)}+E_m^{(e)}\right)\delta_{n'n}\delta_{m'm}+\langle\chi_{n'}\psi_{m'}|\Delta\hat{V}_e|\chi_n\psi_m\rangle \\ &\quad -\frac{i}{2}\langle\chi_{n'}\psi_{m'}|\hat{\Gamma}|\chi_n\psi_m\rangle\end{aligned} \tag{11}$$

and

$$\Delta\hat{V}_e=\hat{V}_e(\mathbf{r},\mathbf{R})-\hat{V}_e(\mathbf{r},\mathbf{R}_0) . \tag{12}$$

The operator, $\hat{\Gamma}$, is represented as

$$\langle\chi_{n'}\psi_{m'}|\hat{\Gamma}|\chi_n\psi_m\rangle=\sum_i\langle\chi_{n'}\psi_{m'}|\chi_i^{(2s)}\psi^{(2s)}\rangle\Gamma_i\langle\chi_i^{(2s)}\psi^{(2s)}|\chi_n\psi_m\rangle . \tag{13}$$

The matrix element, $\langle \chi_{n'}\psi_{m'}|\chi_i^{(2s)}\psi^{(2s)}\rangle$, is the overlap integral between the diabatic basis, $\chi_{n'}(\mathbf{R})\psi_{m'}(\mathbf{r}, \mathbf{R}_0)$, and the adiabatic basis, $\chi_i^{(2s)}(\mathbf{R})\psi^{(2s)}(\mathbf{r}, \mathbf{R})$, where $\psi^{(2s)}$ is the electronic wave function of the 2s state and $\chi_i^{(2s)}$ is the i'th vibrational wave function on the 2s surface.

Note that we did not assume the Born-Oppenheimer approximation in the present treatment.

3.2 Results and Discussion

The wave packets were propagated about 10 ps. The initial vectors were the vibrational ground state with a positive electronic energy (i.e. a "free" electron): the s-state (A_1') with 1.5×10^{-2} eV and the p-state (E') with 0.6×10^{-2} eV. These electronic bases were selected from the eigen functions of $\hat{H}_e$, where the finite volume effect causes the energy difference between the s and p states.

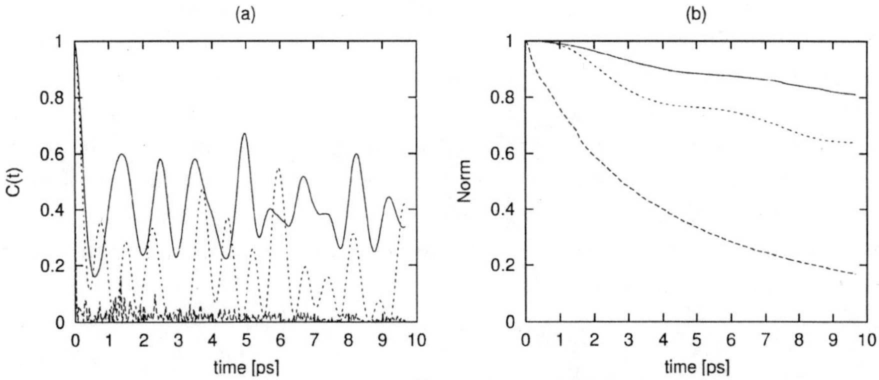

Figure 2.(a) Absolute values of autocorrelation functions C(t). (b) Norms of propagated vectors. The solid lines are for the s-state, the dotted lines for the p-state, and the dashed lines for the 3s state.

The absolute values of the autocorrelation functions $|\langle \Psi(t) | \Psi(0) \rangle|$ and the norms of the vectors are shown in Fig. 2. In the Figure, the result for a vibrationally excited 3s initial state is also shown. The decreases of the norms are slow compared to the predissociation of the vibrationally excited 2s states (see Fig. 1). The s-state initial wave packet transited to the 6s state with one A_1' vibration followed by a transition to the 4d state. A part of the wave packet went to the $n = 3$ state but a significant fraction returned to 4d, 6s and then the initial state. This recursion can be seen as the autocorrelation function in Fig 2. The evolution of the initially p-state wave packet shows similar oscillatory behavior as for the s-state. Compared to the s-state, the decay of p-state is slightly faster, though the dissociation rate is still slower than the average rate of the 2s state (see Fig. 1). As seen in Fig. 2, the predissociation from a vibrationally excited 3s state is much faster than the s- or p- state in the vibrational ground level, indicating that the relaxation to $n = 3$ levels is slow.

The decay of the wave packets seems to be slow. Two reasons can be considered for this behavior. First, we assumed that predissociation of each state was caused by coupling to vibrationally excited 2s states. Since the transitions between vibrationally excited Rydberg states were slow, direct couplings to the 2p dissociative states could affect the decay rate of each state. Second, the slow transitions to the 2s states was the main cause of the slow decay. An additional coupling other than vibronic coupling could change this slow relaxation.

4. SUMMARY

The transition dynamics of vibrationally excited H_3 Rydberg states were investigated. The H_3 system was treated as a Rydberg electron plus H_3^+ core vibration, and the vibrationally excited Rydberg states were taken as basis functions. We constructed the effective Hamiltonian including the predissociation widths of vibrationally excited 2s electronic states. The results indicate a slow transition between Rydberg states and thus a slow decay of the wave packets. These results indicate that we should consider a more elaborate model in which extra couplings or direct coupling to the 2p dissociative states are included.

ACKNOWLEDGMENTS

We thank Prof. T. Oka for calling our attention to this interesting problem and Dr. T. Yamamoto for collaboration at the early stage of this work.

REFERENCES

1 I. F. Schneider and A. E. Orel, and A. Suzor-Weiner, Phys. Rev. Lett. **85**, 3785 (2000).
2 V. Kokoouline, C. H. Greene, and B. D. Esry, Nature, **412**, 891 (2001).
3 B. J. McCall, T. R. Geballe, K. H. Hinkel, and T. Oka, Science **279**, 1910 (1998).
4 A. J. C. Varandas and H.G. Yu, J. Chem. Soc. Faraday Trans. **93**, 819 (1997).
5 MOLPRO is a package of *ab initio* programs written by H.-J. Werner and P. J. Knowles, with contributions from R.D. Amos, A. Bernhardsson, A. Berning, P. Celani, D. L. Cooper, M. J. O. Deegan, A. J. Dobbyn, F. Eckert, C. Hampel, G. Hetzer, T. Korona, R. Lindh, A. W. Lloyd, S. J. McNicholas, F. R. Manby, W. Meyer, M. E. Mura, A. Nicklass, P. Palmieri, R. Pitzer, G. Rauhut, M. Schütz, H. Stoll, A. J. Stone, R. Tarroni, and T. Thorsteinsson.
6 H. G. Yu, Chem. Phys. Lett. **281**, 312 (1997).
7 I. Dabrowski and G. Herzberg, Can. J. Phys. **58**, 1238 (1980).
8 S. Hara, J. Phys. Soc. Japan, **22**, 710 (1967).

Recombination of H_3^+ and D_3^+ with Electrons

Low Limit of the Recombination Rate Coefficient

R. Plašil, J. Glosík, V. Poterya, P. Kudrna, M. Vicher, A. Pysanenko
Charles University Prague, Mathematics and Physics Faculty, Czech Republic
glosik@plasma.troja.mff.cz

1. INTRODUCTION

The ions H_3^+ and D_3^+ have been the subject of a number of studies. The H_3^+ ion and its deuterated analogues play important roles in the kinetics of media of astrophysical interest (interstellar molecular clouds,[1,2] and planetary atmospheres[3]) and also in laboratory produced plasmas (discharges, fusion plasmas at walls,[4] etc.). The recent discovery of interstellar H_3^+ in molecular clouds has confirmed the presence of H_3^+ and opened a problem of observed high column density, which critically depends on the rate of recombination of H_3^+ with electrons at low energies.[1,2,5] The kinetics of formation of H_3^+ and D_3^+ ions in hydrogen (deuterium) containing plasmas is well understood.[6,7] Equally important for the hydrogen and deuterium containing plasmas is to understand the processes of destruction of these ions in collisions with electrons. However, despite enormous efforts, the results of experimental studies determining the rate of recombination of H_3^+ and D_3^+ ions with electrons give values that vary by at least one order of magnitude.[8-13] Studies of D_3^+ recombination reported smaller rate coefficient values, generally by a factor of between 1-10 (see discussion and Ref.'s 11 and 14-16). All experimental recombination rate coefficients obtained for both H_3^+ and D_3^+ are significantly larger than the values predicted by theory.[17-19] The agreement of theory with experimental data was obtained only for electron energies close to 10 eV.[17] The astrophysical problems (observed large column densities of H_3^+) connected with dissociative recombination of H_3^+ are reported in the present volume by Takeshi Oka.

Dissociative Recombination of Molecular Ions with Electrons
Edited by S. Guberman, Kluwer Academic/Plenum Publishers, 2003

In order to solve the "enigma" of recombination of H_3^+ and D_3^+ we built a new afterglow experiment, the **Advanced Integrated Stationary Afterglow (AISA).**[20-22]

In our recent studies of the recombination of H_3^+ we observe in the He-Ar-H_2 afterglow the dependence of the "apparent" recombination rate, α_{eff}, on the number density of H_2, $[H_2]$. Here we will report recent results of our study of recombination of H_3^+ and D_3^+ including new mass spectrometric studies and studies at high number densities of H_2 and D_2.

2. EXPERIMENT

The apparatus is of the Stationary Afterglow type with several improvements to measure recombination rate coefficients as low as 10^{-9} cm^3 s^{-1}. The main part of **AISA** is a large cylindrical stainless-steel UHV discharge chamber[20-22] (40 cm long with diameter 40 cm) cooled to below -40°C during the measurements (see Fig. 1). Purified He was used as a buffer gas (impurity level < 0.1 ppm). The plasma was periodically generated by pulses of microwave power (~ 1 kW, 0.2-2 ms pulses and repetition period 40-100 ms). The electron number density (n_e) was determined from Langmuir probe (18 µm in diameter and 8 mm in length) measurements. For experimental details, see e.g. Ref's. 21 and 23. The plasma composition was monitored by a mass spectrometer[23] working in "integral" and "time resolved" regimes.

The H_3^+ and D_3^+ ions are produced in a sequence of ion molecule reactions, which follow formation of the parent ions during the period of the active discharge. We use "suitable" mixtures of He-Ar-H_2 and He-Ar-D_2 to optimize this formation. This mixture contains mostly of He (~ 99%), with Ar (up to 1%) and traces of H_2 or D_2. Below, we describe mostly hydrogen but experiments were done for both hydrogen and deuterium. We distinguish between these two isotopomers only when discussing results of measurements and mass spectra. During the discharge in He-Ar (100:1), Ar^+ ions are probably preferentially produced. This follows from comparison of the ionization frequencies for electron temperatures of the order of 2-3 eV and reflects the large difference in ionization potentials and cross sections of ionization (σ_ι) for He and Ar. Traces of H_2 will not change this ionization process but will substantially change processes during the afterglow. The partial formation of He^+, He^m, H_2^+ or H^+ during the discharge will not substantially change the production of H_3^+. We modeled the kinetics of formation of H_3^+ during the early afterglow assuming many initial conditions for the overall pressure and composition. For simplicity in the data analyses, we will assume that Ar^+ is the dominant ion formed during the microwave

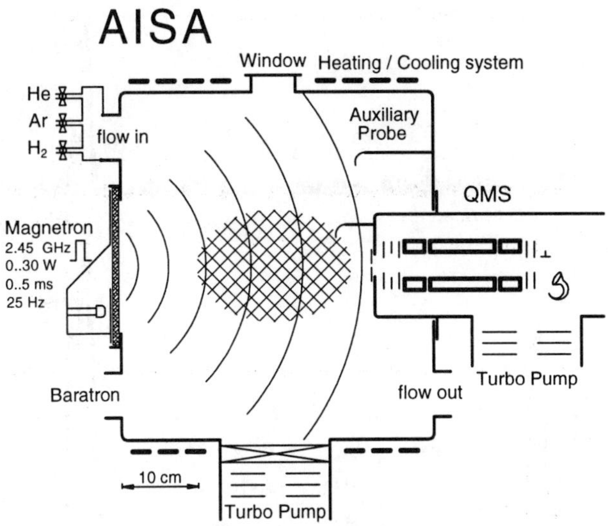

Figure 1. Schematic view of the Advanced Integrated Stationary Afterglow (AISA).

discharge. This assumption is supported by the mass spectrometric observation (see the upper panel in Fig. 2).

The integral spectra are obtained by sampling ions over several cycles. These spectra provide information about ions present in the plasma. Note that Ar^+, ArH^+ and H_3^+ ions are major components of the spectra. In some spectra, traces of H_3O^+ and also N_2H^+ and ArH_3^+ were observed. Here we have to realize the relationship between the spectra and the real ionic composition in the center of the discharge vessel. The ions, prior to the entering the sampling orifice, have to diffuse through the discharge chamber. The characteristic diffusion time is ~ 40ms. During this time, the relative population of H_3O^+ is increasing, particularly for diffusion through the low-density plasma near the wall. We believe that H_3O^+ is enhanced in the observed spectra in comparison with ionic composition in the center of the chamber. Impurities in the buffer gas enhance the decay of the plasma and together with diffusion, they are the source of a lower bound to the measurable recombination rate coefficients.

After the identification of the ions the mass spectrometer is synchronized with the microwave generator and the "time resolved spectra", i.e. the time evolution of the relative populations of the ions can be measured.

In Fig. 3 are plotted time resolved spectra obtained at different partial pressures of H_2. The fast decay of Ar^+ and ArH^+ corresponds to fast proton transfer reactions converting Ar^+ to ArH^+ and then to H_3^+.

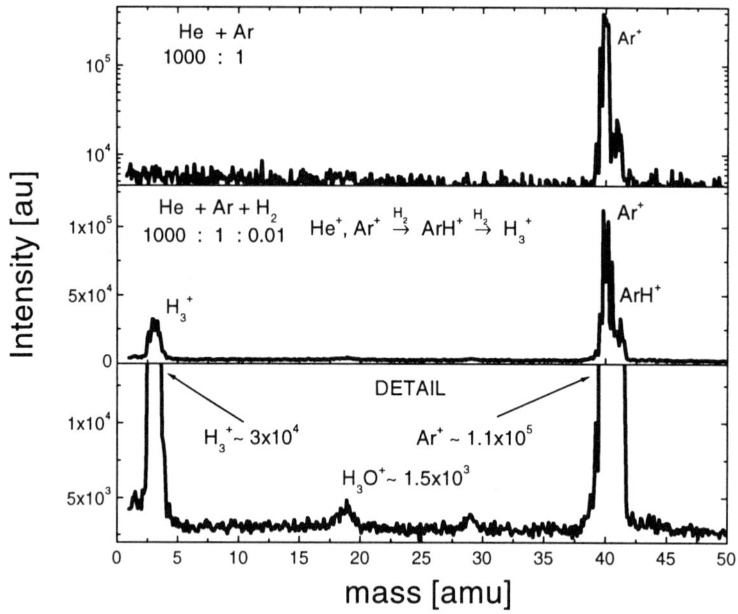

Figure 2. The integral mass spectra of ionic composition. In the upper panel is the plotted spectrum obtained in the He-Ar mixture. The lower panels are for the He - Ar - H_2 mixture.

3. COMPUTER SIMULATION

The mass spectrometer provides important but qualitative information about the ionic composition in the recombination dominated region of the plasma. To obtain further information and to confirm the mass spectrometric data we constructed a kinetic model for calculation of the ionic composition during the afterglow. There is one very uncertain parameter in these calculations: initial conditions. It is difficult to evaluate these conditions without going into details such as the form of the electron energy distribution function (EEDF) at energies close to the ionization potentials of the species present in the mixture. Rough comparison of integrals $<v\cdot \sigma_{iAr}(v)>$ and $<v\cdot \sigma_{iHe}(v)>$ for Maxwellian EEDFs with an electron temperature of ~ 2-3 eV gives comparable ionization frequencies. This can indicate substantial production of Ar^+, in agreement with our mass spectrometric observation (see Fig. 2). Because of these difficulties, we used several initial conditions including very "unfavorable" conditions for the formation of H_3^+. We assumed in these calculations that in the active discharge, He^+ ions are formed and are converted to He_2^+ in three-body association. In the sequence of reactions these ions lead to H_3^+. The plots in Fig. 4 include calculations

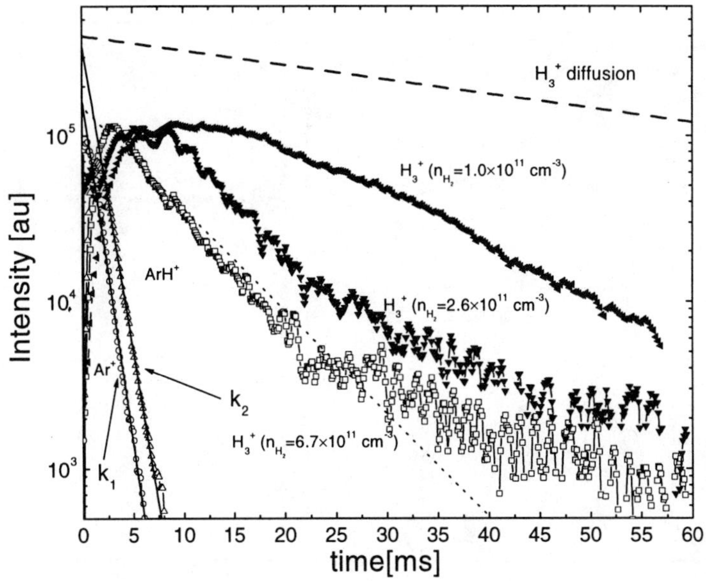

Figure 3. The time resolved mass spectra obtained at different concentration of H_2. For $[H_2] = 6.7\times10^{11}\ cm^{-3}$ the corresponding decays of Ar^+ and ArH^+ are also plotted. The dashed straight line indicates pure diffusive losses. The dotted line indicates the measured decay of ArH^+ for $[H_2] = 1.0\times10^{11}\ cm^{-3}$

made under the assumption that in the discharge the helium ions and the helium metastables are formed with number densities $[He^+] = 10^{10}\ cm^{-3}$ and $[He^m] = 10^{10}\ cm^{-3}$. If the Ar^+ ions are formed during the discharge then the time of the formation of H_3^+ will be just shorter. The difference between calculated and measured integral spectra indicates substantial production of Ar^+ ions during the microwave discharge.

The simulations were made by numerical solution of the set of differential equations describing the reactions and/or by Monte Carlo simulation (see the example in Fig. 4). The measured time resolved mass spectra confirm the validity of the model for our experiment. After 5-15 ms, for nearly all of the H_2 flows employed, the chemistry forming H_3^+ is complete and the plasma can be characterized as an H_3^+/e^- plasma. For the lowest H_2 flows, complete conversion to H_3^+ ions is slower. The eventual influence of this "non complete conversion" on measured rate coefficient will be discussed below.

4. DATA ANALYSIS

In order to describe the decay of the plasma during the afterglow, we can write the balance equation for electrons. The electrons are removed from

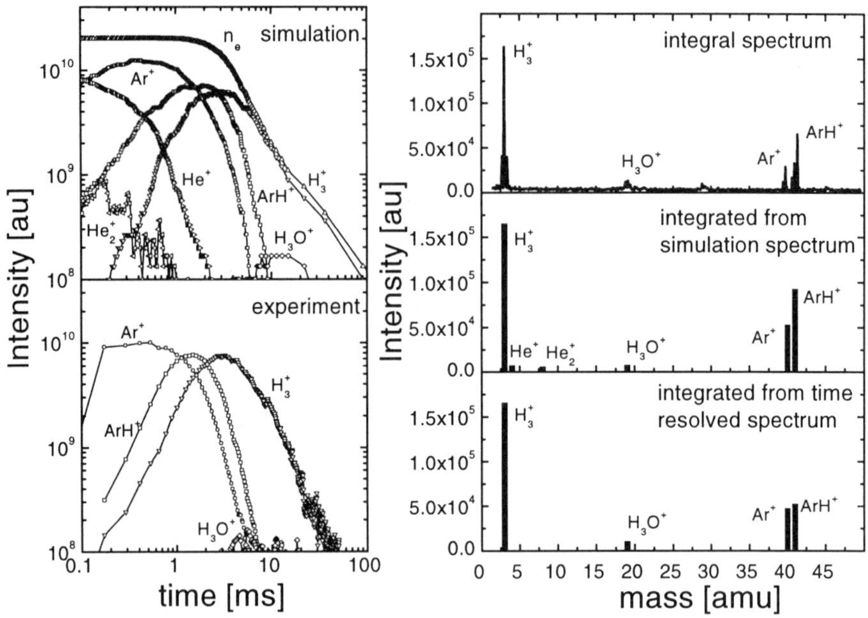

Figure 4. The calculated and measured time evolutions of the ionic composition during the afterglow in He-Ar-H_2 plasma. The corresponding integral spectra are plotted in the right panel.

the cold afterglow plasma (starting at a few milliseconds after termination of microwave pulse) mainly by recombination and ambipolar diffusion. Some very fast processes (e.g. reaction of metastables He^m with Ar for [Ar] ~ 10^{14} cm^{-3}), are completed within < 1 ms and do not have to be considered. Because of quasi-neutrality of the plasma, the RHS of the balance equation can be written as a sum of terms corresponding from left to right to recombination of H_3^+ ions, ambipolar diffusion, and the fast recombination of "impurity ions":

$$\frac{dn_e}{dt} = -\alpha\left[H_3^+\right]n_e - \nu_D n_e - \nu_R n_e \qquad (1)$$

Because of the very low number density of "impurity molecules" the formation of "impurity ion" is the rate determining process. If for simplicity we assume that there is just one type of impurity Y (e.g. H_2O), which is reacting with the ions with comparable rate coefficients, then the rate of destruction of ions is $k_Y([H_3^+] + [Ar^+]+[ArH^+]+[He_2^+]+[He^+])[Y]=k_Y n_e [Y] \sim \nu_{RY} n_e$. For several impurities (in the spectrum we observed just N_2H^+ and H_3O^+) we obtain: $\Sigma(k_Y n_e [Y]) = n_e \Sigma(k_Y [Y]) \sim \nu_R n_e$. In the fundamental diffusion mode approximation the diffusion losses are characterized[24] by ν_D. Because we cannot measure absolute values of $[H_3^+]$, it is convenient introduce the fraction of H_3^+ ions in plasma, $f = f(t) = [H_3^+]/n_e$. The balance equation can be rewritten in the form:

$$\frac{dn_e}{dt} = -\alpha f n_e^2 - \nu_D n_e - \nu_R n_e = -\alpha_{eff} n_e^2 - \nu_D n_e - \nu_R n_e \qquad (2)$$

The fraction, $f(t)$, can be estimated from the ionic composition as determined by the calculations and/or time resolved mass spectra. For $[H_2] > 10^{11}$ cm^{-3} the H_3^+ ions are formed within 5-20 ms and $f \sim 1$ can be used to obtain α. For $[H_2] < 10^{11}$ cm^{-3}, the formation of H_3^+ is slow and $f < 1$ at small t. In such a case, the time dependence of f has to be included when analyzing data (see discussion below). The effective rate coefficient of the processes in which loss of charged particles is proportional to the n_e^2 is given by $\alpha_{eff}(t) = \alpha f(t)$. Because of the high precision of our data, dn_e/dt can be calculated and Eq. (2) can be rewritten to the form suitable for evaluation of α_{eff} (using n_e' for dn_e/dt):

$$-\frac{n_e'}{n_e^2} = \alpha_{eff} + \frac{(\nu_D + \nu_R)}{n_e}$$

If $\alpha_{eff}(t)$ can be considered as a constant, the plot of $-n_e'/n_e^2$ versus $1/n_e$ should be linear with slope given by $(\nu_D + \nu_R)$. Fig. 5 shows an example of measured spectra and $f(t)$.

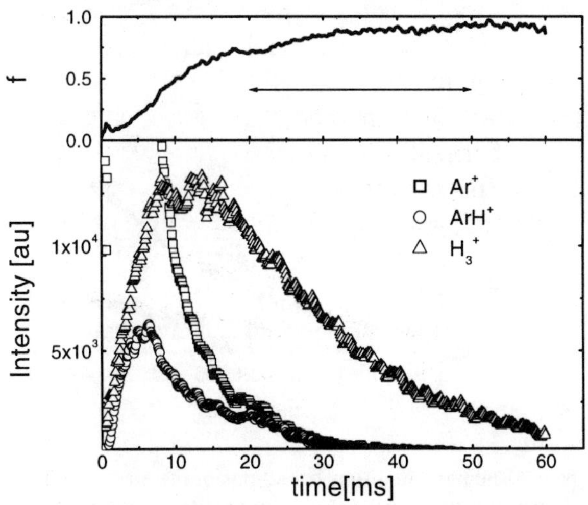

Figure 5. Upper panel: H_3^+ relative population. The f factor is calculated from measured time evolution of ionic composition. The arrow indicates time interval used for determination of α_{eff}. Lower panel: The measured time evolution of the ionic composition during the afterglow. Helium pressure $p_{He} = 2$ Torr, $[H_2] = 9\times10^{10}$ cm^{-3}.

5. RESULTS

In a He-Ar mixture (~ 0.1-1% of Ar) we observed a slow decay of the plasma that corresponded solely to the diffusive loss of Ar^+ (see Fig. 2). The measurement in He-Ar mixture was regularly repeated to confirm the purity of the system and the low level of "reactive losses". In a He-Ar-H_2 mixture we measured the deionization rate constant, α_{eff}, in the plasma containing H_3^+ ions. Examples of measured values of $n_e(t)$ are plotted in the upper panel of Fig. 6. The dependence of the decay of the plasma on [H_2] is evident. In the lower panel of Fig. 6, $1/n_e(t)$ is plotted. These plots are used in standard analyses for calculation of α_{eff}. Fig. 6 also includes the decay curve for O_2^+ (used for the calibration of the system) and the decay curve for Ar^+ (used to estimate diffusive losses). Note that between curves corresponding to O_2^+ (recombination rate coefficient ~ 2×10^{-7} cm^3 s^{-1}) and Ar^+ there is "space" for decay curves covering 1-100 times slower recombination, i.e. we can study recombination with a rate coefficient as small as 2×10^{-9} cm^3 s^{-1}.

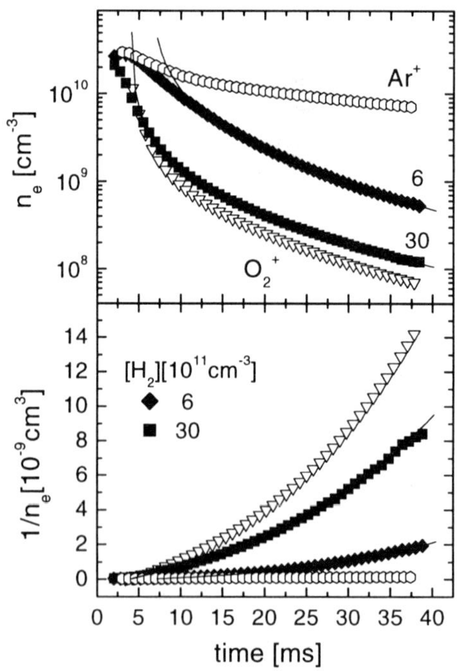

Figure 6. Upper panel: The measured time evolutions of the electron number density, $n_e(t)$, in an H_3^+ dominated afterglow. O_2^+ and Ar^+ indicate the decay curves corresponding to plasma dominated by O_2^+ and Ar^+ ions, respectively. The helium pressure p_{He} = 2 Torr. The full lines indicate the fit by function including diffusion and recombination losses. Lower panel: The corresponding values of $1/n_e(t)$.

From the measured $n_e(t)$, n_e'/n_e^2 was calculated. Examples of $(-n_e'/n_e^2 - \nu_D/n_e)$ versus $1/n_e$ for two values of $[H_2]$ are given in Fig. 7. From the model, calculated values of f were used to account for the ionic composition.

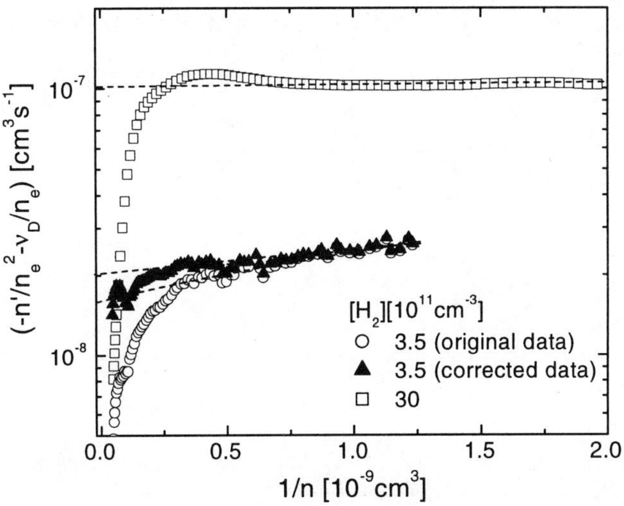

Figure 7. Decay of H_3^+ dominated afterglow. The dependence of $(-n_e'/n_e^2 - \nu_D/n_e)$ versus $1/n_e$ (open symbols) and "f corrected" values, $(-n_e'/n_e^2 - \nu_D/n_e)/f$ versus $1/n_e$, for $[H_2] = 3.5\times10^{11}$ cm^{-3} (full symbols).

Fig. 7 shows good linearity in the plot of "f corrected" values $(-n_e'/n_e^2 - \nu_D/n_e)/f$ versus $1/n_e$. The values for ν_D, calculated from the decay at low n_e and/or from the decay of Ar^+, are not critical for the evaluation of α_{eff}. It is important that ν_D and ν_R are nearly constant in the time interval for which α_{eff} is evaluated. To ensure the proper evaluation of α_{eff} we made a systematic study of the dependence of the function $f(t)$ on $[H_2]$. We also realized that to a good approximation, we can use the similarity law $f(t, [H_2]) = f(t/2, 2[H_2])$. Having this information we decided to skip the first 5-20 ms of the decay time as a formation time. For $[H_2] > 10^{11}$ cm^{-3}, this condition gives $f(t > 20$ ms$) > 0.75$, so for α_{eff} we can write $0.75\ \alpha < \alpha_{eff}(t > 20 \text{ ms}) < \alpha$. Because the measurements are made at least up to 60 ms, this is enough to assume that for $[H_2] > 10^{11}$ cm^{-3}, $\alpha_{eff} \approx \alpha$. For $[H_2] = 0.5\text{-}1\times10^{11}$ cm^{-3}, the proper interval for the determination of α_{eff} was chosen with the respect to the form of the function $f(t)$, but because of the already mentioned similarity law we believe that even for this $[H_2]$, $\alpha_{eff} \geq \alpha/2$. Fig. 8 has $(-n_e'/n_e^2 - \nu_D/n_e)$ versus $1/n_e$ for the decay curves measured in the He-Ar-H_2 afterglow. Dotted lines indicate the linear fit. Fig. 9 has decay curves obtained in He-Ar-D_2 afterglow for several $[D_2]$. Fig.s 10a and 10b have semilogarithmic and linear plots of $(-n_e'/n_e^2 - \nu_D/n_e)$ versus $1/n_e$. Note the good linearity of the plots in Fig. 10b even for $[D_2] =$

10^{11} cm^{-3}. Dotted lines indicate the linear fit for α_{eff}. The intervals for the fit were determined by using time resolved spectra.

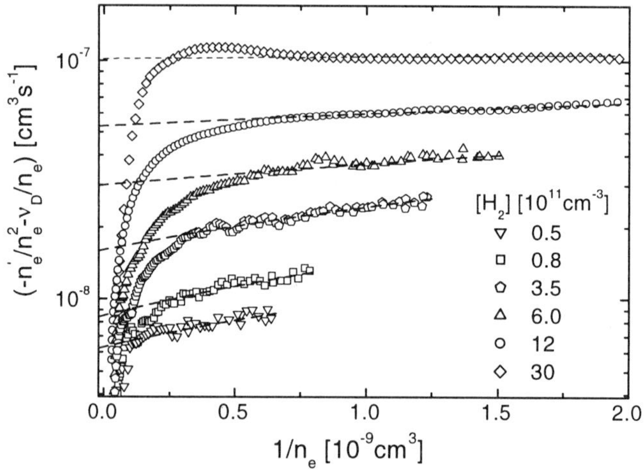

Figure 8. Dependence of the decay of H_3^+ dominated afterglow on concentration of H_2.

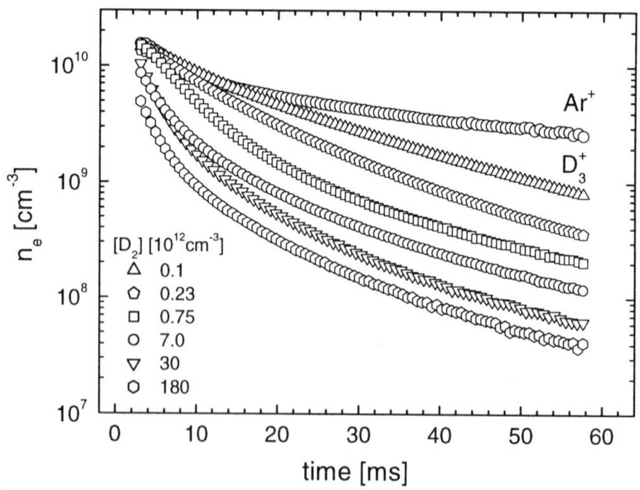

Figure 9. Dependence of the decay of D_3^+ dominated afterglow on concentration of D_2.

α_{eff} for the recombination of H_3^+ and D_3^+ is plotted in Fig.s 11 and 12, respectively. We stress once again, as it follows from the discussion above, that $\alpha_{eff} \sim \alpha$ for [H_2], [D_2] > 10^{11} cm^{-3}. For [H_2], [D_2] = 0.5-1×10^{11} cm^{-3}, α_{eff} has to be corrected to obtain α.

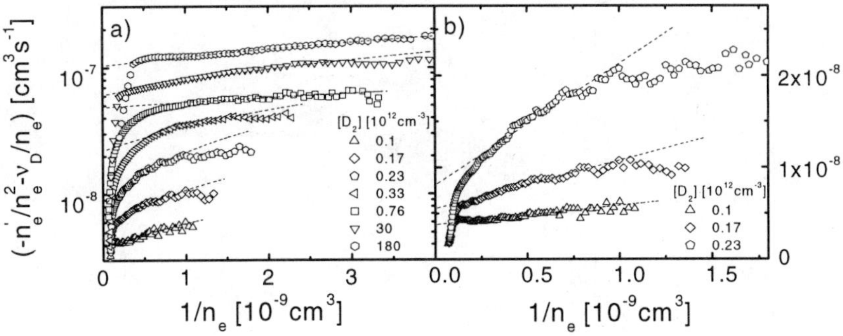

Figure 10. Recombination of D_3^+ : a) Semilogarithmic scale, b) Linear scale.

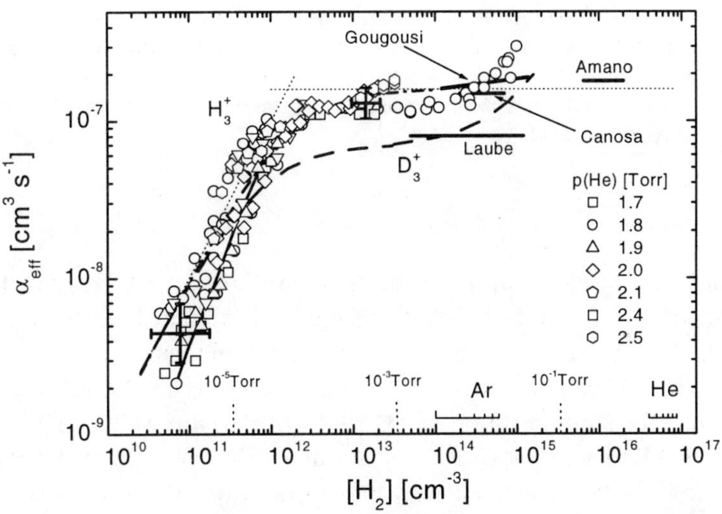

Figure 11. Recombination of H_3^+ and the dependence of the observed α_{eff} on [H_2]. Measurements were performed at several helium pressures, $p_{He} = 1.5\text{-}3$ Torr and several Ar number densities (as indicated). In these experiments, the He temperature was ~ (260 ± 40) K. Also included are previous results.[11-13,25] The crosses indicate estimated error bars. The dashed line indicates the dependence of α_{eff} on [D_2] for the recombination of D_3^+ (see Fig. 12).

6. DISCUSSION

We see that the recombination rate coefficients for H_3^+ and D_3^+ with electrons are dependent upon [H_2] and [D_2], respectively. We can make one very important statement at the outset without going into details of the

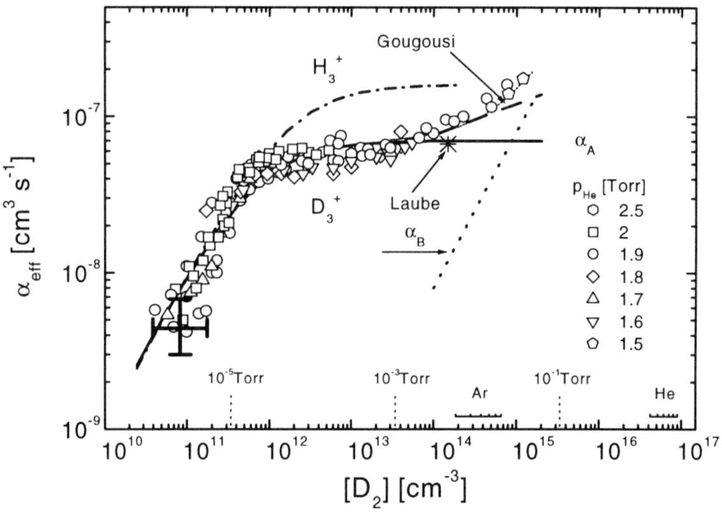

Figure 12. The recombination of D_3^+ and the dependence of the observed α_{eff} on D_2 number density. Pressures of He and Ar are indicated. In this experiment, the He temperature was ~ (230 ± 40) K. Included are also previous results.[11,12] The cross indicates estimated error bars. The dash-dotted line indicates the dependence of α_{eff} on $[H_2]$ for the recombination of H_3^+ with electrons (see Fig. 11).

recombination mechanism. Because the binary dissociative recombination is just one of the possible reaction channels that contribute to α_{eff} we can conclude that the rate of binary dissociative recombination, at nearly thermal temperatures, is small and we can place a limit on its value. The rate coefficients for recombination of H_3^+ and D_3^+ ions are $\alpha_{DR} < 3\times10^{-9}\,\mathrm{cm^3\,s^{-1}}$ (at ~ (260 ± 40) K) and $\alpha_{DR} < 6\times10^{-9}\,\mathrm{cm^3\,s^{-1}}$ (at ~ (230 ± 40) K), respectively. The difference is probably given just by the extent of the measurements.

Because of the similarity of the observed phenomena, we will discuss the recombination of H_3^+ and only at the end will we discuss observed differences between H_3^+ and D_3^+ recombination. The dependence of α_{eff} on $[H_2]$ indicates that not only electrons (e^-) and H_3^+ ions but also the H_2 molecules participate in the observed deionization process. In general we can consider two possible mechanisms that differ in the sequence of collisions:

$$A: \qquad H_3^+ + e^- \underset{k_{-1}}{\overset{k_1}{\rightleftarrows}} H_3^* \xrightarrow{k_2[H_2]} neutral \quad products$$

$$B: \qquad H_3^+ + H_2 \underset{k_{-3}[He]}{\overset{k_3[He]}{\rightleftarrows}} H_5^+ \xrightarrow{\alpha_5[e^-]} neutral \quad products$$

Because both processes can be considered as independent, the overall rate coefficient is the sum of the rate coefficients: $\alpha_{eff} = \alpha_A + \alpha_B$, where α_A and α_B are overall recombination rate coefficients of channels *A* and *B*, respectively. As we will demonstrated below, under our experimental conditions, at $[H_2] < 1\times10^{13}$ cm^{-3}, α_B can be neglected and only channel *A* need be considered. By application of the steady state approximation,[28] we obtain for reaction channel *A* the following description of the overall effective recombination rate coefficient (see also Ref. 21):

$$\alpha_A = \frac{k_1 k_2 [H_2]}{(k_{-1} + k_2[H_2])} \tag{3}$$

where k_1 is the binary rate coefficient of H_3^* formation and k_{-1} is the rate coefficient of unimolecular autoionization corresponding to the H_3^* lifetime, $\tau_{-1} = 1/k_{-1}$. The binary rate coefficient, k_2, describes the stabilization process. The dash-dotted line plotted in Fig. 11 indicates the fit of measured data by this formula for $[H_2] < 1\times10^{13}$ cm^{-3}, where α_B can be neglected. From the data plotted in Fig. 11, we see "high-pressure" and "low-pressure" limits (of the channel *A*) for $k_{-1} < k_2 [H_2]$ and $k_{-1} > k_2 [H_2]$, respectively. In the high-pressure limit we obtain $\alpha_{eff} = \alpha_A = k_1$, i.e. the rate determining process is the formation of H_3^*. The fit of the data gives $k_1 = 1.6\times10^{-7}$ cm^3 s^{-1}. In the low pressure limit, we obtain $\alpha_{eff} = \alpha_A = k_1 k_2[H_2]/k_{-1}$. From the linearly increasing part of the plot in Fig. 11 we obtained $k_2/k_{-1} = 6\times10^{-13}$ cm^3. From this value it will be possible to calculate the lifetime of H_3^*, $\tau_{-1} = 1/k_{-1}$, if k_2 can be measured or estimated.

From the plot in Fig. 12, $k_1 = 7\times10^{-8}$ cm^3 s^{-1} and $k_2/k_{-1} = 1.5\times10^{-12}$ cm^3 are obtained for the recombination of D_3^+. Note that for H_3^+, k_1 is approximately two times higher than for D_3^+. The ratios k_2/k_{-1} differ by a factor of 0.4.

At $[H_2] > 5\times10^{13}$ cm^{-3}, α_{eff} is again increasing with increasing $[H_2]$. A similar increase was observed also for D_2. To explain the increase of α_{eff} we considered that the loss of H_3^+ ions proceeds via formation of the H_5^+ ion and its successive recombination (scheme *B*). In channel *B*, k_3 is the rate coefficient of the three-body association reaction, and k_{-3} is the rate coefficient for collision induced dissociation of H_5^+ in collisions with He. α_5 is the rate coefficient for the recombination of H_5^+ with electrons[26,27] ($k_3 = 5\times10^{-30}$ cm^6 s^{-1}, $\alpha_5 = 3.6\times10^{-6}$ cm^3 s^{-1}). For loss of H_3^+, we can write: $d[H_3^+]/dt = -\alpha_B[H_3^+][e^-]$; naturally α_B can depend on [He], $[H_2]$ and $[e^-]$. By application of the steady state approximation[28] we obtain:

$$\alpha_B = \frac{k_3}{k_{-3}} \alpha_5 [H_2] \frac{1}{1 + \alpha_5 [e^-] / k_{-3} [He]} \tag{4}$$

The contribution of the channel *B* is proportional to $[H_2]$ and it is influenced by values of $[e^-]$ and [He] (for more details of the calculation see Refs. 23 and 29). From[26,30] k_3 and ΔG we found (for our experimental conditions) the relation: $\alpha_B \sim 8\times10^{-23}$ $[H_2]$ $cm^3\,s^{-1}$, where $[H_2]$ is given in units cm^{-3}. This contribution is identical for hydrogen and for deuterium. The calculated values of α_B relevant to our experiment are indicated by the dotted line in Fig. 12. The calculated α_B is in very good agreement with the observed increase of α_{eff} at $[H_2]$, $[D_2] > 5\times10^{13}$ cm^{-3}. Note that α_B can be neglected at $[H_2]$, $[D_2] < 5\times10^{13}$ cm^{-3}. We considered the possibility of fast recombining vibrationally excited ions but because of the long formation-relaxation time before actual recombination controlled decay, the presence of H_3^+ (v>0) can be neglected in our experiment.

7. CONCLUSIONS

We measured the variation of the rate coefficients of the overall recombination of H_3^+ and D_3^+ with electrons (α_{eff}) in a decaying plasma with the addition of H_2 or D_2, respectively. Since only three-body recombinations were observed, we concluded that the binary dissociative recombination channels are negligible. The fundamental binary dissociative recombination of H_3^+ ions with electrons (at ~ 270 K) has the rate coefficient, $\alpha_{DR} < 3\times10^{-9}$ $cm^3\,s^{-1}$, and that for D_3^+ (at ~ 230 K) has the rate coefficient, $\alpha_{DR} < 6\times10^{-9}$ cm^3 s^{-1}. The results are in accordance with theoretical predictions.[17-19] We estimated the contribution of "effective recombination" proceeding via formation of H_5^+ (D_5^+) that is only important at large $H_2(D_2)$ and He densities. This is important for the description of processes in the atmospheres of the large planets.

ACKNOWLEDGEMENTS

Thanks for financial support are due to GACR (202/00/1689, 202/99/D061 and 202/99/M004), GAUK (146/2000 and 171/2000 B FYZ MFF) and MSM 1132000002. The experiments were carried out with support of the EU in frame of the ETR network (HPRN-CT-2000-00142) and with support from EURATOM.

REFERENCES

1 T. R. Geballe, Phil. Trans. R. Soc. London A **358,** 2503 (2000).
2 T. J. Millar, H. Roberts, A. J. Markwick, S. B. Charnley, Phil. Trans. R. Soc. London A **358,** 2535 (2000).

3 S. Miller, N. Achilleos, G. E. Bellester, T. R. Geballe, R. D. Joseph, R. Prange, D. Rego, T. Stallard, J. Tennyson, L. M. Trafton, J. H. Waite, Phil. Trans. R. Soc. London A **358,** 2485 (2000).
4 R.K. Janev in *Dissociative Recombination: Theory, Experiment and Applications IV*, ed. by M. Larsson, J. B. A. Mitchell and I. F. Schneider (World Scientific, Singapore, 1999), p. 40.
5 J. H. Black, Phil. Trans. R. Soc. London A **358**, 2515 (2000).
6 D. Smith and P. Španěl, Acc. Chem. Res. **25**, 414 (1992).
7 V. G. Anicich, J. Phys. Chem. Ref. Data **22**, 1469 (1993).
8 M. Larsson, Phil. Trans. R. Soc. London A **358**, 2433 (2000).
9 M. J. Jensen, H. B. Pedersen, C. P. Safvan, K. Seiersen, X. Urbain and L. H. Andersen, Phys. Rev. A **63**, 052701 (2001).
10 D. Smith and P. Španěl, Int. J. Mass Spectr. Ion Proc. **129**, 163 (1993).
11 S. Laube, A. Le Padellec, O. Sidko, C. Rebrion-Rowe, J. B. A. Mitchell and B. R. Rowe, J. Phys. B **31**, 2111 (1998).
12 T. Gougousi, R. Johnsen and M. F. Golde, Int. J. Mass Spectr. Ion Proc. **149/150**, 131 (1995).
13 A. Canosa, J. C. Gomet, B. R. Rowe, J. B. A. Mitchell and J. L. Queffelec, J. Chem. Phys. **97**, 1028 (1992).
14 D. Smith and P. Španěl, Chem. Phys. Lett. **211**, 454 (1993).
15 A. Le Padellec, M. Larsson, H. Danared, A. Larson, J. R. Peterson, S. Rosen, J. Semeniak and C. Stromholm, Phys. Scripta **57**, 215 (1998).
16 M. Larsson, H. Danared, A. Larson, Le Padellec, J. R. Peterson, S. Rosen, J. Semeniak and C. Strömholm, Phys. Rev. Lett. **79**, 395 (1997).
17 A. E. Orel, I. F. Schneider and A. Suzor-Weiner, Phil. Trans. R. Soc. London A **358**, 2445 (2000).
18 M. R. Flannery in *Atomic and Molecular Physics*, ed. by I. Alvarez, C. Cisneros and T. J. Morgan (World Scientific, Singapore, 1995), p. 329.
19 V. Kokoouline, C. H. Greene and B. D. Esry, Nature **412**, 891 (2001).
20 P. Kudrna, R. Plašil, J. Glosík, M. Tichý, V. Poterya and J. Rusz, Czech. J. Phys., **50/S3**, 329 (2000).
21 J. Glosík, R. Plašil, V. Poterya, P. Kudrna, M. Tichý and A. Pysanenko, J. Phys. B **34**, 485 (2001).
22 J. Glosík, R. Plašil, V. Poterya, P. Kudrna and M. Tichý, Chem. Phys. Lett. **331**, 209 (2000).
23 R. Plašil, J. Glosík, V. Poterya, P. Kudrna, J. Rusz, M. Tichý and A. Pysanenko, Int. J. Mass Spectr., **218**, 105 (2002).
24 E. W. McDaniel, J. B. A. Mitchell, M. E. Rudd, *Atomic Collisions, Heavy Particle Projectiles* (Wiley Interscience, New York, 1993).
25 T. Amano, J.Chem. Phys. **92**, 6492 (1990).
26 W. Paul, B. Lucke, S. Schlemmer and D. Gerlich, Int. J. Mass Spectr. Ion Proc. **149/150**, 373 (1995).
27 M. T. Leu, M. A. Biondi and R. Johnsen, Phys. Rev. A **8**, 413 (1973).
28 P. W. Atkins, *Physical Chemistry*, (Oxford University Press, Oxford, 1988).
29 V. Poterya, J. Glosík, R. Plašil, M. Tichý, P. Kudrna and A. Pysanenko, Phys Rev. Lett. **88**, 044802 (2002).
30 K. Hiraoka and T. Mori, J. Chem. Phys. **91**, 4821 (1989).

Breakup Dynamics in H_3^+ and D_3^+ Dissociative Recombination

D. Zajfman,[1,2] D. Strasser,[1] L. Lammich,[2] H. Kreckel,[2] A. Naaman,[1] S. Krohn,[1,2] D. Schwalm,[2] and A. Wolf[2]
[1] *Department of Particle Physics, Weizmann Institute of Science, Rehovot, 76100, Israel;*
[2]*Max-Planck-Institut für Kernphysik and Physikalisches Institut der Universitat Heidelberg, D-69029 Heidelberg, Germany*
fndaniel@wicc.weizmann.ac.il

1. INTRODUCTION

The triatomic hydrogen positive ion is the dominant positively charged molecular ion in a hydrogen plasma. As such, it plays an important role in diverse fields from chemistry to astronomy. Also, being the simplest polyatomic system, it serves as a benchmark system in quantum chemistry calculations.

The dissociative recombination (DR) of H_3^+ plays an important role in many plasma environments, both as a destruction mechanism and as a pathway for the creation of neutral molecules. In particular the DR of H_3^+ plays a critical role in models of the interstellar medium. A small uncertainty in the H_3^+ DR rate translates to a large uncertainty in the size of the diffuse interstellar clouds where H_3^+ infrared spectra have been detected.

Despite the intensive experimental and theoretical study of the DR of H_3^+, the mechanism that drives the process remains a mystery and the reaction rate coefficient is an issue of long-standing controversy. Laboratory measurements of the recombination cross section differ by four orders of magnitude and although in the last years results from the majority of experiments performed at storage rings seem to converge towards a high recombination cross section, theory predictions are still well below the measured values.

In this work, we present an experimental study of the dissociation dynamics and isotope effect in the DR of H_3^+ and D_3^+. The DR of these species with zero energy electrons and at low excitation energy (below ~0.9 eV), can yield either two ($H_2(v)$ + H(1s)) or three (H(1s) + H(1s) + H(1s)) fragments. In the latter channel, 4.76 eV of kinetic energy is released, while in the two-body channel, the kinetic energy release is 9.23 eV (4.63 eV and 9.19 eV respectively for the triatomic deuterium breakup channels); this energy is distributed over the ro-vibrational excitation of the H_2 (or D_2) molecule and the translational degrees of freedom of the fragments. Here we present results for the energy distribution among the different particles for both the two and three body channels. Our data also indicates that H_3^+ and D_3^+ are rotationally hot before DR. More details related to the H_3^+ experiments can be found in Ref. 9.

2. EXPERIMENTAL SETUP

The experiments were carried out at the Test Storage Ring (TSR) located at the Max-Planck-Institut für Kernphysik, Heidelberg, Germany. The ring, which has already been used for previous DR experiments,[1-5] allows for the storage of ion beams with kinetic energies in the MeV range. In the present case, both H_3^+ and D_3^+ ions are produced in a gas discharge ion source, accelerated to 1.41 MeV using a radio-frequency quadrupole accelerator and injected into the storage ring. After each injection, the ring (55.4 m circumference) typically holds about 10^6 ions, in an average vacuum of ~3 x 10^{-11} mbar. At each pass, the ion beam (~0.8 mm in diameter) is merged, over 1.5 m, with the 'cold' electron beam, which is produced by the electron cooler.[6] The electron beam is guided in and out of the ring by magnetic fields. In the D_3^+ experiments, two settings of the electron beam were used with 14.8 mm and 10.4 mm electron beam radius and electron densities of 3 x 10^6 cm^{-3} and 6 x 10^6 cm^{-3}, respectively. The electron beam has typically a 0.1 meV longitudinal and 20 meV transverse energy spread in its comoving reference frame. The electron beam energy is tuned so that the electrons and the ions have the same mean velocities. In this specific case, the center-of-mass collision energies between the ions and electrons are mainly determined by the transverse electron energy spread, the relative velocities being given by a flattened Maxwell distribution.[7]

During the storage time, the initially 'hot' H_3^+ and D_3^+ molecular ions are given the opportunity to cool down. Vibrational relaxation of H_3^+ and D_3^+ has been demonstrated previously using the Coulomb Explosion Imaging technique. The measured lifetime of the most long lived vibrational mode was ~0.4 s for H_3^+ and ~1.4 s for D_3^+, and no further changes of the vibrational state distribution of the ions were found after 2 s for H_3^+ and 5 s for D_3^+.[8]

In the interaction region of the ion and the electron beam, neutral fragments are produced due to DR reactions. These fragments exit the ring and are detected using an imaging detector that is mounted straight ahead of the interaction region, at a distance of 6 m. The projected relative distances between the fragments hitting the surface of the detector, which is normal to the direction of the beam, are measured using a two-dimensional imaging detector, which is similar to the one used in previous experiments.[3] The use of the imaging detector allows for the measurement of the kinetic energy release, as well as the kinematical correlation for the DR of a single molecule at a time.

3. RESULTS

3.1 Two-fragment breakup analysis

In case of two body fragmentation, at low initial energy above the ionic ground state, an H(1s) and an H_2 molecule (for the DR of H_3^+) or a D(1s) and a D_2 molecule (for the DR of D_3^+) are produced. Careful center-of-mass cuts are made in the data to allow for the separation between a real two-body fragmentation and an event where three particles are produced, but one is missed by the detector because of its final detection efficiency (~50%).

In the two body breakup channel, 9.23 eV (for H_3^+) or 9.19 eV (for D_3^+) is distributed among the ro-vibrational excitation of the molecular fragment and the kinetic energy between the atomic and molecular fragments. Figs. 1(a) and (b) show the observed distribution, $P(D)$, of projected distances D on the surface of the detector for the two body channel recombination of H_3^+ and D_3^+, respectively, for a measurement time $3 < t < 10$ s (for H_3^+) and $6 < t < 20$ s (for D_3^+). As in earlier experiments with diatomic-ion recombination, *P(D)*, for a given kinetic energy release (KER), has a characteristic shape (see dashed line in Figures 1(a) and (b)) resulting from the average over the dissociation angles and the extent of the interaction region.[4] The contributions of different KER, depending on the vibrational state of the molecular fragment, add up to the total $P(D)$ distribution. Rotational excitation can either increase or decrease the KER, depending on whether it is present in the original molecular ion or in the molecular fragment. However, even for considerable excitation, the associated shifts in the KER are small compared to the effect of the vibrations.

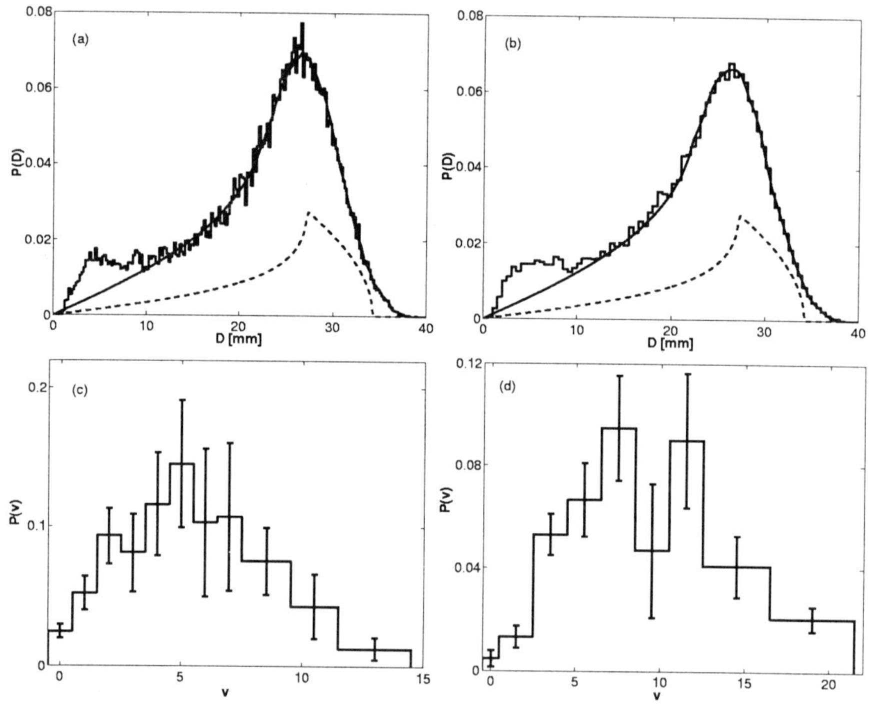

Figure 1. Top panels: Projected distance distributions for the two body DR of (a) H_3^+ and (b) D_3^+. The dashed lines are the characteristic distributions for v=5 (H_3^+) and v=9 (D_3^+), respectively. Bottom panels: Vibrational distributions of (c) H_2 and (d) D_2 products.

Neglecting the rotations, the observed $P(D)$ spectra (for $D > 10$ mm) were fitted with a superposition of projected distributions for various $H_2(v)$ or $D_2(v)$ final states, assuming isotropic dissociation and using the relative vibrational state populations as fit parameters. The resulting vibrational distributions are shown in Figures 1(c) (for H_2) and 1(d) (for D_2). The neglected effects of rotation contribute an uncertainty of f 1 in the vibrational quantum number v.

In both Figs. 1(a) and (b), the fits are very good for projected distances $D > 10$ mm. At small distances, a "shoulder" appears in the data, which cannot be reproduced by the analysis. This shoulder is most likely due to the presence of high rotational states in the H_3^+ and D_3^+ ions (see also Sec. 3.2). Such states, with an energy of ~1 eV above the ground state, can (probably very efficiently) recombine to the next available channel which is $H(2s)+H_2(v)$ (or $D(2s)+D_2(v)$), which lies 0.97 eV for H_3^+ (1.01 eV for D_3^+) above the ro-vibrational ground state of the molecular ion. Measurements made with a D_3^+ beam with two different electron beam sizes (and thus electron densities) show no variations in the projected distance spectrum, thus ruling out the possibility (also disfavored by quantitative estimates) that the observed shoulder could be due to collisions taking place in the toroid regions of the electron cooler, at increased electron energies of > 1 eV.[4]

On the other hand, the vibrational population distribution extracted from the fit of Figs. 1(a) and (b) is not affected by the presence of this additional channel, and the results displayed in Figs. 1(c) and (d) show that the H_2 or D_2 molecules produced by the DR of H_3^+ or D_3^+ are highly excited. No isotope effect is apparent, except for the trivial shift due to the different vibrational level spacings of H_3^+ and D_3^+.

3.2 Three-fragment breakup analysis

In the case where the DR leads to the production of three atomic fragments, the total energy difference between the initial state of H_3^+ or D_3^+ and the asymptote of H(1s)+H(1s)+H(1s) (or D(1s)+D(1s)+D(1s)), 4.76 eV for H_3^+ or 4.634 eV for D_3^+, is distributed as kinetic energy release among the three atomic fragments. To represent the momentum correlation in three body systems, the so-called "Dalitz plot" representation[10] is common in high energy physics. It offers the advantage that an uncorrelated momentum distribution (restricted only by energy and momentum conservation) would produce a uniform distribution in the special coordinates used. Keeping in mind possible symmetry properties due to invariance of the system under exchange of fragments, a possible definition of the Dalitz coordinates η_1, and η_2 is

$$\begin{aligned} \eta_1 &= \frac{(E_2 - E_1)}{\sqrt{3}E_k} \\ \eta_2 &= \frac{2E_3 - E_2 - E_1}{3E_k}, \end{aligned} \quad (1)$$

where E_k is the total KER in the DR process:

$$E_k = E_1 + E_2 + E_3. \quad (2)$$

Fig. 2 shows the η_1, η_2 representation, where each point in this space represents a unique dissociation geometry. The symmetry under exchange of the indices of the three fragments is represented by the three reflection axes, marked with dashed lines. The circle of radius 1/3 indicates the border defined by energy and momentum conservation - the maximum energy carried by a single particle has to be $\leq (2/3)E_k$. It is easy to observe that radius

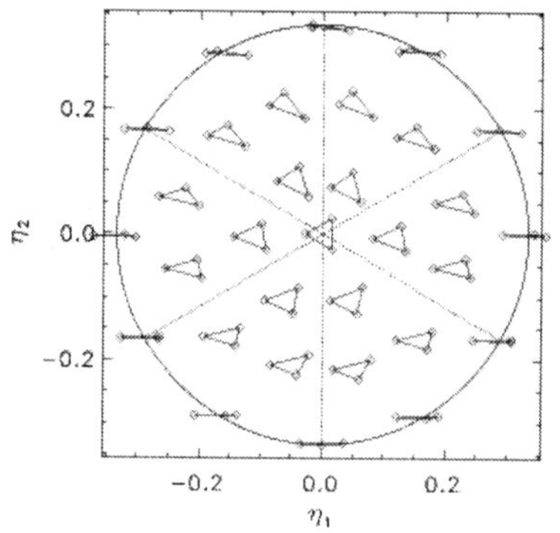

Figure 2. Dalitz map of all the possible dissociation geometries in momentum space.

and angle, the polar coordinates in this representation, have a geometrical meaning regarding the shape of the triangle (independent of its overall size and, in fact, its orientation in space). Increasing radius represents the `approach to linearity', as defined by the momenta of the three fragments, where the center corresponds equilateral triangles and the circumference to linear geometries. The angular coordinate in this plot represents shearing deformations, with coordinates on the three reflection axes corresponding to isosceles triangles.

As the 2D imaging technique records distances in the detector plane, transverse to the beam velocity, only the transverse momenta and energies can be measured. Thus, we defined a new set of coordinates, defined on the basis of the transverse distances (which are proportional to the transverse kinetic energy of each fragment):

$$\begin{aligned} Q_1 &= \frac{(R_2^2 - R_1^2)}{\sqrt{3}R^2} \\ Q_2 &= \frac{2R_3^2 - R_2^2 - R_1^2}{3R^2}, \end{aligned} \qquad (3)$$

where R_i is the distance of particle i from the CM position on the surface of the detector, and R^2 is defined as $R^2 = \sum_i R_i^2$ and is directly related to the total transverse kinetic energy release.

Similar to the η_1, η_2 representation, each point in the Q_1, Q_2 plane represents the geometry of the three impact positions on the detector plane, which is in fact the projection of the η_1, η_2 geometry in the dissociation plane onto the detector plane. However, there is no unique correspondence between the geometry in the dissociation plane and the projected geometry. Each

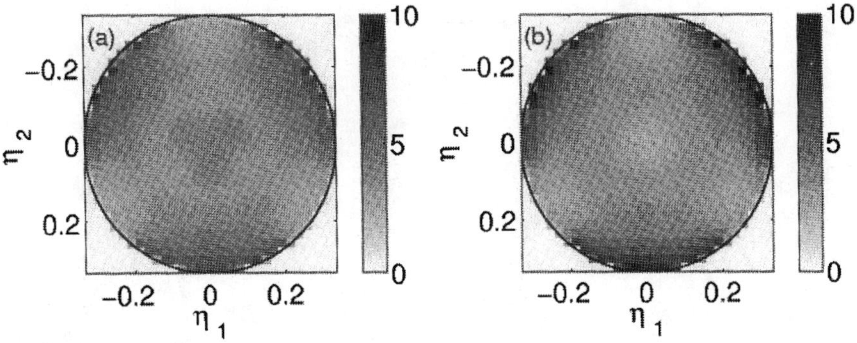

Figure 3. Dalitz plot obtained after Monte Carlo reconstruction for the three-body channel of the DR of (a) H_3^+ and (b) D_3^+.

specific dissociation geometry is transformed into a variety of different projected geometries, depending on the orientation of the dissociation plane. Monte Carlo simulations of the experimental setup were used to study the projection of dissociation geometries on to the detector plane. The simulation samples events with specific dissociation geometries and projects them onto the Q_1, Q_2 plot by rotating them randomly and calculating the transverse coordinates. The simulation also takes into account the fact that overlapping hits on the detector cannot be resolved if their distance is less than 1 mm.

Using this Monte Carlo simulation, a response function $S(Q_1, Q_2)$, which allows for the connection of the (η_1, η_2) space to the (Q_1, Q_2) space, can be built and used in a Monte-Carlo restoration technique.[11] This allows the (η_1, η_2)-space distribution to be retrieved from the measured (Q_1, Q_2)-space distribution. The reconstruction technique was tested on simulated data, with similar statistics, and was found to be reliable.

Figures 3(a) and (b) show the results of the Monte Carlo reconstruction procedure for the DR of H_3^+ and D_3^+, respectively. These results can be directly compared to the geometry map provided in Fig. 2. For both H_3^+ and D_3^+, the central region, which corresponds to near-equilateral triangles (i.e., the initial geometry of H_3^+ and D_3^+), is depleted. On the other hand, there are maxima close to the rim of the circle defined by momentum conservation in the Dalitz plots, a region corresponding to linear structures. Comparison of the results for H_3^+ and D_3^+ shows an isotope effect: the results for D_3^+ are more smeared than for H_3^+.

In the three body dissociation channel, the KER is equal to the relative kinetic energy of the electron (E_e) plus the initial excitation energy of H_3^+ or D_3^+ relative to the limit of the three dissociated atoms. For an electron

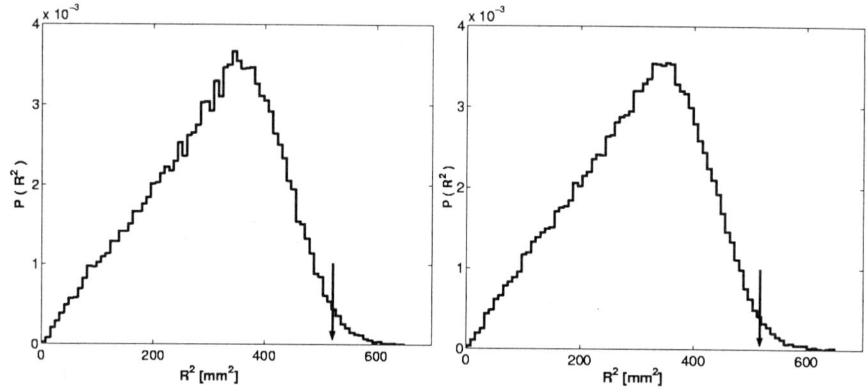

Figure 4. Distribution of the transverse kinetic energy release *(R^2)* as measured for (a) H_3^+ and (b) D_3^+. The arrows indicated the calculated endpoint of the distributions assuming ro-vibrationally cold H_3^+ and D_3^+, respectively. The final detector resolution contributes an uncertainty of only about ±5 mm² to R^2 at ~500 mm².

energy, $E_e = 0$, and assuming that the molecular ions are in their ro-vibrational ground state, the KER should be strictly 4.76 eV for H_3^+ and 4.63 eV for D_3^+. As the 2D imaging technique records distances in the detector plane, transverse to the beam velocity, only the transverse part of the KER can be measured and is simply related to R^2. Figs. 4(a) and (b) show the R^2 spectra of the three body breakup events for H_3^+ and D_3^+, respectively. The spectra are found independent of the storage time up to the maximum of 40 s. The overall shape of these spectra is a function of the kinematical correlation existing in the dissociation process. A complete reconstruction of these distributions can be achieved by the Monte Carlo reconstruction technique, and results will be presented in a future publication.[12] The arrows on Figs. 4(a) and (b) mark the calculated endpoint for an R^2 distribution, assuming that (1) the DR events took place at the far side of the e-cooler, (2) the dissociation occurred in the detector plane, so that the full energy released was deposited in the transverse KER, and (3) the molecular ions were ro-vibrationally cold. Thus, the positions indicated by these arrows are the largest expected values for the R^2 spectra in the case of internally completely relaxed ions. However, for both H_3^+ and D_3^+, the measured KER distribution exceeds these positions with a tail extending to an excess energy up to ~0.8 eV (where still ~1% of the intensity at the calculated end point is found).

Since the three atomic fragments produced by the DR cannot be in an internally excited state, this excess energy is interpreted as being due to a substantial population of initial excited states in the H_3^+ or D_3^+, prior to recombination. The *vibrational* excitation of H_3^+ and D_3^+ as a function of storage time has been previously probed[8] with the CEI technique, which is sensitive to the vibrational but not the rotational excitation of the molecular ion beam; the DR measurements always started after the vibrational relaxation had come to an end according to the CEI measurements (see Sec. 2). In view of these precautions, together with the fact that the tail in the R^2 distribution appears to be unchanged after different storage periods, we conclude that the excess KER measured in the dissociating system originates from the *rotational* degrees of freedom. Indeed, it is well known that H_3^+ in gas discharges, such as in the ion source, are dominantly produced by a reaction of H_2^+ with H_2, in which the H_3^+ products can receive substantial rotational excitation. On the other hand, neither H_3^+ nor D_3^+ have a permanent dipole moment, so that no allowed radiative rotational transitions can take place; hence, very slow rotational cooling is expected and rotational states, excited in the ion production process, can remain in the beam even after a long storage time. When full reconstructions of the KER spectra are performed,[12] it is found that a Boltzmann distribution of ~ 0.3 eV has to be added to the value of the KER in order to reproduce the distributions shown in Figs. 4(a, b). Also, the indications of high rotational excitation found in these distributions are in good agreement with those found in Sec. 3.1 (Figs. 1a and b) for the two-body channel.

4. CONCLUSIONS

The results presented in this paper demonstrate, for the two-body channel in the DR of H_3^+ and D_3^+, that vibrationally excited molecular species are produced. For the three-body channel, it is found that the three atoms like to dissociate with their relative momenta aligned to each other. These two elements of information can be expected to help in understanding the structure of the potential surface on which the DR of H_3^+ and D_3^+ takes place. From our data we also conclude that the molecular ions, prior to recombination, were rotationally excited under the conditions of the present (and probably also the earlier) storage ring experiments. We did not measure the absolute DR rate coefficients here, but rather propose to perform dedicated measurements of these quantities only after the initial rotational population of the molecular ions is under control. We also propose to check whether the large disagreements between the reported rate coefficients for the DR of H_3^+ and D_3^+ could be due to the different initial rotational population distributions produced in the various experimental environments.

ACKNOWLEDGMENTS

This work has been funded in part by the German Federal Minister of Education, Science, Research and Technology (BMBF) within the framework of the German-Israeli Project Cooperation in Future Oriented Topics (DIP), and by the European Community within the Research Training Network "Electron Transfer Reactions."

REFERENCES

1 P. Forck, M. Grieser, D. Habs, A. Lampert, R. Repnow, D. Schwalm, A. Wolf, and D. Zajfman. Phys. Rev. Lett. **70**, 426 (1993).

2 P. Forck, C. Broude, M. Grieser, D. Habs. J. Kenntner, J. Liebmann, R. Repnow, D. Schwalm, A. Wolf, Z. Amitay, and D. Zajfman. Phys. Rev. Lett. **72**, 2002 (1994).

3 D. Zajfman, Z. Amitay, C. Broude, P. Forck, B. Seidel, M. Grieser, D. Habs, D, Schwalm, and A. Wolf. Phys. Rev. Lett. **75**, 814 (1995).

4 Z. Amitay, D. Zajfman, P. Forck, U. Hechtfischer, B. Seidel, M. Grieser, D. Habs, R. Repnow, D. Schwalm and A. Wolf, Phys. Rev. A **54**, 4032 (1996).

5 Z. Amitay, A. Baer, M. Dahan, L. Knoll, M. Lange, J. Levin, I. F. Schneider, D. Schwalm, A. Suzor-Weiner, Z. Vager, R. Wester, A. Wolf and D. Zajfman. Science **281**, 75 (1998).

6 D. Habs et al., Nucl. Instr. and Meth. B **43**, 390 (1989).

7 G. Kilgus et al., Phys. Rev. A **46**, 5730 (1992).

8 H. Kreckel, diploma thesis, University of Heidelberg, unpublished; and H. Kreckel et al., in preparation.

9 D. Strasser et al., Phys. Rev. Lett. **86**, 779 (2001).

10 R. H. Dalitz, Phil. Mag. **44**, 1068 (1953).

11 B. R. Frieden, in *Picture Processing and Digital Filtering*, ed. by T. S. Huang (Springer-Verlag, Berlin, 1979), p. 240.

12 D. Strasser et al., in preparation.

Coupling of Bound States to Continuum States in Neutral Triatomic Hydrogen

H. Helm, U. Galster, I. Mistrík, U. Müller, and R. Reichle
Department of Molecular and Optical Physics, Universität Freiburg, Fakultät für Physik, Hermann-Herder-Str. 3, D-79104 Freiburg, Germany
Hanspeter.Helm@physik.uni-freiburg.de

1. INTRODUCTION

Rydberg states of the neutral triatomic hydrogen molecule can be viewed as an electron attached to a tightly bound H_3^+ core.[1-4] These states are embedded in the dissociation continuum of the repulsive electronic ground state surface as seen from Fig. 1. As a consequence, all excited bound states of H_3 are subject to predissociation. The metastable $2pA_2''$(N=K=0) state is a notable exception. Here N refers to the total angular momentum apart from spin and K to the projection of N on the molecular top axis. This lowest rotational level is long-lived (700 ns)[5] owing to symmetry forbidden coupling with the electronic ground state by either radiation or predissociation. Neutral H_3 in this rotational level can be selectively prepared in a fast beam by charge exchange techniques. This state has been used as a platform for laser-excitation experiments[3-14] to explore the structure and dynamics of highly excited states of H_3.

Some of these investigations have addressed the dissociation dynamics of state-selected low principal quantum number (n=2,3) Rydberg states into the fragments H + H_2[13-16] as well as their three-body breakup into three ground state hydrogen atoms.[17,18] In the latest version of these experiments correlation maps of the ground state atom momentum vectors are obtained. These maps provide the most direct view of how the neutral triatomic species enters into the neutral dissociation continuum at molecular distances and how it emerges from this continuum at the separated atom limit.

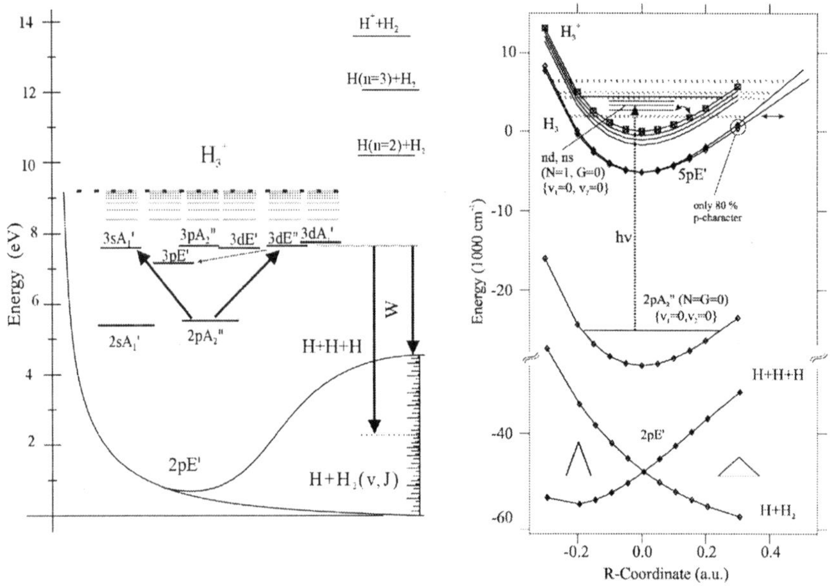

Figure 1. The excimer states of the H_3 molecule, embedded in the repulsive ground state are shown on the left. Also indicated is the translational energy release, W, which appears in the center-of-mass motion of the fragments when a selected state predissociates into neutral products. The Figure on the right shows a cut through the potential energy surfaces. The symbols give calculated energies of the H_3^+ ion and of the 2pE′, 2pA$_2$", and 5pE′ states of H_3. The R coordinate corresponds to a set of asymmetric stretch geometries.[14]

In separate studies high principal quantum number s-, p-, d-, and f-Rydberg states were excited by one- and two-color. The rotational, vibrational, and vibronic coupling mechanisms between the Rydberg electron and the ion core were studied theoretically in MQDT models.[14,19,20] Quantitative comparisons between experiment and theory indicate that with only few exceptions the level-positions and level-widths are precisely accounted for by an MQD theory, which is based solely on ab initio parameters from quantum-chemistry calculations, without regard of the neutral dissociation continuum. Exceptions appear for selected electronic states in that their lifetime is shortened due to predissociation. The experiments indicate that selective coupling to the dissociation continuum is effective for even very high principal quantum number Rydberg states (for example at the energy of n=61). The exact predissociation mechanisms of Rydberg states of H_3 with principal quantum numbers n>3 are currently not well understood.

We had previously pointed out[21] that predissociated Rydberg states may play an important role in supporting the dissociative recombination (DR) process of H_3^+ ions. The puzzling magnitude of the DR cross section which numerous experimenters have observed at low electron energy is still at odds with theory, after substantial efforts from theory and experiment to clarify the situation. The very latest measured cross section for DR[22] is approximately two to four orders of magnitude higher than current theory, which includes direct and indirect DR channels[23] or the Jahn-Teller coupling of p-states.[24]

While this paper cannot pinpoint the exact origin of the failure of theory to account for the huge cross section through which low energy electrons see the open neutral continuum, it discusses several examples of state-selective predissociation and competition between predissociation with autoionization. These results are equally lacking theoretical explanation. Obviously these resonances are examples of intermediate neutral states, which are active in the DR-process. We are confident that a common explanation for their predissociation and for the fast rate of DR of low energy electrons exists.

2. POTENTIAL ENERGY SURFACES

In its ground vibrational level the electronic state of H_3^+ is best described in D_{3h} geometry. The corresponding neutral excimer states of H_3 are shown in the center of Fig. 2. What is important for DR is that all excited state dissociation limits of the neutral molecule lie energetically *above* the lowest energy of the molecular ion. It is evident from Fig. 1, that any DR process involving electrons with energies less than 0.5 eV can only access the neutral electronic ground states of $H+H_2$ or H+H+H. In linear geometry the repulsive $1^2\Sigma_g^+$ state of H_3 intersects all bound excimer states, hence providing a convenient path for predissociation (Fig. 2, left). This channel is closed in D_{3h} geometry. Approximately 2 eV of vibrational excitation are required to open the triangular configuration of the ion into linear geometry in order to directly enter the continuum. Obviously tunneling will permit access at significantly lower energy, however no quantum chemistry calculations are currently available to estimate the likelihood of tunneling. Quantum chemistry calculations of the potential energy surfaces at geometries intermediate to these shown in Fig. 2 would help estimate such effects.

Fig. 2 only shows the energy levels of doublet H_3, as appropriate for addition of one electron to ground state singlet H_3^+. Quartet states of H_3 also exist, however the vertical energy separation of triplet H_3^+ from the ionic ground state geometry is about 20 eV.[25] Hence quartet states should play no role in the dynamics of H_3 near its lowest ionization threshold. The electronically excited doublet states (except for the repulsive 2pE' ground state surface) can be viewed as Rydberg states with a tightly bound H_3^+ core.[4] The equilibrium geometry of the ion, as well as the neutral excimers, is that of an equilateral triangle. Owing to the fermionic nature of particles forming H_3^+ only odd rotational levels exist for K = 0 for the nondegenerate vibrational levels.[26] The Rydberg states of H_3 are labelled[19] by $nL\Gamma\,(N,G)\{\nu_1,\nu_2^l\}$. These states are characterized by the principal quantum

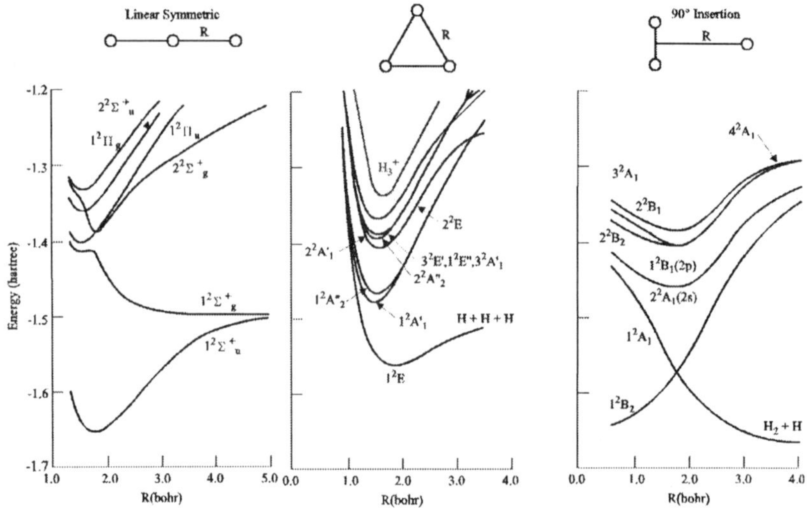

Figure 2. Energy levels of the H_3 molecule for three configurations are compared on an absolute energy scale (taken from Ref. 27). The meaning of the R -coordinate is indicated in the Figure above each diagram.

number n, the electronic angular momentum L of the Rydberg electron, and the electronic symmetry Γ in the D_{3h} group. The nuclear vibrational motion is labeled by the quantum numbers of the symmetric stretch (v_1) and degenerate (v_2) modes, and the vibrational angular momentum ι. The molecular rotation is described by the total angular momentum N disregarding electron spin, its projection on the molecular top axis K, and Hougen's convenient quantum number,[28] $G=\iota+\lambda-K$.

3. PREDISSOCIATION

Experimentally known lifetimes of the lower-lying Rydberg states are collected in the table below. For references see Table I of Ref. 29. As is evident from the magnitude of the lifetimes predissociation is a major decay channel of these states.

Experiments observe a most dramatic variation in lifetime among the s-states at low values of n. The lifetime increases by almost four orders of when going from n=2 to n=3, but decreases again by two orders of magnitude from n=3 to n=4. A recent theoretical study based on vibronic coupling between ns states with the ground 2pE' state[30] is able to explain the ratio of lifetimes between 2s to 3s in terms of the reduced overlap of the 3s potential energy surface with the repulsive ground state. However the experimental result that the coupling is again more efficient for the higher lying 4s and 5s states cannot be accounted for. The surfaces of the higher n-states are further

Table 1. Rydberg state lifetimes.

electronic state label $nL\Gamma$	rovibrational state label $(N,G)\{v_1,v_2^l\}$	experimental lifetime [ps]	theoretical lifetime [ps]	state energy relative to H_3^+ [eV]
$2sA_1'$	$(1,0)\{0,0^0\}$	0.2	1.7	-3.777
$3sA_1'$	$(1,0)\{0,0^0\}$	1000	8000	-1.595
$4sA_1'$	$(1,0)\{0,0^0\}$	7.6	-	-0.885
$5sA_1'$	$(1,0)\{0,0^0\}$	4.4	-	-0.562
$3pE'$	$(1,0)\{0,1^1\}$	>66	39	-1.737
$3sA_1'$	$(1,0)\{1,0^0\}$	>66	1400	-1.197
$4pE'$	$(1,0)\{0,1^1\}$	21	-	-0.714
$5pA_2''$	$(2,0)\{0,2^2\}$	6	-	0.034
$20pE'$	$(2,0)\{1,0^0\}$	5	-	0.369
$6dE''$	$(1,0)\{1,0^0\}$	2	-	0.014

removed from the repulsive ground state, hence having weaker overlap. We note that the avoided crossing between the repulsive $1^2\Sigma_g^+$ state of H_3 and the bound excimer states (see Fig. 2 left) falls energetically between the lowest vibrational levels of the 3s and the 4s state and may play a role in the predissociation of the higher lying s-states.

The npE' states are homogeneously coupled to the ground state surface. No theoretical data on their predissociation lifetimes exist. Such results could be obtained using the formalism presented by Greene et al.[24] The lifetimes of the npA_2'' levels are governed by rotational coupling with the ground electronic state. The lifetime should decrease with N(N+1)-K(K+1), as has been observed for the $2pA_2''$ state.[2] For the vibrationally excited 5p, 20p and 3d states autoionization competes with predissociation, both operating in the time scale of a few picoseconds.

3.1 Predissociation of High-n Rydberg States

High principal quantum number Rydberg states (n>30) were investigated by monitoring both the field ionization signal concurrent with the photofragment yield.[29] Fig.'s 3a and 3b have the photofragment yield and Fig. 3c has the field ionization signal for one-photon excitation from the H_3 $2pA_2''$ state in the energy range just below the lowest ionization threshold.

The transitions in the ionization spectrum were assigned to the d-Rydberg series converging to the $H_3^+(1,0)\{0,0^0\}$ threshold[3], which is marked by a dashed line. Up to the highest resolved principal quantum numbers, contributions also appear in the photofragment signals.

The windows in the field ionization intensity close to n=44 and n=61 correlate with enhanced fragment yield. An earlier laser depletion experiment[13] showed that the absorption strength is a monotonic function of principle quantum number. The intensity modulations that appear in Fig. 3c

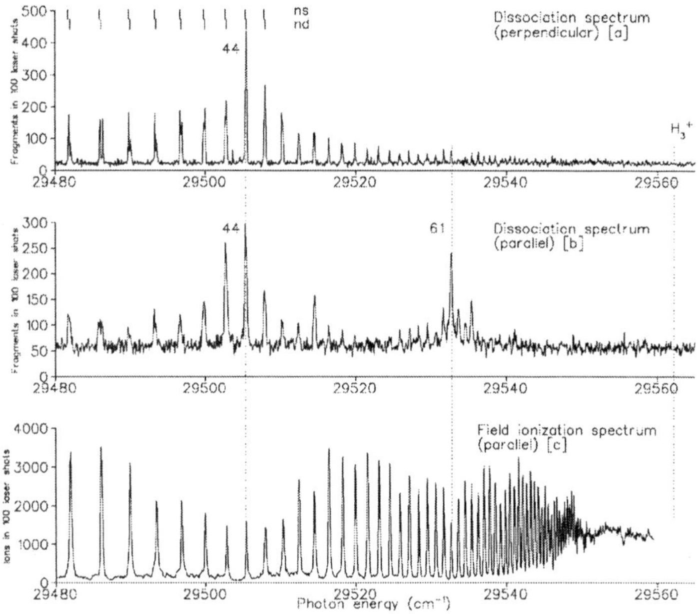

Figure 3. Laser-excitation of H_3 in the vicinity of the lowest ionic limit. The photofragment yield [a,b] is compared with the field ionization spectrum [c]. In [a], the laser polarization was perpendicular, in [b] parallel to the axis defined by the lateral displacement of the detector for neutral fragments. The ns- and nd-Rydberg series converging to the $H_3^+(1,0)\{0,0^0\}$ threshold are indicated by tick marks.

are caused by a shortening of the lifetime due to predissociation. In the experiment only an upper limit can be ascertained for the lifetime of the states, which appear in the spectra of Fig. 3. The Rydberg states predissociate on a time scale faster than about 2μs. The significantly different magnitude of the photofragment signal observed for the two orientations of the laser polarization relative to the detection-axis (Fig.'s 3a and 3b) is caused by the state specific angular distribution of photofragments. When perpendicular laser polarization is chosen, the photodissociation signal in the vicinity of n=44 is significantly larger than that at n=61. This suggests that the doorway states mediating the predissociation process are different in nature at n=44 and n=61. This is consistent with the observation of a profound electric field dependence of the intensity window at n=61. Predissociation of the 61d-state is due to Stark mixing[12] with a vibrationally excited member of the npE′ series. In contrast the predissociation doorway around n=44 is insensitive to the presence of external electrical fields, hence this state must find access to the neutral continuum via another doorway state. In addition to the interlopers at n=44 and n=61, increased predissociation is observed in the vicinity of the n=14 and n=21 states.

It is remarkable that ns-Rydberg members are observed in the photofragment signal in Fig. 3, while they are absent in the ionization signal. This shows that the s-series members are much more strongly predissociated than are higher L-states. The energetic location of most members of the s- and d-series are in agreement with the Rydberg formula using the established quantum defects to a precision of 0.2 cm^{-1}. However, perturbations which shift the line positions by more than 0.5 cm^{-1} appear at 13d, 14d, and 24s. A number of additional predissociating resonances built on vibrationally excited cores of H_3^+ have also been identified (see Table II of Ref. 29).

3.2 Predissociation of autoionizing Rydberg states

In the pE'-series built on a symmetric-stretch excited H_3^+ core, the positions, relative amplitudes, and widths of most of the quasi-discrete lines are generally understood from an MQD-theory which does not take into account the existence of the neutral dissociation continuum. For a few members of the series, the measured relative intensities are observed to be lower and the line widths broader than those predicted by MQDT. This observation was taken as indication that predissociation competes with autoionization.

The autoionization and predissociation spectrum in the range from 369.46 to 370.20 meV above the ionization threshold is shown in Fig. 4e and 4f, respectively. This range covers the vicinity of the 20p$\{1,0^0\}$ states with N=1 and N=2, excited via the 3sA_1'(1,0)$\{1,0^0\}$ intermediate. Taking into account the geometric collection efficiency of the off-axis detector which we find that the branching between predissociation and autoionization for the N=2 state amounts to at least 25%. Apparently, the N=2 state is strongly predissociated while the decay of the N=1 state is dominated by autoionization.

As an example for a d-state, we have investigated the symmetric-stretch excited 6dE"(1,0)$\{1,0^0\}$ state of H_3 which can be excited in a one-photon transition. The corresponding autoionization and predissociation spectra are shown in Fig. 4a and 4b. The Fano-type shape of the line profile is caused by the interference with the underlying ionization continuum. The decay of the 6dE" state is dominated by autoionization, the branching into predissociation is of the order of only a few percent. A third example where predissociation competes with autoionization is the 5pA_2"(2,0)$\{0,2^0\}$ level shown in Fig. 4c and 4d. Also here the branching to predissociation amounts to a few percent. As is evident from the energy scales in Fig. 4, low energy electrons involved in DR will find such states in the collision continuum and thereby gain access to the neutral ground state continuum.

3.3 Three-body decay

Predissociation lifetimes and the branching between predissociation and

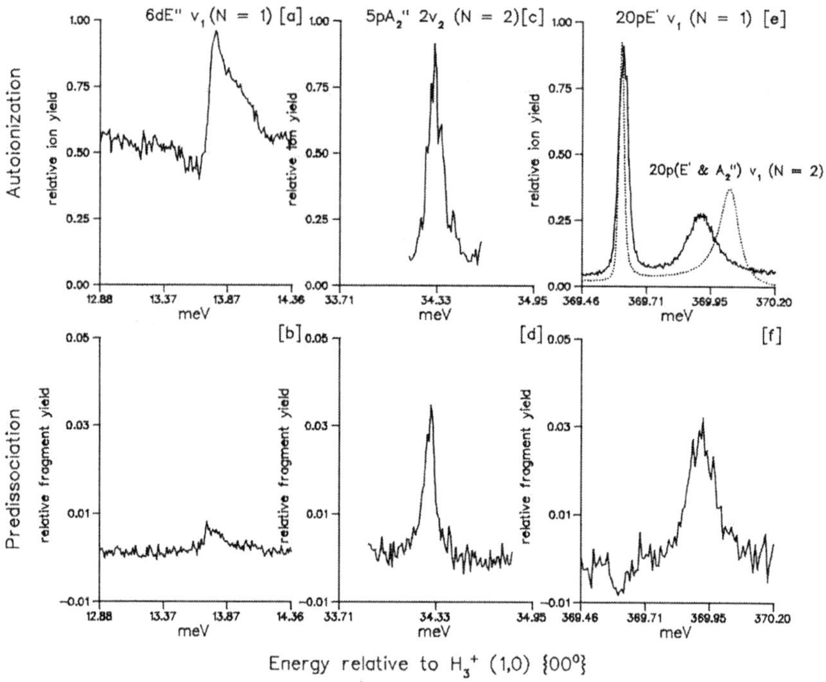

Figure 4. Examples of spectra, which reveal that predissociation competes with autoionization above the first ionization threshold of H_3. The dotted curve gives the MQDT prediction.

other decay channels are important experimental data to elucidate the dynamics of H_3. Additional experimental information can be gained on how the neutral molecule finds access into the dissociative continuum. Such information is contained in the dissociation products. For the states investigated here, as well as for the products of low energy DR, the final states are always in their ground electronic state, H+H_2(v, J) or else three hydrogen atoms, H(1s)+H(1s)+H(1s). In the former case the total energy available is shared between the translational energies of the atomic and the molecular fragment and the rovibrational excitation of the H_2 product.

It is obvious from Fig. 5 that the degree of rovibrational excitation is governed by the exact shape of the trough of the repulsive ground state surface and the initial position of the wave packet when it enters this surface. In this sense the shape of the initial wave packet disperses among the trajectories which populate specific (v, J) combinations.[30] In the case of two-body decay the information on the actual coupling to the neutral continuum, combined with the dynamic evolution on the ground state surface, is cast into a complex set of (v, J) variables, which bar us from a direct view of the predissociation mechanism.

The situation is intrinsically different in case of three-particle decay. Here,

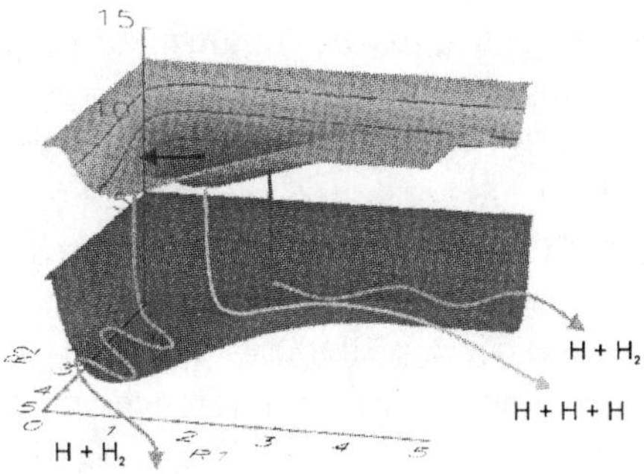

Figure 5. Schematic paths of predissociation into the repulsive ground state surface of H_3. The upper Jahn Teller sheet of the ground state surface, which branches to the H+H+H limit (see Fig. 2, left), is not shown. The origin of vibrational excitation of the H_2 product is apparent from the decay path on the left.

the sum of the kinetic energy of the three fragments is fixed at the energy difference W between the excited state and the H(1s)+H(1s)+H(1s) limit. In this case, the balanced center-of-mass momentum vectors in the molecular state are directly imaged into macroscopic observables: the correlated set of the three momentum vectors at which the fragments appear in the laboratory frame. A direct analysis of the three correlated momentum vectors is possible with the photofragment spectrometer developed at Freiburg.[18] With this instrument, triple coincidences of H+H+H are collected on a multi-hit time- and position-sensitive detector. The individual vectorial momenta of the three hydrogen atoms $m\mathbf{u}_1$, $m\mathbf{u}_2$, and $m\mathbf{u}_3$ in the center-of-mass frame are determined separately and a consistency check that $W = (m\mathbf{u}_1^2 + m\mathbf{u}_2^2 + m\mathbf{u}_3^2)/2$ can be carried out for each individual molecule investigated. Momentum conservation requires that $\mathbf{u}_1+\mathbf{u}_2+\mathbf{u}_3=0$. This implies that the three vectors have only six independent components. Given the experimental information on the three individual vectors we can determine the 6-fold differential cross section, completely characterizing the final state of the three-body decay process.

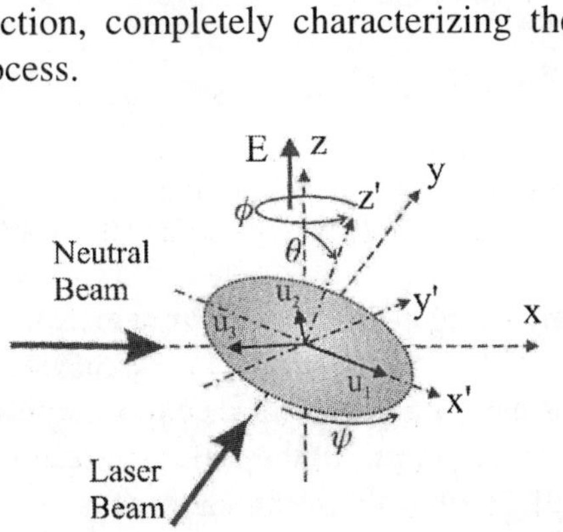

Figure 6. Orientation of the fragment momentum vectors in a three-body decay process.

In order to gain insight into the fragmentation pattern, we consider suitable projections of this six-dimensional set of information. We introduce the parameters in Fig. 6, which uniquely describe the three fragment momentum vectors. The $\mathbf{u}_i$ are contained in a plane. For each event, we define a new coordinate system (x', y', z') by the normal vector on this plane (z'-axis) and the direction of the largest of the three momentum vectors as x'-axis. Three Euler angles (ψ,θ,ϕ) describe the orientation of the (x', y', z')-coordinate system within the laboratory reference system (x, y, z), which is defined by the electric vector of the laser beam (z-axis) and the direction of the neutral beam (x-axis). The spatial orientation of the (x', y', z')-coordinate frame is determined by the spatial anisotropy of the photodissociation process.

The remaining three parameters describe the arrangement of the three momenta in the (x', y')-plane. We introduce the individual fragment energies $\varepsilon_i = m\mathbf{u}_i^2$ and use the total kinetic energy W and two parameters to show the vector correlation among the fragment momenta within the (x', y') plane.

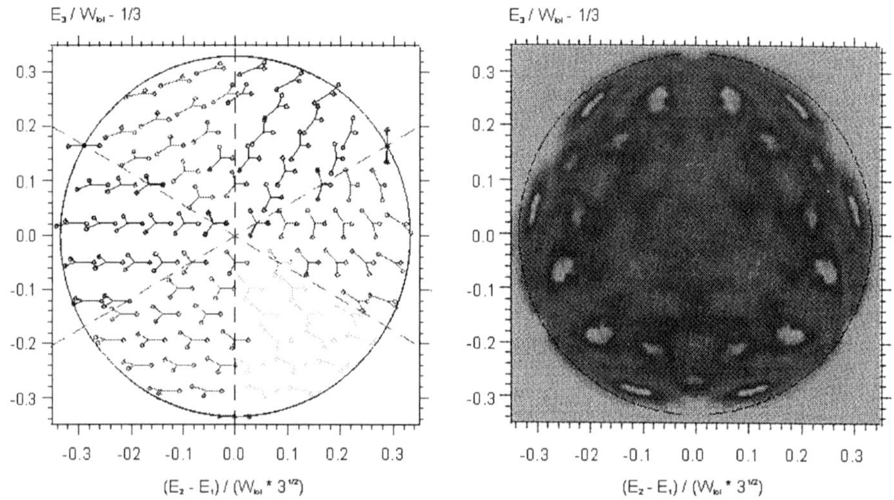

Figure 7. The generalized Dalitz plot on the left shows the correspondence between the location of a data point in the plot and the fragmentation configuration. An experimental distribution of fragmentation vectors is shown on the right (the 3dE'(1,0){1,0^0} state).

This is possible by using a Dalitz plot[18] in which we represent each event by a point in the (ε_3/W-1/3) vs. (ε_1-ε_2)/($\sqrt{3}$ W) plane. Energy and momentum conservation require the data points in this plot to fall inside a circle with radius 1/3, centered at the origin. In the Dalitz representation the phase space density is conserved. This means that a process with a matrix element independent of the fragmentation configuration leads to a homogeneous distribution in the kinematically allowed circular region. Preferred fragmentation pathways can immediately be recognized from the event density in the Dalitz plot as may be seen from Fig.'s 7 and 8.

In Fig.'s 8 (a) and (b), the triple-coincident events following three-body breakup of the $3sA_1$' and 3dE' Rydberg states (both in their ground vibrational level) are shown in Dalitz plots. Since the three hydrogen atoms are indistinguishable, points are drawn in Fig. 8 for the six permutations of the fragment energies, ε_i, measured for each dissociation event.

In order to understand the meaning of the very pronounced islands of correlation appearing in the experimental data in Fig.'s 7 and 8, the geometric and electronic collection efficiency of the detector has to be considered. The efficiency was determined in a Monte Carlo simulation by generating a uniform distribution of fragmentation configurations and calculating the fragment propagation to the detector.[31] Based on these Monte Carlo results, we can definitely exclude experimental artifacts as contributing to the patterning in the Dalitz maps. Nevertheless, in the Figures shown we have corrected the measured distributions by weighting the data points with the geometric collection efficiency.

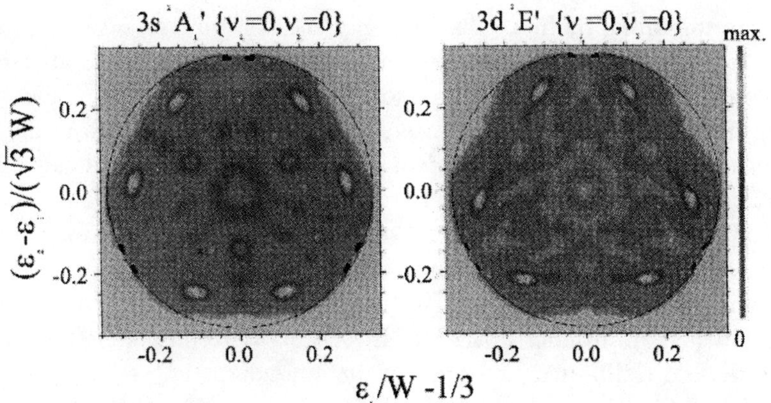

Figure 8. Dalitz plots of the three-body decay of the $3sA_1$'(1,0){$0,0^0$} and 3dE'(1,0){$0,0^0$} initial states of H_3. The geometric collection efficiency was calculated by a Monte-Carlo simulation and the measured data were corrected.

Although the initial molecular states are characterized by high symmetry (D_{3h}), rather asymmetric fragmentation configurations are much preferred in finding a path into the three-particle continuum. Neither the totally symmetric configuration (center of the plot in Fig. 7) nor isosceles configurations (dashed lines in Fig. 7, left) show preferred population. It may surprise, that the preferred fragmentation configurations sensitively depend on the initial state. The absolute energies of the states investigated in Fig. 8 differ by as little as 75 meV and the nuclear equilibrium configuration for each initial state is extremely close to that of the vibrationless H_3^+ ion in its ground electronic state and its lowest rotational level. Since the first publication of such maps,[18] these patterns have been studied for about twenty different rovibronic Rydberg-states of H_3. Each state shows individual maps with significant vibrational dependence (compare for example Fig. 7 with Fig. 8b). Also, when comparing maps of H_3 and D_3 substantial isotope effects are discovered.

The striking difference of the final state distributions in Fig. 8a and 8b reflects the different coupling mechanism active between the initial state and the two sheets of the repulsive ground state potential energy surface. In these two cases nearly identical vibrational wave packets attempt to access the continuum, at nearly equal total energy. The suggestion is that the zero point vibration, while exploring practically identical regions of internuclear configuration, finds individual regions of access to the continuum. This initial correlation among the fragment momentum vectors is propagated into the final state observables seen in the Dalitz plot.

From a symmetry point of view, the breakdown of the Born-Oppenheimer approximation for the $3sA_1$' state is mediated by the zero-point motion in the degenerate vibration. In case of the 3dE' state, the coupling is induced by the rotational tumbling motion.[17] In a semi-classical picture, the vibrational motion explores on a very short time scale the geometries where transitions between the initial state and the ground state surface may occur. Trajectories starting from these geometries interfere, producing a rich pattern in the final state distribution. While the coupling between the bound states and the continuum mediates the first entry of the quasi-bound system into the continuum, a series of avoided crossings between the upper sheet of the repulsive ground state surface and the s- and d-Rydberg states of $1^2\Sigma_g^+$ symmetry in linear geometry (see Fig. 2) will influence the appearance of the continuum state at infinite separation. Efforts are currently being undertaken to interpret the Dalitz maps in terms of the non-adiabatic couplings between the Rydberg states and the neutral continuum.[32]

4. CONCLUSIONS

The investigation of the dynamics of state-selected neutral states of H_3 reveals complex patterns of preferred predissociation. Most notable is an apparently irregular dependence of the predissociation rate on the energy and on the rovibronic nature of the state. These details have currently no firm theoretical understanding. For numerous predissociated states of H_3 a complete microscopic view of the half collision from a well characterized initial state to the momentum set of the three correlated ground state hydrogen atoms at infinite separation has been obtained with the Freiburg coincidence spectrometer.[18,31] In the case of DR a Dalitz-type map was recently also obtained.[33] The map obtained in DR is the superposition from a multitude of electronic and rotational angular momenta of the e+ H_3^+ complex formed as intermediate, even if the rovibrational state of H_3^+ is well defined. The state-selective information provided in our half collision approach is a stringent test of the quality of any theoretical treatment of the e+ H_3^+ problem. It is also a test to its applicability to predict a reliable DR cross section. We are confident that the unrestricted approach to this problem, currently explored by Tashiro and Kato,[34] will be able to account for all experimental details documented for this fundamental system of three protons and three electrons.

ACKNOWLEDGEMENT

This research was supported through generous funding by the Deutsche Forschungsgemeinschaft in the Sonderforschungsbereich SFB276, TP-C13.

REFERENCES

1 G. Herzberg, J. Chem. Phys. **68**, 5298 (1978).

2 I. Dabrowski and G. Herzberg, Can. J. Phys. **58**, 1238 (1980).

3 H. Helm, Phys. Rev. Lett. **56**, 42 (1986); H. Helm, Phys. Rev. A **38**, 3425 (1988).

4 J. K. G. Watson, in *From Atoms to Polymers: Isoelectronic Analogues, Vol. 11 Chap. 4* ed. by J. F. Liebman and A. Greenberg (VCH Publishers, 1989).

5 M. C. Bordas, P. C. Cosby, and H. Helm, J. Chem. Phys. **93**, 6303 (1990).

6 A. Dodhy, W. Ketterle, H. Messmer, and H. Walther, Chem. Phys. Lett. **151**, 133 (1989).

7 W. Ketterle, H.-P. Messmer, and H. Walther, Europhys. Lett. **8**, 333 (1989).

8 L. J. Lembo and H. Helm, Chem. Phys. Lett. **163**, 425 (1989).

9 L. J. Lembo, H. Helm and D. L. Huestis, J. Chem. Phys. **90**, 529 (1989).

10 L. J. Lembo, C. Bordas, and H. Helm, Phys. Rev. A, **42**, 6660 (1990).

11 C. Bordas, L. J. Lembo, and H. Helm, Phys. Rev. A, **44**, 1817 (1991).

12 C. Bordas and H. Helm, Phys. Rev. A **45**, 387 (1992); *ibid.* **47**, 1209 (1993).

13 C. Bordas and H. Helm, Phys. Rev. A **43**, 3645 (1991).

14 I. Mistrík, R. Reichle, U. Müller, H. Helm, M. Jungen and J. A. Stephens, Phys. Rev. A **61**, 033410 (2000).

15 P. C. Cosby and H. Helm, Phys. Rev. Lett. **61**, 298 (1988).

16 U. Müller and P. C. Cosby, J. Chem. Phys. **105**, 3532 (1996).

17 U. Müller and P. C. Cosby, Phys. Rev. A **59**, 3632 (1999).

18 U. Müller, Th. Eckert, M. Braun and H. Helm, Phys. Rev. Lett. **83**, 2718 (1999).

19 R. Reichle, I. Mistrík, U. Müller and H. Helm, Phys. Rev. A **60**, 3929 (1999).

20 J. A. Stephens and C. H. Greene, J. Chem. Phys. **102**, 1579 (1995).

21 H. Helm, in Dissociative Recombination: Theory, Experiment and Applications, edited by B. R. Rowe, A. Mitchell, A. Canosa, (Plenum Press N.Y. 1993) p. 145-153.

22 M. J. Jensen, H. B. Pedersen, C. P. Safvan, K. Seiersen, X. Urbain and L. H. Andersen, Phys. Rev. A **63**, 052701 (2001).

23 I. F. Schneider, M. Larsson, A. E. Orel and A. Suzor-Weiner, in *Dissociative Recombination: Theory, Experiments and Applications IV*, ed. by M. Larsson, J. B. A. Mitchell and I. F. Schneider (World Scientific, Singapore, 1999), p. 131-141.

24 V. Kokoouline, C. H. Greene, and B. D. Esry, Nature **412**, 891 (2001).

25 K. Kaufmann, M. Jungen, and H.-J. Werner, J. Phys. Chem. **87**, 3806 (1983).

26 J. K. G. Watson et al., Can. J. Phys. **62**, 1875 (1984).

27 I. D. Petsalakis, G. Theodorakopoulos, and J. S. Wright, J. Chem. Phys. **89**, 6850 (1988).

28 J. T. Hougen, J. Chem. Phys. **37**, 1433 (1962).

29 I. Mistrík, R. Reichle, H. Helm, and U. Müller, Phys. Rev. A **63**, 042711 (2001).

30 I. F. Schneider and A. E. Orel, J. Chem. Phys. **111**, 5873 (1999).

31 U. Galster et al., Europ. J. Phys. B accepted for publication (2002).

32 U. Müller et al., in preparation.

33 D. Strasser, L. Lammich, S. Krohn, M. Lange, H. Kreckel, J. Levin, D. Schwalm, Z. Vager, R. Wester, A. Wolf, and D. Zajfman, Phys. Rev. Lett. **86**, 779 (2001).

34 M. Tashiro and S. Kato, this volume, p. 243.

Rotationally Resolved Photoelectron Spectra from Vibrationally Autoionizing Rydberg States of NO

Ian Konen, Runchuan Zhao, and Richard N. Zare
Department of Chemistry, Stanford University, Stanford, CA 94305.
rnz@stanford.edu

1. INTRODUCTION

The usual starting point for modeling bound electronic states in molecules is the Born-Oppenheimer approximation. The approximation rests on the assumption that the electrons in a bound state move appreciably faster than the nuclei. Thus the electrons experience a field arising from the "frozen" positions of the nuclei at any given instant, and the nuclei experience a field from an effective electron density equal to the time averaged positions of the electrons.

Highly excited Rydberg states provide an obvious counter example. The Born-Oppenheimer approximation generally remains valid for the particles of the ionic core (all the nuclei and all the electrons except the Rydberg electron), but Rydberg electrons with even a moderate degree of excitation (around n = 10) have characteristic orbital periods longer than a typical vibrational period. Although this situation represents a breakdown of the Born-Oppenheimer approximation, the motion of the Rydberg electron can still be separated from the motion of the nuclei and the core electrons because of its slower speed and its distance from the core. A simplifying approximation that suffices in place of the Born-Oppenheimer approximation is that the potential the Rydberg electron experiences is adequately described by the total charge of the core without the need to consider higher multipole moments.

Rydberg states with low angular momenta complicate the picture further. The classical orbit corresponding to a high n, low l Rydberg state is a highly elliptical orbit. The electron still spends most of its time far away from the core, experiencing only the monopole coulomb potential. But it does approach close to the core (quantum mechanically speaking, the radial wave function has lobes at short radii) and interacts with the other electrons and the nonspherical potential from the core. When it does so, it is moving rapidly, owing to the depth of the potential well, and thus should at least temporarily obey the Born-Oppenheimer approximation. Each region has a different simplifying approximation. The complexity comes from tying them together. Multichannel quantum defect theory (MQDT) treats this connection by using two different basis sets for the total wave function, one that is appropriate for small distances and one that is appropriate for large distances. Phase shifts arising from the coulomb potential and boundary value constraints at infinity are described in the long-range basis set, and phase shifts and channel interactions arising from the core are described in the short-range basis set. A frame transformation[1] that connects the two basis sets is used to solve the entire problem for bound and continuum states.

Vibrational autoionization, which can be viewed as part of the reverse pathway for dissociative recombination, provides a particularly interesting opportunity to study a breakdown of the Born-Oppenheimer approximation. In vibrational autoionization, a vibrationally and electronically excited molecule undergoes a crossing to a vibrationally relaxed ion plus a free electron. The transfer of energy from the nuclei to the electron runs counter to the premise that the nuclei only experience an averaged field from the electron cloud. Energy transfer in the quasibound state can also go from electronic to nuclear motion, in a process known as predissociation. The existence of both autoionization and predissociation pathways at a given energy facilitates dissociative recombination.

Departures from the Born-Oppenheimer approximation are observed frequently in spectroscopy and dynamics studies, but are often adequately modeled by a small perturbative correction to a treatment that uses the approximation as a starting point. The mixing between bound and continuum states is governed by the same physics as the mixing between bound states but the observation of the former is much more direct. By detecting electrons ejected by vibrational autoionization, we are sensitive to very weak coupling, because we are not merely looking for a small change in an otherwise large signal but we are measuring a signal that is only present because of the coupling.

2. EXPERIMENTAL

Vibrationally excited, high-lying (n = 12-18) Rydberg states of nitric oxide (NO) were prepared using two-photon double resonance excitation. Two pulsed lasers were counter-propagated into the vacuum chamber without focusing, and overlapped onto a pulsed molecular beam of pure NO. No attempt was made to cool the beam, as high rotational states were needed for this study in order to achieve rotational resolution. The output of a dye laser using Excalite 428 laser dye was doubled in a BBO crystal to produce 2 μJ pulses of light at 213 nm. The 213 nm light was used to excite the $R_{21}(17.5)$ line of the A $^2\Sigma^+$-X $^2\Pi$ (1,0) band in order to select a single rotational level in the vibrationally excited A state (J = 18.5, N = 19). A second dye laser, running a mixture of DCM and R640, was doubled in KDP to produce 100 μJ pulses of 330-340 nm light, which were used to excite the Rydberg-A (1,1) transitions.

2.1 Double Resonance Scans

Rydberg state resonance maxima were located by holding the wavelength of the first laser constant and scanning the wavelength of the second laser. Ions produced by the laser excitation were collected and amplified in a channeltron electron multiplier held at negative voltage relative to the rest of the vacuum chamber. The field strength in the ionization region was estimated to be no higher than 2 V/cm, insufficient to significantly perturb the Rydberg states we studied.

2.2 Circular Dichroism

Circular dichroism scans were performed to help distinguish R, P, and Q branches in the second excitation step. Quarter wave plates were used to circularly polarize both laser beams. Double resonance scans were performed over regions of interest, once with the lasers using the same helicity, and once with opposite helicity. Because the lasers were counter-propagated, opposite helicity resulted in parallel angular momenta for the two photons and same helicity resulted in anti-parallel angular momenta. Recall that we use the $R_{21}(17.5)$ line in the first step. Consequently, an R-branch transition in the second step is enhanced for the opposite helicity, a P-branch transition is enhanced for the same helicity, and a Q-branch transition is not changed between the two polarization configurations.

2.3 Photoelectron Spectra

Photoelectron spectra were taken by measuring electron flight times down a 50 cm magnetically shielded, field-free flight tube. Electrons were timed by monitoring the current from a set of MCPs after a laser trigger pulse. Signal levels were kept low enough to count individual electrons, and electron impacts were recorded using a LeCroy 3377 time-to-digital converter and summed on a personal computer. The lack of acceleration optics and a conical baffle halfway down the tube ensured that only electrons ejected initially in the direction of the detector were observed.

2.4 Angular Distributions

Angular distributions were measured by rotating the laser polarizations using half-wave plates mounted on computer-controlled stepper motors. The laser polarizations were stepped through 15° increments from 0° to 90°. The wave plates were rotated frequently (every 20 seconds) and cycled through all the angles until complete TOF spectra were recorded for each angle. Numerous repetitions of the whole cycle and careful monitoring of signal levels ensured a minimum level of systematic bias from angle to angle.

3. RESULTS AND DISCUSSION

3.1 Circular Dichroism Double Resonance

Fig. 1 shows an example of the circular dichroism double resonance scans. The expected ratio of peak intensities can be calculated with a simple application of angular momentum algebra. The counter-propagating circularly polarized laser beams preserve cylindrical symmetry of the system, allowing the total signal from a particular Rydberg state and laser geometry to be calculated by summing over an incoherent mixture of initial M states:

$$\mathrm{I}(\Delta J_1, \Delta J_2, \mu_1, \mu_2) = C \sum_{M''} \begin{pmatrix} J'' & 1 & J' \\ M'' & \mu_1 & -M' \end{pmatrix}^2 \begin{pmatrix} J' & 1 & J \\ M' & \mu_2 & -M \end{pmatrix}^2,$$

where $M' \equiv M'' + \mu_1, M \equiv M' + \mu_2, J' \equiv J'' + \Delta J_1, and\ J \equiv J' + \Delta J_2$. Here μ_1 and μ_2 are the spin projections of the absorbed photons onto the quantization axis. For circularly polarized light, they are either equal to 1 or -1 (with the quantization axis being defined along the propagation axis). The proportionality constant, C, depends on laser intensities, initial population of the J'' ground state, oscillator strengths and autoionization quantum yield.

The summation can be evaluated using tabulated formulas for the 3-J symbols[2] and it can be shown that for each Rydberg state, the ratio of signals from the two geometries depends only on ΔJ_1 (always an R branch in this experiment) and ΔJ_2. In the high-J limit, the ratios are

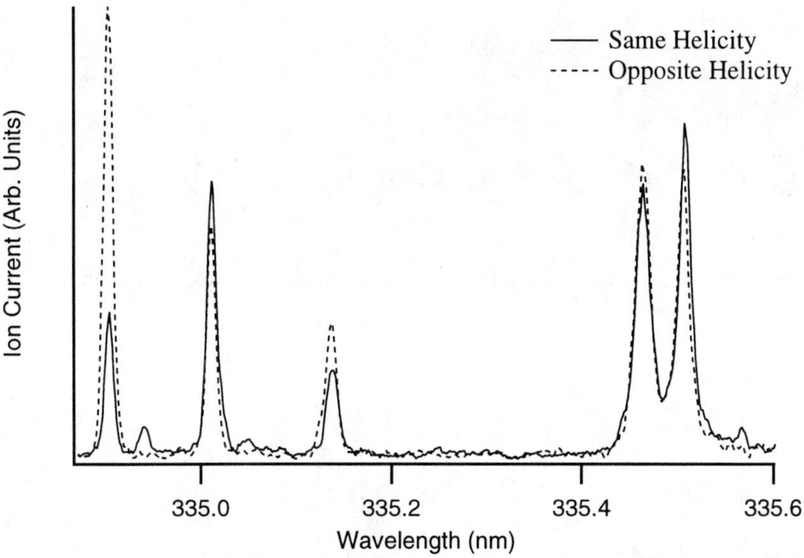

Figure 1. Circular dichroism double resonance scan .

$$\frac{I(R,R,1,1)}{I(R,R,1,-1)} \approx 6,$$
$$\frac{I(R,Q,1,1)}{I(R,Q,1,-1)} \approx 1,$$
$$\frac{I(R,P,1,1)}{I(R,P,1,-1)} = \frac{1}{6} \text{ (for all J).}$$

We were unable measure a ratio greater than 4.5 (or less than 1/4.5) probably owing to residual ellipticity in the polarizations of the two laser beams. Nonetheless, the signal difference was sufficient to distinguish branches. In Fig. 1, the peak at 334.91 nm can be safely attributed to an R branch, the small peak at 334.95 nm is clearly a P branch, and the peak at 335.02 is a Q branch. The peak at 335.13 is probably an overlap of two or more transitions.

3.2 Photoelectron Spectra

The Rydberg states used in this study have been identified in previous work.[3,4] Although there is known to be appreciable mixing, especially between s and d states,[5,6] a single-channel treatment of the Rydberg levels

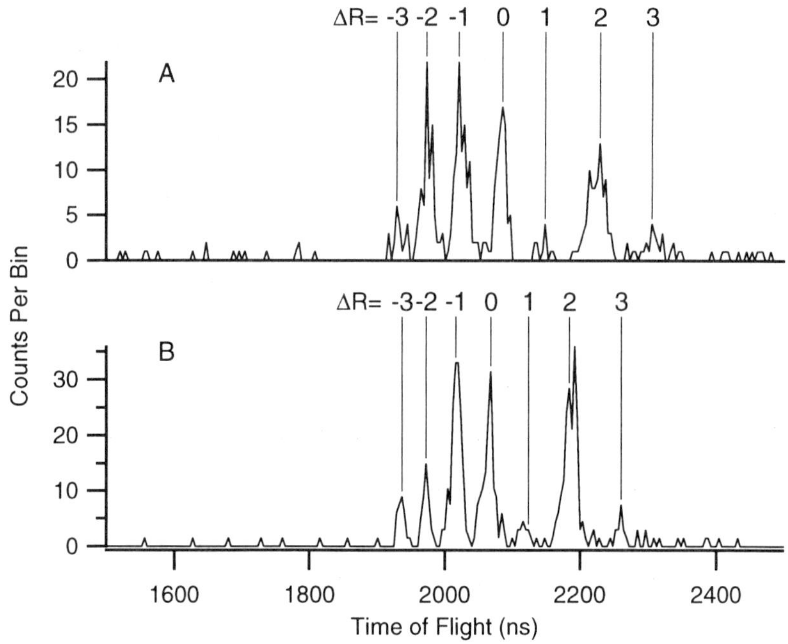

Figure 2. 14s Fixed-angle TOF photoelectron spectra, detection axis parallel to both laser polarizations: (A) 14s(R = 20); and (B) 14s(R = 18).

proved adequate for assigning resonances.

Fig. 2 shows TOF photoelectron spectra from the 14s(R = 18) and 14s(R = 20) states, taken with both laser polarizations parallel to the flight axis. Peaks correspond to the rotational states of the ions, and are labeled by the change in rotational quantum number from the Rydberg state to the ion ($\Delta R \equiv R^+ - R$). The rotational state of the core in the Rydberg state should be thought of only as a label, and its use here is not meant to imply that it is necessarily a good quantum number. The good quantum number is N, the total angular momentum of all particles, discounting spin. In NO, with its $^1\Sigma^+$ ion core, N is simply the vector sum of R and l. For a theoretically pure s state, $N = R$. Visual inspection of these two spectra shows the similarity between the decays from the two 14s levels, but it should be noted that they only represent detection in one direction, and not actual rotational branching ratios. For example, the $\Delta R = 0$ decay channel is the strongest, even though its contribution in the $\theta = 0°$ direction is comparable to the ΔR = -1 and -2 channels. Proper branching ratios are evaluated from the angular distributions, and tabulated in the first two rows of Table 1.

The strong presence of $\Delta R = 0$ peaks would seem to be consistent with the idea that the rotational distributions simply reflect the rotational composition of the bound Rydberg state. Even relatively high contributions from other even ΔR peaks are not ruled out by this picture because it is not clear exactly how much d character these s states have mixed in, and a thorough MQDT analysis is needed to determine it. However, the presence of the odd ΔR peaks indicates exchange of angular momentum in the autoionization process.[7] By parity conservation, odd rotational peaks must come from odd l partial waves, and the abundance of odd partial waves (26% of the total signal for the 14s(20) and 31% for the 14s(18)) is significantly larger than any mixing terms are thought to be for NO Rydberg states.[6,8]

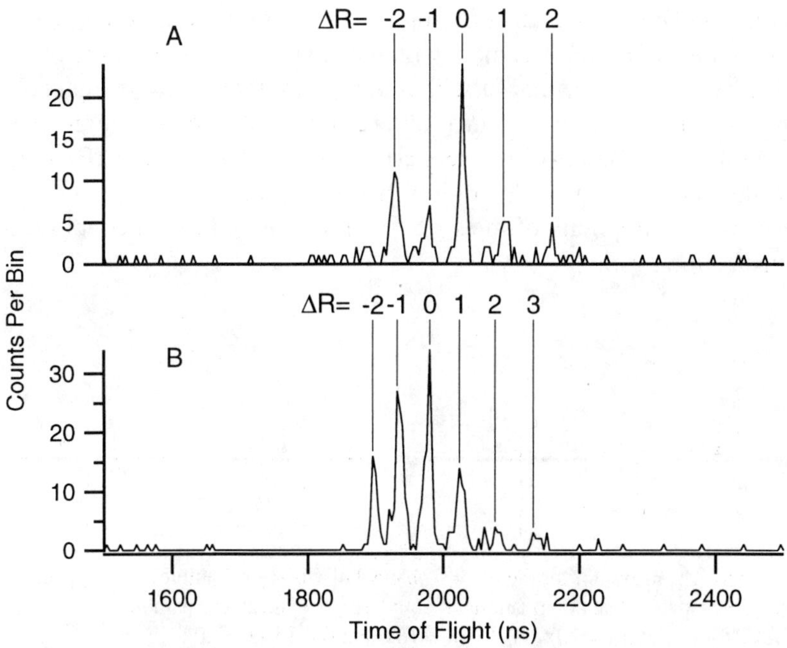

Figure 3. Fixed-angle TOF photoelectron spectra, detection axis parallel to both laser polarizations: (A) 15s(R = 20); and (B) 15s(R = 18).

Fig. 3 shows a similar comparison for the 15s(R = 20) and 15s(R = 18) states. Unlike the 14s states, we see a striking difference in the appearance of these spectra. The branching ratios, shown in Table 1, bare out this apparent difference. Upon inspection of the branching ratios, the 15s(R = 20) state would appear to have the most s character in the ejected electrons. The ΔR = 0 peak comprises the 73% of the detected electrons. However, higher, even valued angular momentum partial waves can contribute to this peak.

Table 1. Normalized rotational branching ratios for selected Rydberg states

ΔR=	-3	-2	-1	0	1	2	3
14s(20)	4.4	11.1	21.7	50.6	<1	12.1	<1
14s(18)	2.4	6.8	24.8	41.3	2.3	20.2	2.1
15s(20)	3.1	8.7	6.9	72.6	5.0	3.6	<1
15s(18)	6.1	21.8	21.8	43.6	3.6	3.1	<1

3.3 Angular Distributions

Angular distributions of the photoelectron spectra offer a more detailed look at the dynamics of the autoionization process. The absorption of the two photons of linearly polarized light results in an alignment of the total angular momentum, *J*. The angular distributions of the different rotational peaks are the convolutions of this *J* alignment with the angular distribution relative to *J*.

Fig. 4 shows the angular distributions from the $\Delta R = 0$ channels of autoionization from the same four Rydberg states. A more thorough analysis of the angular distributions (as was performed for the 14s($R = 20$) state by Park et al.[9]) is required to determine the actual partial wave contribution. Nonetheless, the simplicity of the s waves allows us to reach a conclusion by visual inspection again.

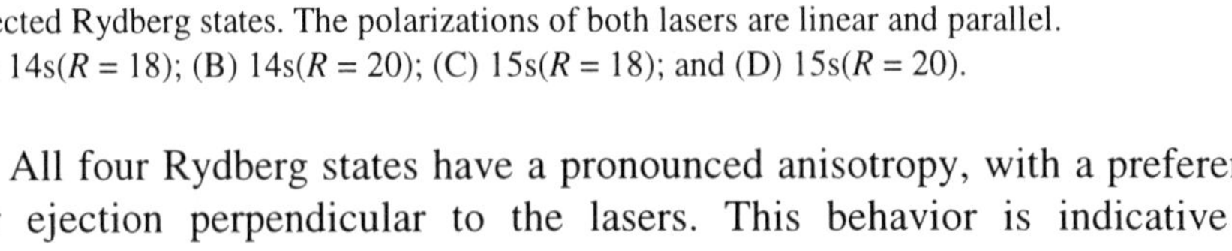

Figure 4. Angular distribution for the ΔR = 0 channel of vibrational autoionization from selected Rydberg states. The polarizations of both lasers are linear and parallel. (A) 14s(R = 18); (B) 14s(R = 20); (C) 15s(R = 18); and (D) 15s(R = 20).

All four Rydberg states have a pronounced anisotropy, with a preference for ejection perpendicular to the lasers. This behavior is indicative of significant contributions from higher angular momentum components to the $\Delta R = 0$ peak because s waves are isotropic.

Another set of angular distributions is compared in Fig. 5. The $\Delta R = -1$ peaks result from odd angular momentum partial waves. There seem to be more variation in these from state to state than is seen in the $\Delta R = 0$ angular distributions. Similar plots can be made for the rest of the autoionization transitions from $\Delta R = -3$ to $\Delta R = 3$, but little more can be learned from a simple visual inspection.

In treating Rydberg states and autoionization with MQDT, we usually make the approximation that processes that occur near the core should not

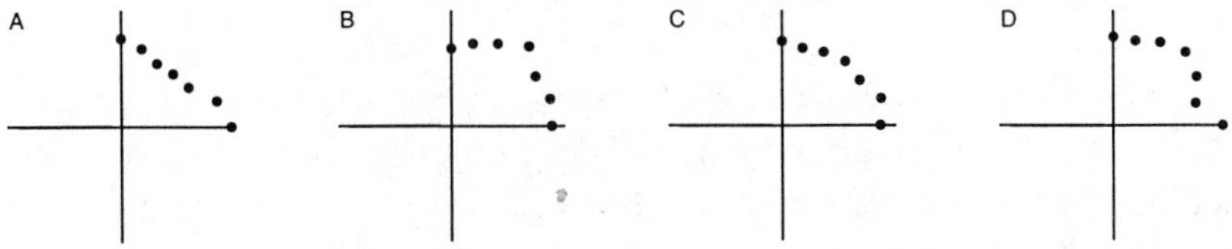

Figure 5. Angular distribution for the ΔR = -1 channel of vibrational autoionization from selected Rydberg states. (A) 14s(R = 18); (B) 14s(R = 20); (C) 15s(R = 18); and (D) 15s(R = 20).

depend heavily on the energy of the Rydberg state. The justification for this approximation is that coulomb wave functions vary slowly with energy close to the core. The reaction matrix, **K**, which describes the short-range interactions of the Rydberg orbitals, is assumed to be constant over short ranges in energy.[10] The importance of this is that departures from simple, hydrogen-like behavior, either from the presence of core electrons or from the nonspherical nature of the core, are caused primarily by short-range effects. This constancy may still be valid, but it is clear that the process does depend heavily on the identity of the Rydberg state. If not the energy, then it seems likely that the channel composition of the Rydberg states is the determining factor. Different autoionization efficiencies for different channels could account for the variations observed in the photoelectron angular distributions.

A look at the characteristic time scales involved can provide some insight into the reason why the composition of these bound states would vary significantly among these supposedly similar states. With a rotational constant of 1.969 cm^{-1},[11] the characteristic rotational period for an (R = 20) rotor is 0.41 ps, whereas the characteristic period for a classical orbit corresponding to a 15s Rydberg state is 0.40 ps. We are observing this phenomenon right in the region of transition from Hund's case b to case d coupling as we increase the Rydberg energy. In this intermediate region, where the rotational period is comparable to the orbital period of the electronic motion, we expect facile coupling between nuclear and electronic motions. This coupling might account for the dramatic differences between nearby Rydberg states. Indeed, whenever the rotational and electronic periods are commensurate, interesting stroboscopic effects occur.[12]

As a simple demonstration, we can examine the energies of the four states discussed here. The difference in energy between the 14s(R = 18) state and the 14s(R = 20) state is 153.2 ± 0.3 cm^{-1}, within experimental error of the calculated difference in rotational energy for the ion of 153.5 cm^{-1}. The 15s levels, on the other hand, are separated by 155.7 ± 0.3 cm^{-1}. The 2.2 cm^{-1} discrepancy is probably indicative of some kind of perturbation, which could result in a significantly different composition of channels.

4. CONCLUSIONS/FUTURE WORK

The vibrational autoionization of even a simple molecule like nitric oxide can prove to be dauntingly complex. This study demonstrates how sensitive branching ratios and angular distributions are to the identity of the autoionizing Rydberg state.

Two major treatments of the data are needed to illuminate this subject. A complete MQDT analysis of the line positions will help determine the composition of the Rydberg states. The study performed by Park et al.[9] determined matrix elements that represent the coupling of a specific Rydberg state to Hund's case-b partial waves in the continuum. In a sense, the matrix elements determined were only one row from an autoionization Hamiltonian that couples a "Rydberg state" basis set to a case-b continuum basis set,

$$\left\langle v=0, l_e, \lambda_e \middle| \mathrm{H}^{\mathrm{AI}} \middle| 14s(R=20), v=1 \right\rangle .$$

Further analyses of the angular distributions to determine the partial wave composition for the rest of the Rydberg states studied, as was done for the 14s(R = 20) state, may allow us to evaluate autoionization matrix elements within the short-range basis set. This short-range Hamiltonian should then only depend weakly on the overall energy, and show a clearer picture of the dynamics behind vibrational autoionization,

$$\left\langle v=0, l_e, \lambda_e \middle| \mathrm{H}^{\mathrm{AI}} \middle| v=1, l_{ryd}, \lambda_{ryd} \right\rangle .$$

ACKNOWLEDGMENTS

The authors would like to thank Wim J. van der Zande for helpful suggestions on the significance of the stroboscopic effect to this study. This work was supported by the National Science Foundation.

REFERENCES

1 U. Fano, Phys. Rev. A **2**, 353-65 (1970).
2 R. N. Zare, Angular Momentum (Wiley-Interscience, New York, 1988).
3 H. Park and R. N. Zare, J. Chem. Phys. **106**, 2239-47 (1997).
4 S. T. Pratt, J. Chem. Phys. **108**, 7131-7140 (1998).
5 S. Fredin, D. Gauyacq, M. Horani, C. Jungen, L. Georges, and M.-S. Francoise, Mol. Phys. **60**, 825-66 (1987).

6 D. Gauyacq, A. L. Roche, M. Seaver, S. D. Colson, and W. A. Chupka, Mol. Phys. **71**, 1311-1331 (1990).
7 H. Park, D. J. Leahy, and R. N. Zare, Phys. Rev. Lett. **76**, 1591-1594 (1996).
8 S. T. Pratt, C. Jungen, and E. Miescher, J. Chem. Phys. **90**, 5971-5981 (1989).
9 H. Park, I. Konen, and R. N. Zare, Phys. Rev. Lett. **84**, 3819-3822 (2000).
10 C. H. Greene and C. Jungen, Adv. Atom. Mol. Phys. **21**, 51-121 (1985).
11 K.-P. Huber and G. Herzberg, Constants of Diatomic Molecules (Van Nostrand, New York, 1979).
12 P. Labastie, M. C. Bordas, B. Tribollet, and M. Broyer, Phys. Rev. Lett. **52**, 1681-1684 (1984).

Predissociation versus Autoionization for High Rydberg States of H_2

Laura Dinu[1], Y.J. Picard[2], and Wim J. van der Zande[1]
[1]FOM-AMOLF Institute for Atomic and Molecular Physics, Kruislaan 407, 1098 SJ Amsterdam, The Netherlands; [2]Laboratoire des Collisions Atomiques et Moleculaires (UMR C8625) Universite Paris-Sud, 91405 Orsay Cedex, France
zande@amolf.nl, ldinu@amolf.nl

1. INTRODUCTION

Dissociative Recombination (DR) has been a subject of research for more than 50 years.[1] In a DR event, a molecular ion captures an electron forming an intermediate electronically excited neutral molecule that rapidly dissociates into fragments: $AB^{+}+e(E)\rightarrow AB^{**}\rightarrow A+B^{*}+KER$. The symbol 'KER' stands for the kinetic energy released to the fragments during the reaction and E is the electron's energy. Among the numerous species studied so far, the molecular ions H_2^+, H_3^+, and HeH^+ and their isotopomers have been studied in most detail. DR of molecular hydrogen serves as the prototypical DR reaction. In spite of its apparent simplicity, being a one-electron system, DR of H_2^+ is theoretically challenging. Although theory is making progress,[2,3,4] theory cannot yet be used as an *a priori* reliable standard. In H_2 the direct process, as formulated by Bates, is prominent. The lowest neutral doubly excited state, $Q_1\,{}^1\Sigma_g^+$, crosses the ionic state close to the outer turning point of the ground vibrational level. The lowest triplet doubly excited state, $Q_1\,{}^3\Pi_g$, does not have such a favorable crossing.[5] As a consequence, theory generally assumes that the direct process is very slow for triplet collisions. In systems in which the direct process is impossible, such as HeH^+, an alternative mechanism has been formulated in which the ion plus free electron is coupled with bound Rydberg states through the radial part of the kinetic energy operator.[6,7] In HeH^+, this mechanism explains a substantial DR cross section at low electron collision energies. Since the coupling between the singlet and triplet manifolds in molecular hydrogen is very small, triplet collisions contribute independently to the cross section. It is of interest to note that the structure of triplet H_2 and HeH have many similarities, as witnessed by their respective ground states and the dynamics of the lower lying states.[8,9] The following contribution aims at presenting *indirect* evidence that triplet molecular hydrogen may contribute to the total DR cross section in H_2^+.

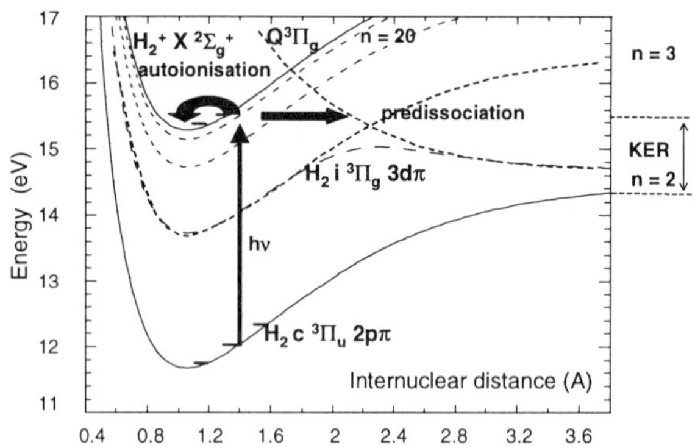

Figure 1. Relevant potential curves of H_2^+ and H_2 (from Ref. 5 and Ref.'s therein).

Highly excited triplet hydrogen is selectively excited by laser excitation starting from the long-lived metastable $c^3\Pi_u^-$ state. By a one photon transition, Rydberg states with principal quantum numbers, $10 < n < 20$, are prepared. Here, we invoke the concepts of quantum defect theory. The electron dynamics within these triplet Rydberg states is highly similar to the collision physics of a near 0 eV free electron with the ionic ground state. The binding energy of $n = 20$ equals only 34 meV. The necessary condition for DR to be important in this system is the occurrence of both autoionization as well as predissociation in this system. We show that upon excitation of $v = 1$ Rydberg states, autoionization as well as predissociation are observed. The predissociation is found to be sensitive to a small electric field in the interaction region. In triplet hydrogen, both the $h^3\Sigma_g^+$ and $i^3\Pi_g$ states are possible candidates as intermediate states for the non-crossing DR mechanism.[6,7]

2. EXPERIMENT

The experimental set-up is shown in Fig. 2. The H_2^+ ion beam is produced in a hollow cathode discharge ion source. The ion beam is extracted at 5 kV. After approximately 170 cm it passes through a cell of Cs vapor where charge exchange processes take place. Part of the H_2^+ ion beam is neutralized and metastable excited states of molecular hydrogen are populated ($H_2^+ (X^2\Sigma_g^+) + Cs \rightarrow H_2 (c^3\Pi_u) + Cs^+$). The residual ion beam is deflected after the collision cell. The neutral beam travels undisturbed along the set-up axis and 50 cm away from the cell it passes a 1 mm diaphragm and enters the interaction region with the laser. The molecular beam and the laser cross each other at 90 degrees. The laser system used for this experiment consists of a 50 Hz Spectra Physics Nd-YAG laser pumping a Lambda Physik ScanMate dye laser. The output of the ScanMate (around 690 nm) is frequency doubled by a KDP-C non-linear crystal. The UV laser light produced this way has 15 mJ per pulse and it is focused into the set-up. The length of the line focus (1 cm) determines the overlap region between the molecular beam and the laser. Photons are absorbed and high Rydberg states are populated. The lifetime of these Rydberg states is short compared with the time-of-flight of the molecules in the laser region (20 ns). Therefore we can consider that predissociation and autoionization occur inside the laser focus.

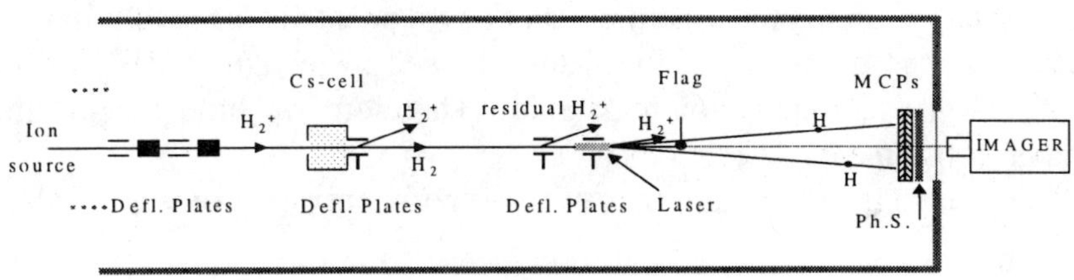

Figure 2. Schematic of the apparatus.

If autoionization takes place, the trajectory of the ions is slightly bent by a set of deflection plates located around the laser region such that ions escape the flag and they arrive on the detector in a defined region. This electric field affects the degree of predissociation.

If predissociation occurs, two hydrogen atoms fly apart with a kinetic energy corresponding to the internal state of the Rydberg molecule. After a time-of-flight of 1.8 μs, the atoms hit the time and position sensitive detector. By measuring the kinetic energy release in the predissociation process, useful information can be extracted concerning the molecular Rydberg state such as the energy of the state and the angular orientation of the molecule, etc. (see Fig.1).

If at the moment of dissociation the molecule is oriented perpendicular to the set-up axis, the kinetic energy released is encoded in the distance between fragments. If the molecule is oriented along the set-up axis, the fragments arrive in the same spot on the detector, but at different times. When the molecule is randomly oriented with respect to the set-up axis, the kinetic energy release can be expressed as $KER = E_0(R^2 + v_0^2\Delta t^2)/4L^2$, where E_0 is the kinetic energy of the beam (5 keV), L is the distance between laser region and detector (120 cm), R is the distance between the fragments on the detector, v_0 is the velocity of the beam ($7*10^5$ m/s) and Δt is the arrival time difference of the fragments.

In order to estimate the kinetic energy release, both fragments have to be detected. Two particles are recognized as fragments of the same molecule if their center of mass coincides with the center of mass of the ion beam. It is necessary to record not only the position of the fragments, but also the arrival time difference.

The detector consists of two microchannel plates and a P46 phosphor screen. A CCD based double exposure camera is placed behind the phosphor screen. The camera retrieves the (x,y) coordinates of the spots (80 μm resolution) and the arrival time of the particles[12,13] (1 ns resolution). The camera allows us, in principle, to detect hundreds of particles per image (the only limitation on the maximum number of fragments is the resolution of the camera). In the treatment of the data, we need to recognize the coincident fragments (via the center of mass rule). Therefore we limit the maximum number of fragments detected to 10.

The camera is the master clock of the experiment and it runs at 25 Hz.

3. RESULTS

By scanning the laser over 5 nm (343.3 – 348.3 nm), we induce transitions from the initial state, $c^3\Pi_u^-$, to $n = 10$ to $n = 20$ Rydberg states. Since these states are high Rydberg states, we use Hund's case d to describe them. The conserved quantities are $\vec{l}$ (the orbital angular momentum of the exited electron; $l = 2$), $\vec{R}$ (the rotational angular momentum of the ion core) and $\vec{N}$ (the vector sum; $\vec{N} = \vec{R} + \vec{l}$). The electron spin $\vec{S} = 1$ gives rise to an unresolved splitting. The initial state, $c^3\Pi_u^-$, is described according to Hund's case b. The rotational states are indexed according to N_c, the total angular momentum exclusive of spin.

The $v = 0$ Rydberg levels lie below the ground state of the ion and can only predissociate (see Fig.1). For vibrationally excited Rydberg states, autoionization becomes an open decay channel competitive with predissociation. Since the lowest triplet unbound state, $Q_1\ {}^3\Pi_g$, leading to the $n = 2$ dissociation limit crosses the Rydberg states around $v = 5$,[5,10,11] direct

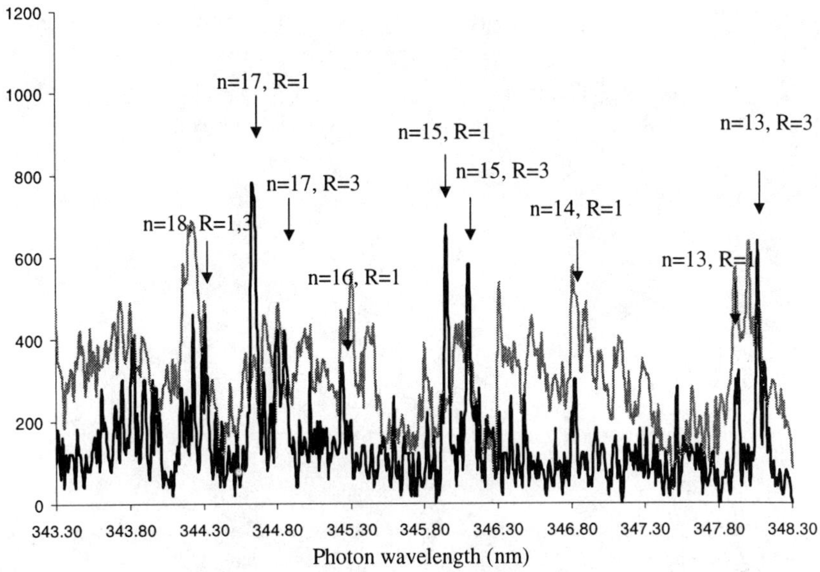

Figure 3. The black curve represents the predissociation spectrum for $v = 1$ transitions. The gray curve represents the autoionization spectrum and it includes transitions from H_2 $c^3\Pi_u^-$ ($v = 1, ... v = 5$) initial states; n is principal (Rydberg) quantum number; R is the rotational quantum number of the core of the Rydberg state. The amplitude of the signal is expressed in arbitrary units.

predissociation of $v = 1$ Rydberg states is very unlikely. Contrary to this fact, we observed a predissociation signal from $v = 1$, $n = 13$ to $n = 18$ Rydberg states (see Fig.3).

The predissociation spectrum of Fig.3 was obtained by selecting events with a KER between 600 meV and 800 meV, i.e. by selecting only events starting from the $c^3\Pi_u^-$ ($v = 1$) state that predissociate forming an $H(n = 2) + H(n = 1)$ fragment pair. For the assignment of the rotational lines, we followed the calculation by Eyler and co-workers[14,15] and Knight and Wang'[16].

Observed transitions are characterized by $\Delta v = 0$; $\Delta N = 0,\pm1$; $R = N_c$. The orthohydrogen levels dominate (odd R). Parahydrogen lines ($R = 2$) are very weak and hard to distinguish from the background.

An ionization spectrum was obtained by counting the deflected laser-induced-ions on the detector. A continuous background was subtracted. The ionization spectrum contains many more lines than the predissociation spectrum mainly because the ionization spectrum has contributions from all vibrational states that are excited in that photon region, while the predissociation spectrum shows only $v = 1$ transitions.

An enlargement of the spectrum in the region of $n = 15$ $v = 1$ transitions is shown below. The two broad peaks represent the $R = 1$ and $R = 3$ transitions

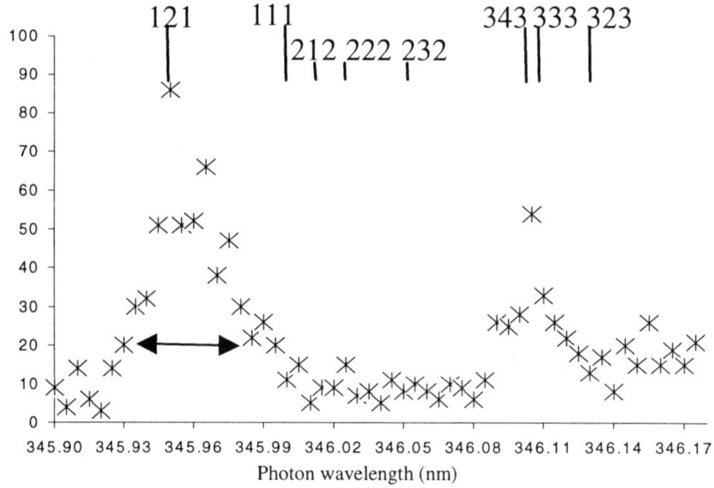

Figure 4. Predissociation spectrum for $n = 15$ $v = 1$ Rydberg states. The assignment (i j k) refers to R, N and N_{c^+}. The line width ~ 4 cm^{-1} is indicated by the arrow.

(with an unresolved N structure). In this Figure, the arrow represents the width of the 121 line. We estimated the major contributions to the width of the lines as follows. The laser bandwidth is 0.14 cm^{-1} and Doppler broadening is 0.2 cm^{-1} (FWHM) as calculated from the very small divergent angle ($\varphi = 2*10^{-3}$ rad) in combination with the fast beam ($v = 7*10^5$ m/s). Stark broadening is estimated to give the most significant contribution. For the 121 line, we calculated a maximum energy shift of 3.35 cm^{-1} (for 250 V/cm).

To our surprise, dissociation is found to compete with autoionization. Since the direct predissociation cannot explain the strength of the dissociation signal, we have estimated the effect of a coupling of the inner turning part of the Rydberg state with the $i^3\Pi_g$ state (see Fig. 1). We calculated the Franck-Condon factors as a function of principal quantum number, n. A comparison between experimental data and this calculation is shown in Fig. 5.

Interestingly, the predissociation signal seems to be an oscillatory function of the odd/even principal quantum number. This must reflect an accidental energy degeneracy with either even or odd principal quantum numbers of a perturber state (which we have not yet identified).

Even if we cannot express the absolute predissociation efficiency, it is still possible to estimate the competition between the two decay channels (predissociation and autoionization). For odd principal quantum numbers, the measured predissociation versus autoionization rate is between 5% and 15%.

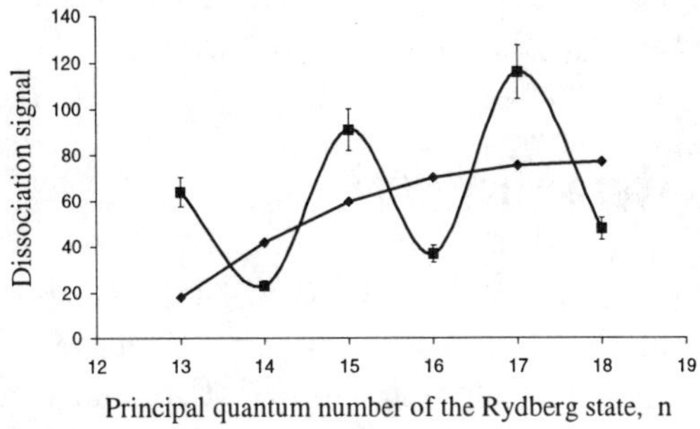

Figure 5. Comparison between the expected behavior of dissociation signal versus principal quantum number n (the smooth curve) and the experimental data (the oscillatory curve)

4. DISCUSSION AND CONCLUSIONS

This is a preliminary report on the observation of competition between autoionization and predissociation of triplet molecular hydrogen. We believe this competition to be the necessary condition for triplet hydrogen to have a finite DR cross section at small electron collision energies. This work was motivated by three reasons. Firstly, HeH was found to have a small but significant DR rate. This rate was calculated by Guberman[7] and by Sarpal and Tennyson[6] by invoking a $\partial/\partial R$ coupling between the ion plus free electron and the inner repulsive walls of several dissociative states. The former calculation[7] found that the C state is the dominant dissociative route. Secondly, triplet H_2 has many similarities with HeH. The triplet nature forces the $1\sigma_g$ orbital to be partially unfilled, which is occupied in HeH with the atomic like 1*s* orbital around the helium core. The ground state of HeH ($^2\Sigma^+$) is repulsive as is the ground triplet state of hydrogen ($^3\Sigma_u^+$). The first excited states of HeH are the $A^2\Sigma^+$, $B^2\Pi$ and $C^2\Sigma^+$ states which have similar configurations as the $a^3\Sigma_g^+$, $c^3\Pi_u$ and $h^3\Sigma_g^+$ states in H_2 and similar decay mechanisms. Thirdly, if the coupling is present, due to the triplet character this channel has a statistical factor of 3. Although we do not give direct evidence for the presence of a $\partial/\partial R$ coupling, we have made some significant observations. Recently Stolyarov and Child have provided a method to derive the necessary $\partial/\partial R$ coupling based on the R-dependent quantum defects only[17]. This procedure should be applicable to triplet hydrogen for which accurate calculations of potential curves have been done. Finally we want to mention the effect of the electric field in our experiments. We have observed that the electric field has an influence on the predissociation behavior. This suggests that the *nd* character of the outer electron might not be sufficiently effective in driving the predissociation. The effect of the static electric field is to mix higher and lower angular momentum character.

ACKNOWLEDGEMENTS

The work described in this paper is part of the research program of the Stichting Fundamenteel Onderzoek van de Materie (Foundation for Fundamental Research on Matter) and it was financially supported by the Nederlandse Organisatie voor Wetenschappelijk Onderzoek (The Netherlands Organization for Scientific Research). We are grateful for the support from the EU-Network "Imagine".

REFERENCES

1 D. R. Bates, Phys. Rev. **78**, 492 (1950).
2 T. Tanabe, I. Katayama, H. Kamegaya, K. Chida, Y. Arakaki, I. Watanabe, M. Yoshizawa, M. Saito, Y. Haruyama, K. Hosono, K. Hatanaka, T. Honma, K. Noda, S. Ohtani and H. Takagi, Phys. Rev. Lett. **75**, 1066 (1995).
3 I. F. Schneider, C. Strömholm, L. Carata, X. Urbain, M. Larsson and A. Suzor-Weiner, J. Phys. B **30**, 2687 (1997).
4 H. Takagi, in *Dissociative Recombination: Theory, Experiment and Application IV* , ed. by M. Larsson, J. B. A. Mitchell and I. F. Schneider, (World Scientific, Singapore, 1999), p.180.
5 S. L. Guberman, J. Chem. Phys. **78**, 1404 (1983).
6 B. K. Sarpal, J. Tennyson and L. A. Morgan, J. Phys. B **27**, 5943 (1994).
7 S. L. Guberman, Phys. Rev. A **49**, R4277 (1994).
8 Al-Khalili, H. Danared, M. Larsson, A. Le Padellec, R. Peverall, S. Rosen, J. Semaniak, M. af Ugglas, L. Vikor and W. J. van der Zande, Hyperfine Interact. **114**, 281 (1998).
9 W. J van der Zande, W. Koot, D. P. de Bruijn and C. Kubach, Phys. Rev. Lett. **57.** 1219 (1986).
10 N. Bjerre, S. R. Keiding, L. L. Lembo and H. Helm, Phys. Rev. Lett. **60**, 2465 (1988).
11 L. J. Lembo, N. Bjerre, D. L. Huestis and H. Helm, J. Chem. Phys. **92**, 2219 (1990).
12 D. Strasser, X. Urbain, H. B. Pedersen, N. Altstein, O. Heber, R. Wester, K. G. Bhushan and D. Zajfman, Rev. Sci. Instr. **71**, 3092 (2000).
13 L. Dinu, A. T. J. B. Eppink, F. Rosca-Pruna, H. L. Offerhaus, W. J. van der Zande, M. J. J. Vrakking, *in preparation.*
14 E. E. Eyler and F. M. Pipkin, Phys. Rev. A **27**, 2462 (1983).
15 M. D. Lindsay, A. W. Kam, J. R. Lawall, P. Zhao, F. M. Pipkin and E. E. Eyler, Phys. Rev. A **41**, 4974 (1990).
16 R. D. Knight and L. Wang, Phys. Rev. Lett. **55**, 1571 (1985).
17 A. V. Stolyarov and M. S. Child, Phys. Rev. A **63**, 2510 (2001).

Vibrational Autoionization in Polyatomic Molecules

S. T. Pratt, J. A. Bacon, and C. A. Raptis
Argonne National Laboratory, 9700 South Cass Avenue, Argonne, Illinois 60439
stpratt@anl.gov

1. INTRODUCTION

Molecular Rydberg states converging to vibrationally excited states of a given electronic state of the ion and having total energy greater than the lowest vibrational level of that electronic state of the ion can spontaneously ionize through the conversion of vibrational energy of the ion core into electronic, and ultimately translational energy, of the Rydberg electron. This process is known as vibrational autoionization.[1] Vibrationally autoionizing resonances can be thought of as vibrationally excited Feshbach resonances in the scattering of electrons off molecular ions, and the decay process is closely related to the superelastic scattering of slow electrons off vibrationally excited ions. If they are coupled to a dissociative continuum, vibrationally autoionizing resonances can also play a role in dissociative recombination.

Vibrational autoionization is a particularly well defined example of a nonadiabatic process involving the coupling between the vibrational and electronic degrees of freedom. By using double resonance excitation techniques, it is possible to excite well characterized resonances in which the nominal electronic, vibrational, and rotational quantum numbers are known. By using high resolution photoelectron spectroscopy, it is also possible to determine the electronic, vibrational, and sometimes even rotational quantum numbers of the product ion. Thus, the process can be characterized in great detail allowing for an understanding of the mechanism. Considerable work has been performed on diatomic molecules with this approach. This approach has also been extended to the study of vibrational autoionization in polyatomic molecules, where it is particularly interesting to investigate how the vibrational autoionization rate depends on the different normal modes of the molecule. In this paper, recent work on the mode dependence of vibrational autoionization in ammonia is discussed, and the results are explained in terms of a simple qualitative model for the autoionization process.

Vibrational autoionization was first discussed theoretically by Berry,[2] who considered terms in the molecular Hamiltonian that are left out when making the Born-Oppenheimer approximation. These terms involve first and second derivatives of the vibrational and electronic wave functions with respect to the internuclear distance in diatomic molecules, or to the normal coordinate in polyatomic molecules. By considering the dominant term, Berry derived a propensity rule: vibrational autoionization involving the smallest possible change in the vibrational quantum number(s) will be most efficient.[2,3] Subsequent experimental studies have confirmed this rule in most cases, with exceptions occurring when other ionization mechanisms (e.g., electronic autoionization) are also operative.

The next significant theoretical advance in understanding vibrational autoionization came in a study of molecular hydrogen by Herzberg and Jungen,[4] who based their analysis on multichannel quantum defect theory (MQDT), which Fano[5] had recently extended to molecules. In their paper, Herzberg and Jungen derived the following two channel formula for the vibrational autoionization rate:

$$\Gamma_{va} = \frac{(2\pi)2Ry}{n^{*3}}\left[\frac{d\mu}{dR}\right]_{R_e}^{2} \left\langle \chi^{v^+} \middle| R - R_e \middle| \chi^{v^{+'}} \right\rangle^{2} \tag{1}$$

Here, n* is the effective principal quantum number, μ is the quantum defect (n-n*), Ry is the Rydberg constant, and χ^{v+} is the vibrational wave function of the vth ion level. This formula can be generalized to polyatomic molecules by using the normal coordinate rather than the internuclear distance, R. Eq.(1) is particularly useful because the important physical points are broken into three factors. The first includes several constants and a factor of $1/n^{*3}$, which arises from the normalization factor in the Rydberg wave function and determines how the rate scales with n. The second factor is the derivative of the quantum defect with respect to the internuclear distance (or normal coordinate) and shows how the rate depends on the electronic character of the Rydberg state. The third factor is similar to a term in Berry's analysis and gives rise to the $|\Delta v|$ = minimum propensity rule.

Eq.(1) was developed for the two channel problem and, in most cases, the situation is considerably more complex. Since the early 70's, the full machinery of MQDT has been applied to a substantial number of diatomic molecules and several small polyatomic molecules.[6] Most notable for the present discussion, Stephens and Greene[7,8] have performed detailed calculations on the spectroscopy and dynamics of Rydberg states of H_3, and

Child and coworkers[9,10] have performed calculations on autoionizing states of H_2O. In both cases, the agreement with experiment is quite good. While such studies will ultimately be possible for much larger molecules, at present there is insufficient input data (e.g., quantum defect functions and transition moments) to perform high quality calculations. Considerable data on the relative rates of vibrational autoionization as a function of normal mode have been obtained by Grant and coworkers[11-13] on NO_2 and HCO, and new data also exist for ammonia[14] and aniline.[15] In the absence of full MQDT calculations, a qualitative model for these results would be quite informative.

Aside from the matrix element (which can be readily evaluated) of the displacement along the normal coordinate, Q_i, the factor in Eq. 1 that determines the dependence on the normal mode and the electronic character is the derivative of the quantum defect with respect to the normal coordinate. As an example, if the Rydberg potential curve is parallel to the ion curve (that is, simply displaced to lower energy), the quantum defect obtained from:

$$\mu(Q_i) = n - \sqrt{Ry/(U^+(Q_i) - U(Q_i))} \tag{2}$$

is constant with Q_i. Thus, the derivative with respect to the normal coordinate is zero, with the result that the vibrational autoionization rate is zero. The effect of the Rydberg electron is to lower the energy of the system, with no effect on the potential in which the nuclei move. Thus, there is no influence of the electronic structure on the vibrational motion of the nuclei, and thus no vibrational autoionization. On the other hand, if the ionic and Rydberg curves are not parallel, this indicates that the Rydberg electron is influencing the motion of the nuclei, and the vibrational autoionization rate will be nonzero. Vibrational autoionization will be fastest for electronic states and normal modes for which the quantum defect varies strongly near the equilibrium geometry of the molecule, because this is the geometry sampled in the Rydberg state. Qualitatively, the dependence of the quantum defect on geometry reflects how the bonding or antibonding character of the Rydberg orbital changes, and is thus closely connected to the orbital energy geometry dependence. This dependence is described in correlation[16] and Walsh[17] diagrams. In addition, vibrational autoionization will be fastest from penetrating Rydberg states, because an electron at close range interacts with the other electrons and can affect the motion of the much heavier nuclei. In general, penetrating Rydberg states are those that have precursors in the core,[18-20] i.e., lower occupied orbitals of the same Rydberg series as the Rydberg state. These ideas are used to rationalize the mode dependence of vibrational autoionization observed in the spectra of ammonia discussed below.

2. EXPERIMENT

The details of the experiment have been described previously.[14] The focus of the present work is on understanding how the vibrational autoionization rates depend on the character of the Rydberg electron and on the normal mode that drives the autoionization process. Perhaps the most clean-cut approach for studying the normal mode dependence is to compare the rates for two (or more) different modes directly, and it is this approach that is used in the studies of NH_3 presented here. Specifically, two different Rydberg vibrational modes are excited, and the photoelectron spectra resulting from the autoionization of these states display the relative efficiency of $\Delta v_i = -1$ autoionization. In the experiments discussed here, double resonance excitation techniques are used to excite Rydberg series converging to selected vibrational levels of the ion. Specifically, by using the two-photon pump transition, $C'\,^1A_1'$ (1300) $\leftarrow$ $X\,^1A_1'$ (0000), it is possible to excite predominantly ns and nd Rydberg series converging to the $X\,^2A_2''$(1300) level of NH_3^+. Here, (1300) denotes 1 quantum of symmetric stretch, 3 quanta of umbrella bend, and 0 quanta of the asymmetric stretch and bend, respectively. These Rydberg states and the ground state of the ion are all planar. By choosing the pump laser wavelength, it is also possible to select the rotational level in the $C'\,^1A_1'$ (1300) state, which significantly restricts the allowed rotational quantum numbers of the Rydberg states excited by the probe transition.

The experiments were performed using a seeded, supersonically cooled molecular beam of NH_3 in He. The first step in these experiments is to map out the pump transition; this was achieved by monitoring the total electron signal produced by two-photon resonant, three-photon ionization using a single laser. This laser was then tuned to the pump transition of interest, and the probe laser was introduced. This second laser was then scanned to drive transitions from the selected intermediate level to Rydberg series converging to the $X\,^2A_2''$(1300) threshold. The spectrum was mapped out by monitoring the photoelectrons resulting from the two-color excitation process. An example of such a spectrum is shown in Fig. 1. Finally, the probe laser was tuned to the autoionizing resonance of interest and a high resolution photoelectron spectrum was recorded. Photoelectron spectra for a number of resonances in Fig. 1 are shown in Fig. 2.

3. DISCUSSION

The most striking aspect of Fig. 2 is that all of the photoelectron spectra show an intense doublet corresponding to the population of two rotational levels of the $X\,^2A_2''$(1200) level along with an extremely weak feature

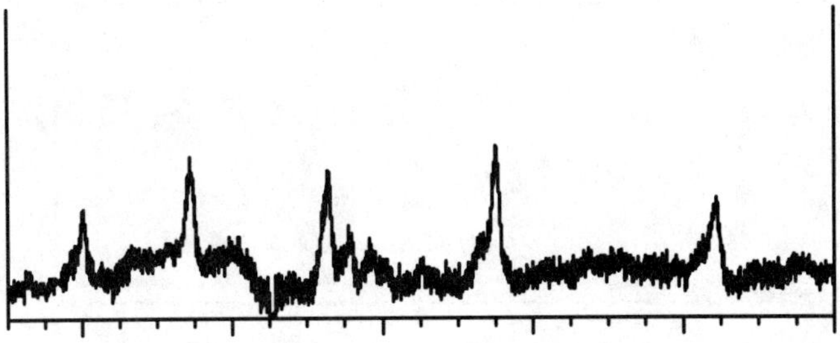

Figure 1. The double-resonance spectrum obtained by pumping the C'(1300) ← X(0000) transition and probing between the X(1300) and (1200) thresholds of the ion. The spectra were recorded by monitoring the two-color photoelectron signal as the probe laser was scanned. The pump laser was tuned to the $Q_a(0,0)$ transition at 69712±1 cm^{-1}, This Figure was adapted with permission from Fig. 1 of Ref. 14.

corresponding to the population of the $X\,^2A_2''$(0300) level. This observation indicates that, for the Rydberg series studied, the Δv_2 = -1 process, i.e. vibrational autoionization via the umbrella motion, is considerably faster than the Δv_1 = -1 process involving the symmetric stretching motion. In the remainder of this paper, these observations will be explained in terms of how the Rydberg orbital energies and quantum defects change along the two normal coordinates of interest. To a first approximation, the short-range radial wave functions of a given Rydberg series have very similar shapes,[18-20] differing only by a normalization factor of $1/n^{*3/2}$. It is assumed that the normal coordinate dependence of a given Rydberg series is independent of n. Thus, if the Rydberg series with precursors in the core can be identified, the normal coordinate dependence of those precursors should provide insight into which orbitals and normal modes will have the most rapid vibrational autoionization. Autoionization rates for nonpenetrating orbitals without precursors are expected to be small.

Based on the 3p character of the $C'\,^1A_1'$ intermediate state, the dominant probe transitions are expected to be to ns and nd Rydberg states, and the intense resonances in Fig. 1 are believed to correspond to nd Rydberg states. The arguments presented below are based on this assignment, but the same conclusions are reached if the assignment is changed to the ns series. There are two allowed nd series accessible from the $C'\,^1A_1'$ state, corresponding to nda_1' and nde" series. The forms of the angular wave functions for these two series are shown in Fig. 3. Fig.'s 4 and 5 show the Walsh diagrams for the occupied and unoccupied valence orbitals[20] of planar NH_3 along the umbrella and symmetric stretching coordinates, respectively. These diagrams show qualitatively how the orbital energies change with geometry, and can be used to provide insight into the behavior of the quantum defect functions. Fig. 4 was drawn simply based on considerations discussed in Walsh's original paper.[20] Fig. 5 was drawn using N, H, Ne, and NH_3 orbital energies from the

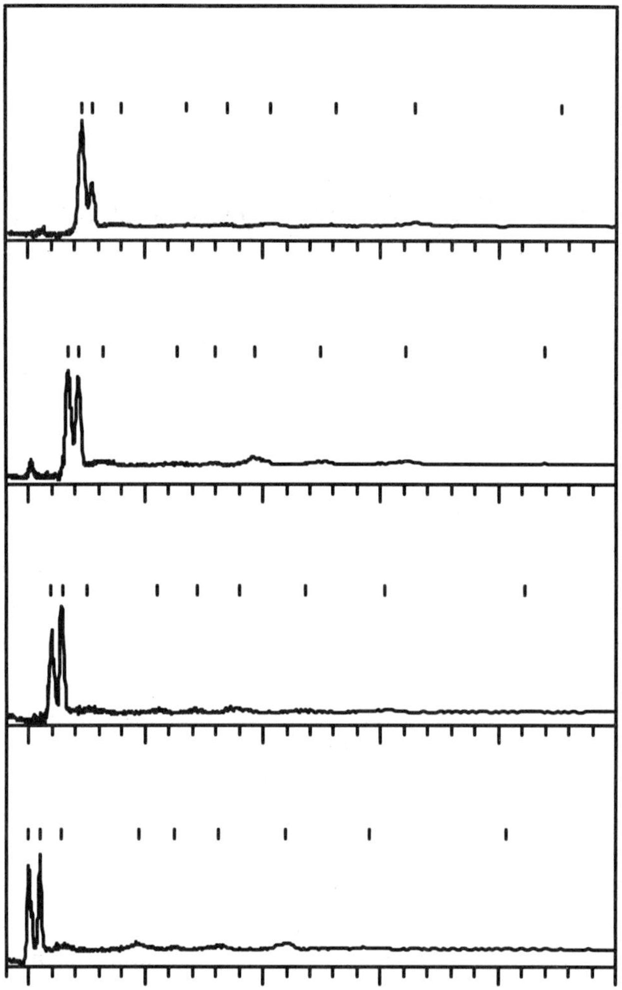

Figure 2. The photoelectron spectra obtained for the (a) n = 14, (b) n = 13, (c) n=12, and (d) n=11 resonances in the spectrum of Fig. 1. The doublet splitting in the X(1200) band is due to two rotational levels. This Figure is from Ref. 14 and is used with permission.

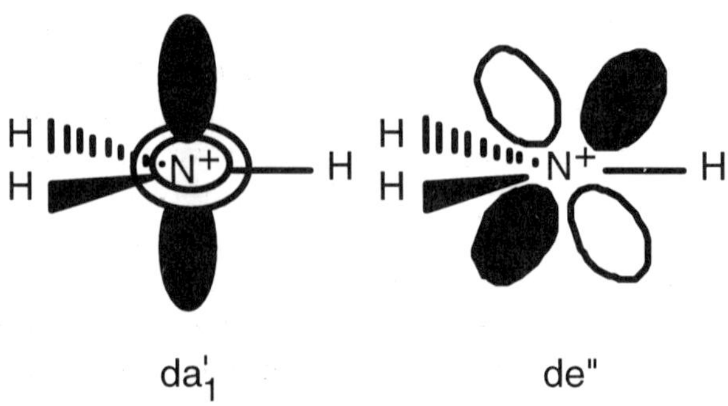

Figure 3. A schematic diagram of the relevant nd angular wave functions.

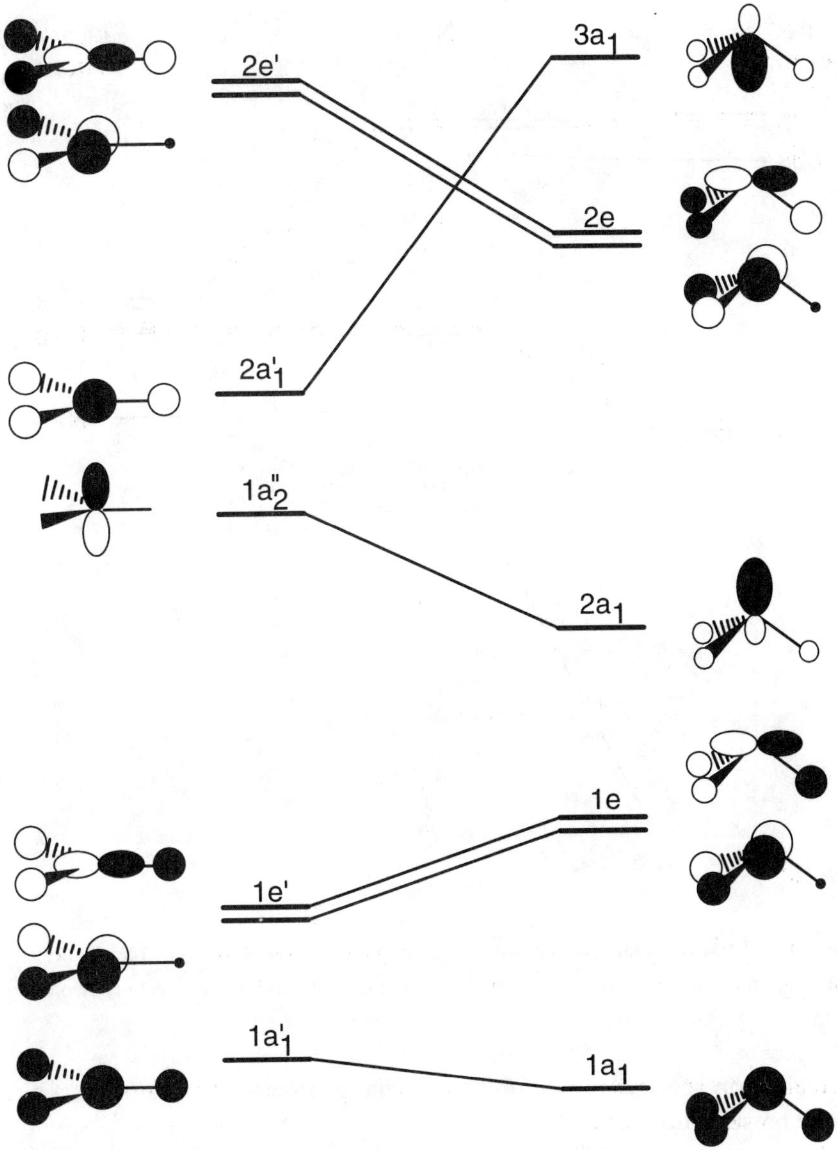

Figure 4. A Walsh diagram for the umbrella bending coordinate of NH3, with qualitative pictures of the corresponding molecular orbitals. Adapted from An Introduction to Molecular Orbitals *by Y. Jean and F. Volatron, ed. by J. Burdett, translated by J. Burdett, (Oxford University Press, Inc., 1993). Used by permission of Oxford University Press, Inc.*

literature.[21,22] For the atoms, the energies were averaged over the multiplets. For NH_3, the two highest orbital energies came from Rydberg state spectroscopy of the planar molecule. The lowest three energies are from

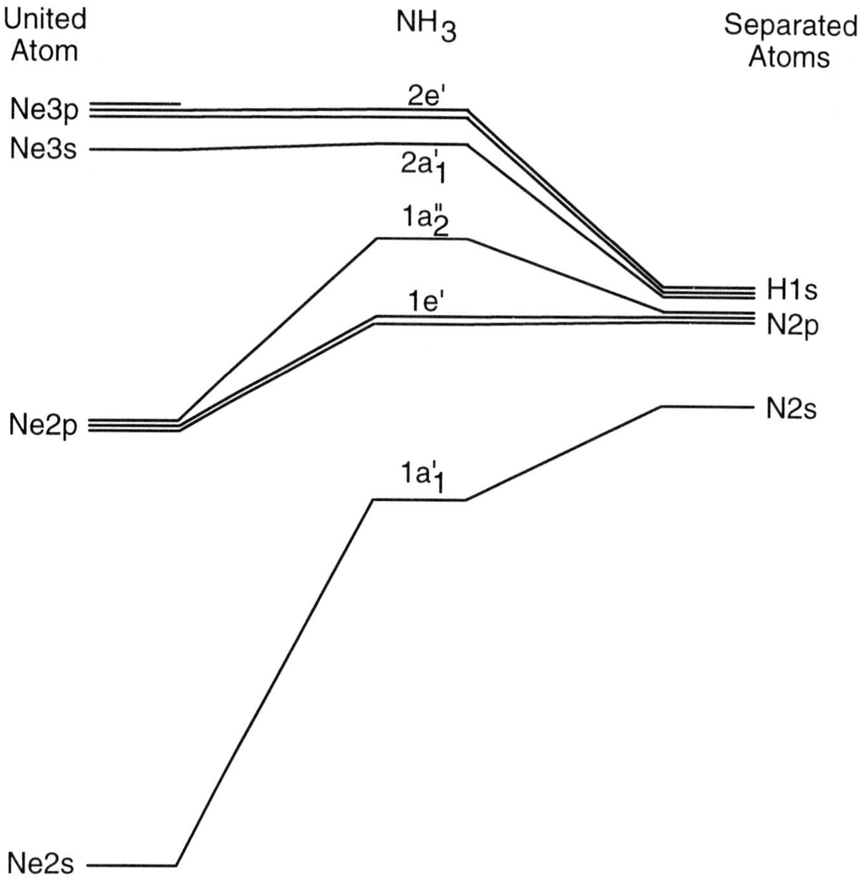

Figure 5. A Walsh diagram for the symmetric stretching vibration of NH_3. The shapes of the orbitals are the same as those for the corresponding orbitals in the planar geometry of NH_3 in Fig.4.

photoelectron spectroscopy in the pyramidal geometry. This is not a concern for the present discussion.

As an example, consider the nde" orbital in the planar geometry. Fig. 4 shows that there are no precursor orbitals with e" symmetry. However, as the molecule is distorted away from the planar geometry, the symmetry breaks, and the e" orbital becomes e in C_{3v}. As can be seen in Fig. 4, the e' orbital also become e symmetry in C_{3v}, and the form of the wave function clearly gains an admixture of what would be de" character in the planar molecule. The energy of this orbital also decreases significantly upon distortion. Thus, it is expected that as the molecule distorts away from planarity, the de" orbital becomes more penetrating and the orbital energy and the quantum defect vary considerably. In contrast, for the symmetric stretch, there are never any precursor orbitals with de" symmetry, and the orbital energy is not expected to change much along this normal coordinate. Put another way, the motion of the nuclei in the symmetric stretch is in a nodal plane of the nde" orbital, and thus the quantum defect should remain relatively constant for this distortion. These considerations indicate that autoionization via the symmetric stretch should be considerably slower than via the umbrella mode, as is observed in the data. Similarly, Fig. 4 shows the energy of the precursor $2a_1'$ or $3a_1$ orbital changes significantly with distortion along the umbrella coordinate. The energy of the $2a_1'$ orbital also changes significantly along the stretching coordinate between the separated atom and NH_3 geometries. However, the $2a_1'$ orbital changes little between the NH_3 and united atom geometries. It appears likely that the increase in energy at larger NH distances occurs at relatively long range and that, near the NH_3 geometry, the change in energy is small. In this case, vibrational autoionization via the umbrella mode would be expected to be faster than via the symmetric stretch. More detailed calculations are necessary to put these arguments on a more quantitative footing.

While the present results can be rationalized in terms of the Walsh diagrams of Fig.'s 4 and 5, more extensive studies are necessary to evaluate the general utility of this approach. It would be particularly interesting to study the behavior of the third, in-plane component of the nd Rydberg states (the nde' component). Although this component is forbidden in optical transitions from the $C'\,^1A_1'$ state,[23] it can be accessed from the 3pe' B E" state. It would also be interesting to compare the ν_1 and ν_2 vibrational autoionization rates with those of ν_3 and ν_4. Although combination bands involving ν_2 and ν_3 or ν_4 are not known for the $C'\,^1A_1' \leftarrow X\,^1A_1'$ transition, they are well characterized for the $B\,E'' \leftarrow X\,^1A_1'$ transition.[24] For this reason, we are currently performing a number of studies with the B E" intermediate state. The first of these studies involves pumping selected rotational levels of the B E"(0301) state and probing Rydberg series converging to the $X\,^2A_2''(0301)$ threshold. Preliminary photoelectron spectra suggest that the relative efficiencies of $\Delta\nu_2 = -1$ and $\Delta\nu_4 = -1$ processes are much more similar than those for $\Delta\nu_2 = -1$ and $\Delta\nu_1 = -1$ processes. Future studies to compare ν_2 and ν_3 are also planned, as are studies of some related compounds.

The discussion of vibrational autoionization presented above is based on a two channel description of the process, and it is clear that the real situation is considerably more complex. For example, l uncoupling in the Rydberg states will mix the nda_1', nde', and nde" components, and the nonspherical symmetry of the molecule can result in considerable l mixing. While such effects can be included in a full-blown MQDT analysis, they are obviously not included in any systematic way here. Nevertheless, it is hoped that the present treatment will provide insight into more complex molecules for which detailed theoretical treatments are in short supply.

The photoelectron spectra in Fig. 2 show a number of weak peaks corresponding to the population of a wide range of vibrational levels of the X^2A_2'' state. Some of these may be populated through direct photoionization of the C'^1A_1' (1300) level, but some may be populated through the autoionization process. Particularly interesting are those peaks that involve excitation of the asymmetric stretching vibration, which are hard to rationalize in terms of simple vibrational autoionization. However, it is well known that many of the high Rydberg states of NH_3 are predissociated, and that electronic autoionization from the state responsible for the predissociation can be an important process.[25] The vibrational distributions for such a process should reflect the Franck-Condon factor between the state responsible for the predissociation and the ionic state. Because dissociation is expected to occur through the lengthening of an NH bond, the observation of the asymmetric stretching vibration (and perhaps the asymmetric bending vibration) would be expected in the photoelectron spectrum. Elucidation of this decay mechanism will require further work, but the resulting process is closely related to dissociative recombination, in which there is a competition between the ionizing and dissociating pathways in the highly excited molecule.

4. SUMMARY

In conclusion, the study of vibrational autoionization in polyatomic molecules can provide considerable information on the quantum state dependence of the flow of energy between the vibrational and electronic degrees of freedom in isolated molecules. Although a detailed theoretical understanding of these processes requires high level MQDT calculations, some insight can be gained by using Mulliken's ideas about precursor orbitals[18-20] along with qualitative correlation or Walsh diagrams that describe how the orbital energies change along the normal coordinates. This approach appears to provide a reasonable interpretation of the results on NH_3 presented here, but more comprehensive experimental data is necessary to determine its overall consistency and general utility.

ACKNOWLEDGMENTS

This work was supported by the U.S. Department of Energy, Office of Science, Office of Basic Energy Sciences, Division of Chemical Sciences, under Contract W-31-109-Eng-38.

REFERENCES

1 See, for example, J. Berkowitz, *Photoabsorption, Photoionization, and Photoelectron Spectroscopy* (Academic Press, New York, 1979).
2 R. S. Berry, J. Chem. Phys. **45**, 1228 (1966).
3 J. N. Bardsley, Chem. Phys. Lett. **1**, 229 (1967).
4 G. Herzberg and Ch. Jungen, J. Mol. Spectrosc. **41**, 425 (1972).
5 U. Fano, Phys. Rev. A **2**, 353 (1970).
6 Ch. Jungen, *Molecular Applications of Quantum Defect Theory* (Institute of Physics, Philadelphia, 1996).
7 J. A. Stephens and C. H. Greene, Phys. Rev. Lett. **72,** 1624 (1994).
8 J. A. Stephens and C. H. Greene, J. Chem. Phys. **102**, 1579 (1995).
9 M. S. Child and Ch. Jungen, J. Chem. Phys. **93**, 7756 (1990).
10 M. S. Child and W. L. Glab, J. Chem. Phys. **112**, 3754 (2000).
11 H. Matsui and E. R. Grant, J. Chem. Phys. **104**, 42 (1996), and references therein.
12 E. E. Mayer, H. G. Hedderich, and E. R. Grant, J. Chem. Phys. **108**, 1886 (1998).
13 E. E. Mayer, H. G. Hedderich, and E. R. Grant, J. Chem. Phys. **108**, 8429 (1998).
14 J. A. Bacon and S. T. Pratt, J. Chem. Phys. **113**, 7188 (2000).
15 C. A. Raptis and S. T. Pratt, J. Chem. Phys. **113**, 4190 (2000).
16 G. Herzberg, *Spectra of Diatomic Molecules.* (Van Nostrand Reinhold, New York, 1950).
17 A. D. Walsh, J. Chem. Soc. London **1953**, 2260 (1953).
18 R. S. Mulliken, J. Am. Chem. Soc. **86**, 3183 (1964).
19 R. S. Mulliken, J. Am. Chem. Soc. **88**, 1849 (1966).
20 R. S. Mulliken, J. Am. Chem. Soc. **91**, 4615 (1969).
21 C. E. Moore, *Atomic Energy Levels, Vol. I.* (NSRDS, Washington, D.C., 1971).
22 M. N. R. Ashfold, R. N. Dixon, and R. J. Stickland, Chem. Phys. **88**, 463 (1984).
23 G. Herzberg, *Molecular Spectra and Molecular Structure, III* (Van Nostrand Reinhold, New York, 1966).
24 J. M. Allen, M. N. R. Ashfold, R. J. Stickland, and C. M. Western, Mol. Phys. **74**, 49 (1991).
25 A. Giusti-Suzor and Ch. Jungen, J. Chem. Phys. **80**, 986 (1984).

State-Selected Associative Ionisation as a Probe of the Molecular Dissociative Channels

Xavier Urbain, Pierre-Célestin Karangwa, Driss Nehari, Vola Andrianarijaona, Jozo Jureta*, and Freddy Brouillard

Unité de Physique Atomique et Moléculaire, Université Catholique de Louvain, Chemin du Cyclotron 2, 1348 Louvain-la-Neuve, Belgium; * *On leave from the Institute of Physics, Beograd, P.O.Box 68, Yugoslavia*

urbain@fyam.ucl.ac.be

1. INTRODUCTION

Associative ionization (AI): $A + B^* \rightarrow AB^+ + e^-$ and dissociative recombination (DR) share the same reaction channels, although the precise dynamics depend on the actual collision taking place. The large angular momentum put into the system by heavy particle collisions does indeed obliterate the subtle role of closed Rydberg channels, as these resonances are displaced by the centrifugal energy. However, the systematic study of AI processes at low energy provides us with additional information on the reaction channels, their branching into atomic products, and the preferred symmetries of DR. The effect of isotope substitution on total cross sections, together with the development of a new diagnostic tool to measure the internal energy of AI products, give an unambiguous identification of the reaction pathways.

The experiments performed on hydrogen atom collisions are presented in the following section, and their interpretation is put into perspective with the most recent data gathered on the DR of HD^+ and H_2^+, i.e. the branching among the electronic channels, $H + D(n\ell)$, the ion pair formation, and the vibrational dependence of the DR rate.

2. ASSOCIATIVE IONIZATION IN HYDROGEN ATOM COLLISIONS

2.1 Total cross sections

Our laboratory has been involved in measurements of the total cross sections for AI between hydrogenic species since 1978, with the work of Poulaert et al.[1] on the reaction :

$$H^+ + H^- \rightarrow H_2^+ + e^- \quad (1)$$

A merged beam set-up, based on a single ion source, was used for that purpose. Protons were partly converted into negative ions by two-step electron capture on a xenon target, and appropriate voltages were applied to the charge exchange cell to create a velocity difference between the positive and negative beams.

More recently, a consistent set of data has been collected on AI between hydrogen atoms:[2,3]

$$H(1s) + H(ns) \rightarrow H_2^+ + e^- \quad (2)$$

A generic apparatus is represented in Fig. 1, the actual design of the ion analyzer having changed with time from dipole magnet to triple cylindrical condenser, the latter being best suited to block the huge flux of Lyman-α radiation produced by electric-field quenching of the metastable atoms. Excited atoms are obtained by neutralization of protons with cesium vapor, and subsequent excitation is made by laser irradiation in a region where a carefully chosen electric field is present, in order to open optical transitions among the whole multiplet of Stark substates.[4]

Populations are monitored by quenching the metastable atoms in front of a bare channel electron multiplier. Special care is taken to simulate the actual process of laser excitation in order to infer the actual ns-state population from the observed depopulation of the 2s state. Ground-state atoms are produced in an electrically biased charge-exchange cell containing argon or xenon in order to generate the second neutral beam that is naturally merged with the laser-excited one, to allow for low-energy collisions in the center-of-mass frame.

The collision energy is continuously swept by varying the bias voltage, yielding reliable relative cross sections (Fig. 2), that are put on an absolute scale after evaluation of the form factor (beam overlap) and the fluxes of the different neutral species, as derived from secondary emission current, known metastable fraction and excitation efficiency.

A common feature of all results is the large value of all cross sections up

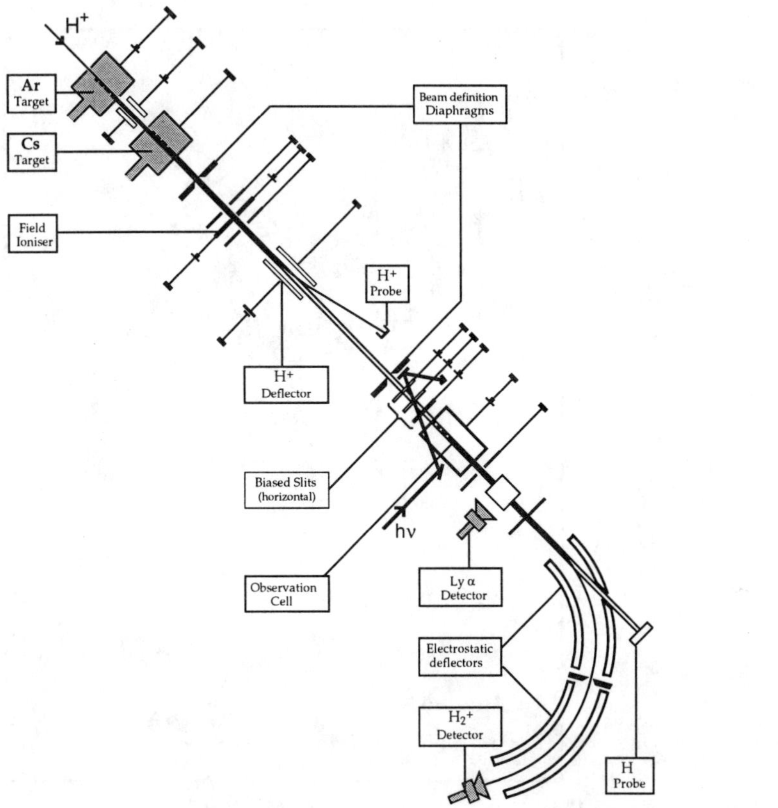

Figure 1. Merged-beam apparatus for the measurement of AI cross sections between neutrals. The CW laser beam is produced by a dye laser or an argon-ion laser, for 3s and 4s excitation.

to a few electron volts. Also visible is the rapid drop above the ionization limit of the excited reactant, corresponding to the Penning ionization channel that competes efficiently with AI. The E^{-1} behavior of the cross section at the lowest collision energies for processes involving H(3s) and H(4s) is remarkable, as these reactions involve neutrals and only the dipole-dipole interaction is attractive at long range. Such a behavior is expected as long as the cut-off in angular momentum (or impact parameter) does not arise from a centrifugal barrier, but rather from the lack of accessible bound states of H_2^+ at a given total energy.

Similar measurements have been performed with deuterium in place of hydrogen, and reveal the quantal nature of the oscillations observed in all cross sections at intermediate energy. These Rosenthal oscillations originate from the fact that all substates with the same principal quantum number are quasi-degenerate, and become equally populated at long internuclear distance. The amplitudes developing along the different channels may

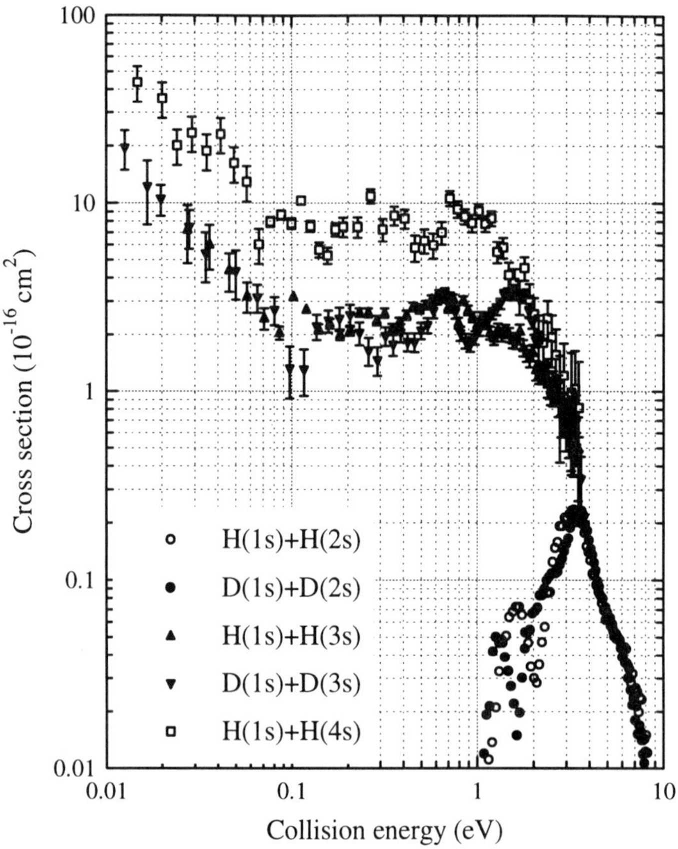

Figure 2. Absolute cross sections for reaction (2) as a function of collision energy and principal quantum number. Error bars are statistical only.

interfere and modulate the total reaction probability.

A quasidiabatic picture of the potential energy curves of all states calculated up to now in ${}^1\Sigma_g^+$ symmetry is given in Fig. 3. The hump of the O state is responsible for the absence of interference in the H(1s) + H(3s) and D(1s) + D(3s) cross sections below 0.08 eV, since a single channel is open below that limit, namely the P state. Furthermore, one may suspect the $H\overline{H}$ state, of ion-pair character, to contribute to the cross section at very low energies, somewhat below 10 meV, which is the size of the adiabatic splitting of the avoided-crossing located at 36 a_0.[5]

A semiclassical calculation of the associative ionization along the O, P and $H\overline{H}$ ${}^1\Sigma_g^+$ states has been performed (Fig. 4), which reproduces the overall behavior of the cross section, but is smaller than the experiment by an order of magnitude, as a result of the neglect of other symmetries.

This fact will be discussed in connection with the DR results. Finally,

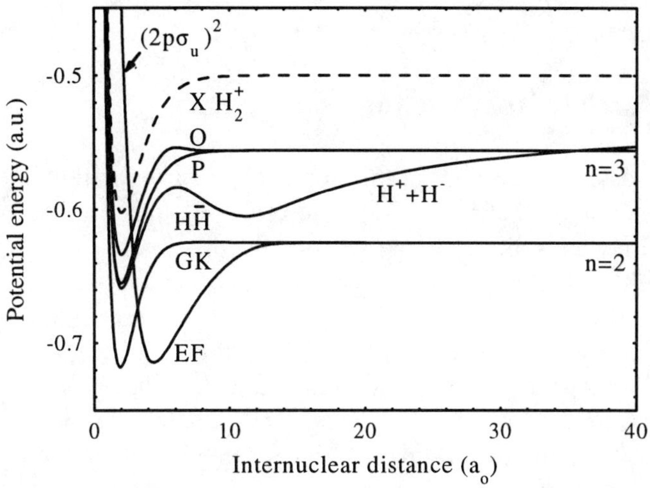

Figure 3. Relevant potential energy curves of ${}^1\Sigma_g{}^+$ symmetry for AI between ground state and excited hydrogen atoms (quasidiabatic picture).

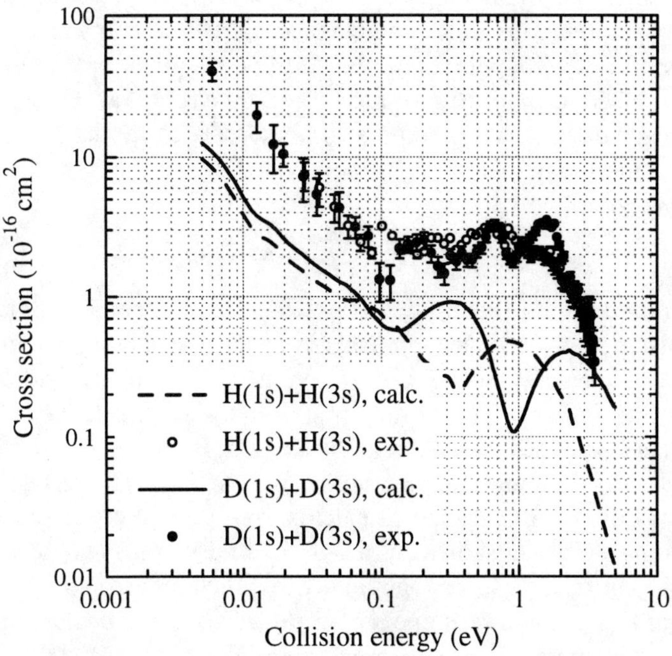

Figure 4. Semiclassical calculation (lines) compared with experimental results.

none of the calculations includes the collisional ionization channel, although methods exist that were applied to $He(2^{1,3}S) + X$ collisions[6] (e.g. the optical potential method).

2.2 Partial Cross Sections

Partial cross sections refer in this context to the rovibrational distribution of the molecular products. State-to-state cross sections are out of reach of such experiments, while the internal energy distribution can be accessed either by ejected electron spectroscopy[7] or by photodissociation of the molecular ions in flight.[8] An alternative to the photodissociation method was extensively developed and used at FOM[9] for H_2^+ beams, namely the dissociative charge exchange with alkali targets. This process leaves the molecule in a predissociative state (c or d $^3\Pi_u$), and careful detection in time and space of the departing atoms suffices for the determination of the initial energy content of the neutral molecule. This distribution deviates very little from the ionic distribution, provided the intermediate states have a strong Rydberg character.

To implement this method, we have combined our merged beam set-up with a detection stage consisting of an effusive potassium beam emerging from a capillary array, a long flight tube and a pair of position-sensitive detectors operating in time coincidence. Kinetic-energy-release (KER) measurements allow the vibrational levels to be resolved for rotationally cold ions, as produced in an electron-impact ion source (Fig. 5a).

The power of this new method was demonstrated by the analysis of the products of another reaction:

$$H(2s) + H(2s) \rightarrow H_2^+ + e^- \tag{3}$$

From the distribution of internal energy (Fig. 5b), one can immediately infer that the distance of closest approach governing the vibrational distribution lies well outside the Franck-Condon region of the H_2^+ ground state (R_{min} ~$7a_0$) as shown in Fig.6. This observation rules out the existence of a strongly binding state correlated to the doubly excited atomic limit, as tentatively introduced[10] to account for an additional feature in the electron-loss spectroscopy of H_2. A restricted CI calculation[11] gives a better account of the actual turning point, as shown by the dashed curve on Fig. 5b obtained by vertical projection of the continuum wave function onto the rovibrational spectrum of H_2^+ and summation over partial waves of the nuclear motion.

Similar measurements are in preparation for Reactions (2) in order to identify the dominant association routes.

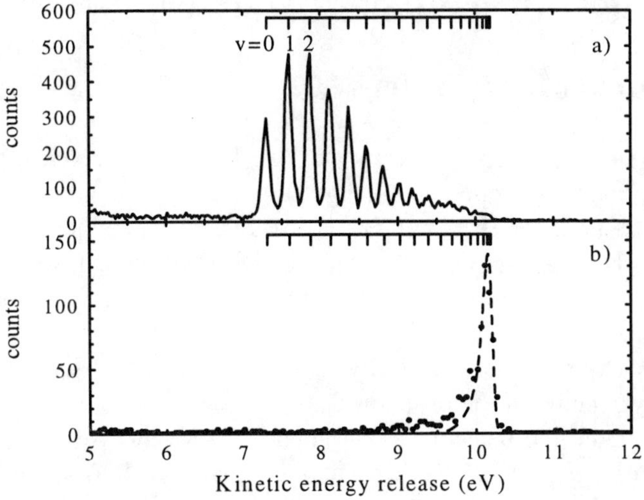

Figure 5. KER spectrum - a) direct H_2^+ beam produced by electron impact ; b) ions produced by associative ionization between H(2s) atoms - dashed line : calculated spectrum.

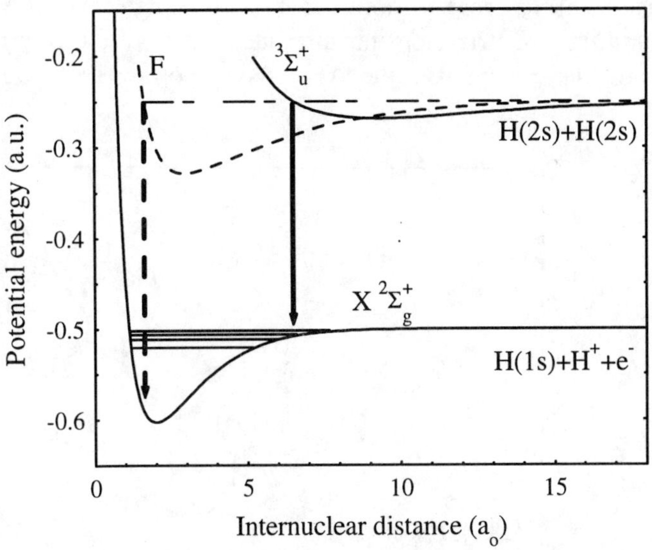

Figure 6. Potential energy curves relevant to H(2s) + H(2s) AI - dashed line: F state;[10] full line: quasidiabatic state correlated to the 2s + 2p limit (□□u and ı□u potential energy curves are assumed identical).[11]

3. THRESHOLD BEHAVIOUR OF H_2^+ AND HD^+ DISSOCIATIVE RECOMBINATION

3.1 Cold Molecular Ions

Hydrogen molecular ions with v<5 are known to recombine along the $(2p\sigma_u)^2\ {}^1\Sigma_g^+$ doubly excited state (Fig. 3), and v-specific cross sections were recently measured for HD^+ and H_2^+ at the TSR in Heidelberg.[12] The precise location of the crossing point with the H_2^+ ground state is responsible for the lower DR rate of v=0 ions, although the v-dependent rate does not follow the Franck-Condon factors. Off-the-energy-shell effects (non-local) have been invoked to explain this peculiarity,[13] but none of the calculations gives a satisfactory account of the observations.

At this point, the behavior of the total cross section measured for the AI reaction H(1s)+H(2s) at threshold may be instructive. A close look at the cross section below 1 eV reveals that almost no signal is present, with an apparent threshold at 1.05 eV, which coincides with the opening of the $H_2^+(v=1) + e^-$ channel. The number of contributing partial waves depends on the total energy minus that of the rotationless vibrational state, explaining the continuous rise of the cross section in that energy range.

Assuming the *S*-matrix elements are independent of the (low) centrifugal distortion of the potentials, the AI cross section takes the following

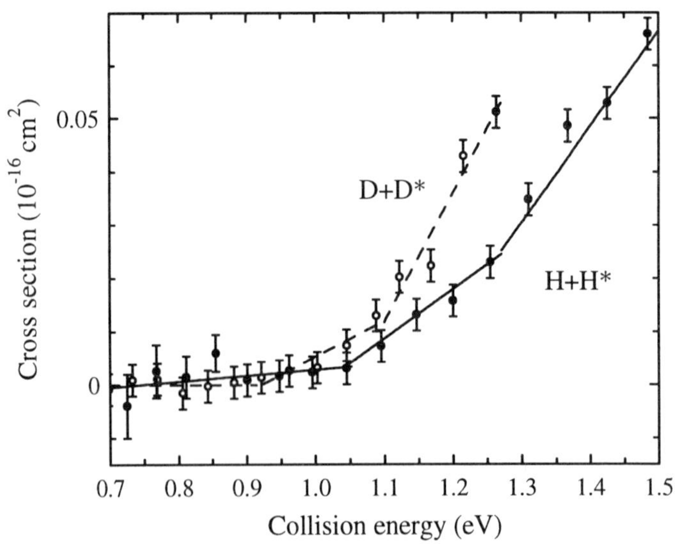

Figure 7. H(2s) + H(2s) AI in the threshold region - full lines : linear fit between vibrational thresholds. The scatter in the data below 0.9 eV is caused by a large background subtraction.

approximate form:

$$\sigma = \frac{\pi}{2\mu E_{coll}} \sum_J (2J+1) \sum_v \left|S_v^J\right|^2 \approx \frac{\pi}{2\mu E_{coll}} \sum_v J_v (J_v+1) \cdot \left|S_v\right|^2 \qquad (4)$$

where μ is the relative mass, E_{coll} the collision energy, and J_v the maximum angular momentum for which level v is accessible. The factor $J_v(J_v+1)$ is proportional to the amount of energy above threshold, so we obtain the following energy dependence:

$$\sigma \propto \sum_v \frac{E_{as} + E_{coll} - E_v}{E_{coll}} \cdot \left|S_v\right|^2 \qquad (5)$$

where E_{as} is the asymptotic energy of the pair of atoms and E_v the energy of the (v, J=0) level. This analytical form allows a linear fit from threshold to threshold, and the increasing slope reflects the increasing sum of S-matrix elements. The relative slope (proportional to $|S_v|^2$) can be compared with the relative DR rate coefficients of Nakashima et al[14] (Table 1), showing reasonable agreement, in view of the experimental uncertainties and the oscillation of the total cross section occurring on a broader energy scale.

Table 1. Relative DR rate coefficients for H_2^+. All results are normalized to v=1.

v	H_2^+ Theory (100 K)	H_2^+ Experiment (AI)	D_2^+ Theory (100K)	D_2^+ Experiment (AI)
0	0.13	0.14±0.11	0.03	<0.1
1	1	1	1	1
2	0.74	0.84±0.08	1.29	2.5±0.7

3.2 Hot Molecular Ions

As demonstrated by Amitay et al,[12] a sudden transition occurs in the HD^+ DR rate for v>5, i.e. the H(1s) + H(3ℓ) threshold. This limit corresponds to v>4 in H_2^+, and v>6 in D_2^+. The lowest doubly excited states, apart from the $(2p\sigma_u)^2\ {}^1\Sigma_g^+$, all cross the ground state potential energy curve of the ion in the vicinity of the outer turning point of these high-lying vibrational levels, enhancing the recombination process by an order of magnitude.

One may expect the AI reaction involving n=3 atoms to be faster than the reaction with n=2 atoms by the same factor. However, the total AI cross section is a sum over all accessible rovibrational levels of the molecular ion, and cross sections have to be compared at the same excess energy above the vibrational levels considered. This would correspond, for an excess energy of 0.1 eV, to 1.15 eV (0.1 above v=1 in H_2^+) for the H(1s) + H(2s) reaction and

0.1 eV for H(1s)+H(3s) reaction (above v=5). In order to eliminate a possible effect of the interference pattern on the results, an average of the hydrogen and deuterium measurements is considered. Furthermore, the cross sections are multiplied by the collision energies in order to compare reaction probabilities. With σ_{3s}=2.25 10^{-16} cm^2 and σ_{2s} = 1.54 10^{-18} cm^2, one obtains a 12:1 ratio after rescaling, in good agreement with Amitay et al. This result is not expected, since H(1s) + H(ns) collisions are limited to Σ states, while Amitay et al. refer to the lowest Π state to account for their experimental results. Our semiclassical calculations indicate the dominant contribution of additional doubly excited states that do not have the $^1\Sigma_g^+$ symmetry.

4. TOWARDS A COMPLETE S MATRIX

The advent of three-dimensional imaging has made possible the measurement of state-specific branching ratios, and forced the theoretical community to tackle the long-range dynamics that governs the fate of the recombined molecule. Highly satisfactory results have already been obtained regarding the detailed branching in the low-energy region,[15] where the $(2p\sigma_u)^2\ ^1\Sigma_g^+$ state is the only recombination route.

With the imaging technique described in section 2.2, one has access to the internal energy distribution of the molecular ions. Such results were obtained for the $H^+ + H^-$ reaction (Fig. 8), although simulations based on the theoretical model of Urbain et al.,[16] in which the reaction proceeds along the $(2p\sigma_u)^2\ ^1\Sigma_g^+$ state, do not fit the experimental spectrum. The deviation most probably arises from a second reaction channel, namely the B" $\overline{\mathrm{B}}\ ^1\Sigma_u^+$ state.

For a definitive understanding to be gained, one needs a consistent set of potential energy curves and couplings, and an appropriate treatment of non-

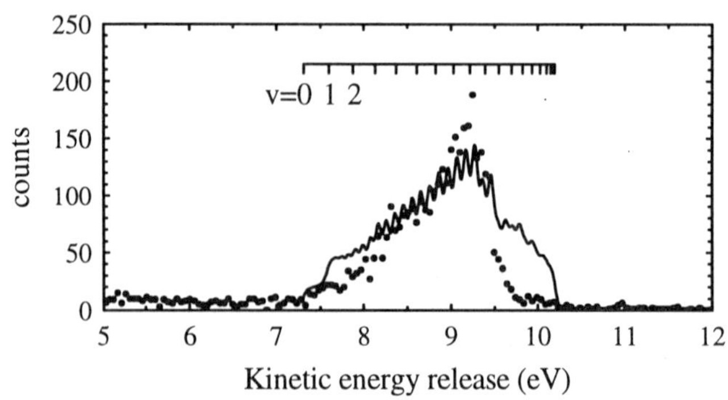

Figure 8. Kinetic energy release obtained by dissociative charge exchange of H_2^+ formed by AI between H^+ and H^- - full line: simulated spectrum.

adiabatic transitions at all internuclear distances. In that respect, the challenge of reproducing the complicated interference pattern modulating the AI cross sections presented in section 2.1 may prove beneficial. As an example, the energy at which a sudden transition occurs, in the H + H(3s)/D + D(3s) cross sections, from monotonic to oscillatory behavior, is an exact measure of the height of the hump present in one of the potential energy curves leading to AI.

5. CONCLUSIONS

Associative ionization experiments were shown to be closely related to dissociative recombination studies, provided one takes into account the heavy-particle nature of the collisional process. The total cross sections for H(1s) + H(2-4s) associative ionization measured at low energy reveal some of the dynamics of the reverse process. Threshold behavior and vibrational dependence of the DR cross section appear to be connected with the energy dependence of state-specific AI cross sections.

Such measurements may give an additional check of the existing theories, particularly in cases where the reaction mechanism is uncommon, as in HeH^+ DR.[17] The sudden switching from He + H(n=2) to He + H(n=3) products[18] should manifest itself in the reverse process. Would these reactions be observed, the measurement of the internal energy of the products would provide the theoretician with an additional clue of the vibronic nature of the coupling. Such investigations are in preparation in our laboratory.

ACKNOWLEDGMENTS

The present research has been supported by the 'Institut Interuniversitaire des Sciences Nucléaires', the Euratom-Belgian State Association, and the IHP program of the EC under contract HPRN-CT-2000-00141. XU is a research fellow of the Belgian National Fund for Scientific Research (FNRS).

REFERENCES

1 G. Poulaert, F. Brouillard, W. Claeys, J. W. McGowan and G. Van Wassenhove, J. Phys. B: *At. Mol. Opt. Phys.* **11**, L671 (1978).

2 X. Urbain, A. Cornet, F. Brouillard and A. Giusti-Suzor, Phys. Rev. Lett. **66**, 1685 (1991).

3 F. Brouillard and X. Urbain, *Rearrangement Processes Involving Hydrogen and Helium Atoms and Ions*, in Atomic and Molecular Processes in Fusion Edge Plasmas, ed. by R. Janev (Plenum Press, New York and London, 1995), p. 309.
4 V. Lorent and P. Antoine, J. Phys. B: *At. Mol. Opt. Phys.* **24**, 227 (1991).
5 V. Sidis, C. Kubach and D. Fussen, Phys. Rev. A **27**, 2431 (1983).
6 P. E. Siska, Rev. Mod. Phys. **65**, 337 (1993) and references therein.
7 J. Lorenzen, H. Hotop, M.-W. Ruf and H. Morgner, Z. Phys. A **297**, 19 (1980).
8 J. J. Blangé, J. M. Zijlstra, A. Amelink, X. Urbain, H. Rudolph, P. van der Straten, H. C. W. Beijerinck and H. G. M. Heideman, Phys. Rev. Lett. **78**, 3089 (1997).
9 D. P. de Bruijn, J. Neuteboom, V. Sidis and J. Los, Chem. Phys. **85**, 215 (1984).
10 T. Odagiri, N. Uemura, K. Koyama, M. Ukai, N. Kouchi and Y. Hatano, J. Phys. B : *At. Mol. Opt. Phys.* **29**, 1829 (1996).
11 L. D. A. Siebbeles and C. Le Sech, J. Phys. B : *At. Mol. Opt. Phys.* **27**, 4443 (1994).
12 Z. Amitay, A. Baer, M. Dahan, J. Levin, Z. Vager, D. Zajfman, L. Knoll, M. Lange, D. Schwalm, R. Wester, A. Wolf, I. F. Schneider and A. Suzor-Weiner, Phys. Rev. A **60,** 3769 (1999); S. Krohn, Z. Amitay, A. Baer, D. Zajfman, M. Lange, L. Knoll, J. Levin, D. Schwalm, R. Wester and A. Wolf, Phys. Rev A **62**, 032713 (2000).
13 H. Takagi, Theoretical Problems in the Dissociative Recombination of H_2^+ + e, in *Dissociative Recombination : Theory, Experiment and Applications*, ed. by B. R. Rowe, J. B. A. Mitchell and A. Canosa, (Plenum Press, New York,1993), p. 75.
14 K. Nakashima, H. Takagi and H. Nakamura , J. Chem. Phys. **86**, 726 (1987).
15 D. Zajfman, Z. Amitay, M. Lange, U. Hechtfischer, L. Knoll, D. Schwalm, R. Wester, A. Wolf and X. Urbain, Phys. Rev. Lett. **79**, 1829 (1997).
16 X. Urbain, A. Giusti-Suzor, D. Fussen and C. Kubach, J. Phys. B : *At. Mol. Opt. Phys.* **19**, L273 (1986).
17 S. L. Guberman, Phys. Rev. A **49**, R1531 (1994).
18 J. Semaniak, S. Rosén, G. Sundström, C. Strömholm, S. Datz, H. Danared, M. af Ugglas, M. Larsson, W. J. van der Zande, Z. Amitay, U. Hechtfischer, M. Grieser, R. Repnow, M. Schmidt, D. Schwalm, R. Wester, A. Wolf and D. Zajfman, Phys. Rev. A **54**, R4617 (1996).

Ultracold Molecules: Formation, Detection

Olivier Dulieu
Laboratoire Aimé Cotton, Bat. 505, Université Paris-Sud, F-91405 Orsay Cedex, FRANCE
dulieu@perceval.lac.u-psud.fr

1. INTRODUCTION

During the last decade, laser cooling and trapping of atoms has allowed researchers to obtain gaseous samples of atoms at ultralow temperatures, i.e. well below 1 milliKelvin. The most spectacular result, recently honored by the Nobel Prize, is probably the achievement of Bose-Einstein condensation (BEC) in a sample of ultracold alkali atoms (^{87}Rb,[1] ^{23}Na,[2] ^{7}Li[3]) and more recently in Helium.[4,5] In contrast, direct laser cooling of molecules, which may appear as a simple extension of laser cooling of atoms, encounters major difficulties. Indeed, laser cooling of atoms relies on the existence of a closed energy level scheme, which allows an efficient cycling of the level population without loss. Such a closed scheme cannot be easily found in molecules, due to their much more complicated energy level structure (Fig. 1). Direct laser cooling would necessitate a large number of lasers,[6] and has not been realized up to now.

A major advance was performed[7] in our group in 1997 using the photoassociation[8] of ultracold cesium atoms to generate ultracold Cs_2 molecules in their lowest triplet (metastable) electronic state. Similar observations of metastable or ground state molecules have been presented later with Cs_2,[9] K_2,[10, 11] and Rb_2.[12] Ensembles of ultracold molecules with typical density of $\approx 10^{11}$ cm^{-3}, and temperature in the 20-100 μK range are now available, opening the way to further studies.

The present article is organized as follows: the basics of photoassociation are recalled first, and a survey of the main results is given. The various

Dissociative Recombination of Molecular Ions with Electrons
Edited by S. Guberman, Kluwer Academic/Plenum Publishers, 2003

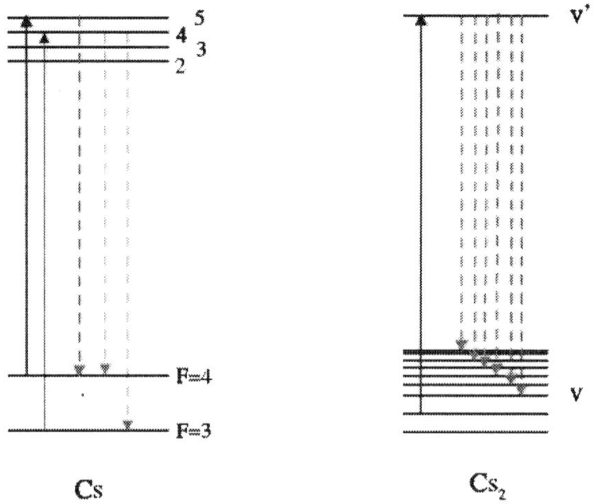

Figure 1. Schematic representation of transitions involved in laser cooling of cesium atoms (left) and cesium dimers (right). F is the atomic hyperfine quantum number, and v, v' are vibrational levels of ground and excited Cs_2 electronic states

mechanisms that have been proposed or demonstrated for ultracold molecule formation are described in detail in Section 3. In Section 4, an efficient procedure for detection of ultracold molecules is presented, relying on photoionization and/or molecular autoionization and providing a link with the central topic of this conference. Finally we discuss the main prospects for ultracold molecule research that are currently under investigation.

2. PHOTOASSOCIATION OF ULTRACOLD ATOMS

The photoassociation (PA) process is simply the inverse of photodissociation (PD): a pair of colliding atoms absorbs a photon with frequency ν_L detuned to the red of an atomic transition frequency ν_a, in order to create a diatomic molecule in an excited electronic bound state. In contrast with PD at thermal energies, the initial state for PA does not have a well defined energy, due to the large thermal distribution of kinetic energies: the PA process is actually competing with many other free-free absorption processes, and is hardly observable experimentally.[13]

Using cold colliding atoms offers a unique opportunity: the kinetic energy distribution is now so sharp (at 100 μK, $k_BT \approx 5$ MHz, or 10^{-4} cm^{-1}) that it is smaller than any other characteristic energy of the system. The PA process acts now as a resonant free-bound transition, which is depicted in Fig. 2 for cesium atoms: the radial motion of the colliding atoms is represented by a stationary continuum wave function with a large amplitude at large interatomic distances, determined by the R^{-6} variation of the long-range potential correlated to Cs(6s)+Cs(6s). In a typical experiment, the photon

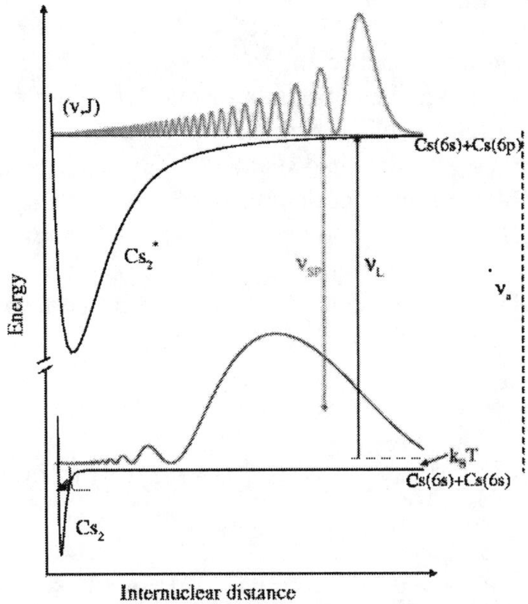

Figure 2. Schematic view of photoassociation of cold cesium atoms.

frequency, ν_L, is detuned to the red of the 6s → $6p_j$, (j = 1/2, 3/2) transition frequency. In contrast with the initial state, the R^{-3} long-range behaviour of the electronic excited state implies the existence of bound levels with a vibrational motion reaching distances of $100a_o$ (where a_o is the Bohr radii), or more. Using quasiclassical terms, the photon is mainly absorbed at large distances, and creates an excited molecule in a bound level close in energy to the dissociation limit. Such excited molecules are actually ultracold, as the absorption does not modify significantly the initial kinetic energy of the atom pair. But they have a short lifetime, as they decay within a few tens of nanoseconds back into a pair of hot free atoms, which escape from the cold atom trap. This loss process is actually the simplest way to observe the PA process.[14-16]

Several review papers have been devoted to the PA process and its various applications,[17-20] and include extensive lists of references that cannot be listed here. In particular, the remarkable features of cold atom PA described above, has resulted in the development of a new spectroscopic technique, the so-called photoassociative spectroscopy, which is now a well-established tool for cold atom studies. Most results concern alkali dimers from Li_2 to Cs_2, and only a few concern other systems such as H_2,[21] metastable He_2,[22] and Ca_2.[23] PA spectroscopy gives access to the determination of the long-range part of the electronic potential curves. The so-called pure long-range molecules,[24] i.e. molecules with excited electronic states exhibiting a potential well at interatomic distances well beyond the standard chemical bond length, have been observed for all alkali dimers except Li_2. Long-range coefficients, C_3, governing the R^{-3} variation of the excited electronic states, have been extracted, providing new determinations of the radiative lifetime of the first excited P_j (j = 1/2, 3/2) levels of alkali atoms with a better accuracy than standard atomic spectroscopy experiments. Determination of the scattering length for elastic collisions, which is a key parameter for BEC studies, also benefited from PA studies in several cases. Other effects such as tunneling through a potential barrier excited in photoassociated Cs_2 molecules, or spin-orbit perturbations and pre-dissociation in photoassociated Rb_2 and Cs_2 molecules have also been investigated.

3. FORMATION OF ULTRACOLD MOLECULES

As mentioned in the previous section, the PA process creates only short-lived molecules, decaying back to a pair of free atoms, as shown in Fig. 3a. This is due to the poor overlap between the vibrational wave function of the PA level, and the vibrational wave functions of the lowest electronic states correlated to 6s + 6s. The way to stabilize the photoassociated molecules into long-lived ultracold molecules was first demonstrated with cesium dimer in our group in 1997,[7] and is illustrated in Fig. 3b. The existence of a long-range potential well (see the pure long-range molecules quoted above) in an excited electronic state of 0_g^- symmetry (in Hund's case c notation), correlated to the ($6s + 6p_{3/2}$) dissociation limit, makes the radiative decay towards bound vibrational levels very efficient because of a turning point in the 15-20a_o range of the PA vibrational level. Such molecules are actually created in vibrational levels of the lowest triplet state in Cs_2, and are metastable. They are not trapped by the standard magnetooptical trap used for the cesium atoms in the experiment, so they escape from the trap due to gravity within 10 ms. The ultracold molecules are detected using an ionizing probe laser (see the next section) below the atom trap. The ballistic expansion of the molecular cloud provides an estimate for their temperature, found in the 20-100 μK range, i.e. similar to the atom trap temperature.

More recently, we proposed[25] a formation scheme for ground state ultracold molecules (see Fig. 3c). These molecules have also been observed for Cs_2.[25] Due to spin-orbit coupling, two excited electronic states (of 0^+ symmetry here, correlated to ($6s + 6p_{1/2}$) and ($6s + 6p_{3/2}$) respectively) can interact through an avoided crossing at short distance (around 10a_o in Cs_2). The vibrational series of each electronic state are resonantly coupled to each other, inducing mixing of vibrational wave functions, according to the

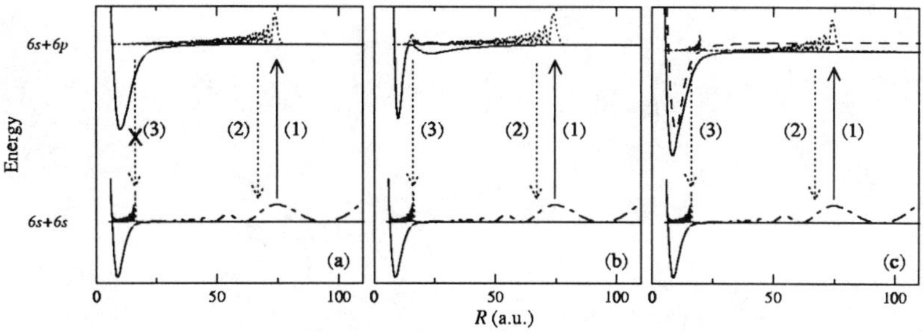

Figure 3. Photoassociation [arrow (1)] from the cesium 6s + 6s continuum to (a) a typical potential; (b) a double-well potential [e.g. 0_g^- (6s + $6p_{3/2}$)]; (c) two coupled states [e.g. 0_u^+ (6s + $6p_{1/2,3/2}$)]. The system decays by spontaneous emission either back to the continuum [arrow (2)] or to a bound level of the ground state [arrow (3)]. For case (a), reaction (3) is usually unlikely.

example shown in the Figure. As in the previous case, a turning point in the 15-20a_o range for the PA vibrational level enhances the radiative decay toward bound levels of the ground state (of $^1\Sigma_g^+$ symmetry). We have also investigated the possibility to achieve PA into vibrational levels lying within shallow potential wells located around 17ao in K_2 and Rb_2, and embedded into the dissociation continuum of the (6s+$6p_{3/2}$) dissociation limit.[26]

Ultracold molecules have also been found using PA of cold Rb and K atoms. The Rb experiment in Pisa[12,27] relies on the same processes described by Fig. 3b. The K experiment at University of Connecticut implies a two-photon PA scheme, populating vibrational levels of a Rydberg electronic state that provides a good overlap between the PA and ground state vibrational wave functions.[11]

In all the above experiments, the rate for ultracold molecule formation has been found quite important, i.e. 1 to 5 molecules per second per atom. We will see in the next section how such rates can be determined and modelled.

4. DETECTION OF ULTRACOLD MOLECULES

The detection procedure is a key point of ultracold molecule experiment. When PA is performed in a cold atom trap, the excited molecule can decay back into a pair of hot atoms (Fig. 3a) or into a bound level of a stable molecular state (Figs. 3b and 3c). In both cases, the products escape from the trap, and dips are observed in the fluorescence signal of the trap. This trap loss detection has been extensively used for photoassociation studies, but cannot be used to discriminate molecules from atoms when they are produced. A powerful technique has been demonstrated,[7] that consists of pulsed resonant two-step ionization of the Cs_2 molecules into Cs_2 molecular

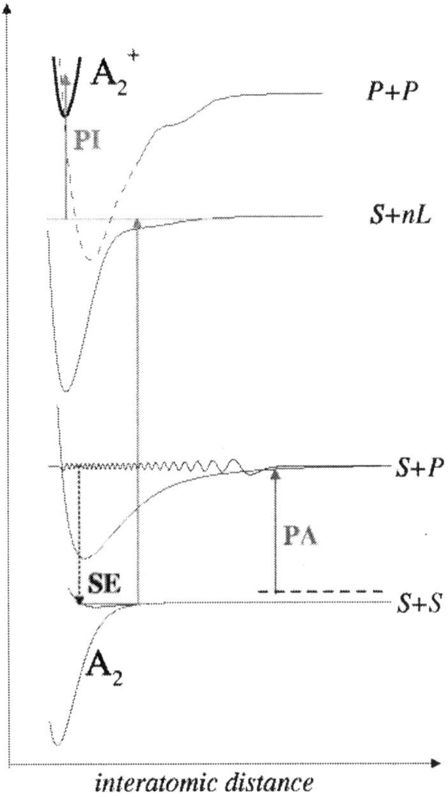

Figure 4. Schematic view of the resonant two-photon ionization of ultracold molecules formed after photoassociation (PA) followed by spontaneous emission (SE).

ions, via an intermediate electronic state (see Fig. 4). In a typical experiment, the atom trap is illuminated continuously by the PA laser, while an intense laser pulse of ≈1 ns duration is applied at a 10 Hz repetition rate. All ultracold molecules accumulated within the trap between two pulses are then ionized in a single shot.

Equivalent perturbative models for the PA rate have been proposed by several authors,[28-30] and for the cold molecule formation rate and ionization rate in a more recent paper.[31] The rate calculation mainly involves, for the PA step, the overlap between the radial continuum wave function of the initial atom pair with the vibrational wave function of the PA bound level, and for the cold molecule rate the overlap between the latter wave function with the vibrational wave functions of the lowest electronic states. A detailed comparison of this model with experimental data has been presented[25,32] and shows the validity of such a perturbative approach.

A similar detection method has been applied for ultracold Rb_2[12] and K_2.[11] It is also worthwhile to mention that an ionization step has been also used for the detection of photoassociated molecules. The competition between direct photoionization, and ionization through autoionizing molecular levels has been demonstrated in Na_2 photoassociation.[33,34] In a few cases, the spectroscopy of molecular autoionizing states has been achieved using a two color photoassociation process in Na_2[35] and K_2.[36]

5. THEORETICAL APPROACHES FOR ULTRACOLD MOLECULE FORMATION

As suggested by Fig. 3, ultracold molecule formation involves both the large distance region (well beyond $50a_o$) governed by weak interactions and low kinetic energies, and the short distance region (around $20a_o$, or below), which is dominated by strong chemical interactions, and large kinetic energies. In other words, the local de Broglie wavelength associated to the radial motion changes dramatically from large R towards small R. The theoretical description of these processes should treat both situations with equivalent accuracy. For this purpose, we developed in our group the Mapped Fourier Grid Hamiltonian (MFGH) method.[37,38] Most of the calculations performed in the papers from our group quoted above use the MFGH approach.

In the standard Fourier Grid Hamiltonian method, wave functions are represented on a grid of N equally-spaced points, which is equivalent to an expansion of wave functions in a basis of N plane waves.[39-41] The Hamiltonian for a given electronic state is expressed as a $N \times N$ matrix, whose diagonalization provides eigenvalues and eigenvectors.

The MFGH method involves *an adaptive grid step, which is controlled by the local de Broglie wavelength of the system,* leading to a drastic reduction of the required number of grid points, particularly suitable when large distances are studied. The MFGH method is easily generalized to the calculation of eigenvalues of a system with p coupled electronic states (like in Fig. 3c), which are obtained after diagonalization of the corresponding $pN \times pN$ Hamiltonian matrix. When relevant, predissociation lifetimes can be deduced by adding an absorbing complex potential to the Hamiltonian.[42] The diagonalization of the associated $pN \times pN$ complex matrix provides complex eigenvalues, the imaginary part giving the predissociation width.

6. CONCLUSIONS, PERSPECTIVES

The aim of the present overview is to suggest that the research field of ultracold molecule is rapidly evolving, stimulating a growing number of teams to start new studies. Once those dense and cold molecular samples are produced, the next step is to trap them in particular configurations of electromagnetic fields. This has been demonstrated with a small number of ultracold cesium molecules,[43] and is presently under investigation in our group.[44] Another improvement is to use a coherent decay process (like stimulated Raman emission) to control the final state of the formed ultracold molecule.[45]

Such achievements will open a way for a better control of these molecules, including the possibility to increase their number density and temperature to reach the quantum degenerate regime. An alternate way to obtain molecular BEC, which is currently more studied, is to perform PA within an atomic BEC, which could transform it into a molecular BEC using suitable transitions like for example those described in the present article.

It is worthwhile also to note that the present discussion on photoassociative spectroscopy, which focuses on bound levels close to the dissociation limit of a given molecular state, has some similarities with the study on near-threshold photodissociation spectroscopy of cold CH^+ and OH^+ in a storage ring,[46] which was presented at the Nasslingen conference on dissociative recombination in 1999. Moreover, cold molecular ions can also be trapped, as demonstrated on MgH^+ and MgD^+,[47] and could be used for further DR studies.

ACKNOWLEDGMENTS

The work presented in this article has been performed within the "Cold Molecule" group at Laboratoire Aimé Cotton, involving both the theoretical team (C. Amiot, A. Crubellier, O. D., and F. Masnou-Seeuws) and the experimental team (D. Comparat, S. Guibal, and P. Pillet). Several Ph.D. students and post-docs have also been strongly involved in this work, including C. Dion, C. Drag, A. Fioretti, R. Gutterres, V. Kokoouline, B. Laburthe-Tolra, N. Vanhaecke, M. Vatasescu.

REFERENCES

1 M. H. Anderson, J. R. Ensher, M. R. Matthews, C. E. Wieman, and E. A. Cornell, Science **269**, 198 (1995).

2 K. B. Davis, M.-O. Mewes, M. R. Andrews, N. J. van Druten, D. S. Durfee, D. M. Kurn and W. Ketterle, Phys. Rev. Lett. **75**, 3969 (1995).

3 C. C. Bradley, C. A. Sackett, J. J. Tollett and R. G. Hulet, Phys. Rev. Lett. **75**,1687 (1995).

4 A. Robert, O. Sirjean, A. Browaeys, J. Poupard, S. Nowak, D. Boiron, C. I. Westbrook, and A. Aspect, Science **292**, 461 (2001).

5 F. Pereira Dos Santos, J. Leonard, J. Wang, C. J. Barrelet, F. Perales, E. Rasel, C. S. Unnikrishnan, M. Leduc, and C. Cohen-Tannoudji, Phys. Rev. Lett. **86**, 3459 (2001).
6 J. T. Bahns, P. L. Could, and W.C. Stwalley, J. Chem. Phys. **104**, 9689 (1996).
7 A. Fioretti, D. Comparat, A. Crubellier, O. Dulieu, F. Masnou-Seeuws, and P. Pillet, Phys. Rev. Lett. **80**, 4402 (1998).
8 H. R. Thorsheim, J. Weiner, and P. S. Julienne, Phys. Rev. Lett. **58**, 2420 (1987).
9 T. Takekoshi, B. M. Patterson and R. J. Knize, Phys. Rev. A **59**, R5 (1999).
10 A. N. Nikolov, E. E. Eyler, X. T. Wang, J. Li, H. Wang, W. C. Stwalley and P. L. Gould, Phys. Rev. Lett. **82**, 703 (1999).
11 N. Nikolov, J. R. Enscher, E. E. Eyler, H. Wang, W. C. Stwalley, and P. L. Could, Phys. Rev. Lett. **84**, 246 (2000).
12 C. Gabbanini, A. Fioretti, A. Lucchesini, S. Cozzini, and M. Mazzoni, Phys. Rev. Lett. **84**, 2814 (2000).
13 P. Backhaus, B. Schmidt, and M. Dantus, Chem. Phys. Lett. **306**, 18 (1999).
14 J. D. Miller, R. A. Cline, and D. J. Heinzen, Phys. Rev. Lett. **71**, 2204 (1993).
15 E. R. I. Abraham, W. I. McAlexander, C. A. Sackett, and R. G. Hulet, Phys. Rev. Lett. **74**, 1315 (1995).
16 D. Comparat, C. Drag, A. Fioretti, O. Dulieu, and P. Pillet, J. Molec. Spect. **195**, 229 (1999).
17 P. D. Lett, P. S. Julienne, and W. D. Philips, Annu. Rev. Phys. Chem. **46**, 423 (1995).
18 J. Weiner, V. S. Bagnato, S. C. Zilio, and P. S. Julienne, Rev. Mod. Phys. **71**, 1 (1999).
19 W.C. Stwalley and H. Wang, J. Molec. Spect. **195**,194 (1999).
20 F. Masnou-Seeuws and P. Pillet, Adv. At. Mol. Opt. Phys. **47**, 53 (2001).
21 A. P. Mosk, M. W. Reynolds, T. W. Hijmans and J. T. M. Walraven, Phys. Rev. Lett. **82**, 307 (1999).
22 N. Herschbach, P. J. J. Tol, W. Wassen, W. Hogervorst, G. Woestenenk, J. W. Thomsen, P. Van der Straten, and A. Niehaus, Phys. Rev. Lett. **84**, 1874 (2000).
23 G. Zinner, T. Binnewies, F. Riehle, and E. Tiemann, Phys. Rev. Lett. **85**, 2292 (2000).
24 W. C. Stwalley, Y. H. Uang, and G. Pichler, Phys. Rev. Lett. **41**, 1164 (1978).
25 C.M. Dion, C. Drag, O. Dulieu, B. Laburthe Tolra, F. Masnou-Seeuws and P. Pillet, Phys. Rev. Lett. **86**, 2253 (2001).
26 M.-L. Almazor, O. Dulieu, F. Masnou-Seeuws, R. Beuc, and G. Pichler, Eur. Phys. J. D **15**, 355 (2001).
27 A. Fioretti, C. Amiot, C. M. Dion, O. Dulieu, M. Mazzoni, G. Smirne, and C. Gabbanini, Eur. Phys. J. D **15**, 189 (2001).
28 P. S. Julienne, J. Res. Natl. Inst. Stand. Technol. **101**, 487 (1996).
29 P. Pillet, A. Crubellier, A. Bleton, O. Dulieu, P. Nosbaum, I. Mourachko, and F. Masnou-Seeuws, J. Phys. B **30**, 2801 (1997).
30 R. C6te and A. Dalgarno, Phys. Rev. A **58**, 498 (1998).
31 M. Dion, O. Dulieu, D. Comparat, W. de Souza Melo, N. Vanhaecke, P. Pillet. R. Beuc, S. Milosevic, and G. Pichler, Eur. Phys. J. D, in press (2002).
32 C. Drag, B. Laburthe Tolra, O. Dulieu, D. Comparat, M. Vatasescu, S. Boussen. S. Guibal, A. Crubellier, and P. Pillet, IEEE J. Quant. Electron. **36**, 1378 (2000).
33 P. A. Molenaar and P. Van der Straten, Phys. Rev. Lett. **77**, 1460 (1996).
34 K.M. Jones, S. Maleki, L.P. Ratliff, and P. D. Lett, J. Phys. B: Atomic Molecular and Optical Physics **30**, 289 (1997).
35 A. Amelink, K.M. Jones, P.D. Lett, P. van der Straten, and H.G.M. Heideman, Phys. Rev. A **61**, 42707 (2000).
36 H. Wang, X.T. Wang, P. L. Could, and W.C. Stwalley, Phys. Rev. Lett. **78**, 4173 (1997).
37 V. Kokoouline, O. Dulieu, R. Kosloff, and F. Masnou-Seeuws, J. Chem. Phys. **110**, 9865 (1999).

38 V. Kokoouline, O. Dulieu, and F. Masnou-Seeuws, Phys. Rev. A **62**, 022504 (2000).
39 R. Kosloff, J. Phys. Chem., **92**, 2087-2100 (1988).
40 C. C. Martson and G. G. Balint-Kurti, J. Chem. Phys. **91**, 3571 (1989).
41 O. Dulieu and P. S. Julienne, J. Chem. Phys. **103**, 60 (1995).
42 V. Kokoouline, O. Dulieu, R. Kosloff and F. Masnou-Seeuws, Phys. Rev. A **62**, 032716 (2000).
43 T. Talcekoshi, B. M. Patterson and R. J. Knize, Phys. Rev. Lett. **81**, 5105 (1998).
44 B. Laburthe Tolra, D. Comparat and P. Pillet, submitted.
45 B. Laburthe Tolra, C. Drag and P. Pillet, Phys. Rev. A **64**, 61401 (R) (2001).
46 U. Hechtfischer, Z. Amitay, P. Forck, M. Lange, J. Linkemann, M. Schmitt, U, Schramm, D. Schwalm, R. Wester, D. Zajfman, and A. Wolf, Phys. Rev. Lett. **80**, 2809 (1998).
47 K. Moolhave and M. Drewsen, Phys. Rev. A **62**, 011401 (R) (2000).

Merged Beam Studies of Associative Ionization

A. Le Padellec[1], X. Urbain[2], T. Nzeyimana[2], and E.A. Naji[2]
[1]LCAR UMR 5589 - Université Paul Sabatier-Toulouse III - 118, Route de Narbonne Bât. IIIR1b4 31062 Toulouse Cedex 4, France; [2]Unité FYAM - Département de Physique - Université Catholique de Louvain (LLN) - Chemin du cyclotron 2 - B-1348 Louvain la Neuve, Belgium
lepadellec@yosemite.ups-tlse.fr, urbain@fyam.ucl.ac.be

1. INTRODUCTION

Carbon monoxide was detected in emission in the supernova 1987A spectrum and it is likely that its ion, CO^+, has also been detected.[1] The radiative association process was considered[2] for formation of CO^+, from C^+ + O, but not the associative ionization (AI) process from charged reactants $C^+ + O^-$. Bertrand and Van Tiggelen[3] found that the process, $N(^2D, ^2P)+O(^3P)\rightarrow NO^+ + e$, is the primary source of nitrogen oxide cations in ammonia and hydrogen-oxygen-nitrogen flames. The AI process that arises from the charged reactants $N^+ + O^-$, was not considered. The hydroxyl cation, OH^+, was detected in interstellar clouds[4] (where its destruction leads to H_3O^+), in comets and in planetary atmospheres[5]. It is important to characterize the formation pathways for OH^+, and among them the AI process. Here, we provide AI cross sections for the association of charged reactants that leads to the production of CO^+, NO^+, O_2^+ and OD^+.

2. EXPERIMENTAL TECHNIQUE

The whole apparatus can be divided into four sections pumped at ultra high vacuum. The first one includes the ion sources, the acceleration, and the mass selectors as well as the beam optics. In the second section, the two ionic beams are merged. The third section has the 7.8 cm (length)

Dissociative Recombination of Molecular Ions with Electrons
Edited by S. Guberman, Kluwer Academic/Plenum Publishers, 2003

interaction region. The last section contains the magnetic analyzer where the intensities of the primary beams as well as that of the reaction products are recorded. In a merged beam set-up, the number of reactions, N, that occur within a certain time, T, is connected to the absolute cross section σ by the following expression:

$$N(T)=\sigma\frac{v_r}{q_1 q_2 v_1 v_2}F\int_0^T I_1(t)I_2(t)dt \qquad (1)$$

where v_r, v_1, v_2, q_1, q_2, I_1 and I_2 depict the relative velocity, the laboratory velocities of the two beams, their charges and their intensities, respectively. The form factor, F, is assumed to be time independent. It is a non-trivial parameter to establish, since it involves the measurement of the density profiles of the two interacting beams. However, one can overcome this experimental difficulty. Indeed, if the two beams display a sharply characterized cross section but also a perfect overlap over a known region L, the form factor would be simply expressed by:

$$F=\frac{L}{S_>} \qquad (2)$$

where $S_>$ denotes the section of the largest beam. In practice, one only needs to strongly diaphragm the two beams just before they interact and make use of the cores of the two reactant beams. L is defined cautiously, by applying a certain observation voltage, V, in the interaction region. The molecular cations formed in this region experience an increase of their kinetic energy by an amount +eV and are therefore completely distinguishable, after magnetic analysis, from other molecular cations produced anywhere else. The interaction length is therefore equal to that of the region where the electric potential is equal to V. The kinematics of the merged beam experiment give the center-of-mass energy, E_{cm}, as

$$E_{cm}=\mu\left[\sqrt{\frac{q_2(A_2-V)}{m_2}}-\sqrt{\frac{q_1(A_1-V)}{m_1}}\right] \qquad (3)$$

with μ, A_1, A_2, m_1 and m_2, the reduced mass, accelerating voltages and masses, respectively. Thus, a fine adjustment of the center of mass energy can be achieved while changing the observation voltage. Cross sections are measurable down to 10meV.

The C^+, N^+, O^+ and D^+ beams were produced within the ECR ion source, located about 2.6 m upstream to the interaction region, whereas the anionic beams O^- and D^- were produced within a duoplasmatron ion source. The electronic excitation of the O^- and D^- target beams is not a problem since they display a unique stable electronic level. However, the excitation of the cationic counterparts is far from being simple. For C^+ below 10 eV, in addition to the ^2P ground state, there are two excited states, the ^4P and ^2D lying at 5.33 and 9.29 eV, respectively. For N^+ below 6 eV, there are three states above the ^3P ground state, the ^1D, ^1S and ^5S at 1.90, 4.05 and 5.85 eV, respectively. For O^+ below 6 eV, there are two states above the ^4S ground state, namely the ^2D and ^2P at 3.32 and 5.02 eV, respectively. Evidence of metastable contamination was demonstrated by Harrison et al.[6] in their N^+ ion beam, in particular with the ^1D and ^1S states. Moreover, the presence of electronically excited cations N^+ and O^+ was explicitly highlighted by Hamdan and Brenton[7] on one hand, and by Reid[8] on the other hand, who show that about 45 % of the O^+ cations produced by electron impact of O_2 were in the ^2P and ^2D metastable states. It is necessary to evaluate the residence time of the cations within the ECR source plus their time-of-flight and compare these with the different possible transition lifetimes of the cations.

The residence times are about 1.4, 2.0 and 2.1 μs, for C^+, N^+ and O^+, respectively. The time-of-flights are 8.4, 8.7 and 9.3μs, respectively.

For the C^+ cations, the radiative lifetimes for the $^2D \rightarrow {}^2P$ and $^4P \rightarrow {}^2P$ transitions, are 3.5 ns and at least 6.7 ms, respectively. Therefore, the ^2D state had time to decay, and the C^+ beam was comprised of a mixture $C^+(^2P$ and $^4P)$. For the N^+ cations, the lifetimes for the $^1S \rightarrow {}^1D$ and $^1D \rightarrow {}^3P$ transitions, are 855ms and 275s, respectively. If produced, the ^1S and ^1D states of the target N^+ ions are populated, in addition to the ^3P ground state. For the O^+ $^2P \rightarrow {}^4S$ and $^2D \rightarrow {}^4S$ transitions, the lifetimes are at least 19 and 6290 s, respectively. The ^2D and ^4S states are populated in the interaction region.

The energy resolution is an important issue, especially for the low center-of-mass energy data that we will present below. The AI process from charged particles is, indeed, a pure Coulomb interaction case for which the cross section should behave like E_{cm}^{-1} in the low energy limit. Together with each set of our low energy cross section data, we display this E_{cm}^{-1} dependence, folded with our experimental energy resolution.

3. EXPERIMENTAL RESULTS

3.1 $C^+ + O^-$

The AI cross sections are in Fig. 1a. The full line below 1 eV fits our data. extremely well. At low CM energy, the AI process populates rovibrational levels of several electronic states of CO^+: X $^2\Sigma^+$, A $^2\Pi$ and B $^2\Sigma^+$. Therefore, this measurement refers to total AI cross sections. It is difficult to give a qualitative and quantitative description of the AI process due to the scarcity of the relevant molecular data. Indeed, the correlation rules gives twelve states that correlate to the $C^+(^2P)$ + $O^-(^2P)$ limit. These are $^{1,3}\Sigma^+(2)$, $^{1,3}\Sigma^-(1)$, $^{1,3}\Pi(2)$, and $^{1,3}\Delta(1)$, all poorly documented. The numbers in parentheses represent the number of electronic states of that particular symmetry. Depending upon the CM energy of the C^+ + O^- system, several channels are open in the studied energy range (<20 eV):

$$
\begin{aligned}
&\rightarrow CO(X^1\Sigma^+) + 20.89 eV \\
&\rightarrow C(^3P) + O(^3P) + 9.80 eV \\
&\rightarrow CO^+(X^2\Sigma^+) + e + 6.88 eV \\
C^+(^2P) + O^-(^2P) &\rightarrow C^+(^2P) + O(^3P) + e - 1.46 eV \\
&\rightarrow C(^3P) + O^+(^4S) + e - 3.82 eV \\
&\rightarrow C^+(^2P) + O^+(^4S) + 2e - 15.08 eV \\
&\rightarrow CO^{2+}(X^3\Pi) + 2e - 20.38 eV
\end{aligned}
$$

One might expect an interplay between AI and the four endothermic processes. Below about 1 eV, only a structureless E^{-1} dependence can be seen in Fig. 1a. The opening of the $C^+(^2P)$ + $O(^3P)$ + e channel, i.e. collisional detachment, nearly coincides with the rapid fall-off of our measured cross sections above 2.5 eV. For higher energies, the cross sections decrease to a non-measurable level just below 15 eV where abruptly decrease again. This may be due to the opening of the $C^+(^2P)$ + $O^+(^4S)$ + 2e channel to which $CO^{2+}(X\ ^3\Pi)$ correlates.

3.2 $N^+ + O^-$

The AI cross sections are in Fig. 1b. The line below 1 eV represents the E_{cm}^{-1} dependence and again, the model fits reasonably well. Several electronic states and rovibrational levels of NO^+ can be populated : $X^1\Sigma^+$, $a^3\Sigma^+$, $b^3\Pi$, $w^3\Delta$, $b'^3\Sigma^-$, $A'^1\Sigma^-$, $w^1\Delta$ and $A^1\Pi$. Once again there is a paucity of molecular

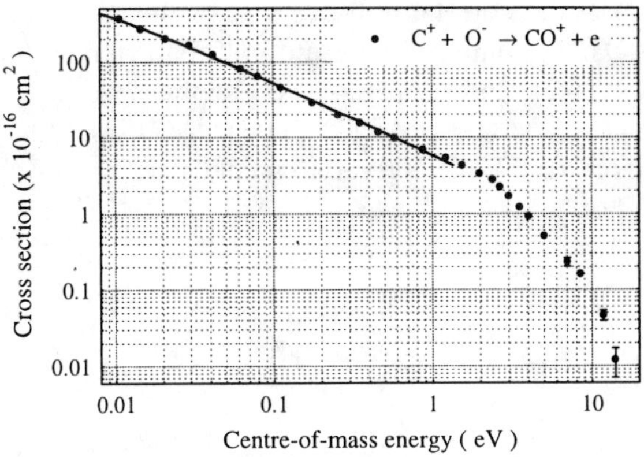

Figure 1a. The filled circles represent the AI cross sections for C^+ + O^- whereas the full line shows the normalized data with the 1/E dependence.

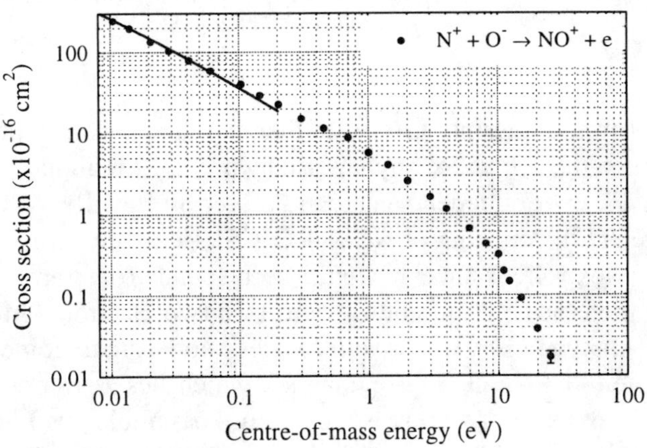

Figure 1b. The filled circles represent the AI cross sections for N^+ + O^- whereas the full shows the normalized data with the 1/E dependence.

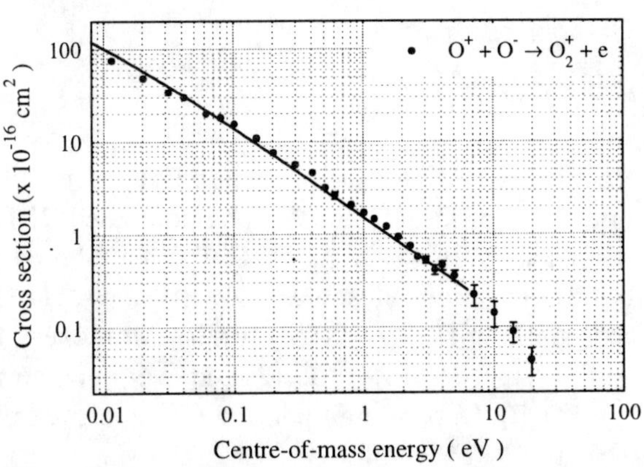

Figure 1c. The filled circles represent the AI cross sections for O^+ + O^- whereas the full line shows the normalized data with the 1/E dependence.

data. We know from the correlation rules that twelve states correlate to the lowest $N^+(^3P) + O^-(^2P)$ limit and most are not characterized. These are the $^{2,4}\Sigma^+(1)$, $^{2,4}\Sigma^-(2)$, $^{2,4}\Pi(2)$, and $^{2,4}\Delta(1)$ states. For the next limit, $N^+(^1D) + O^-(^2P)$, there are nine states: $^2\Sigma^+(2)$, $^2\Sigma^-(1)$, $^2\Pi(3)$, $^2\Delta(2)$, and $^2\Phi(1)$. Depending upon the CM energy of $N^+ + O^-$, several channels are open over the studied energy range (<30 eV):

$$
\begin{aligned}
& \rightarrow NO(X^2\Pi_r) + 19.55eV \\
& \rightarrow N(^4S) + O(^3P) + 13.07eV \\
& \rightarrow NO^+(X^1\Sigma^+) + e + 10.29eV \\
N^+(^3P) + O^-(^2P) & \rightarrow N(^4S) + O^+(^4S) + e - 0.54eV \\
& \rightarrow N^+(^3P) + O(^3P) + e - 1.46eV \\
& \rightarrow N^+(^3P) + O^+(^4S) + 2e - 15.07eV \\
& \rightarrow NO^{2+}(X^2\Sigma^+) + 2e - 18.85eV
\end{aligned}
$$

AI and the four endothermic processes that can all interact with each other, have never been the subject of any reported work. Below about 0.7 eV, only a structureless E^{-1} energy dependence can be seen in Fig. 1b. At 0.54eV, it is difficult to relate the opening of $N(^4S) + O^+(^4S) + e$ to any structure in the AI curve. This is expected since it would correspond to a transfer ionization process in which the oxygen anion would lose two electrons. Moreover, the rapid fall-off that we observe above 6 eV does not seem to coincide with the opening of the $N^+(^3P) + O(^3P) + e$ channel, which lies at 1.46 eV. At higher energies, the cross sections decrease to a non-measurable level just below 20 eV. This may be caused by the opening of the $N^+(^3P) + O^+(^4S) + 2e$ channel to which the $NO^{2+}(X^2\Sigma^+)$ correlates.

3.3 $O^+ + O^-$

The AI process from $O^+ + O^-$ is presented in Fig. 1c. At low cm energies, three electronic states of the O_2^+ ions can be populated: $X^2\Pi$, $a^4\Pi_u$ and $A^2\Pi_u$. Excited states such $b'^4\Pi_g$, $b^4\Sigma_g^-$ and $C^2\Phi_u$, may be populated as well. According to the correlation rules, the $^{3,5}\Sigma^+_{u,g}(1)$ and $^{3,5}\Pi_{u,g}(1)$ states correlate to the ground state $O^+(^4S) + O^-(^2P)$ limit. Moreover, thirty six states, $^{3,5}\Sigma^+_{u,g}(2)$, $^{3,5}\Sigma^-_{u,g}(1)$, $^{3,5}\Pi_{u,g}(3)$, $^{3,5}\Delta_{u,g}(2)$ and $^{3,5}\Phi_{u,g}(1)$, correlate to the next limit, $O^+(^2D) + O^-(^2P)$. Depending upon the CM energy, several channels are open (at <20 eV):

$$O^+(^4S)+O^-(^2P)\;\begin{array}{l}\rightarrow O_2(X^3\Sigma_g^-)+17.27eV\\ \rightarrow O(^3P)+O(^3P)+12.15eV\\ \rightarrow O_2^+(X^2\Pi_g)+e+5.20eV\\ \rightarrow O(^3P)+O^+(^4S)+e-1.46eV\\ \rightarrow O^+(^4S)+O^+(^4S)+2e-15.07eV\\ \rightarrow O_2^{2+}(X^1\Sigma_g^+)+2e-18.65eV\end{array}$$

On purely energetic grounds, there should be interplay between the AI and the three endothermic processes at large CM energies. It is remarkable that below 7 eV, only a structureless E^{-1} energy dependence can be seen in Fig. 1c, and that the opening of the $O(^3P)$ + $O^+(^4S)$ + e channel that occurs at 1.46 eV, does not give rise to any structure in the cross section curve. On the other hand, the shallow fall off above 10 eV might be caused by the opening of the $O^+(^4S)$ + $O^+(^4S)$ + 2e channel to which the O_2^{2+} (X $^1\Sigma_g^+$) correlates.

3.4 O^+ + D^- and D^+ + O^-

The AI cross section are in Fig. 2. The low energy dependence for the O^+ + D^- channel is remarkable because it does not follow the E_{cm}^{-1} law. This is not due to an experimental artifact but rather to a not yet elucidated physical effect. Few electronic states of the OD^+ ions can be populated : X $^3\Sigma^-$, a $^1\Delta$, b $^1\Sigma^+$ and A$^3\Pi$ states. The paucity of the molecular data is once again a problem and is surprising since the hydroxyl cation is quite well documented.

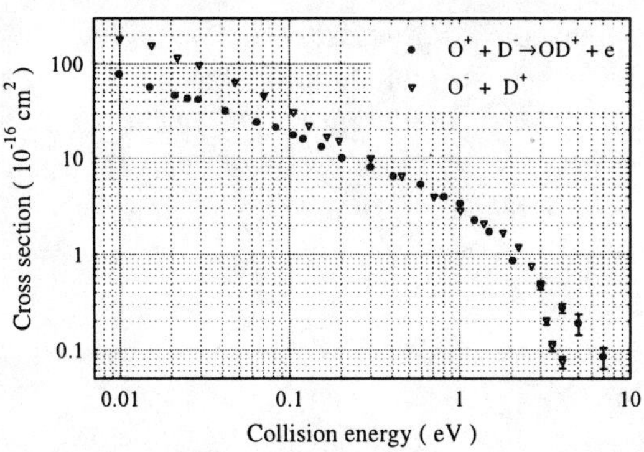

Figure 2. The open triangles represent the AI cross sections for D^+ + O^- and the filled circles represent the data for O^+ + D^-.

We know from the correlation rules that two states arise from the D^++$O^-(^2P)$ limit and to the best of our knowledge, they are not characterized. These are the $^2\Sigma^+(1)$ and $^2\Pi(1)$ states. From the three $O^+(^4S$, ^2D, $^2P)$+$D^-(^1S)$ limits, six states arise: $^4\Sigma^-(1)$, $^2\Sigma^-(1)$, $^2\Pi(1)$, $^2\Delta(1)$, $^2\Sigma^+(1)$ and $^2\Pi(1)$. The first state is from $O^+(^4S)$. The next three are from $O^+(^2D)$ and the last two are from $O^+(^2P)$.

Depending upon the center-of-mass energy in the $D^+ + O^-$ and $O^+ + D^-$ systems, several channels are open over the studied energy range (<10 eV):

$O^-(^2P) + D^+$	$\rightarrow OD(X^2\Pi) + 16.53$ eV
	$\rightarrow O(^3P) + D(^2S) + 12.14$ eV
	$\rightarrow OD^+(X^3\Sigma^-) + 3.51$ eV
	$\rightarrow O^+(^4S) + D^-(^1S) - 0.72$ eV
	$\rightarrow O(^3P) + D^+ - 1.46$ eV
	$\rightarrow O^+(^4S) + D(^2S) - 1.48$ eV
	$\rightarrow O^+(^4S) + D^+ - 15.08$ eV
$O^+(^4S) + D^-(^1S)$	$\rightarrow OD(X^2\Pi) + 17.25$ eV
	$\rightarrow O(^3P) + D(^2S) + 12.86$ eV
	$\rightarrow OD^+(X^3\Sigma^-) + 4.23$ eV
	$\rightarrow O^-(^2P) + D^+ + 0.72$ eV
	$\rightarrow O(^3P) + D^+ - 0.74$ eV
	$\rightarrow O^+(^4S) + D(^2S) - 0.76$ eV
	$\rightarrow O^+(^4S) + D^+ - 14.36$ eV

Concerning the interplay between the processes at higher energies, the opening up of the $O(^3P)+D^++e$ and $O^+(^4S)+D(^2S)+e$ channels affect the $O^+ + D^-$ AI curve whereas the $O^+(^4S)+D^++2e$ channel does not play any role (where AI is already at a non-measurable level). The $O^+(^4S)+D^-(^1S)$ channel does not affect $D^+ + O^-$ AI whereas the AI cross section is affected by the $O(^3P)+D^++e$ and $O^+(^4S)+D(^2S)+e$ channels. The $O^+(^4S)+D^++2e$ does not play any role.

4. CONCLUSION

The cross sections for the AI processes (C^+, N^+, O^+ and D^+) + O^- and those for $O^+ + D^-$ were measured. Special attention was paid to the internal energy of the target cation. Our most important result concerns the large size of the measured cross sections, which reach values near 1×10^{-14} cm^2 at thermal energies. The lack of molecular data is a limiting factor in formulating qualitative and quantitative interpretations.

REFERENCES

1 P. Bouchet, E. Slezak, T. LeBertre, A. Moneti and J. Manfroid, A. & A. **80**, 379 (1989).
2 A. Dalgarno, M. L. Du and J. H. You, Astrophys. J. **349**, 675 (1990).
3 C. Bertrand and P. J. van Tiggelen, J. Phys. Chem. **78**, 2320 (1974).
4 M. A. Coplan, K. W. Olgilvie, M. F. A'Heare, P. Boschler and J. Geiss, J. Geosphys. Res. **92**, 39 (1987).
5 W. F. Huebner in The Photochemistry of Atmospheres, ed. by J.S. Levine (Academic Press, Orlando, 1985).
6 M. F. A. Harrison, K. Dolder and P. C. Thonemann, Proc. Phys. Soc. **82**, 368 (1963).
7 M. Hamdan and A. G. Brenton, J. Phys. B **22**, 2289 (1989).
8 C. J. Reid, J. Phys. B **25**, 475 (1992).

Dissociative Recombination in Interstellar Clouds

Eric Herbst
Department of Physics and Department of Astronomy, The Ohio State University, Columbus, OH 43210 USA
herbst@mps.ohio-state.edu

1. INTRODUCTION

1.1 The Interstellar Medium

Dissociative recombination has a universality far greater than most researchers in the field realize – it occurs in interstellar matter, the ubiquitous material that exists among the stars of our own galaxy and most others. The interstellar medium is not uniform but is condensed into relatively dense regions known as interstellar clouds, which can be many light years in extent and contain many solar masses of material. The physical conditions in interstellar clouds comprise a wide range of temperature and density; the matter consists of a gaseous component, which contains roughly 99% of the mass, and a particulate component of tiny dust particles perhaps 0.1μ in radius.[1]

In so-called diffuse interstellar clouds, the gas densities range from 10-1000 species cm^{-3} and temperatures are 50-100 K. Radiation from background stars penetrates these clouds and acts as a lamp for absorption spectroscopy in the visible and ultra-violet. Mainly but not totally atomic in nature, the gas is a weak plasma, with a fractional ionization of 10^{-4}. The relative abundances of elements in the gas phase are similar to those found in stellar atmospheres, although the heavier elements are depleted somewhat to form the dust grains. Hydrogen is by far the dominant element, while some appropriate numbers for elemental abundances relative to hydrogen are: He (0.10), C(1×10^{-4}), O(3×10^{-4}) and N(8×10^{-5}). The dominant ion, C^+, is produced by photoionization.

Dissociative Recombination of Molecular Ions with Electrons
Edited by S. Guberman, Kluwer Academic/Plenum Publishers, 2003

In dense interstellar clouds, gas densities range from 10^{4-6} species cm^{-3}, the temperature can be as low as 10 K, and there is sufficient material to block out the visible light of background stars. The gas is warm enough to be detected in emission in the radio, microwave, and millimeter-wave regions of the spectrum, and the detection of discrete emission frequencies caused by quantized rotational motions has led to the identification of many gas-phase molecules.[2] Unlike visible radiation, infra-red radiation can penetrate through dense clouds and be utilized as a lamp for absorption studies of molecular vibrational spectra. As opposed to the situation at longer wave lengths, infra-red studies can be directed both at gas-phase species, which exhibit narrow spectra with rotational fine structure, and at condensed-phase molecules, which exhibit broader spectra.[3]

From high-resolution rotational and vibrational studies, it is known that the gas in dense interstellar clouds is overwhelmingly molecular in nature, with molecular hydrogen the most abundant species by four orders of magnitude, but with more than 100 different species detected containing up to 13 constituent atoms.[4] The many molecules are mainly organic in nature, and are quite abundant in an absolute if not a relative sense. Perhaps half of the molecules can be considered standard in a terrestrial sense, but many others are quite unusual and consist of ions, radicals, 3-membered rings, and isomers. The fractional ionic abundance is approximately 10^{-7}, because photoionization, such as occurs in diffuse clouds, does not occur appreciably. Because the elemental abundances in the gas cannot be directly determined, there is an uncertainty in how closely they resemble those found in diffuse clouds. There is reasonable but ambiguous evidence from broad infra-red features that large carbon-containing species (through 100 atoms in size), known as polycyclic aromatic hydrocarbons, are also present.[5] The dust particles, which prevent visible radiation from penetrating dense clouds, consist of cores of either silicates or carbonaceous material surrounded by mantles of ices such as water, carbon dioxide, and carbon monoxide.[3]

Dense interstellar clouds are far from homogeneous unless they are quite small.[6] The so-called "giant" clouds are generally clumpy in nature, with the densest clumps often in a state of collapse leading to the formation of stars. Other dense clumps show no signs of star formation and are said to be "quiescent." Star formation is one of the main areas of research in astronomy, and is found to be a most complex topic with many manifestations depending on the mass of the collapsing, or "protostellar", object. One of the most interesting phases attendant upon the formation of solar-mass stars is the protoplanetary disk, which is a rotating rather flat object of high gaseous and particulate density. It is no stretch of the imagination to suggest that many protoplanetary disks will soon, at least in astronomical terms, mature into planetary systems! The organic molecules in dense clouds are also detected in protoplanetary disks, and it is not far-fetched to assume that these molecules will also find their way into incipient planets.

1.2 Gas-Phase Chemistry

Interstellar molecules are formed in situ via both gas-phase reactions and reactions on the surfaces of dust particles.[7] The material in clouds derives from the detritus of previous generations of stars; towards the end of their lives stars emit matter, either explosively, as in the case of high-mass stars, or more gently, as in the case of low-mass stars. The material is in the form of dust particles, which are synthesized in the atmospheres of stars, and a mainly atomic gas phase. Any molecules synthesized in stars are photodissociated by the radiation field present in interstellar space. Under the influence of gravity, the stellar debris condenses into clouds. The molecules detected in the clouds must be formed from the initially atomic material, which has elemental abundances similar to those found in stars but with the heavier elements depleted to form the dust particles. Molecules produced in dense clouds are not destroyed rapidly by photodissociation and photoionization since the dust particles protect them to a large extent. Unlike the case of photons, cosmic rays, which are high energy nuclei produced in supernovae and other cataclysmic events, can penetrate through dense clouds, and cause a small degree of ionization.

The slowest step in the synthesis of molecules is the conversion of atomic into diatomic species. Consider the formation of molecular hydrogen, the most abundant molecule. This molecule cannot be formed efficiently in the gas phase because, under the low-density conditions of even the most "dense" clouds, two hydrogen atoms must stick together under radiative conditions, a very inefficient process. Although there are gas-phase formation mechanisms involving H^- and H^+, the formation of molecular hydrogen occurs principally on grain surfaces, most probably by a diffusive (Langmuir-Hinshelwood) mechanism in which one H atom lands on a surface and remains there long enough before evaporation for a second H atom to land and find the first one. Molecular formation here occurs via transfer of chemical energy to the grain, which can be considered a third body. Modern experiments show that the molecular hydrogen formed in this manner has a high probability of being ejected into the gas during the act of formation.[8]

Once gaseous molecular hydrogen is formed, gas-phase processes can produce a large assortment of molecules, even under the low-density, low-temperature conditions of interstellar clouds, in which binary reactions with no activation energy are the only reactions that need to be considered. The major processes are exothermic ion-molecule reactions, since these are known to have zero activation energy in the great majority of instances, and ions are available for reaction. Ion-molecule reactions typically possess a large and temperature-independent (Langevin) rate coefficient of 1-2 x 10^{-9} cm^3 s^{-1} if the neutral species is non-polar, and an even larger rate coefficient that varies inversely with temperature if the neutral reactant is polar.[9] *To produce neutral species from chains of ion-molecule reactions normally requires dissociative recombination reactions as the final step.* Selected neutral-neutral reactions, mainly involving at least one so-called radical, can also occur rapidly at low temperatures.[10] Subsequent to their formation, molecules are depleted by gas-phase reactions. In diffuse clouds, however, photodissociation plays the dominant role in molecular destruction and is rapid enough to prevent the formation of large abundances of polyatomic species except perhaps in relatively dense clumps. In dense clouds, photons are much rarer except in regions near newly-born stars; these "photon-dominated regions" are both heated and chemically affected by large fluxes of photons.

Cosmic rays ionize a small portion of the neutral molecular hydrogen into the ion H_2^+; this ion in turn reacts with H_2 to form the ion H_3^+ through the well-studied reaction

$$H_2^+ + H_2 \rightarrow H_3^+ + H. \qquad (1)$$

The H_3^+ ion, which has been detected via infra-red absorption,[11] is a precursor of many species through reactions with assorted neutral atoms and molecules. It is depleted through these reactions as well as through the dissociative recombination reaction with electrons. This latter process is the major depletion mechanism in diffuse clouds, and can be competitive in dense clouds. Its rate, of course, is still in doubt, with measurements of a normal rate coefficient at 300 K ($\approx$ 1 x 10^{-7} cm^3 s^{-1}) co-existing with measurements of a rate coefficient at least an order of magnitude smaller. The size of the rate coefficient has two major astronomical ramifications: (1) in diffuse clouds, a large rate coefficient may be incompatible with the surprisingly large amount of the ion recently detected (see McCall & Oka in this volume), and (2), in dense clouds, a large rate coefficient (> 2 x 10^{-8} cm^3 s^{-1}) can lead to multiple solutions of the chemical models if certain choices for elemental abundances of the heavier elements are made.[12] The two phases are known as the low and high ionization phases; in the model results discussed here, the elemental abundances used lead only to the low ionization phase.

To see how H_3^+ can be a precursor of other molecules, let us consider the reaction of H_3^+ with atomic C:

$$C + H_3^+ \rightarrow CH^+ + H_2. \qquad (2)$$

This reaction, followed by two reactions with H_2, in which H atoms are transferred, leads to the methyl (CH_3^+) ion. Although the methyl ion cannot react with H_2 in a standard manner, it can radiatively associate with it:

$$CH_3^+ + H_2 \rightarrow CH_5^+ + h\nu, \quad (3)$$

to form the saturated CH_5^+ ion. This process is much more efficient than the association of two H atoms; theory and experiment indicate that at 10 K one out of every 10^4 collisions leads to the associative product.[13] The process is competitive because H_2 is 10^4 times as abundant as CO, the second most abundant species. The CH_5^+ ion is depleted via dissociative recombination and, in dense clouds, by reaction with CO (the synthesis of which has not been discussed). The former reaction leads mainly to the methyl radical (CH_3) while the latter leads to methane.

A similar set of reactions starting with O and H_3^+ leads to water and the OH radical. Once again, the final step in the chain is a dissociative recombination reaction, this time the process

$$H_3O^+ + e \rightarrow H_2O + H; \; OH + 2H; \text{etc.} \quad (4)$$

Although there is still disagreement concerning the product branching fractions, it is certain that OH is the dominant product. The formation of ammonia is somewhat more complex since the reaction between N and H_3^+ to form $NH^+ + H_2$ is endoergic and does not occur under interstellar conditions. Nevertheless, a longer chain of reactions leads eventually to the NH_4^+ ion, from which dissociative recombination leads mainly to the two-body channel $NH_3 + H$.

More complex neutral species are also produced via chains of ion-molecule reactions in which the final step is a dissociative recombination reaction. For example, once methane is formed, it can react with C^+, a relatively abundant ion, to form acetylenic ions:

$$C^+ + CH_4 \rightarrow C_2H_2^+ + H_2; \; C_2H_3^+ + H, \quad (5)$$

which can then react with other heavy species or dissociatively recombine. Condensation reactions, viz.,

$$C_2H_2^+ + C_2H_2 \rightarrow C_4H_3^+ + H \quad (6)$$

can also be competitive. In selected cases, neutral-neutral reactions are more important. For example, the reaction

$$C + C_2H_2 \rightarrow C_3H + H \quad (7)$$

dominates the production of both the linear and cyclic forms of the C_3H radical, whereas the more abundant cyclopropenylidene species (c-C_3H_2) is produced mainly via an ion-molecule/dissociative recombination chain of reactions. Review articles have been written on the many reactions leading to complex interstellar molecules.[14]

It should be mentioned that there is a possibility that the electron concentration is much lower than estimated in dense clouds because efficient radiative attachment reactions may occur between electrons and large, (heretofore undetected) unsaturated neutral molecules with high electron affinities. If the population of negative ions reaches or exceeds that of electrons, then astrochemists will also have to worry about neutralization via positive ion-negative ion reactions.

2. MODELS

Gas-phase chemical models of interstellar clouds consist of coupled differential, or kinetic, equations in which the time derivative of the concentration (abundance) of each species is set equal to the sum of the source and sink terms for that species. The equations are solved as a function of time for either fixed or varying physical conditions with elemental abundances in the gas as a constraint. Chemical models containing both gas-phase and grain-surface chemistry also exist but are less common because of the uncertainties in our understanding of surface chemistry. In all gas-phase models, there must be one surface reaction, however – the formation of molecular hydrogen. Adsorption onto dust particles and desorption from dust particles are also often included, even if surface chemistry is not. The rate of evaporation for most heavy species, however, is so slow that the gas-phase is lost on a time scale of $10^{6\text{-}7}$ yr for dense cloud cores, which either limits the age of the cores or implies that non-thermal desorption processes are clearly important. Although many such processes have been suggested (e.g., grain-grain collisions, photo-desorption, desorption via the energy from surface chemical reactions), these are all poorly understood. Current models often include cosmic ray-induced desorption because rate coefficients for this process have been estimated.[15]

Recently constructed gas-phase model networks consist of ca. 4000 reactions and photodestructive processes, perhaps 10-20% of which have been studied in the laboratory or theoretically in some temperature range.[16, 17] The elements included are H, He, N, O, C, S, Si, Fe, Na, Mg, P, and Cl; abundances for these elements must be estimated based on stellar and diffuse cloud observations. Of the gas-phase reactions, there are 500 dissociative recombination channels, only 5-10% of which have been measured. Early "estimates" of the branching fractions of the neutral products were made based on a combination of thermodynamics, chemical intuition, statistical theories (RRKM, phase-space) and inclusiveness. Recent experimental results have been included in networks, but large-scale extensions of these results have not been attempted.

Rather than discussing the results of the models in some detail, it is more appropriate to consider first their qualitative successes for the simplest of astronomical objects with some complex chemistry – so-called quiescent dense cores, in which little evidence of star formation can be seen. For such objects, the gas-phase models appear to reproduce the sense of the chemistry. In particular, they

(i) show that detectable abundances of ions and radicals are present,
(ii) yield high abundances of unusual isomers (e.g. HNC),
(iii) reproduce the large isotopic fractionation seen for the case of deuterium-containing molecules,
(iv) predict that, despite the large abundance of hydrogen, most large neutral species are unsaturated.

The detectable abundances of ions stem from the role of ion-molecule reactions. Radicals and isomers arise as products of dissociative recombination reactions, although the production of isomers in this manner has not received experimental confirmation.

Isotopic fractionation occurs as a thermodynamic effect via a few exchange reactions in which direction matters at low temperatures. Consider as an example the reactive system[18]

$$H_3^+ + HD \Leftrightarrow H_2D^+ + H_2 \quad . \tag{8}$$

In dense clouds, HD is a prime repository of deuterium, but since there is very little deuterium compared with hydrogen, HD has an abundance of 1-2 x 10^{-5} that of H_2. The left-to-right reaction has an exoergicity ($\Delta E/k$) of over 200 K, so that at low temperature the backward reaction is considerably slowed and the H_2D^+/H_3^+ abundance ratio can build up to a rather large value. Of course, other reactions deplete H_2D^+ in addition to the reaction with H_2; still, the abundance of this ion can reach 10% of the normal ion in cold clouds. Once ions such as H_2D^+ are produced, they lead to other deuterated species via ion-molecule reactions.

The final successful qualitative prediction that of mainly unsaturated species, derives from two facts: (i) reactions leading to complex species tend to reduce the saturation level of the ions, and (ii) many ions with two or more heavy atoms tend to be unreactive with molecular hydrogen.

To compare quantitative model results with astronomical observations, one has to choose specific sources. Perhaps the best-studied quiescent dense core is known as TMC-1 and is found in the Taurus molecular cloud. Of the more than 50 molecules seen in this source, gas-phase model results can reproduce their abundances to within an order of magnitude or better for approximately 80% of the species at a variety of reasonable cloud ages.[17] Yet the model results are dependent upon many uncertain rates (especially involving neutral-neutral reactions) and, in the case of dissociative recombination reactions, many uncertain product branching fractions. So let us now discuss what has been learned about some of the important product channels of interstellar dissociative recombination reactions and what remains to be learned.

3. DISSOCIATIVE RECOMBINATION REACTIONS

3.1 Some Important Results

Table 1 shows some of the recently measured neutral product branching fractions, mainly determined in storage rings, which have been incorporated into chemical networks of interstellar clouds. Perhaps the most interesting system is H_3O^+ + e, because three different sets of measurements have been made on it. The two storage ring results do not quite agree to within the error bars given but are not in strong disagreement either. They show that the dominant channel is OH + 2H but that the two-body channel H_2O + H is non-negligible. The flowing afterglow results, on the other hand, still are very different from the storage ring values – very little if any water is obtained and the OH + H_2 channel is significant! The discrepancy has never been resolved.

Four years ago, Herbst & Lee[28] investigated the sensitivity of models of the quiescent core TMC-1 to changes in branching fractions based on extensions of the then available storage ring and flowing afterglow results for H_3O^+ + e. At that time, few of the other numbers in Table 1 had been measured. They found that extension of the flowing afterglow results has a dramatic effect on some species, especially saturated ones formed in two-body channels with H as the second body, whereas extension of the storage ring results was less disruptive. Now that many more storage-ring results are available and have been put into model networks alongside the H_3O^+ + e results, few great changes have occurred. In all fairness, the result that CH_5^+ + e yields mainly CH_3 rather than CH_4 would have dramatically lowered the calculated abundance of methane were it not for the reaction between CH_5^+ and CO.

Given the many dissociative recombination branching fractions currently in our networks that have not been studied, the time may be propitious for a new attempt to generalize the experimental data we have to the remaining reaction channels in the networks. But is it possible to generalize from these

Table 1. Some measured branching fractions at room temperature. Note that CRY=Cryring storage ring, AST=Astrid storage ring, FL=flowing afterglow

Reactants	Products	Fraction	Source
H_3^+ + e	H+H+H	0.75	CRY[19]
	H_2 + H	0.25	
H_3O^+ + e	H_2O+H	0.25,0.05,0.18	AST,FL,CRY[20,21,22]
	OH+H+H	0.60,0.29,0.67	
	OH + H_2	0.14,0.36,0.11	
	O+H_2+H	0.01,0.30,0.04	
CH_3^+ + e	CH_2 + H	0.4	AST[23]
	CH+H+H	0.16	
CH_5^+ + e	CH_4 + H	0.049	CRY[24]
	CH_3+H+H	0.698	
NH_4^+ + e	NH_3 + H	0.69	CRY[25]
	NH_2+H+H	0.21	
$HCNH^+$ + e	HCN(HNC)+H	0.67	CRY[26]
	CN+H+H	0.33	
$C_2H_2^+$ + e	C_2H + H	0.50	CRY[27]
	C_2 + H + H	0.30	
	CH + CH	0.13	

results? The answer is: only partially. First of all, most of the ions studied are saturated or nearly saturated. While the smaller interstellar ions in interstellar clouds tend to be hydrogen-rich, the larger ones tend to be hydrogen-poor. Secondly, although a study of Table 1 indicates that the more exothermic systems are associated with large three-body product channels, this generalization is not quantitative. It is clear that two-molecule two-body channels seem *not* to be favored (e.g. OH + H_2, CH + CH), and one might also generalize that breaking multiple bonds (e.g. CH + CH) does not occur easily if other channels are present, despite the availability of sufficient energy for this purpose. Of course, these generalizations are not based on much evidence, and we will need a much larger body of data before sound generalizations become possible, if ever. One must remember that generalizations in chemical kinetics are notoriously difficult to pronounce; differences in individual potential surfaces from one system to a supposedly similar one can be large.

Regarding our current network, all measured storage ring results are included. Unmeasured branching fractions are conservatively chosen to reflect both channels in which one or more H atoms come off and those in which a multiple-bond might break. A list of the dissociative recombination reactions/neutral product branching fractions in our model networks can be obtained from the author.

3.2 Some Urgently Needed Branching Fractions

Table 2 contains some branching fractions that would be *especially* useful to know for astronomical purposes. These are divided into several categories. In the first category are isomeric products. It is assumed in interstellar models that dissociative recombination reactions are mainly responsible for the many isomeric structures detected, normally with branching fractions emphasizing the lowest energy forms. Yet, there is very little experimental evidence for this view. Calculations of relevant potential surfaces for the two reactions listed – $HCNH^+$ + e and $HCCCNH^+$ + e – indicate that the isomeric products listed in Table 2 are reasonable outcomes. This is especially true for the $HCNH^+$ + e reaction, where the crossings between ionic and neutral surfaces have been investigated.[29] It is interesting to note that in cold cores of dense clouds, there is often more HNC than HCN. Since it is unlikely that HNC is depleted less rapidly than HCN, the higher abundance of HNC implies a higher rate of formation, which in turn implies that the $HCNH^+$ + e reaction leads to more HNC than HCN.[30] Experimental verification is needed.

Two "small" systems are listed in Table 2 in the second category. For the case of $CH_3OH_2^+$ + e, the extent of any methanol produced is critical since this is the only gas-phase pathway for production of methanol in the models. The other reaction listed – $C_2H_3^+$ + e – is important in clouds since this ion is rather abundant and since measurement of the products will yield another clue as to whether two-body or three-body channels are preferred and whether the breaking of multiple bonds is likely.

The third category consists of H-poor ions; here the problem is that especially for large ions there are many possible neutral fragments (although they are normally limited to two-body channels because the breaking of more than one multiple bond may not be exothermic.) Finally, simple statistical theories suggest that radiative recombination to the parent neutral may be increasingly competitive as ions reach 10-20 atoms in size.[31] Once again, experimental verification or contradiction of this suggestion is necessary. If radiative recombination is rapid, then the growth of far larger molecules than those considered in current models may indeed be likely.

4. SUMMARY

Dissociative recombination reactions play a significant role in chemical models of interstellar clouds. Under the cold and rarefied conditions of these clouds, the chemistry is dominated by positive ion-molecule reactions, which produce polyatomic ions. These ions are in turn neutralized via dissociative recombination reactions, for which modelers of interstellar chemistry must

Table 2. Some urgently needed branching fractions

Reactants	Products
A. Isomers	
$HCNH^+$ + e	HCN + H
	HNC + H
HC_3NH^+ + e	HCCCN + H
	HNCCC + H
	HCCNC + H
	HCNCC + H
B. Small Systems	
$CH_3OH_2^+$ + e	CH_3OH + H ?
$C_2H_3^+$ + e	C_2H_2 + H
	C_2H + H + H
	CH + CH_2
C. H-Poor Species	
C_n^+ + e	Fragments?
D. Radiative Recombination	

know both total rate coefficients and neutral product branching fractions. Although much information concerning these parameters has been obtained in the last 20-30 years, much more remains to be learned. Perhaps the most important rate coefficient concerning which there is ambiguity is that for H_3^+ + e; whether or not this rate coefficient is "normal" or "slow" has still not been determined beyond a reasonable doubt. Furthermore, the very uncertainty in this rate might indicate that different experiments are probing subtly different processes, preparing ions in different states of excitation or, at the least, interfering with the system that is being studied. If the rate coefficient is "normal" (i. e., large), then it becomes all the more difficult to explain the high abundance of H_3^+ in diffuse clouds.

In dense clouds, the major problem is in the determination of neutral product branching fractions. Although quite a few systems have now been successfully studied, it is not clear how to generalize these results quantitatively to the many unstudied systems. Moreover, there are still some critical experiments of a different nature to be done, such as determining the isomeric content of neutral product channels, and elucidating when radiative recombination becomes competitive with the dissociative process.

In conclusion, the study of dissociative recombination reactions by many experimentalists and theoreticians has dramatically helped the study of interstellar chemistry in the past, and will continue to do so in the future.

ACKNOWLEDGEMENT

Financial support for my program in astrochemistry by the National Science Foundation (US) is gratefully acknowledged.

REFERENCES

1 L. Spitzer, Jr., *Physical Processes in the Interstellar Medium*, (Wiley Interscience, New York, 1998).

2 *Millimeter-wave Astronomy: Molecular Chemistry & Physics in Space*, ed. by W. F. Wall, A. Carraminana, L. Carrasco, and P. F. Goldsmith (Kluwer Academic Publishers, New York, 1999).

3 *Molecules in Astrophysics: Probes and Processes*, ed. By E. F. van Dishoeck (Kluwer Academic Publishers, New York, 1997).

4 E. Herbst, Interstellar Chemistry, in *Spectroscopy from Space*, ed. by J. DeMaison et al. (Kluwer Academic Publishers, New York, 2001).

5 L. J. Allamandola, A. G. G. M. Tielens, and J. Barker, Astrophys. J. Suppl. Ser. **71**, 733 (1989).

6 *Astrochemistry: From Molecular Clouds to Planetary Systems*, ed. by Y. C. Minh and E. F. van Dishoeck (Astronomical Society of the Pacific, 2000).

7 E. Herbst, Ann. Rev. Phys. Chem. **46**, 27 (1995).

8 N. Katz, I. Furman, O. Biham, V. Pirronello, and G. Vidali, Astrophys. J. **522**, 305 (1999).

9 E. Herbst, Gas Phase Reactions, in *Atomic, Molecular, & Optical Physics Handbook*, ed. by G. W. F. Drake (American Institute of Physics, New York,1996).

10 I. R. Sims and I. W. M. Smith, Ann. Rev. Phys. Chem. **46**, 109 (1995).

11 T. R. Geballe and T. Oka, Nature **384**, 334 (1996).

12 H.-H. Lee, E. Roueff, G. Pineau des Forets, O. M. Shalabiea, R. Terzieva, and E. Herbst, Astron. Astrophys. **334**, 1047 (1998).

13 D. Gerlich and S. Horning, Chem. Reviews **92**, 1509 (1992).

14 D. Smith, Chem. Reviews **92**, 1473 (1992).

15 T. I. Hasegawa and E. Herbst, Mon. Not. R. Astron. Soc. **261**, 83 (1993).

16 T. J. Millar, P. R. A. Farquhar, and K. Willacy, Astron. Astrophys. Suppl. Ser. **121**, 139 (1997).

17 R. Terzieva and E. Herbst, Astrophys. J. **501**, 207 (1998).

18 E. Herbst, Astron. Astrophys. **111**, 76 (1982).

19 S. Datz, G. Sundström, Ch. Biedermann, L. Broström, H. Danared, S. Mannervik, J. R. Mowat, and M. Larsson, Phys. Rev. Lett. **74**, 896 (1995).

20 M. J. Jensen, R. C. Bilodeau, C. P. Safvan, K. Seiersen, L. H. Andersen, H. B. Pedersen, and O. Heber, Astrophys. J. **543**, 764 (2000).

21 T. L. Williams, N. G. Adams, L. M. Babcock, C. R. Herd, and M. Geoghegan, Mon. Not. R. Astron. Soc. **282**, 413 (1996).

22 A. Neau, A. Al Khalili, A. Le Padellec, A. M. Derkatch, W. Shi, L. Vikor, M. Larsson, J. Semaniak, R. Thomas, M. B. Någård, K. Andersson, H. Danared, and M. af Ugglas, J. Chem. Phys. **113**, 1762 (2000).

23 L. Vejby-Christensen, L. H. Andersen, O. Heber, D. Kella, H. B. Pedersen, H. T. Schmidt, and D. Zajfman, Astrophys. J. **483**, 531 (1997).

24 J. Semaniak, Å. Larson, A. Le Padellec, C. Strömholm, M. Larsson, S. Rosen, R. Peverall, H. Danared, N. Djuric, G. H. Dunn, and S. Datz, Astrophys. J. **498**, 886 (1998).
25 L. Vikor, A. Al-Khalili, H. Danared, N. Djuric, G. H. Dunn, M. Larsson, A. Le Padellec, S. Rosen, and M. Af Ugglas, Astron. Astrophys. **344**, 1027 (1999).
26 J. Semaniak, B. F. Minaev, A. M. Derkatch, F. Hellberg, A. Neau, S. Rosen, R. Thomas, and M. Larsson, Astrophys. J. **135**, 275 (2001).
27 A. M. Derkatch, A. Al-Khalili, L. Vikor, A. Neau, W. Shi, H. Danared, M. af Ugglas, and M. Larsson, J. Phys. B **32**, 3391 (1999).
28 E. Herbst and H.-H. Lee, Astrophys. J. **485**, 689 (1997).
29 Y. Shiba, T. Hirano, U. Nagashima, and K. Ishii, J. Chem. Phys. **108**, 698 (1998). See also D. Talbi and Y. Ellinger, Chem. Phys. Lett. **288**, 155 (1998).
30 E. Herbst, R. Terzieva, and D. Talbi, Mon. Not. R. Astron. Soc. **311**, 869 (2000).
31 R. P. A. Bettens and E. Herbst, Int. J. Mass Spectrom. Ion Proc. **149/150**, 321 (1995).

Enigma of H_3^+ in Diffuse Interstellar Clouds

Benjamin J. McCall[1,2] and Takeshi Oka[1]
[1]Department of Chemistry, Department of Astronomy & Astrophysics, and the Enrico Fermi Institute, University of Chicago, 5735 South Ellis Ave., Chicago, IL 60637; [2]Present address: University of California at Berkeley, 601 Campbell Hall, Berkeley, CA 94720
t-oka@uchicago.edu

1. INTRODUCTION

H_3^+ is of fundamental importance to interstellar chemistry both because it is a hydrogenic species (and >90% of all nuclei in the universe are hydrogen nuclei; see Fig. 1) and because it serves as the universal protonator, initiating a chain of ion-neutral reactions that is responsible for the formation of most interstellar molecules.

The chemistry of H_3^+ in dense (molecular) clouds is now well understood,[1] and observations of H_3^+ in these environments now permit measurements of the path lengths, number densities, and kinetic temperatures of dense clouds. However, as discussed in more detail in a separate paper[2] in this volume, the situation in diffuse clouds is much less clear. H_3^+ has now been observed[3,4] in several diffuse cloud sources with unexpectedly large absorption strength. This enigma requires at least one of the model parameters (cosmic ray ionization rate, electron fraction, and the rate constant for dissociative recombination of H_3^+) to be changed by at least one to two orders of magnitude!

In this paper we describe in more detail the observations of H_3^+ in diffuse interstellar clouds, and the simple chemical model which is so obviously in error. We appeal to the specialists of dissociative recombination to resolve the current order-of-magnitude controversy in the rate constant for dissociative recombination of H_3^+, which is hindering further progress in the use of H_3^+ as an astronomical probe of the diffuse interstellar medium.

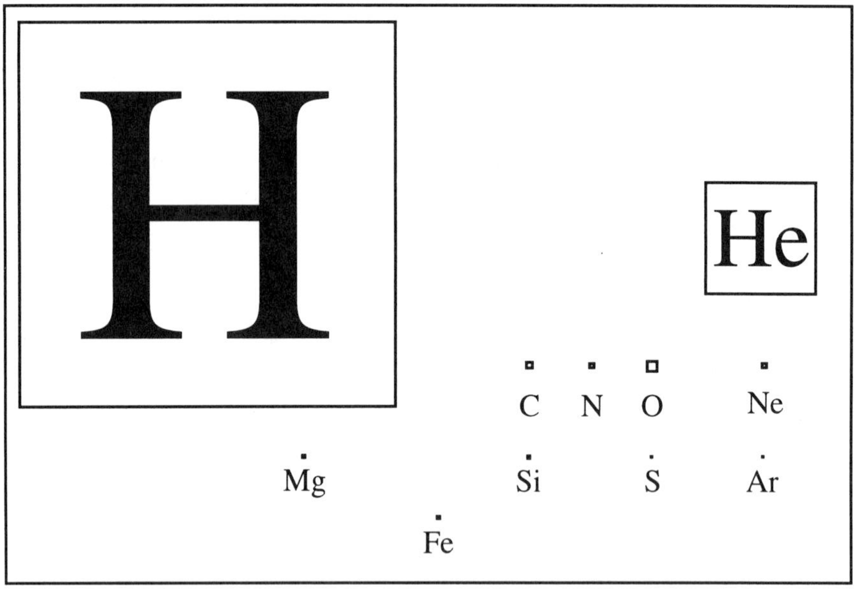

Figure 1. "The Astronomer's Periodic Table". The area of each element is proportional to its cosmic abundance.

2. OBSERVATIONAL TECHNIQUE

Because H_3^+ has no permanent dipole moment (its equilibrium structure is an equilateral triangle), it possesses no allowed rotational spectrum. This means that the sensitive techniques of radioastronomy cannot be brought to bear on the study of this key molecule. Furthermore, H_3^+ does not have any allowed electronic transitions, so visible spectroscopy (the most well-established form of spectroscopy in astronomy) is also useless. Thus our only tool for studying interstellar H_3^+ is its infrared rotation-vibration spectrum, first detected in the laboratory by Oka.[5]

The band origin of the ν_2 fundmanental vibration of H_3^+ lies at 2521.31 cm^{-1}, around 4 μm. Fortunately for us, few other molecules have strong vibrational transitions in this frequency range, and as a result the atmosphere is fairly transparent there – hence we can study interstellar H_3^+ from the ground rather than needing space-based telescopes. Because of the large rotational constant of H_3^+ the individual ro-vibrational transitions are generally well-separated in frequency. Another consequence of the large rotational constant is that at the low temperatures of the interstellar medium, only the lowest two allowed rotational states (in the vibrational ground state) are populated. A diagram of the energy levels and transitions of H_3^+ is shown in Fig. 2. A detailed explanation of the notation for H_3^+ transitions is given elsewhere.[6]

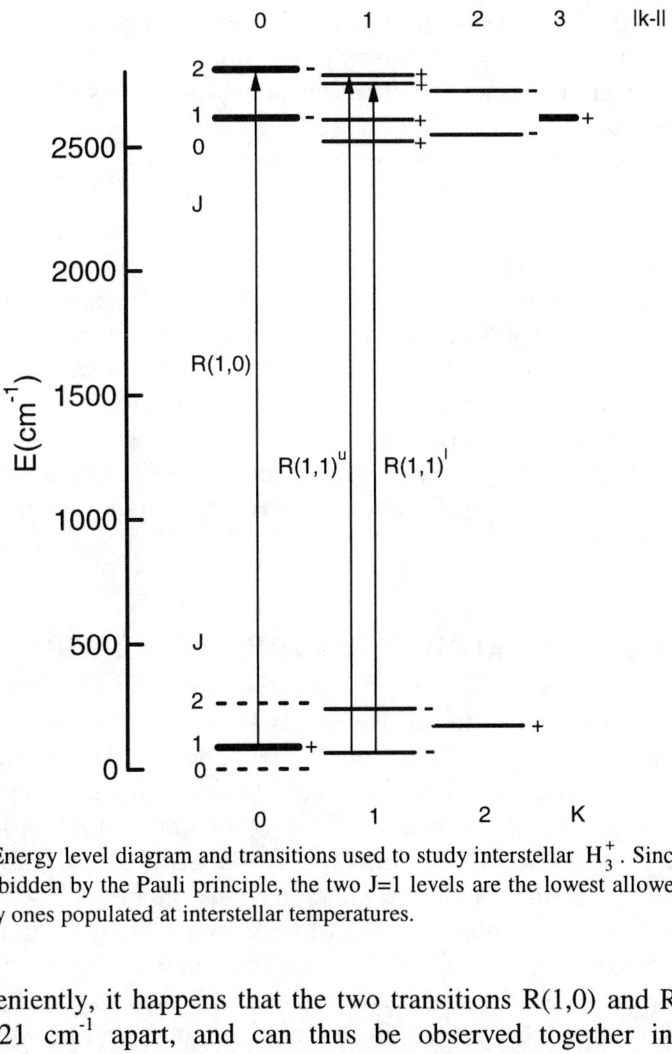

Figure 2. Energy level diagram and transitions used to study interstellar H_3^+. Since the J=K=0 level is forbidden by the Pauli principle, the two J=1 levels are the lowest allowed levels and are the only ones populated at interstellar temperatures.

Conveniently, it happens that the two transitions R(1,0) and R(1,1)^u are only 0.321 cm^{-1} apart, and can thus be observed together in a single observation. This "doublet" is convenient for recognizing the spectral signature of H_3^+ in low signal-to-noise spectra, and also provides measurements of both the *ortho* and *para* spin configurations of H_3^+. The downside of this doublet is that it lies near a fairly strong (~30% deep) line of atmospheric methane, so it can only be observed from high mountains such as Mauna Kea in Hawaii. Elsewhere (for example, at Kitt Peak in Arizona) we use the R(1,1)^l transition, which is less affected by atmospheric absorption lines.

We study H_3^+ using absorption spectroscopy, with a hot star (which is essentially a blackbody lamp with T>10,000 K) as the continuum source. The light from the star passes through the interstellar cloud, is collected by a large telescope (such as the United Kingdom Infrared Telescope in Hawaii or the Mayall Telescope at Kitt Peak), dispersed by a cryogenically cooled echelle spectrometer (such as CGS4 at UKIRT or Phoenix at Kitt Peak), and finally detected (in two-dimensional form) by a cooled InSb array. Image processing techniques then yield a one-dimensional absorption spectrum of the interstellar cloud.

H_3^+ is not very abundant in interstellar clouds, because it is formed slowly by cosmic-ray ionization of H_2 and destroyed quickly by chemical reactions (to form new molecules) or by dissociative recombination. As a result, the H_3^+ column densities (column density is essentially the number density times the path length) in interstellar clouds are only $\sim 10^{14}$ cm^{-2}, which yield absorptions of order 1% deep. Measuring such weak absorption signals from interstellar clouds requires long integration times and careful data reduction techniques. On the other hand, the fact that the lines are weak means that they are not saturated, so that measurement of the integrated area of the absorption lines (what the astronomers call "equivalent width") directly yields the column density of H_3^+. Fig. 3 shows a few representative spectra of H_3^+ in diffuse clouds.

3. THE H_3^+ PROBLEM IN DIFFUSE CLOUDS

The chemistry of H_3^+ in diffuse clouds is described elsewhere[2] in this volume in detail, so here we give only a brief summary. H_3^+ is produced by cosmic-ray ionization of H_2 to form H_2^+, followed by a fast ion-neutral reaction with another H_2 to form H_3^+. The cosmic-ray ionization is the rate limiting step, so the rate of H_3^+ formation can be given as ζ [H_2], where $\zeta \sim 10^{-17}$ s^{-1} is the cosmic-ray ionization rate. In diffuse clouds, all carbon atoms (which are about 10^{-4} as abundant as hydrogen atoms) are ionized to form C^+ and e^-, so electron recombination is the dominant destruction mechanism for H_3^+. Consquently, the H_3^+ destruction rate can be expressed as $k_e[e^-][H_3^+]$. The steady state number density of H_3^+ can then be found by equating the formation and destruction rates: $[H_3^+] = (\zeta/k_e)([H_2]/[e^-])$.

If we substitute the accepted values $([H_2]/[e^-]) \sim 2000$, $\zeta \sim 3\times10^{-17}$ s^{-1}, and $k_e \sim 5\times10^{-7}$ cm^3 s^{-1}, we find that $[H_3^+] \sim 10^{-7}$ cm^{-3}. Then we can estimate the path length of the cloud as the column density divided by the number density(from the definition of column density), so L $\sim (10^{14}$ cm$^{-2})/(10^{-7}$ cm$^{-3})$ $\sim 10^{21}$ cm. This is an enormous distance of about 1000 light-years, or ~300 parsecs, and is in fact a substantial fraction of the distance to the background

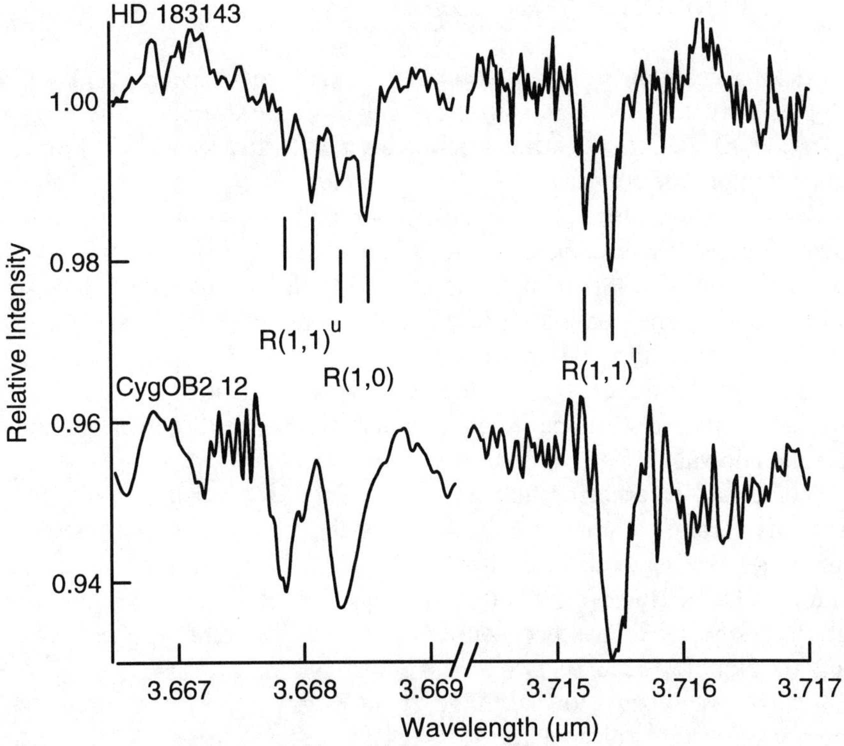

Figure 3. Spectra of H_3^+ in two diffuse cloud sightlines HD 183143 and CygOB2 12. HD 183143 has two velocity components, so each transition appears as a doublet.

stars! If true, this would imply that much of the volume of the galaxy is filled with H_3^+.

A simple calculation shows that this is unreasonable. Since we know the column density of H_2 is of order 10^{22} cm^{-2}, we can use our inferred length to estimate the average number density of hydrogen along the line of sight: $[H_2] \sim (10^{22}\ \text{cm}^{-2})/(10^{21}\ \text{cm}) \sim 10\ \text{cm}^{-3}$. This is a ridiculously low hydrogen number density; at such low densities, hydrogen would not be predominantly in molecular form, as we have assumed in the model. In addition, observations of other molecular species suggest the density is much higher. This suggests that we can exclude the possibility that the path lengths are this large, so it must be the case that H_3^+ is much more abundant in diffuse clouds than our model suggests. Evidently the H_3^+ number density needs to be increased by one to two orders of magnitude to bring the cloud path lengths into accord with other observations.

4. POSSIBLE SOLUTIONS

Looking again at the equation for the H_3^+ number density, $[H_3^+] = (\zeta/k_e)([H_2]/[e^-])$, we see that there are really only three parameters: the cosmic-ray ionization rate ζ, the electron fraction $[e^-]/[H_2]$, and the H_3^+ dissociative recombination rate constant k_e.

The first parameter (ζ) is considered unlikely to be different in diffuse clouds than in dense clouds (where it is reasonably well-constrained), but there is a possibility that there could be a large flux of lower energy cosmic-rays that could penetrate into diffuse clouds but not dense clouds. If this were the case, it would be extremely interesting and have widespread ramifications. However, this parameter is very difficult to get a handle on observationally, so we hope it can be determined once the other parameters are nailed down.

The second parameter (the electron fraction) is governed largely by the form of carbon atoms. The conventional wisdom in the astronomy community is that in diffuse clouds all carbon is ionized, so the electron fraction is large. However, this has not been directly confirmed in any of the sightlines where H_3^+ has been observed – we are hoping to pin this down using the Hubble Space Telescope in the next year or so.

Finally we come to the biggest suspect of all, the dissociative recombination rate constant for H_3^+. This "constant" has varied by at least four order of magnitude over the past 20 years, and different experimental techniques continue to give different results, as reviewed elsewhere[2] in this volume. Complicating matters is that theoretical calculations do not yet provide any support for the large values of k_e currently being measured. The need for a "solid" value of k_e cannot be overemphasized!

5. CONCLUSIONS

H_3^+ has demonstrated its promise as a powerful probe of the physical conditions in dense (molecular) clouds. Because H_3^+ is now being routinely observed in the diffuse interstellar medium, one would naïvely expect that it should also serve as a useful probe of the conditions in diffuse clouds. Instead, due to the uncertainty in the value of the dissociative recombination rate constant k_e, we are at an impasse — at this stage, all we know for sure is that "something is wrong" with the chemical model for diffuse clouds. Perhaps k_e is wrong, in which case the cosmic-ray flux and electron fraction in diffuse clouds are what we expect them to be. But perhaps the current value of k_e is correct! Then there is exciting new astrophysics waiting to be explored...either carbon is mostly not ionized in diffuse clouds or the cosmic-ray flux is much higher than generally thought.

Once the dissociative recombination community is finally able to nail down the value of k_e, H_3^+ will assume its rightful place as a "barometer, thermometer, and ruler" for diffuse clouds. Until then, we are left groping in the darkness for an explanation of the observed H_3^+ column densities.

ACKNOWLEDGEMENTS

It is a pleasure to acknowledge the contributions of our many collaborators in the study of interstellar H_3^+ and related species, especially Tom Geballe (Gemini) and Ken Hinkle (NOAO). The observational work described in this paper has been supported by NSF Grant PHY-0099442, NASA grant NAG5-9583, and a Sigma Xi Grant-in-aid of Research. B.J.M. has been supported by the Fannie and John Hertz Foundation and the Miller Institute for Basic Research in Science.

REFERENCES

1 B. J. McCall, T. R. Geballe, K. H. Hinkle, and T. Oka, Astrophys. J. **522**, 538 (1999).
2 T. Oka, Help!!! Theory for H_3^+ Recombination Badly Needed, this volume.
3 B. J. McCall, T. R. Geballe, K. H. Hinkle, and T. Oka, Science, **279**, 1910 (1998).
4 B. J. McCall, K. H. Hinkle, T. R. Geballe, G. H. Moriarty-Schieven, N. J. Evans II, K. Kawaguchi, S. Takano, V. V. Smith, and T. Oka, Astrophys. J., submitted.
5 T. Oka, Phys. Rev. Lett. **45**, 531 (1980).
6 B. J. McCall, and T. Oka, J. Chem. Phys. **113**, 3104 (2000).

Dissociative Recombination and Deuterium Fractionation in Interstellar Clouds

Franck Le Petit and Evelyne Roueff
DAEC and UMR 8631, Observatoire de Paris, 5 place Jules Janssen, F-92190 Meudon, France ; franck.lepetit@obspm.fr ; evelyne.roueff@obspm.fr

1. INTRODUCTION

Deuterium is remarkable in astrophysics since its formation occurred only during the primordial nucleosynthesis in the Big Bang. The amount of deuterium that was formed is directly connected to the baryonic density Ω_B of the early universe, a fundamental cosmological parameter since it controls in part the fate of the universe. With the baryonic and dark matter densities, one can expect to learn if the universe will expand forever or if it will finally collapse on itself.

The nuclear binding energy of deuterium is very weak. It is destroyed when exposed to temperatures higher than 10^5 Kelvin. These temperatures can be found in star interiors. Thus, any deuterium atom that has been incorporated in a star has been destroyed. The D/H ratio is then expected to decrease from its primordial value since no deuterium atom has been formed since the Big Bang. The D/H ratio in the interstellar medium (ISM) is thus a significant probe of the stellar evolution and its determination is an outstanding problem in astrophysics. The presently accepted value of D/H in the ISM is about $1.6 \cdot 10^{-5}$.[1]

1.1 Interstellar Deuterium

In the neutral ISM, depending upon the molecular fraction of the cloud, D or HD can be the reservoir of deuterium. However, direct observations are not easy. For atomic deuterium, the 92 cm radio line (corresponding to the 21

Dissociative Recombination of Molecular Ions with Electrons
Edited by S. Guberman, Kluwer Academic/Plenum Publishers, 2003

cm line of H) has an exceedingly low transition probability and its detection has been performed only in one line of sight[2] and is subject to large uncertainties. The other observing possibilities are in the vacuum ultraviolet via the Lyman transitions that are only accessible in space. Such observations are performed in absorption and require the presence of a background star with UV continuum.

HD is also observable in ultraviolet absorption via the Lyman and Werner band systems. The FUSE satellite has recently been launched for such observations. These observations are limited to the so-called diffuse and translucent interstellar clouds with moderate densities ($n \approx 100$ - $1000\ cm^{-3}$) where photo-processes are playing an important role. In dense molecular clouds, HD has been detected via infrared rotational emissions[3] with the Infrared Space Observatory (ISO). But, the interpretation of such observations requires the understanding of the excitation conditions of the molecule. Much more straightforward are the millimeter observations of polyatomic molecules containing deuterium atoms. These have been detected since the beginning of millimeter astronomy and there are presently more than 20 singly deuterated molecules detected. Two doubly deuterated molecules, D_2CO and ND_2H have even been found in dark cold molecular clouds and protostar clouds. Table 1 displays the list of the deuterated species detected towards several lines of sight as well as the deduced XD/XH ratio.

1.2 Chemical deuterium fractionation

The obvious fact displayed in Table 1 is that the deuterium fractionation ratio is much larger than the $1.6 \cdot 10^{-5}$ value deduced from other means. This is directly explained by the low temperatures prevailing in interstellar clouds. Chemical fractionation is a direct consequence of the differences in zero point energies of the hydrogenated and deuterated molecules.

The first step of deuterium fractionation occurs in the reaction between H_3^+ and HD, the deuterium reservoir of cold molecular clouds (but not yet observed in such objects):

$$H_3^+ + HD \longrightarrow H_2D^+ + H_2$$

with a rate of: $k_f = 3.5 \cdot 10^{-10}\ cm^3 \cdot s^{-1}$.[4]

The actual value of the endothermicity is still under debate. But in cold interstellar clouds, where $T \approx 10$ K, the backward reaction is not efficient. As a result, the H_2D^+ / H_3^+ ratio is much larger than HD / H_2, resulting in a deuterium enrichment. H_2D^+ is subsequently more likely destroyed in reactions with other abundant interstellar species, as O and CO, leading to the formation of other deuterated species:

Table 1. Observed deuterium fractionation (XD/XH) in the direction of several lines of sight

Molecule	TMC1	L134N	Hot core	NGC133 IRAS 4A	IRAS 16293-2422	ρ OphE
H_2D^+				0.02 [22]		
DCO^+	0.02 [1]	0.08 [1]		0.01 [22]	0.0086 [12]	0.02 [14]
N_2D^+	0.08 [1]	0.35 [1]				
C_2D	0.048 [2]				0.18 [12]	
DCN	0.023 [3]		0.0015 [17]	0.06 [16]	0.013 [12]	
DNC	0.015 [4]				0.03 [12]	
HDO			$2\text{-}6\ 10^{-4}$ [18]		> 0.01 [12]	
HDS			< 0.001 [19]		0.1 [12]	
NH_2D	0.01 [1,5]	0.05 [1]	0.06 [20]		0.1 [12]	0.18 [15]
ND_2H		0.05 [13]				0.4 [14]
HDCO	0.005 – 0.11 [4]		0.14 [20]		0.15 [12]	
D_2CO			0.02 [20]		0.33 [12]	> 0.40 [14]
HDCS	0.02 [6]					
C_4D	0.04 [7]					
DC_3N	0.03-0.1 [8]					
C_3HD	0.08-0.16 [9]	0.10 [9]				
CH_2DOH			0.04 [21]			
CH_3OD			0.01-0.06 [11]		≤ 0.1 [12]	
DC_5N	0.08-0.16 [9]					
CH_2DCCH	0.06 [10]					

[1] S. Tiné et al., A&A **356**, 1039 (2000).
[2] B. E. Turner, ApJ supp. **136**, 579 (2001).
[3] A. Wooten, IAU Symp. **120**, 131 (1987).
[4] M. Guélin et al., A&A **107**, 107 (1982).
[5] R.Y. Shah et al., ApJ **554**, 933 (2001).
[6] H. Minowa et al., ApJ **491**, L63 (1997).
[7] B. E. Turner, ApJ **347**, L39 (1989).
[8] D. A. Howe et al., MNRAS **267**, 59 (1994).
[9] M. B. Bell et al., ApJ **326**, 924 (1988).
[10] M. Gerin et al., A&A **253**, L29 (1992).
[11] R. Mauersberger et al., A&A **194**, L1 (1988).
[12] E. van Dishoeck et al., ApJ **447**, 760 (1995).
[13] E. Roueff et al., A&A **354**, L63 (2000).
[14] L. Loinard et al., ApJ **552**, L163 (2001).
[15] S. Saito et al., ApJ **535,** 227 (2000).
[16] G. A. Blake et al., ApJ **441**, 689 (1995).
[17] J. Hatchell et al., A&A **332**, 695 (1998).
[18] F. P. Helmich et al., A&A **313**, 657 (1996).
[19] J. Hatchell et al., A&A **346**, 227 (1999).
[20] B. E. Turner, ApJ **362**, L29 (1990).
[21] T. Jacq et al., A&A **271**, 276 (1993).
[22] R. Stark et al., ApJ **521**, L67 (1999).

$$H_2D^+ + CO \longrightarrow \begin{array}{l} HCO^+ + D \\ DCO^+ + H \end{array} \quad \text{and}$$

$$H_2D^+ + O \longrightarrow \begin{array}{l} OH^+ + HD \\ OD^+ + H_2\,. \end{array}$$

These destruction processes compete with dissociative recombination:

$$H_2D^+ + e^- \longrightarrow \begin{array}{l} H_2 + D, \\ HD + H, \\ H + H + D. \end{array}$$

The adopted rate for this last reaction is $k = 6 \cdot 10^{-8} \cdot (T/300)^{-0.65}$ $cm^3 \cdot s^{-1}$.[5] It can be shown that the DCO^+ / HCO^+ ratio is directly linked to the H_2D^+ / H_3^+ ratio. Depending upon the fractional ionization in dark clouds, one or the other process of destruction prevails resulting in the production of many or few deuterated molecules.

1.3 Bistability

The phenomenon of bistability in dark clouds was first raised by Le Bourlot *et al.*[6] in the context of interstellar chemistry. In dark clouds, ionization is induced by cosmic rays. The differential equations involving the time variations of the molecular abundances as well as the energy gains and losses are intrinsically non linear.[6,7,8] A complex dynamical behavior may then occur and bistable solutions may be found in a range of control parameters relevant to interstellar chemistry. We have performed numerical models of interstellar clouds relevant to deuteration studies by including 138 chemical species containing H, D, He, C, O, N, S and a metal, linked by about 1700 chemical reactions. The chemical network is an updated version of previous studies.[9] The right side of Fig. 1 displays the time evolution of the fractional ionization for an interstellar cloud with $n_H = 10^4$ cm^{-3}, at a temperature of 10 K, exposed to a cosmic ray ionization rate of $5 \cdot 10^{-17}$ s^{-1}. The two curves correspond to different initial conditions in the phase space. After less than 1 million years, the solutions become steady state, but the steady state values differ by more than one order of magnitude. The two types of solutions have been labeled as High Ionization Phase (HIP) and Low Ionization Phase (LIP). The left Figure displays the set of *steady state* solutions of the fractional ionization found for densities between 500 and 10^6 cm^{-3}. The various solutions belong to a so-called *hysteresis* curve where *both* stable and unstable solutions are displayed. The vertical line corresponds to the density condition for which the time evolution equations have been solved. As a test of the consistency of the calculations, one may verify that steady state values obtained in the time dependent model ($t > 1$ Myr) and the stable solutions of the hysteresis curve for $n_H = 10^4$ cm^{-3} are the same.

However, whether bistability is really taking place in the interstellar medium is still an open question. All variables are affected by bistability. In particular, the deuterium fractionation of molecular species is strongly dependent on the chemical phase, giving a potential way to test

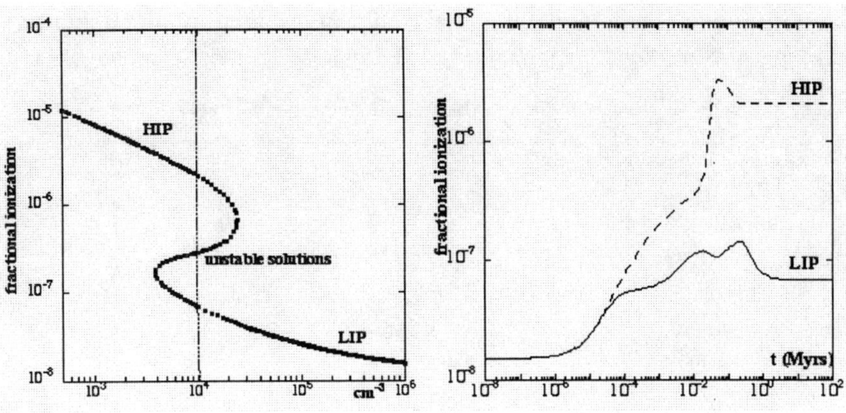

Figure 1. Left: Fractional ionization as a function of the density of the cloud. Each point corresponds to a stationary model at 10 Kelvin. The adopted depletion is δ=10. The region where several solutions are found corresponds to the bistability phenomenon. *Right:* Temporal evolution of the fractional evolution for a cloud of density 10^4 cm^{-3}, temperature T=10 K and a depletion of δ=10. Physical parameters and elemental abundances are the same between the two models. Only the distribution of elements between atomic and molecular species differs in the two models.

observationally the occurrence of bistability in the interstellar medium. The steady state results for deuterium are given in Table 2 below for the two phases.

It can be noticed that the deuterium fractionation is higher in the LIP than in the HIP. As discussed above, in the HIP, H_2D^+ is mainly destroyed by dissociative recombination, whereas in the LIP, its reactions with O and CO contribute to the formation of deuterated ions and molecules.

2. INFLUENCE OF THE RECOMBINATION RATE OF H_3^+

The rate of the dissociative recombination of H_3^+ is still under debate (see papers in this volume). Since H_2D^+ is formed in the reaction between H_3^+ and HD, the abundance of H_2D^+ and subsequently of all deuterated species depends on the abundance of H_3^+. The temperature dependence of the corresponding rate coefficient is given by:

$$k = A\,(T/300)^{-0.65}$$

where the exponent is from the storage ring experiment.[10] In order to study the effect of this rate on the chemistry of the cloud we have calculated the 10 stationary states of a dark cloud model of density 10^4 cm^{-3} at a temperature of K taking A as a control parameter[11] and using the most recent chemical

Table 2. Deuterium fractionation for low and high ionization phases.

Deuterium fractionation	LIP	HIP
H_2D^+/H_3^+	$2.1\cdot10^{-2}$	$5.8\cdot10^{-3}$
DCO^+/HCO^+	$9.5\cdot10^{-3}$	$1.5\cdot10^{-3}$
N_2D^+/N_2H^+	$7.4\cdot10^{-3}$	$1.6\cdot10^{-3}$
C_2D/C_2H	$6.9\cdot10^{-2}$	$5.4\cdot10^{-3}$
DCN/HCN	$2.9\cdot10^{-2}$	$3.4\cdot10^{-3}$
HDO/H_2O	$1.2\cdot10^{-3}$	$1.7\cdot10^{-4}$
NH_2D/NH_3	$2.3\cdot10^{-2}$	$1.4\cdot10^{-3}$

data. The results are presented in Fig. 2. We find again the hysteresis curve characteristic of bistability. For the given parameters, the bistable zone occurs for A between 6.6 10^{-8} and 2 10^{-7} cm^3 s^{-1}.

Note that for the H_3^+ DR rate constant, Glosik et al.[12] find an upper limit of $5\cdot10^{-8}$ cm^3 s^{-1} while Sundström et al.[13] find $1.15\cdot10^{-7}$ cm^3 s^{-1} at 300K. As shown above, the chemical abundances are very different in the two phases. We stress here, once more, the importance of identifying the source of disagreement between the experiments and the need to understand the theory of this key reaction.

3. ISOTOPIC EFFECT IN DISSOCIATIVE RECOMBINATION

In the dissociative recombination of ions having hydrogen or deuterium, one has to ask how hydrogen and deuterium are distributed between the different products. In the absence of any further information, one usually assumes that D and H react in the same way and the branching ratios are obtained from statistical arguments.[14] However, in a recent study of the dissociative recombination of HD_2O^+, Jensen *et al.*[15] observed that hydrogen rather than deuterium preferentially leaves the molecular skeleton. This possible isotopic effect is studied by comparing two models. In Model A, branching ratios for dissociative recombination are given by statistical values. In model B, the probability of hydrogen to leave the molecular skeleton is set at 80% whereas the corresponding value for deuterium is 20%. Results are presented in Table 3.

Since the total rate of reactions is unchanged, the fractionation of deuterated ions is the same in both models. For neutral molecules, fractionation can vary up to a factor 2 depending on chemical species. If this isotopic effect would be confirmed, it would help to improve interstellar chemical models.

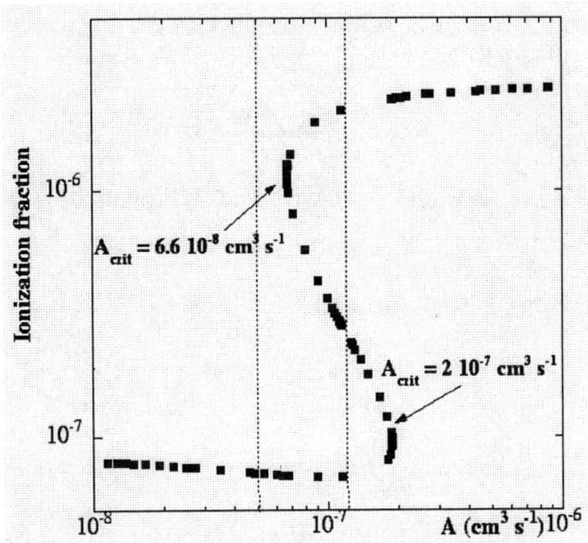

Figure 2. Influence of the dissociative recombination rate of H_3^+ on the fractional ionization. The bistability depends upon the adopted value for A.

Table 3. Deuterium fractionation for models A (same probability that D or H leaves the molecular skeleton) and B (20% probability that D leaves the molecular skeleton).

	Model A	Model B
C_2D/C_2H	$6.9 \cdot 10^{-2}$	$1.3 \cdot 10^{-1}$
DCN/HCN	$2.9 \cdot 10^{-2}$	$4.4 \cdot 10^{-2}$
HDO/H_2O	$1.2 \cdot 10^{-3}$	$1.3 \cdot 10^{-3}$
NH_2D/NH_3	$2.3 \cdot 10^{-2}$	$4.6 \cdot 10^{-2}$

4. BRANCHING RATIOS

Dissociative recombination reactions are often very exothermic and different channels may arise after recombination. Branching ratios leading to the formation of three products are often unknown. A law which predicts these branching ratios would determine if a route of dissociation can be neglected or not. First Vikor *et al.*[16] then Derkatch *et al.*[17] suggested that the greater the enthalpy of the reaction, the higher the branching ratio leading to the formation of three products. The results of eight storage ring experiments[9,16-19] are displayed in Fig. 3. The measured branching ratios of the three body dissociation channel are plotted versus the reaction enthalpy, ΔH. (Enthalpies have been taken at 0 K). In the case of reactions with several routes leading to the formation of three products, we have always selected the channel giving two hydrogen atoms:

$$XH_2^+ + e^- \longrightarrow X + H + H\ .$$

The trend is very clear and we have fitted the points by a polynomial of rank 2 with ΔH in eV :

$$\text{Branch. Ratio} = 0.184 + 0.239\ \Delta H - 0.026\ \Delta H^2\ .$$

The available energy of the reactions is apparently used to dissociate products. Other experiments would be desirable to confirm the trend.

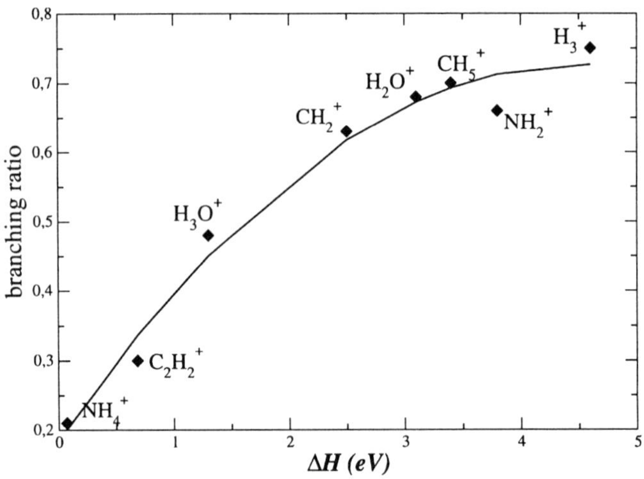

Figure 3. Branching ratio of reactions giving three products as a function of the enthalpy of formation

5. INFLUENCE OF DEPLETION

Millimeter emission spectroscopy allows for the detection of interstellar molecules present in the gas phase. However, some species can also be trapped in dust particles. The ISO satellite has used infrared spectroscopy[20] to clearly detect such species in ice mantles. The amount by which gas phase elements are reduced from their total cosmic abundance is called depletion. This parameter is poorly known in dense clouds and is often discussed in terms of a standard given by the diffuse cloud observations.[21] We discuss in following the influence of this parameter on the chemical models. We present in addition to the previous model, results obtained from a calculation where the gas phase abundance, except Fe, have been divided by ten compared to Hydrogen. Such a model has been labeled δ=10 whereas the standard model corresponds to δ=1.

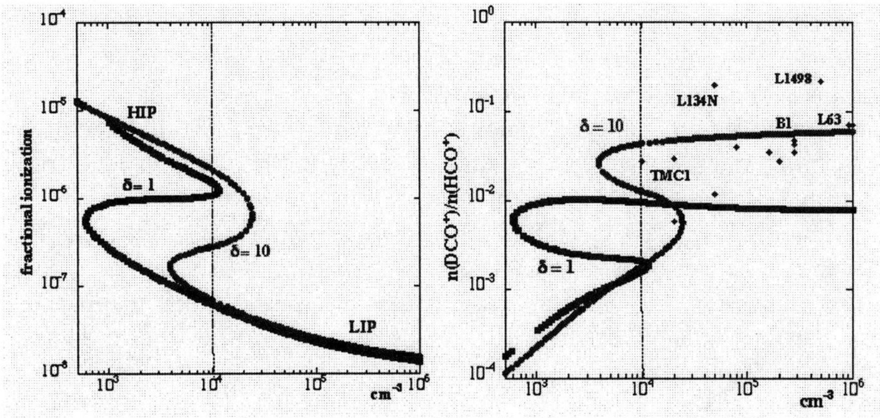

Figure 4: Left: Fractional ionization as a function of the density. Each point represents a stationary model at a temperature of 10 K with δ=1 (black) and δ=10 (gray). *Right*: Fractionation of DCO^+ as a function of density with δ=1 (black) and δ=10 (gray). Each point represents a stationary model at 10 Kelvin. Diamonds correspond to observed values towards different cold dense clouds.

The first effect of depletion is to move the bistable region towards higher densities. As discussed by Le Bourlot et al.,[6,7] bistability is driven by the competition between dissociative recombination of H_3^+ and its destruction by neutral species such as O and CO whose abundances are directly proportional to the total density. These latter reactions are less efficient when the depletion increases and the bistable region is moved towards higher densities. With similar arguments, we may infer that H_2D^+ is less destroyed by O and CO when $\delta = 10$. Then, in the regions where these reactions are preponderant, i.e. in the LIP phase, deuterium fractionation is expected to be larger as shown on the right part of Fig. 4. We have added to the Figures the observed deuterium fractionation toward different cold dark clouds for which densities have also been inferred. It is interesting to note that the observed deuterium fractionation of HCO^+ is better reproduced with the depleted model.

6. CONCLUSION

The deuterium enrichment of interstellar molecules in cold dark clouds is due to fundamental chemical kinetics. Dissociative recombination reactions of H_3^+ and H_2D^+ are driving the chemistry in these regions and the corresponding reaction rate coefficients are urgently needed to fix the actual chemical scenarios. Some experimental trends seem to show that atomic hydrogen rather than atomic deuterium is ejected in the dissociative recombination of deuterated molecular ions. Such an isotopic effect can change the deuterium fractionation of neutral species by a factor 2 and improve the comparison between observations and modeling results. In addition, the available experiments also seem to show a significant correlation between the exoergicity of the reactions and the three body branching ratio. We have extended this result to reactions, which have not yet been studied.

Finally, we have seen that depletion increases the fractionation of deuterated species. Observations are better reproduced in a depleted model. This confirms the well known idea that elements could be more depleted in grains in dark clouds than in diffuse clouds because of the low temperature.

ACKNOWLEDGEMENTS

We thank S. Guberman for his careful reading that improved the English considerably.

REFERENCES

1 J.L. Linsky et al., ApJ **451**, 335 (1995).
2 J. N. Chengalur, R. Braun and W. B. Burton, A&A **318**, L35 (1997).
3 C. M. Wright, E. F. Van Dishoeck, P. Cox, S. D. Sidher, and M. F. Kessler, ApJ **515**, L29 (1999).
4 D. Gerlich, in Planet. Space Sci. (in preparation).
5 M. Larsson et al., A&A **309**, L1 (1996).
6 J. Le Bourlot, G. Pineau des Forêts, E. Roueff, and P. Schilke, ApJ **416**, L87 (1993).
7 J. Le Bourlot, G. Pineau des Forêts, E. Roueff and D. Flower, A&A **302**, 807 (1995).
8 G. Pineau des Forêts and E. Roueff, Phil. Trans. R. Soc. Lond. A **358**, 2549 (2000).
9 G. Pineau des Forêts, E. Roueff and D. R. Flower MNRAS, **240**, 167 (1989).
10 S. Datz, G. Sundström, Ch. Biedermann, L. Broström, H. Danared, S. Mannervik, J. R. Mowat and M. Larsson, Phys. Rev. Lett. **74**, 896 (1995).
11 E. Roueff, J. Le Bourlot, and G. Pineau des Forêts, *Impact of Dissociative Recombination Reactions On Dark Interstellar Cloud Models*, in Dissociative Recombination: Theory, Experiment and Application III ed. by D. Zajfman, J. B. A Mitchell, D. Schwalm and B. R. Rowe (World Scientific, Singapore, 1995) p. 11.

12 J. Glosik, R. Plasil, V. Poterya, P. Kudrna and M. Tichy, Chem. Phys Lett. **331,** 2009 (2000).
13 G. Sundström et al., Science **263**, 785 (1994).
14 H. Roberts, T.J. Millar, A&A **361**, 348 (2000).
15 M. J. Jensen, R. C. Bilodeau, C. P. Safvan, K. Seiersen, L. H. Andersen, H. B. Pedersen and O. Heber, ApJ **543**, 764 (2000).
16 L. Vikor, A. Al-Khalili, H. Danared, N. Djuric, G.H. Dunn, M. Larsson, A. Le Padellec, S. Rosén and M. af Ugglas, A&A **344**, 1027 (1999).
17 A. M. Derkatch, A. Al-Khalili, L. Vikor, A. Neau, W. Shi, H. Danared, M. af Ugglas and M. Larsson, J. Phys. B: At. Mol. Opt. Phys. **32**, 3391 (1999).
18 J. Semaniak, A. Larson, A. Le Padellec, C. Strömholm, M. Larsson, S. Rosen, R. Peverall, H. Danared, N. Djuric, G.H. Dunn and S. Datz, ApJ **498**, 886 (1998).
19 L. Vejby-Christensen, L. H. Andersen, O. Heber, D. Kella, H. B. Pedersen, H. T. Schmidt and D. Zajfman, ApJ **483**, 531 (1997).
20 E. L. Gibb, ApJ **536**, 347 (2000).
21 B. D. Savage, K. R. Sembach, ARA&A **34**, 279 (1996).

Dissociative Recombination ın Cometary Ionospheres

T. E. Cravens
Department of Physics and Astronomy, University of Kansas, Lawrence, KS, USA
cravens@kuplas.phsx.ukans.edu

1. INTRODUCTION

Ionospheres are formed throughout the solar system by the ionization of the neutral constituents of the upper atmospheres of the planets, satellites, and comets. Ionization processes include photoionization by solar extreme ultraviolet (EUV) radiation and impact ionization by superthermal particles such as photoelectrons and energetic auroral electrons. Ion-neutral reactions are important for altering the chemical composition of an ionosphere, and often transform atomic ions into molecular ions. Dissociative recombination (DR) of molecular ions with thermal electrons is the main loss process for ionospheric plasma (see the review by Fox[1]) and is a key process in determining ion and electron densities. The dynamics and energetics of the ionospheric plasma are also strongly affected by DR, mainly due to the density effect but secondarily due to its role in the collisional source terms associated with the momentum and energy conservation equations for the plasma. The role of DR in cometary ionospheres will be emphasized in this paper.

2. FLUID EQUATIONS FOR IONOSPHERIC PLASMA

The fluid description is usually used for both the neutral and ionized species in upper atmospheres and ionospheres, although a kinetic description (i.e., a Boltzmann equation description) must be retained for some particle populations, such as photoelectrons or the fast fragments from DR, which are highly non-Maxwellian and have complex distribution functions. The fluid quantities for species s are the velocity moments of the distribution function of this species. The most important such quantities, as functions of position, **x**, and time, t, are the density of species s, $n_s(\mathbf{x}, t)$, the flow velocity, $\mathbf{u}_s(\mathbf{x}, t)$, and the pressure, $p_s(\mathbf{x}, t)$. In some situations, the pressure is not isotropic and a pressure tensor must be used instead of a scalar pressure.

The fluid conservation equations can be obtained by taking the moments of the Boltzmann equation. The zeroth moment of the Boltzmann equation for a species s gives the continuity equation for this species:

$$\frac{\partial n_s}{\partial t} + \nabla \cdot (n_s u_s) = P_s - L_s \tag{1}$$

where P_s is the production rate of species s and L_s is the loss rate of species s. The loss term associated with dissociative recombination for an ion species s appears as $L_s = \alpha n_s n_e$, where n_e is the electron density and α is the DR rate coefficient. For a quasi-neutral plasma[2,3] the sum of the densities of all ion species equals the electron density:

$$n_e = \sum_{s(ions)} n_s \tag{2}$$

where n_e is the total electron density.

The momentum equation, or fluid equation of motion, results from the first velocity moment of the Boltzmann equation and the energy equation from the second velocity moment. DR enters into the collision terms for both the momentum and energy equations (cf. Cravens[3]; Cravens et al.[4]). The pressure is related to the thermodynamic temperature of species s (T_s) by the equation of state: $p_s = n_s k_b T_s$ where k_b is the Boltzmann constant.

Magnetohydrodynamic (MHD) theory is often a valuable tool for studying planetary plasma environments (e.g., Cravens et al.[5]; Cravens[3]) and the MHD equations can be found from the multi-fluid equations for electrons and ions. The single fluid approximation is usually made and a single mass density, ρ, is employed. In cases where one ion species dominates (s = i), $\rho = n_e m_i$, where m_i is the mass of this major ion species. The electric field is found from the momentum equation for the thermal electron fluid (s = e) and is also called the generalized Ohm's law. The magnetic field can then be determined from Faraday's law, using this form of the electric field. The resulting magnetic induction equation for the magnetic field, **B**, can be written as:

$$\frac{\partial B}{\partial t} = -\nabla \times (u \times B) + \nabla \times (D \nabla \times B) \tag{3}$$

where the magnetic diffusion coefficient is given by $D = \eta/\mu_0$ and μ_0 is the permeability of free space. The plasma resistivity, η, depends on the plasma density (which is affected by DR) and on the electron collision frequency.

The first term on the right-hand side of equation (3) is associated with the motional electric field and, in the absence of the second term, results in magnetic field lines being "frozen into" and carried along with the fluid flow. The second term is the Ohmic dissipation term that becomes larger at low altitudes in planetary ionospheres, in which case the magnetic field can de-couple from the plasma flow.

The momentum equation in the single fluid and MHD approximations can be written as:

$$\rho\left(\frac{\partial \mathrm{u}}{\partial \mathrm{t}}+\mathrm{u}\cdot\nabla\mathrm{u}\right)=\mathrm{J}\times\mathrm{B}-\nabla\mathrm{p}+\rho\mathrm{g}-\left(\rho\nu_{\mathrm{in}}+\mathrm{m}_{\mathrm{n}}\mathrm{P}\right)\left(\mathrm{u}-\mathrm{u}_{\mathrm{n}}\right) \tag{4}$$

where **J** is the current density, P is the total ion production rate, and $\mathbf{u}_n$ is the velocity of the neutral gas. The ion-neutral momentum transfer collision frequency is given by $\nu_{in} = k_{in}\, n_n$, where $k_{in} \approx 10^{-9}$ cm^3 s^{-1} is the rate coefficient and n_n is the neutral number density. The thermal pressure is given by $p = p_e + p_i = n_e k_b(T_e + T_i)$, where T_e and T_i are the electron and ion temperatures, respectively. The gravitational acceleration is **g** (**g** $\approx$ 0 for comets). When the magnetic field lines are not curved, Ampère's law can be used to express $\mathbf{J}\times\mathbf{B}$ as the gradient of the magnetic pressure $p_B = B^2/2\mu_0$.

The plasma flow speed is often low enough, particularly at lower altitudes in an ionosphere, that the inertial terms on the left-hand side of the momentum equation can be neglected. In this case, equation (4) can be rewritten as an ambipolar diffusion equation:

$$\mathbf{u}-\mathbf{u}_n \cong -\frac{1}{\rho\nu_{in}}\nabla\left(p+p_B\right)+\mathbf{g}/\nu_{in} \tag{5}$$

where $m = m_i$. Equation (5) indicates that the total pressure, $p + p_B$, tends to stay constant where the friction term is small.

3. THE COMETARY PLASMA ENVIRONMENT

The ionospheres of some planets, such as Earth and Jupiter, are immersed in strong magnetic fields created by internal planetary dynamos. However, other bodies in the solar system possess only weak intrinsic fields and the ionospheric magnetic fields in these cases are largely induced by the interaction of the external plasma (e.g., the solar wind) with the body.[3] Examples include the planets Venus and Mars, Titan (a satellite of Saturn), and comets.

The solar wind "directly" interacts with a planet's (or a comet's) neutral atmosphere or ionosphere, if it does not possess a strong intrinsic magnetic field. The solar wind slows down and flows around the obstacle created by the planet's atmosphere and ionosphere. At Venus and Mars, the interaction is primarily with the ionosphere itself, whereas for a comet the interaction is mainly with the extensive neutral atmosphere (or coma). A bow shock forms upstream of the obstacle because the solar wind is a supersonic (actually supermagnetosonic) fluid. Fig. 1 is a schematic of the cometary plasma environment. For an active comet like Halley, the bow shock is located at a subsolar distance from the nucleus of $\approx 3 \times 10^5$ km. For comparison, the Venus subsolar bow shock is located at a distance of $\approx$ 2000 km from the surface. The interplanetary magnetic field lines carried by the solar wind pile up outside the obstacle and form a magnetic barrier.

Our knowledge of the cometary plasma environment is mainly a consequence of the spacecraft missions to comets Halley, Giacobini-Zinner, and Grigg-Skjellerup (see the two-volume book *Comets in the Post-Halley Era*[6]). The plasma flow near the comet remains collisionless down to radial distances of roughly 10^5 km (commonly called the cometopause). Charge transfer collisions with cometary neutrals become important inside this distance, as do other collisional processes. The cometary ionosphere can be defined as the region containing cold cometary plasma relatively undisturbed by the solar wind, and it is located within cometocentric distances of $1 - 2 \times 10^4$ km for an active comet like Halley[7]. A field-free region, called the diamagnetic cavity, was observed by instruments onboard the Giotto spacecraft inside a radial distance of $\approx$ 5000 km. The plasma flow inside the cavity is coupled to the neutrals by ion-neutral collisions and is directed outward. The boundary of this cavity is sometimes called the contact surface (CS in short), or sometimes just the cavity boundary surface. A magnetic barrier with enhanced magnetic field strength exists outside the cavity. The plasma is slow moving and almost stagnant in the inner part of this magnetic barrier, but moves faster and roughly antisunward at larger radial distances. The magnetic field lines drape around this inner region because the ends of the field lines remain out in the more rapidly flowing solar wind far from the comet. Cometary plasma is "trapped" by the draped magnetic lines and forms a plasma tail.

The nucleus of a comet is typically a few kilometers in diameter and contains ice and dust. The ice is mainly frozen water (about 80%; see Krankowsky et al.[8]), but other frozen volatiles (e.g., CO_2, NH_3, CH_4,….) are also present. This ice sublimates as a comet travels into the inner solar system and the resulting gas flows away from the nucleus at a speed $u_n \approx 1$ km/s. Dust is carried off the surface by this gas flow. The total neutral density versus radial distance from the nucleus, r, for distances inside the cometopause is given by:

$$n_n = \frac{Q}{4\pi u_n r^2} \tag{6}$$

where Q is the gas production rate of the comet, which depends on the size of the active part of a nucleus and on the distance from the Sun. $Q \approx 10^{30}\ s^{-1}$ for comet Halley near a heliocentric distance of 1 AU.[8]

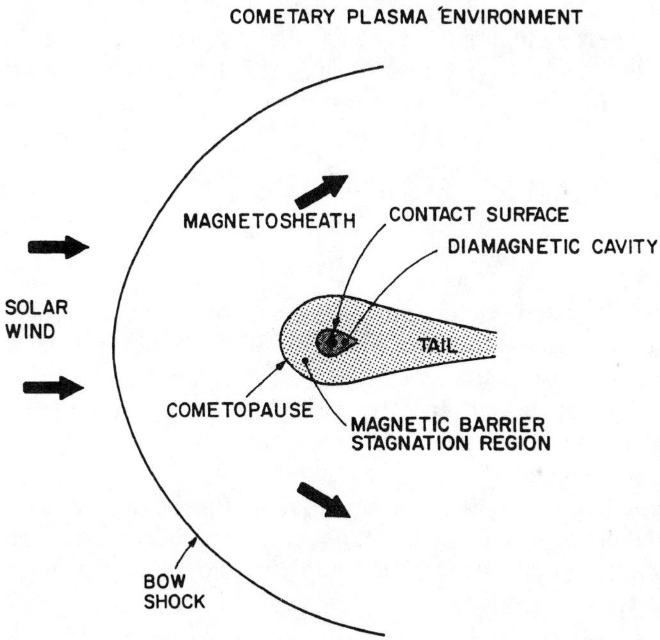

Figure 1. Schematic of the cometary plasma environment. From Cravens.[29]

The primary source of cometary ions is photoionization of neutral molecules and atoms by solar EUV photons. Many ion species are produced but the most important is H_2O^+:

$$
\begin{aligned}
h\nu + H_2O &\rightarrow H_2O^+ + e \qquad (7)\\
&\rightarrow OH^+ + H + e\\
&\rightarrow OH + H^+ + e
\end{aligned}
$$

Most of the water (or other molecular species) is dissociated in the outer coma (beyond a few tens of thousand of km), leaving a neutral coma containing atomic oxygen and hydrogen. The total ion production rate is approximately $P = I\ n_n$, where the total ionization frequency (photoionization and some electron impact ionization) at 1 AU is $I \approx 10^{-6}\ s^{-1}$.

Ionospheric chemistry controls the ion composition and density in the inner coma (the photochemical region). Key reactions include:

$$H_2O^+ + H_2O \rightarrow H_3O^+ + OH \qquad (8)$$

$$H_3O^+ + NH_3 \rightarrow NH_4^+ + H_2O \quad (9)$$

$$NH_3^+ + NH_3 \rightarrow NH_4^+ + NH_2 \quad (10)$$

The rate coefficient for (8) is $k_8 = 2.1 \times 10^{-9}$ cm^3 s^{-1} and this reaction converts H_2O^+ ions to H_3O^+ ions[7]. Plasma is lost primarily via the DR of the ions and thermal electrons. The most important DR reactions are:

$$H_3O^+ + e \rightarrow \text{neutral products} \quad (11)$$

$$NH_4^+ + e \rightarrow \text{neutral products} \quad (12)$$

The H_3O^+ DR rate coefficient most often seen in the cometary literature prior to a few years ago is $\alpha = 7 \times 10^{-7} (300/T_e)^{1/2}$ cm^3 s^{-1} (cf. Mendis[9]), but the Mul et al.[10] rate is now being used. The full cometary photochemistry is more complicated than the brief description given here; however, the behavior of the total plasma density (and hence the dynamics) can be obtained using just H_2O^+ and H_3O^+.

Photochemical equilibrium expressions for the density can be found by setting the net production rate on the right-hand side of equation (1) equal to zero (i.e., $P_s = L_s$). Photochemical expressions for the H_2O^+ and H_3O^+ densities are:

$$n_{H2O+} = I / k_8 \approx 500 \text{ cm}^{-3} \text{ (independent of r)} \quad (13)$$

$$n_e = n_{H3O+} = [I / \alpha]^{1/2} \approx [IQ / (4 \bullet u_n \alpha)]^{1/2} r^{-1} \quad (14)$$

Fig. 2 shows ion densities for mass 18 (H_2O^+), mass 19 (H_3O^+), and mass 33 ($CH_3OH_2^+$) measured in the inner coma by the Giotto ion mass spectrometer. Indeed, the H_2O^+ density is roughly independent of r, and the electron density (and H_3O^+ density) varies inversely with r for the region inside about 10^4 km.

Ion density profiles depart from simple photochemical profiles where transport effects (that is, the left-hand side of the continuity equation) become important. A comparison of a characteristic transport time ($\tau_T \approx r / u$) with a chemical (or collisional) lifetime can be used to determine where these effects begin to come into play for a particular process. In general, transport effects become important at larger distances (or for other planets at higher altitudes) where the neutral and ion densities are lower, and hence the collision times longer. The boundary between the regions has been called the collisionopause.[11,9] For example, a collisionopause distance of $r \approx 40{,}000$ km is found for H_2O^+ with a chemical lifetime of $\tau_c \approx 1/k_8 n_n$.

The chemical lifetime of the major ion in the inner coma, H_3O^+, is determined by the DR process ($\tau_C \approx 1 / \alpha n_e$). Depending on the electron

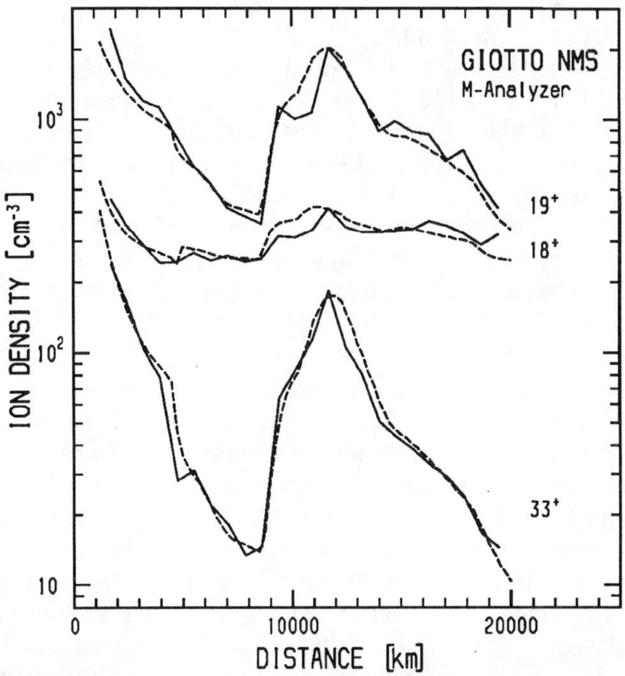

Figure 2. Solid lines are the ion densities versus radial distance measured by the Giotto IMS. Dashed lines are model results.[17]

temperature adopted, this collisionopause is located somewhere between 1 and 4×10^4 km. The observed electron density profile appears to depart from a simple inverse r behavior near $r = 10^4$ km (see Fig. 2, Ip and Axford[12]), suggesting that transport effects are important outside this distance.

The collisionopause concept can also be applied to the electron and ion energy equations[11,13]. Inside the "thermal collisionopause," cooling of the electrons (or ions) due to electron-neutral collisions is more important than transport effects (such as heat conduction), in which case the electron and neutral temperatures are about the same (that is, a few hundred degrees). At larger distances the plasma thermally decouples from the neutral gas, and electron and ion temperatures become much higher than the neutral temperature. Numerical calculations for Halley[13] demonstrated that T_e increases rapidly with r for $r > 1.5 \times 10^4$ km, which has implications for the plasma density due to the temperature dependence of the DR rate coefficient.

4. THE ION PILEUP REGION

An "ion pile-up," or enhancement, region was observed both by the Giotto ion mass spectrometer (IMS)[14,15,16] and the neutral mass spectrometer (NMS) in its ion mode[17] near a radial distance of 10^4 km in the ionosphere of comet Halley. The relative density enhancement was a factor of 4 (Fig. 2). One of the explanations offered for this pile-up was a reduced DR ion loss due to an elevated electron temperature.[18,17,19,20,21] Two groups, Eberhardt and Krankowsky[17] and Häberli et al.[19], used photochemical models (with some transport effects included) to deduce the electron temperature profile which was needed to reproduce the ion density profile. The H_3O^+ DR coefficient measured by Mul et al.[10] (see Fig. 3) was used and is in excellent agreement with the rate recently measured at the heavy-ion storage ring CRYRING by Neau et al.[22] This rate coefficient has a stronger dependence on T_e at high temperatures than does the coefficient previously used by most comet modelers (cf. Mendis et al.[9]). The temperature needed to explain the pile-up increases from $\approx$ 300 K just inside r $\approx$ 8000 km to $\approx 2.5 \times 10^4$ K at r = 12,000 km (see Fig. 4). The electron temperature profile was, unfortunately, not measured by Giotto in this region. Both two-dimensional[21] and three-dimensional[20] numerical MHD calculations of the plasma environment of comet Halley have been carried out. Ion energy equations were solved in these models but not electron energy equations; electron temperature profiles, as shown in Fig. 4, were adopted. Fig.s 5 and 6 show some of the calculated density and magnetic field profiles, respectively, for the Giotto track through the Halley environment (basically in the flanks) from the Lindgren et al.[21] model. The electron density enhancements produced by this MHD model were not as high as the measured enhancement because the plasma flow has positive divergence near the pile-up region which drives the density lower than photochemical values (which are similar to the measured values). This discrepancy is not surprising because the major ion lifetime due to DR is $\approx 10^4$ s, which exceeds the transport time scale by a factor of 2 or more. On the other hand, the ion DR lifetime just inside the pile-up is only about 1000 s, considerably less than the transport time. The reason for the divergent flow can be seen from the MHD momentum relation (5). Due to both the high temperature and the high density, the total plasma pressure is very high in the enhancement region ($p \approx 10^{-8}$ dynes cm^{-2}) in comparison with the regions either inside or outside. This enhanced pressure forces the plasma away from the enhancement region.

The divergent plasma flow near the pile-up region also results in a decrease of the magnetic field strength because plasma flow carries magnetic flux with it (equation (3)). This can be seen by comparing curves A and B in Fig. 6. A large "bite" is taken out of the calculated magnetic field profile right where the density is enhanced. The measured B profile shows a much smaller bite-out than the model profiles. Apparently, the very high electron temperatures demanded by the density data are inconsistent with the magnetic field data. Among the suggested explanations for this inconsistency[21] was that the electron temperature in the pile-up region might only be $\approx$ 10,000 K rather than 25,000 K (this lowers the thermal pressure) and that the DR rate coefficient might then be a factor of 3 less than the Mul et al. value at this temperature (in order to still sustain the high electron density). Vibrational excitation of the parent ion is known to affect DR (cf. Mitchell[23]) and this is perhaps relevant here.

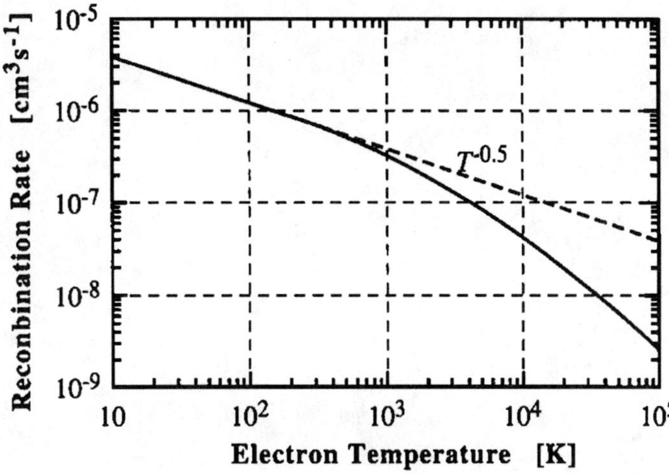

Figure 3. The solid line is the recombination rate coefficient versus electron temperature from Mul et al.[10], and the dashed line is the coefficient used in earlier cometary models. From Gombosi et al.[20]

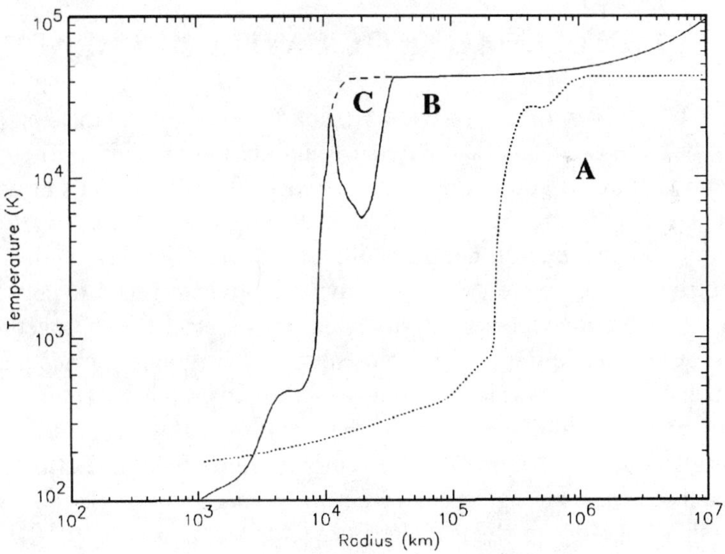

Figure 4. Electron temperature versus cometocentric distance for comet Halley for several cases, as discussed in the text. The temperature for curve A stays low throughout the inner coma. The solid curve labelled B is very similar to the profile used by Eberhardt and Krankowsky[17] for distances less than 20,000 km. The dashed curve labelled C is quite similar to the profile used by Häberli et al.[19] and to one used by Gombosi et al.[20]

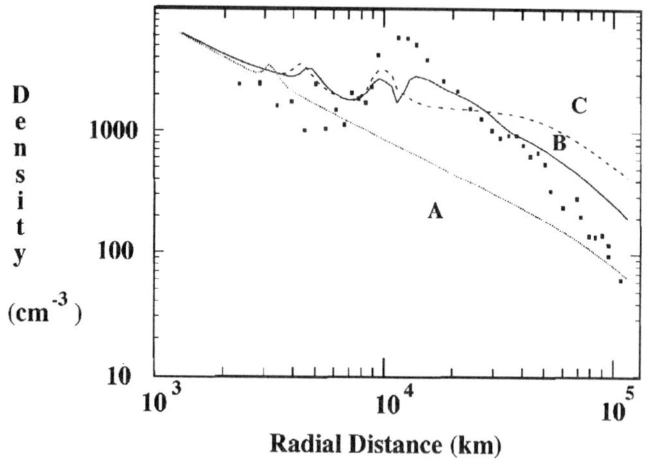

Figure 5. Electron density profiles for comet Halley calculated by a 2D MHD model[21]. The curves are for the cases described earlier. Data points from the Giotto mission are also shown[16]. Adapted from Lindgren et al.[21]

5. THE DIAMAGNETIC CAVITY BOUNDARY

One of the exciting discoveries made by the Giotto mission to comet Halley was the diamagnetic cavity surrounding the nucleus[24]. The cavity can be seen in Fig. 6; B goes to zero at a distance from the nucleus of about 4500 km. The field-free region is caused by the outward force exerted on the cometary plasma by ion-neutral collisions (i.e., friction) in the magnetic barrier region located outside the cavity boundary (cf. Cravens[25]; Ip and Axford[12]). The ionospheric plasma and the neutrals flow together at the neutral speed inside the cavity, whereas in the magnetic barrier the ions are tied to the field lines and are almost stationary (at least for $r < 10^4$ km). Hence, in the region just outside the cavity, neutrals flow past the ions exerting an outward frictional force on them. Equating the friction force with the magnetic pressure gradient force and using the simple photochemical expression (14) for the electron density yields an expression for the magnetic field strength as a function of r that agrees quite well with the measured profile[25,12] except right near the cavity boundary. DR plays an important role in determining the distance to the cavity boundary because the friction force is proportional to the electron density.

The dynamics in the immediate vicinity of the cavity boundary are more

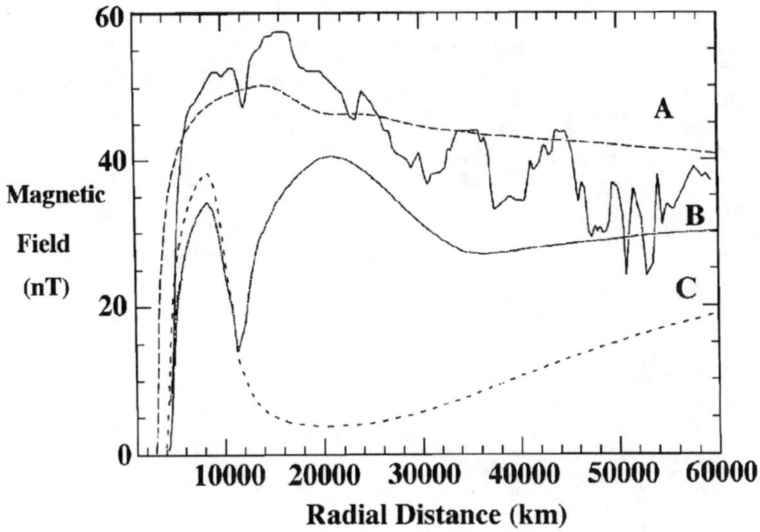

Figure 6. Magnetic field strength profiles for comet Halley calculated by a 2D MHD model[21]. The curves are for the cases described earlier. The measured profile from the Giotto magnetometer is also shown (solid line with structure).[24] Adapted from Lindgren et al.[21]

complicated than the description just given. A narrow ($\Delta r \approx 50$ km) boundary layer with enhanced plasma density (a factor of 3 – 10) was predicted to exist at the boundary[26]. This layer was then seen in Giotto IMS data[27]. A one-dimensional hybrid code[28] was subsequently used to study the layer in more detail. Electrons are treated as a fluid, but ions are treated as particles in a hybrid code. Collisional processes, including DR, can be included in such a code using Monte Carlo methods.

The density increase in the layer, which has been called a "recombination layer" by Goldstein et al.[27], takes place because the ionospheric plasma piles up, as it flows radially outward (initially with the neutral gas at a speed $u_n \approx$ 1 km/s) in the cavity and "collides with" the magnetic barrier. The continual input of plasma from the cavity into the boundary layer is balanced by the enhanced ion loss due to DR. Due to these transport effects, the plasma is not in photochemical equilibrium in this layer. Note that the electron temperature stays low in this region near 5000 km, unlike in the pile-up region found further out near 10^4 km as discussed earlier.

Fig. 7 shows some of the hybrid code results[28]. The plasma flow speed decreases across the layer starting where the density first begins to increase.

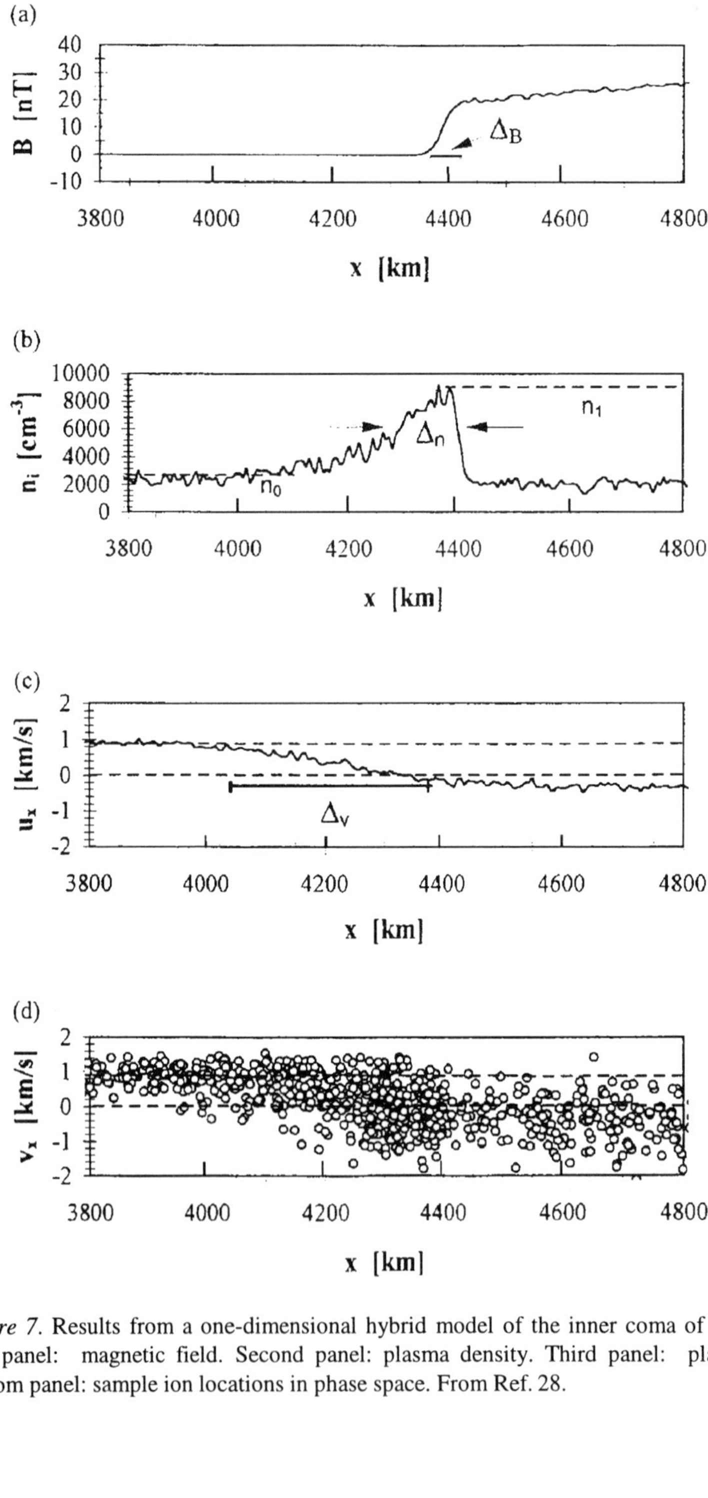

Figure 7. Results from a one-dimensional hybrid model of the inner coma of comet Halley. Top panel: magnetic field. Second panel: plasma density. Third panel: plasma velocity. Bottom panel: sample ion locations in phase space. From Ref. 28.

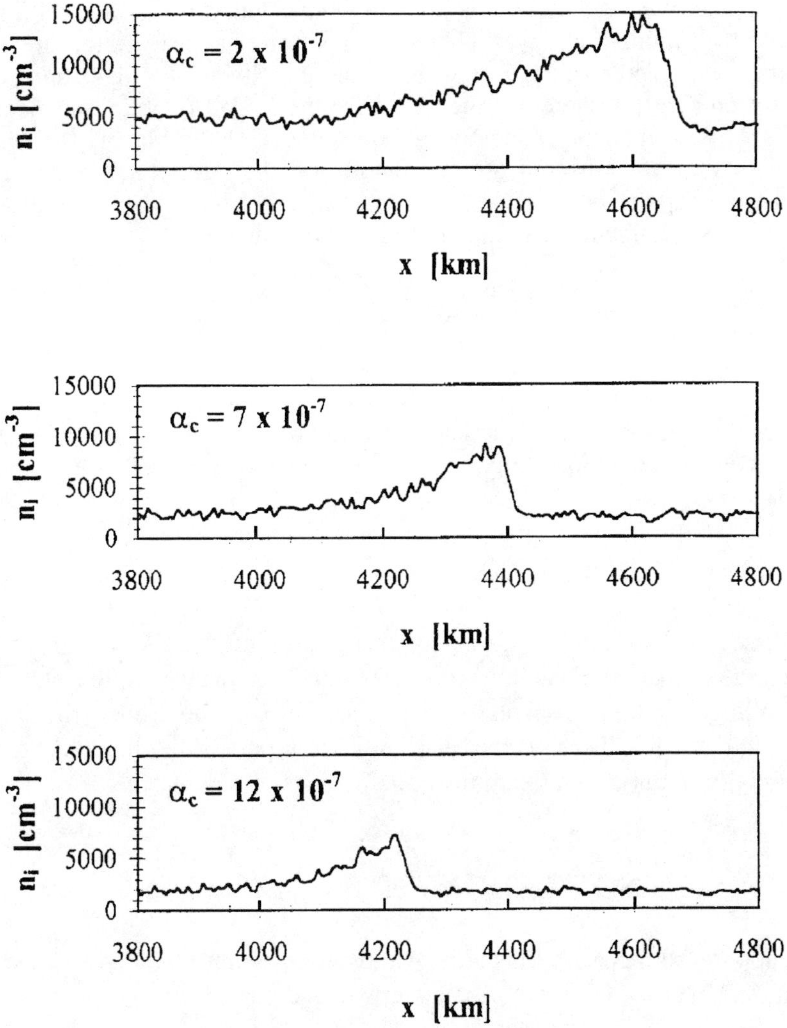

Figure 8. Density profiles from a one-dimensional hybrid model of the inner coma of comet Halley for different dissociative recombination rate coefficients for the major ion. From Ref. 28.

The density decreases in the outer part of the layer where the magnetic field exhibits a sharp increase. The momentum flux, or dynamic pressure (i.e., ρu^2) associated with the flowing plasma inside the cavity is converted to thermal plasma pressure in the layer itself and then finally is balanced by magnetic pressure on the outside of the layer. At even larger radial distances, ion-neutral friction becomes an important force term in the momentum balance, as mentioned earlier.

Dissociative recombination plays an important role in controlling the cavity recombination layer. The momentum balance determines the enhancement factor (n_l/n_0, where n_l is the density in the layer and n_0 the density outside), as demonstrated by Cravens[26]. The relative enhancement increases with increasing incident Mach number of the plasma flow inside the cavity. But DR controls the width of the layer. An approximate expression for the boundary layer width, Δ, can be found by equating the flux of ions into the layer with the total loss rate due to DR[26]:

$$\Delta \approx \frac{u_n}{\alpha n_0} \frac{1}{\left(\frac{n_l}{n_0}\right)^2 - 1} \tag{15}$$

Results from the hybrid model for different values of α also illustrate this DR-dependent behaviour (Fig. 8).

6. CONCLUSIONS

Dissociative recombination plays an important role in determining the structure and dynamics of planetary and cometary ionospheres. By controlling ion loss in an ionosphere, DR helps to determine the ion and electron densities. These densities then form a part of the thermal pressure, which is so important in the momentum balance of any plasma.

ACKNOWLEDGMENTS

Support from NASA Planetary Atmospheres grant NAG5-4358 and NSF grant ATM9815574 is gratefully acknowledged.

REFERENCES

1 J. L. Fox, Dissociative Recombination In Planetary Ionospheres, in *Dissociative Recombination: Theory, Experiment and Applications*, ed. by B. R. Rowe, J. B. A. Mitchell and A. Canosa, (Plenum Press, New York, 1993), p. 219.

2 R. W. Schunk and A. F. Nagy, *Ionospheres: Physics, Plasma Physics, and Chemistry*, Cambridge Univ. Press (2001).

3 T. E. Cravens, *Physics of Solar System Plasmas*, Cambridge Univ. Press (1997).

4 T. E. Cravens, C. J. Lindgren, and S. A. Ledvina, A Two-Dimensional Multifluid MHD Model of Titan's Plasma Environment, Planet. Space Sci. **46**, 1193 (1998).

5 T. E. Cravens, H. Shinagawa, and J. G. Luhmann, Magnetohydrodynamic Processes: Magnetic Fields in the Ionosphere Of Venus, in *Venus II - Geology, Geophysics,*

Atmosphere, and Solar Wind Environment, ed. by S. W. Bougher, D. M. Hunten, and R. J. Phillips, (Univ. of Arizona Press, Tucson, 1997), pp. 61-93.
6 *Comets in the Post-Halley Era*, ed. by R. L. Newburn, Jr. et al., (Kluwer Academic Publishers, The Netherlands, 1991).
7 T. E. Cravens, Theory and Observations Of Cometary Ionospheres, Adv. Space Res. **7**, 147 (1987).
8 D. Krankowsky et al., In Situ Gas and Ion Measurements at Comet Halley, Nature **321**, 326 (1986).
9 D. A. Mendis, H. L. F. Houpis and M. L. Marconi, The Physics of Comets, Fund. Cosmic Phys. **10**, 1 (1985).
10 P. M. Mul, J. W. McGowen, P. DeFrance, and J. A. Mitchell, Merged Beam Experiments, V, Dissociative Recombination of OH^+, H_2O^+, H_3O^+, and D_3O^+, J. Phys. B. **16**, 3099 (1983).
11 T. E. Cravens, Cometary Plasma Boundaries, Adv. Space Res. **9**, 3293 (1989).
12 W.-H. Ip and W. I. Axford, The Formation of a Magnetic Field Free Cavity At Comet Halley, Nature **325**, 418 (1987).
13 L. Gan, and T. E. Cravens, Electron Energetics in the Inner Coma Of Comet Halley, J. Geophys. Res. **95**, 6285 (1990).
14 H. Balsiger, K. Altwegg, F. Bühler, J. Geiss, A. G. Ghielmetti, B. E. Goldstein, R. Goldstein, W. T. Huntress, W.-H. Ip, A. J. Lazarus, A. Meier, M. Neugebauer, U. Rettenmudd, H. Rosenbauer, R. Schwenn, R. D. Sharp, E. G. Shelley, E. Ungstrup, and D. T. Young, Ion Composition and Dynamics at Comet Halley, Nature **321**, 330 (1986).
15 R. Schwenn, W.-H. Ip, H. Rosenbauer, H. Balsiger, F. Bühler, R. Goldstein, A. Meier, and E. G. Shelley, Ion Temperature and Flow Profiles In Comet Halley's Close Environment, in *Proc. 20th ESLAB Symposium on the Exploration of Halley's Comet, ESA SP-250*, ed. by B. Battrick, E. J. Rolfe, and R. Reinhard, (Eur. Space Agency Spec. Publ., 1986), p. 225.
16 K. H. Altwegg et al., The Ion Population Between 1300 and 230,000 Km in The Coma of Comet P/Halley, Astron. Astrophys. **279**, 260 (1993).
17 P. Eberhardt and D. Krankowsky, The Electron Temperature in The Inner Coma of Comet P/Halley, Astron. Astrophys. **295**, 795 (1995).
18 T. E. Cravens, Plasma Processes in the Inner Coma, in *Comets in the Post-Halley Era, Vol. 2*, ed. by R. L. Newburn Jr., et al., (Kluwer Academic Publishers, The Netherlands, 1991) p. 1211.
19 R. M. Häberli, K. Altwegg, H. Balsiger, and J. Geiss, Physics and Chemistry of Ions in The Pile-Up Region of Comet P/Halley, Astron. Astrophys. **297**, 881 (1995).
20 T. I. Gombosi, D. L. DeZeeuw, R. M. Häberli, and K. G. Powell, A 3D Multiscale MHD Model of Cometary Plasma Environments, J. Geophys. Res. **101**, 15233 (1996).
21 C. J. Lindgren, T. E. Cravens, and S. Ledvina, Magnetohydrodynamic Processes in the Inner Coma of Comet Halley, J. Geophys. Res. **102**, 17395 (1997).
22 A. Neau et al., Dissociative Recombination of D_3O^+ and H_3O^+: Absolute Cross Sections and Branching Ratios, J. Chem. Phys. **113**, 1762 (2000).
23 J. B. A. Mitchell, The Dissociative Recombination of Molecular Ions, Physics Reports **186**, 215 (1990).
24 F. M. Neubauer, K. H. Glassmeier, M. Pohl, J. Raeder, M. H. Acuna, L. F. Burlaga, N. F. Ness, G. Musmann, F. Mariani, M. K. Wallis, E. Ungstrup, and H. U. Schmidt, First Results from the Giotto Magnetometer Experiment at Comet Halley, Nature **321**, 352 (1986).
25 T. E. Cravens, The Physics of the Cometary Contact Surface, in *Proc. 20th ESLAB Symposium on the Exploration of Halley's Comet, ESA SP-250*, ed. by B. Battrick, E. J. Rolfe, and R. Reinhard, (Eur. Space Agency Spec. Publ., 1986), p. 244.

26 T. E. Cravens, A Magnetohydrodynamical Model of the Inner Coma of Comet Halley, J. Geophys. Res. **94**, 15025 (1989).

27 B. E. Goldstein et al., Observations of a Shock and a Recombination Layer at the Contact Surface of Comet Halley, J. Geophys. Res. **94**, 1725 (1989).

28 P. Puhl-Quinn and T. E. Cravens, One-Dimensional Hybrid Simulations of the Diamagnetic Cavity Boundary Region of Comet Halley, J. Geophys. Res. **100**, 21631 (1995).

29 T. E. Cravens, The Solar Wind Interaction with Non-Magnetic Bodies and the Role of Small-Scale Structures, in *Solar System Plasma Physics, Geophys. Monograph No. 54,* ed. by J. H. Waite, Jr., J. L. Burch, and R. L. Moore, (American Geophysical Union, Washington, 1989) p. 353.

Absolute Total State-Selected Cross Sections for Ion-Molecule Reactions of Importance in Planetary Ionospheres

Reactions of $O^+(^4S, ^2D, ^2P)$

C. Y. Ng
Department of Chemistry, University of California, Davis, CA, 95616, USA
cyng@chem.ucdavis.edu

1. INTRODUCTION

State-selected and state-to-state ion chemistry has played an important role in our fundamental understanding of elementary bimolecular ion-molecule collision dynamics.[1-3] The progress in state selected ion chemistry depends on the development of experimental techniques for state selection of reactant ions. The formation of O^+ in its ground ^4S states can be achieved when one of a variety of ionization methods is followed by collisional relaxation.[4]

The O^+ ions formed by solar VUV photoionization of O atoms are the most abundant ions in the F-region of the Earth's ionosphere,[5] with the branching ratios for $O^+(^4S):O^+(^2D):O^+(^2P) \approx 0.43:0.29:0.28$.[6-9] Since the $O^+(^2D)$ and $O^+(^2P)$ states are metastable with lifetimes of 3.57 hr and 4.47 s, respectively, the ion-molecule reactions involving $O^+(^4S, ^2D, ^2P)$ and atmospheric gases constitute the most important set of reactions in the Earth's ionosphere.[10-13] Similarly, ion-molecule reactions involving $O^+(^4S, ^2D, ^2P)$ play an important role in modeling the ionospheric chemistry of other planets such as Mars and Venus.[14] The chemistry of $O^+(^4S)$ has been investigated extensively.[15] However, absolute cross sections for ion-molecule processes involving $O^+(^2D)$ and $O^+(^2P)$, which are 3.32 and 5.02 eV above the ground $O^+(^4S)$ state, respectively, have not been examined directly.[16]

It is well known that the rates of radiative recombination of O^+ with electrons are several orders of magnitude smaller than those of dissociative recombination of N_2^+, O_2^+, and NO^+ with electrons.[17,18] The major decay mechanism for $O^+(^4S, ^2D, ^2P)$ involves reactions with N_2 and O_2, leading to the formation of N_2^+, O_2^+, and NO^+ etc.[19-21] The ion-molecule reaction $O^+(^4S, ^2D, ^2P) + H_2$ is relevant to the study of H_2 rich interstellar clouds where OH^+ ions further react with H_2 to yield H_2O^+ and H_3O^+ molecular ions.[22-24] The study of the ion-molecule reaction $O^+(^4S, ^2D, ^2P) + H_2O$ is stimulated by the need to understand the atmospheric plasma environment due to spacecraft atmosphere interactions.[25-27] It is generally recognized that the primary source of observed H_2O^+ ions is due to charge transfer collisions between atmospheric O^+ and H_2O, which are the main constituents of the contaminant cloud surrounding the space shuttle. The modeling of ion density data[28-31] requires accurate cross sections (σ's) or rate coefficients (k's) for the reactions of $O^+(^4S, ^2D, ^2P) + H_2O$.

We have developed an experimental scheme for the preparation of O^+ reactant ions in the selected $O^+(^2P)$ or $O^+(^2D)$ or $O^+(^4S)$ state.[32] This scheme combines the use of dissociative charge transfer reactions,

$$He^+ + O_2 \rightarrow O^+(^4S, ^2D, ^2P) + O(^3P, ^1D, ^1S) + He, \quad (1)$$

$$Ne^+ + O_2 \rightarrow O^+(^4S, ^2D) + O(^3P) + Ne, \quad (2)$$

$$Ar^+(^2P_{3/2,1/2}) + O_2 \rightarrow O^+(^4S) + O(^3P) + Ar, \quad (3)$$

with the radio frequency (RF) octopole ion trap technique.[33] The branching ratios for the formation of $O^+(^4S)$, $O^+(^2D)$, $O^+(^2P)$ by reaction (1) have been measured as a function of collision energy in great detail in previous crossed-beam and RF octopole ion guide experiments.[33,34] The formation of $O^+(^2D)$ and $O^+(^4S)$ by reactions (2) and (3), respectively, are endothermic. We have demonstrated that $O^+(^2D)$ and $O^+(^4S)$ can be produced efficiently by these reactions at collision energies above their respective thermochemical thresholds. The O^+ ions produced by reaction (3) have been found predominantly in the $O^+(^4S)$ ground state.[35] We have shown that state selected $O^+(^4S)$, $O^+(^2D)$, and $O^+(^2P)$ reactant ions, prepared by reactions (1)-(3) at appropriate collision energies and RF trapping conditions of the octopole ion guide, are of sufficiently high intensities for performing absolute total cross section measurements.[32]

2. EXPERIMENT

2.1 The TQDO Apparatus

The experimental arrangement and procedures for cross section measurements using the triple-quadrupole double-octopole (TQDO) apparatus (Fig. 1) have been described in detail.[36] The TQDO apparatus

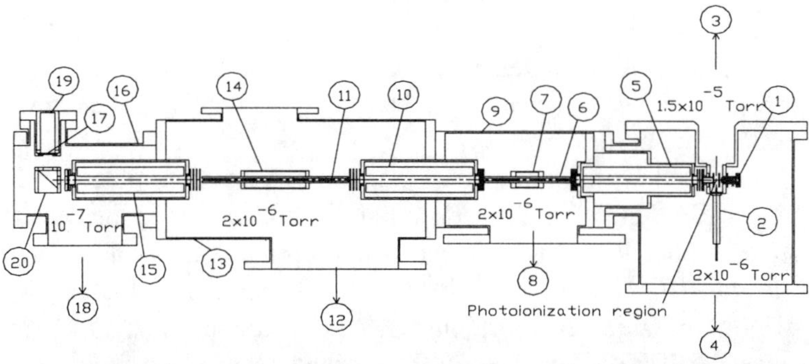

Figure 1. Schematic diagram of the TQDO apparatus: (1) electron impact ionization ion source, (2) atomic or molecular nozzle beam, (3) to freon-trapped 6" diffusion pump (DP), (4) to liquid-nitrogen (LN2)-trapped 6" DP, (5) reactant QMS, (6) lower RF octopole ion guide, (7) lower RFOIGGC, (8) to LN2-trapped 6" DP, (9) the lower RF octopole ion guide chamber, (10) middle QMS, (11) upper RF octopole ion guide, (12) to LN2-trapped 4" DP, (13) upper RF octopole ion guide chamber, (14) upper RFOIGGC, (15) product QMS, (16) detector chamber, (17) plastic scintillator window, (18) to LN2-trapped 2" DP, (19) photomultiplier tube, (20) aluminum ion target.[3]

consists of, in sequential order, a vacuum ultraviolet (VUV) photoionization ion source or an electron impact ionization ion source (1), a reactant quadruple mass spectrometer (QMS) (5), a lower RF octopole ion guide reaction gas cell (RFOIGGC) [(6) + (7)], a middle QMS (10), an upper RFOIGGC [(11) + (14)], a product QMS (15), and a Daly-type scintillation ion detector [(17) + (19) + (20)]. The electron impact ionization ion source used here is of the Brink design and has been added to the TQDO apparatus recently. The TQDO apparatus is partitioned into five chambers, which are separately evacuated by liquid nitrogen or freon trapped diffusion pumps. The differential pumping arrangement is important for the reactant ion preparation scheme and the cross section measurement experiments described here.

The application of an RF octopole field as an ion guide for absolute total cross section measurements was pioneered by Teloy and Gerlich.[33] An electric multipole structure with $2n$ poles is formed by $2n$ rods symmetrically spaced on an inscribed circle with a diameter of $2r_o$. The principle of the RF electric multipole field of $2n$ poles for trapping and guiding ions (charge = q, mass = m) is based on the fact that with a sufficiently high RF field (amplitude = V_o) and an appropriate frequency (ω), a sufficiently high effective potential [$V_{eff}(r)$] is established to confine the ions. Within the limit of certain operating conditions, the expression for V_{eff} of an octopole ($n = 4$) is

$$V_{eff}(r) = 4\,[(q^2\,V_o^2)/(m\,\omega^2\,r_o^2)]\,(r/r_o)^6. \tag{4}$$

This potential is close to a square well, providing a tubular trapping volume with a low potential near the center and a rapidly increasing potential wall near the poles of the octopole.

An ion formed with kinetic energy acquired in an ion-molecule collision process in an RFOIGGC can be trapped radially and guided to the detector with nearly 100% collection efficiency when appropriate values of V_o and ω are applied to the RF octopole ion guide. This is the principle for absolute total cross section measurements of ion-molecule reactions using the (reactant) QMS-(upper) RFOIGGC-(product) QMS arrangement when reactant ions are prepared by VUV or electron impact ionization.

2.2 Preparation of $O^+(^4S, ^2D, ^2P)$

In this experiment,[32] He^+ (Ne^+, Ar^+) ions formed by electron impact ionization are extracted and mass-selected by the reactant QMS before reacting with O_2 in the lower RFOIGGC. Stated-selected $O^+(^4S, ^2D, ^2P)$ reactant ions prepared by processes (1)-(3) in the lower RFOIGGC (using the procedures described below) are mass-selected by the middle QMS and are further guided into the upper RFOIGGC to react with neutral reactants of interest. The intensities of the reactant O^+ and the product ions formed in the upper RFOIGGC are measured using the product QMS and the ion detector. Thus, the measurement of absolute total cross section measurements of ion-molecule reactions involving $O^+(^4S, ^2D, ^2P)$ prepared by reactions (1)-(3) fully utilize the QMS-RFOIGGC-QMS-RFOIGGC-QMS arrangement.

The principle utilized here to form an O^+ ion beam in the nearly pure $O^+(^2P)$ [or $O^+(^2D)$] state is based on the fact that $O^+(^2P)$ [$O^+(^2D)$] ions formed by reaction (1) [reaction (2)] in the lower RFOIGGC possess lower kinetic energies than those for $O^+(^2D, ^4S)$ [$O^+(^4S)$] ions. By choosing appropriate V_o and ω values applied to the lower RF octopole ion guide, $O^+(^2D, ^4S)$ [$O^+(^4S)$] formed with higher kinetic energies can be released, whereas the majority of $O^+(^2P)$ [$O^+(^2D)$] with lower kinetic energies are collected and guided towards the middle QMS. We note that in using this scheme a finite fraction of unwanted $O^+(^2D, {}^4S)$ $[O^+(^4S)]$ ions originally scattered along the lower RF octopole ion guide come through the ion guide together with the trapped $O^+(^2P)$ $[O^+(^2D)]$ ions. However, assuming that the O^+ product ions are scattered isotropically, the number of unwanted $O^+(^2D, {}^4S)$ $[O^+(^4S)]$ ions is determined by the solid angle sustained by the exit of the octopole. For the lower RF octopole (r_o = 0.5 cm, length of rods = 18 cm) used in this experiment, a fraction of ≤ 0.16% of the total $O^+(^2D, {}^4S)$ or $[O^+(^4S)]$ ions originally produced by reaction (1) [reaction (2)] is expected to be transmitted as impurity reactant ions.

We have characterized the $O^+(^2D)$ and $O^+(^2P)$ prepared using reactions (1) and (2) under the conditions described above.[32] In a recent study of the ion-molecule reaction of $O^+(^4S) + CO_2$, we found that the experimental onset or appearance energy (AE) for the Reaction (5) is in accord with its thermochemical threshold of 5.93 eV.

$$O^+(^4S) + CO_2 \rightarrow CO^+ + O + O \tag{5}$$

For reactions (6) and (7), involving $O^+(^2D)$ and $O^+(^2P)$, the experimental onsets should shift lower to corresponding thresholds of 2.63 and 0.91 eV.

$$O^+(^2D) + CO_2 \rightarrow CO^+ + O + O \tag{6}$$

$$O^+(^2P) + CO_2 \rightarrow CO^+ + O + O \tag{7}$$

The absolute cross sections for Reactions (5)-(7) in the center of mass kinetic energy (E_{cm}) range 1-30 eV characteristic of $O^+(^4S)$, $O^+(^2D)$, and $O^+(^2P)$ prepared by the schemes described above are depicted in Fig. 2. These cross sections rise from their onsets and reach a constant value of 0.4 Å^2 for $O^+(^4S)$, 2 Å^2 for $O^+(^2D)$, and 4.7 Å^2 for $O^+(^2P)$. The CO^+ absolute total cross sections observed for $O^+(^4S)$ of Fig. 2 agree with those obtained using $O^+(^4S)$ reactant ions formed by dissociative photoionization of O_2.

As expected, the experimental onsets for reactions (6) and (7) are in good agreement with the respective thermochemical thresholds, assuring that the O^+ ion beams prepared in these experiments using reactions (1) and (2) consist of predominantly $O^+(^2P)$ and $O^+(^2D)$, respectively. On energetic grounds, $O^+(^2P)$ ions cannot be formed by reaction (2) at E_{cm} = 0.73±0.06 eV. However, the $\sigma(CO^+)$ values observed for $O^+(^2D)$ cannot be used to rule out the existence of a finite number of $O^+(^4S)$ ions. Similarly, based upon the $\sigma(CO^+)$ values for $O^+(^2P)$ shown in Fig. 2, we cannot exclude the existence of a finite number of $O^+(^4S, {}^2D)$ ions.

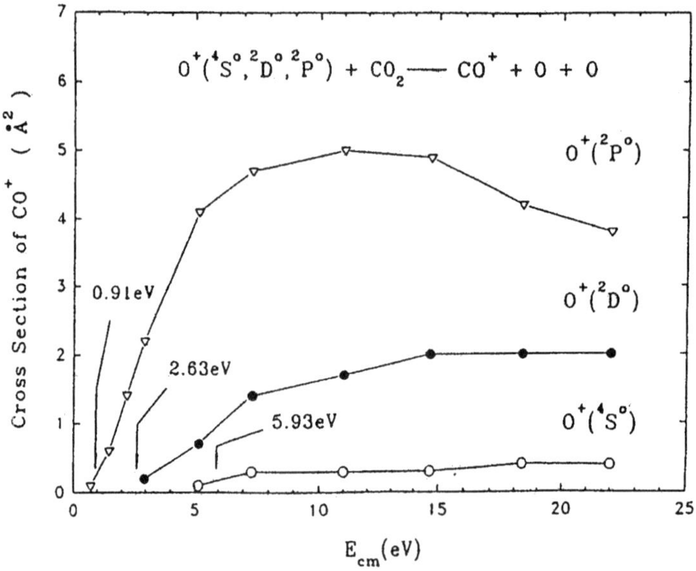

Figure 2. Absolute total cross sections for Reactions (5)-(7) in the E_{cm} range of ≈1-30 eV. The $O^+(^4S^o)$ (•), $O^+(^2D^o)$ (○), and $O^+(^2P^o)$ (∇) are prepared by reactions (1)-(3).[32]

2.3 Differential Retarding Potential Method

Low energy ion-molecule reaction cross sections are most relevant to modeling the ion chemistry in planetary atmospheres. The accuracy of experimental low energy cross sections depends in part on the laboratory kinetic energy spread (ΔE_{lab}) of the reactant ion beam. For O^+ reactant ions prepared by reactions (1)-(3) in the lower RFOPIGGC, the ΔE_{lab} value of the O^+ is determined mainly by the exothermicity of the ion-molecule processes involved. It is well known that in passing through a QMS, the radial kinetic energies of reactant ions may be altered, resulting in a broader ion kinetic energy distribution. The ΔE_{lab} value is proportional to the RF amplitude supplied to the QMS.

To overcome these difficulties, we have developed a simple differential retarding potential (DRP) method,[37] which can be used to reduce the ΔE_{lab} of reactant ions prepared by an ion-molecule reaction ion source. The principle of this method is similar to the retarding potential difference method introduced previously by Fox et al.[38] to obtain a higher electron energy resolution of an electron beam for electron impact ionization studies. In essence, the DRP achieves a narrower kinetic energy distribution by taking the difference between a non-retarded and a retarded kinetic energy distribution of the reactant ions. To implement the DRP method in the application of the TQDO apparatus, two electrostatic lenses situated between the middle QMS (10) and the upper RF octopole ion guide (11) are used as the retarding lenses. We have shown that the ΔE_{lab} of the reactant ion beam can be reduced from 0.5 eV to 2 eV using this method.

3. ABSOLUTE TOTAL CROSS SECTION MEASUREMENTS

Using the experimental method described above, we have conducted detailed absolute total cross section measurements for the reactions $O^+(^4S, ^2D, ^2P) + N_2 (O_2, H_2, D_2, H_2O)$. Selected results are briefly discussed below.

3.1 $O^+(^4S, ^2D, ^2P) + H_2 (D_2)$

Here, we present the results for $O^+(^4S, ^2D, ^2P) + H_2$.[39,40] The discussion also applies to $O^+(^4S, ^2D, ^2P) + D_2$.[40] For $O^+(^4S) + H_2$, the cross section for OH^+ formation is known to be dominant at low collision energies. Fig. 3(a) compares the $\sigma(OH^+)$'s formed by $O^+(^4S, ^2D, ^2P) + H_2$ in the E_{cm} range of 0.02-7 eV. The trend and magnitude for the $\sigma(OH^+)$'s roughly follow the prediction by the Langevin-Gioumousis-Stevenson (LGS) model[41] (see dashed line in Fig. 3). The drop of $\sigma(OH^+)$'s at higher $E_{c.m.}$'s is believed to arise from the further dissociation of excited product OH^+. The lower $\sigma(OH^+)$ from $O^+(^2P) + H_2$ is due to the significantly higher $\sigma(H_2^+)$ and $\sigma(H^+)$ from $O^+(^2P) + H_2$ as shown in Fig. 3(b). As expected, the $\sigma(H_2^+)$ and $\sigma(H^+)$ from $O^+(^4S) + H_2$ are significantly lower than those from $O^+(^2D, ^2P) + H_2$. The AE's for H_2^+ (1.70±0.10 eV) and H^+(4.60±0.10 eV) from $O^+(^4S)$ are found to be in excellent agreement with their thermochemical thresholds. The formation of H_2^+ and H^+ from $O^+(^2D, ^2P) + H_2$ is exothermic. In the $E_{c.m.}$ range of 0.04-5.0 eV, the $\sigma(H_2^+)$ and $\sigma(H^+)$ from $O^+(^2P) + H_2$ are in the range of 16-19 Å^2 and 2-8 Å^2, respectively.

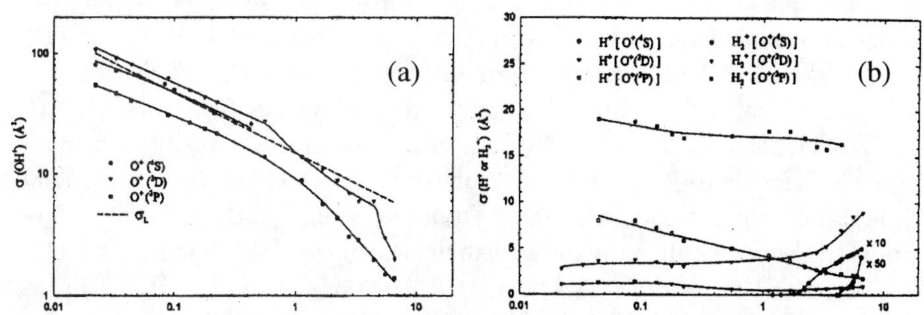

Figure 3. (a) Absolute total cross sections for OH^+ from $O^+(^4S) + H_2$ (solid circles), $O^+(^2D) + H_2$ (solid triangles), and $O^+(^2P) + H_2$ (solid squares). The LGS cross sections are given by the dashed line. (b) Cross sections for H^+ and H_2^+ from the reactions of $O^+(^4S, ^2D, ^2P) + H_2$.[40]

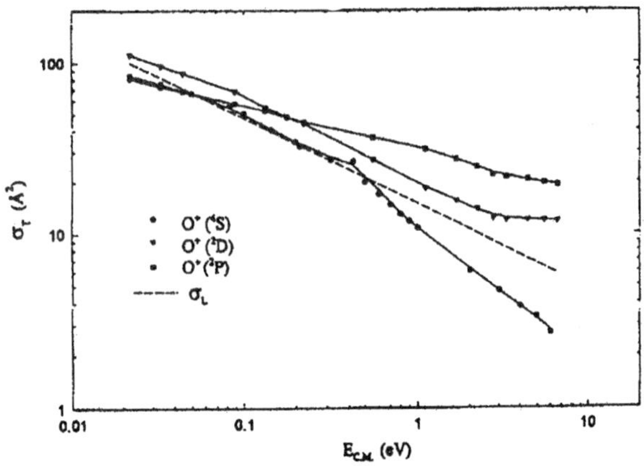

Figure 4. Comparison of σ_T's for the reactions $O^+(^4S) + H_2$ (solid circles), $O^+(^2D) + H_2$ (solid triangles), and $O^+(^2P) + H_2$ (solid squares).[40]

Fig. 4 depicts the sum (σ_T) of the absolute cross sections for all product ions (OH^+, H_2^+, and H^+) in the $E_{c.m.}$ range of 0.02-7 eV for the individual reactions $O^+(^4S, {}^2D, {}^2P) + H_2$.[40] It is interesting to note that the σ_T values for the three $O^+(^4S, {}^2D, {}^2P)$ states at $E_{c.m.} < 4$ eV agree within 25%. The higher σ_T values for $O^+(^2D, {}^2P) + H_2$ at $E_{c.m.} > 4$ eV, compared to those for $O^+(^4S) + H_2$, can be attributed to the greater number of open channels for the $O^+(^2D, {}^2P) + H_2$ reactions.

3.2 $O^+(^4S, {}^2D, {}^2P) + H_2O$

Fig. 5(a) compares the $\sigma(H_2O^+)$ from the reactions of $O^+(^4S, {}^2D, {}^2P) + H_2O$, in the $E_{c.m.}$ range of 0.15-50 eV.[42,43] The charge transfer cross sections for these reactions are found to be ≈100 Å^2 at $E_{c.m.} = 0.15$ eV and to decrease rapidly as $E_{c.m.}$ is increased. This observation is consistent with the conclusion that the ion-molecule reactions of $O^+(^4S, {}^2D, {}^2P) + H_2O$ take place via a long-lived complex mechanism at low $E_{c.m.}$'s. At $E_{c.m.} < 0.5$ eV, the $\sigma(H_2O^+)$ from $O^+(^2D) + H_2O$ are about 10% higher than those from $O^+(^4S) + H_2O$. This difference increases to ≈60-100% in the $E_{c.m.}$ range of 1-30 eV. The higher $\sigma(H_2O^+)$ for $O^+(^2D)$ compared to those for $O^+(^4S)$ is due to the greater number of available product channels for $O^+(^2D) + H_2O$ than for $O^+(^4S) + H_2O$. While the $\sigma(H_2O^+)$ from $O^+(^2P)$ are comparable to those from $O^+(^4S)$ at $E_{c.m.} \geq 1$ eV, the $\sigma(H_2O^+)$ from $O^+(^2P)$ are significantly lower than those from $O^+(^4S, {}^2D)$ at $E_{c.m.} < 1$ eV.

The $\sigma(OH^+)$ from $O^+(^4S) + H_2O$, $O^+(^2D) + H_2O$, and $O^+(^2P) + H_2O$ are compared in Fig. 5(b).[42,43] As shown in the Figure, the $\sigma(OH^+)$ from both $O^+(^2D)$ and $O^+(^2P)$ are significantly higher than those from $O^+(^4S)$. This observation indicates that the formation of OH^+ from $O^+(^2D, {}^2P) + H_2O$ should be included in the analysis of ion density data observed in space shuttle surroundings in the ionosphere. The $E_{c.m.}$ dependencies of the OH^+ cross

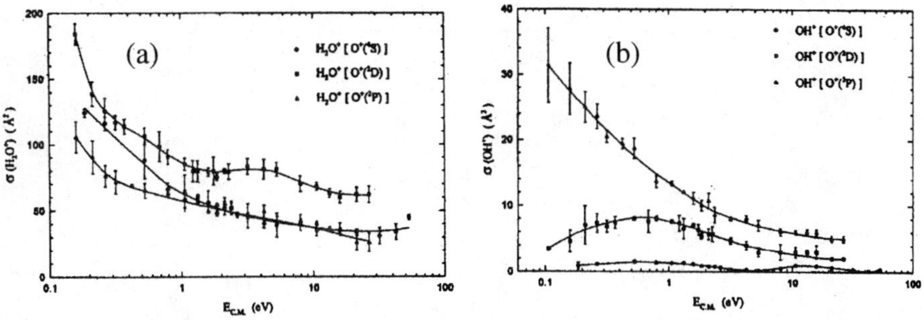

Figure 5. (a) Comparison of the $\sigma(H_2O^+)$ from $O^+(^4S) + H_2O$ (solid circles), $O^+(^2D) + H_2O$ (solid squares), and $O^+(^2P) + H_2O$ (solid triangles). (b) Comparison of the $\sigma(OH^+)$ from $O^+(^4S) + H_2O$ (solid circles), $O^+(^2D) + H_2O$ (solid squares), and $O^+(^2P) + H_2O$ (solid triangles).[42,43]

section curves of the three O^+ reactant states are different. The OH^+ cross section curve associated with $O^+(^2D)$ gradually rises from $E_{c.m.} = 0.1$ eV and shows a broad peak at ≈ 0.8 eV, only to decrease slowly towards higher $E_{c.m.}$'s. Since significant intensities of OH^+ are observed at $E_{c.m.}$ below the thermochemical threshold for $OH^+ + O + H$, we conclude that the exothermic channel $OH^+ + OH$ also plays an important role in the production of OH^+. The decrease in the $\sigma(OH^+)$ at higher $E_{c.m.}$ (>1.74 eV) may be accounted for in part by the further dissociation of internally excited OH^+ into $O^+ + H$.

The $\sigma(OH^+)$ from $O^+(^2P) + H_2O$ are relatively high, reaching a value >30 Å^2 in the thermal energy range. We note that the production of $OH^+ + OH$ and that of $OH^+ + O + H$ from $O^+(^2P) + H_2O$ are both exothermic. The $E_{c.m.}$ dependence for the OH^+ and H_2O^+ cross section from $O^+(^2P) + H_2O$ are similar, and characteristic of an exothermic process occurring via a long-lived complex mechanism. The competition of the $OH^+ + OH$ product channel may also account for the low $\sigma(H_2O^+)$ from $O^+(^2P) + H_2O$.

Since H_2O has a large dipole moment, the two leading terms of the long range potential for the O^+-H_2O system are

$$V(r) = -\alpha q^2/(2r^4) - (\mu q/r^2)\cos\theta, \tag{8}$$

where α is the isotropic dipole polarizability of H_2O, q is the charge of O^+, and θ is the angle between the dipole axis and the line connecting the centers of gravity of the two colliding partners.[41] The first term is the ion-induced dipole potential and the second term is the ion-dipole potential. The capture cross sections corresponding to the ion-induced dipole and ion-dipole potentials are $\pi q(2\alpha/E_{c.m.})^{1/2}$ and $\pi q(\mu/E_{c.m.})\cos\theta$, respectively. For large impact parameter interactions, the ion-dipole potentials dominate and the capture cross section is proportional to $1/E_{c.m.}$. This condition is expected at sufficiently low $E_{c.m.}$'s, where the ion-dipole interaction plays a dominant role in the formation of long-lived collision complexes and the dipole is oriented to maximize the ion-dipole interaction during the collision. Fig. 6 compares the plots of σ_T versus $1/E_{c.m.}$

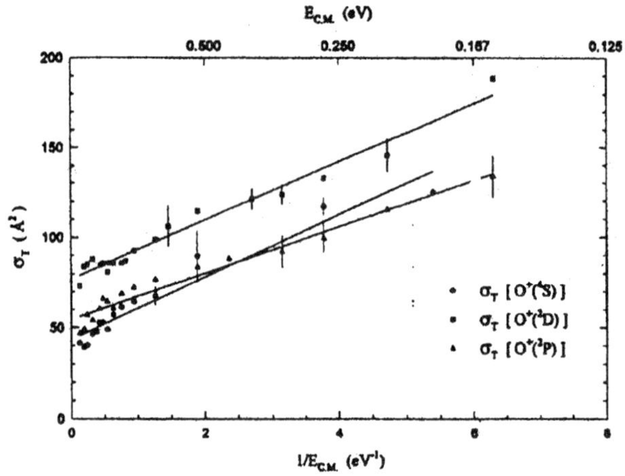

Figure 6. Plots of σ_T versus $1/E_{c.m.}$ in the $E_{c.m.}$ range of 0.15-10 eV. $O^+(^4S) + H_2O$ (solid circles); $O^+(^2D) + H_2O$ (solid squares); and $O^+(^2P) + H_2O$ (solid triangles).[43]

for $O^+(^4S) + H_2O$, $O^+(^2D) + H_2O$, and $O^+(^2P) + H_2O$ in the $E_{c.m.}$ range of 0.15-10 eV. The solid lines are the linear least square fits to the data sets. As shown in the figure, these data sets conform reasonably well with the linear fits after taking into account the experimental uncertainties. However, we note that the fit to the cross section data at $E_{c.m.} > 1$ eV is likely fortuitous because the collision times at these $E_{c.m.}$'s are too short for the molecule to reorient during the collision.

3.3 $O^+(^4S, ^2D, ^2P) + N_2$

We have measured the absolute cross sections for N_2^+, NO^+, and N^+ from the reactions $O^+(^4S, ^2D, ^2P) + N_2$.[15,44] The cross sections for N_2^+, NO^+, and N^+ from these reactions in the $E_{c.m.}$ range of 0.06-40 eV are compared in Figs. 7(a), 7(b), and 7(c), respectively.[44] As shown in Fig. 7(a), it is obvious that the charge transfer efficiencies of $O^+(^2P)$ and $O^+(^2D)$ to N_2 are significantly greater than that for $O^+(^4S)$ to N_2. The ratio of the $\sigma(N_2^+)$ for $O^+(^2D) + N_2$ to that for $O^+(^2P) + N_2$ varies from ≈0.75 at $E_{c.m.} \approx 0.1$ eV to 0.5 at $E_{c.m.} = 40$ eV. Figure 7(b) shows that the profiles for the $\sigma(NO^+)$ from $O^+(^4S) + N_2$ and $O^+(^2D) + N_2$ are similar except that the profile from $O^+(^2D) + N_2$ is shifted to lower $E_{c.m.}$ by ≈ 3 eV. It is interesting to note that the energy level of $O^+(^2D)$ is 3.3 eV higher than that of $O^+(^4S)$. The profile for the $\sigma(NO^+)$ from $O^+(^2P)$ is different from those for $O^+(^4S, ^2D)$, and reveals a decreasing trend as $E_{c.m.}$ is increased. The $\sigma(NO^+)$ from $O^+(^4S, ^2D, ^2P) + N_2$ become negligible at $E_{c.m.} > 25$ eV. The $\sigma(N^+)$ for $O^+(^4S) + N_2$, $O^+(^2D) + N_2$, and $O^+(^2P) + N_2$ shown in Fig. 7(c) all exhibit an increasing trend as $E_{c.m.}$ increases. The profiles for the $\sigma(N^+)$ curves associated with the excited states are shifted to lower $E_{c.m.}$'s with respect to that for the ground $O^+(^4S)$ state. The $\sigma(N^+)$ curves for $O^+(^4S)$ +

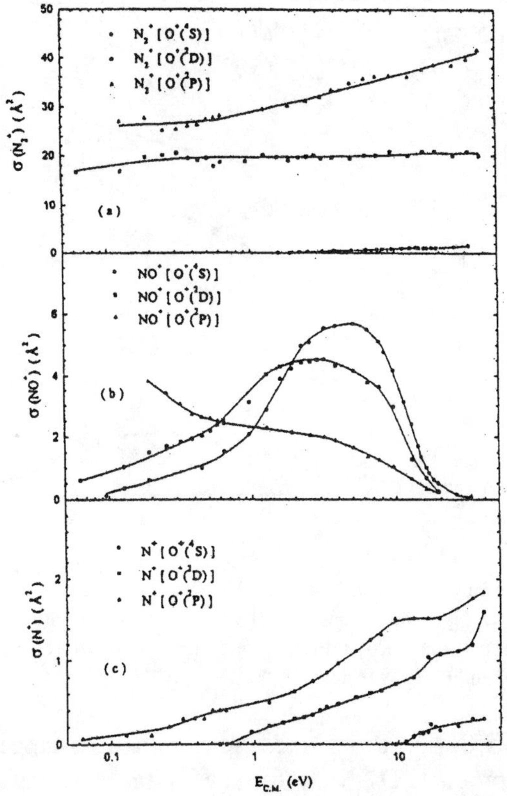

Figure 7. Comparison of (a) N_2^+, (b) NO^+, and (c) N^+ cross sections from reactions involving $O^+(^4S, ^2D, ^2P)$.[44]

N_2 and $O^+(^2D) + N_2$ reveal distinct thresholds for the $N^+ + N + O$ and $N^+ + NO$ thresholds, respectively.

Fig. 8 compares the σ_T, i.e., $\sigma(N_2^+ + NO^+ + N^+)$, from $O^+(^4S) + N_2$, $O^+(^2D) + N_2$, and $O^+(^2P) + N_2$.[44] The total rate constants (k_T's) for these reactions calculated using the equation $(2E_{c.m.}/\mu)^{1/2}$ (where μ is the reduced mass) are also depicted in Fig. 8. This comparison shows that the maximum σ_T values for $O^+(^2D) + N_2$ and $O^+(^2P) + N_2$ are relatively insensitive to $E_{c.m.}$. While the maximum σ_T values for $O^+(^4S) + N_2$ are only about 30% of the σ_T values for $O^+(^2D) + N_2$, the σ_T values for $O^+(^2P) + N_2$ are 50-100% greater than those for $O^+(^2D) + N_2$. The higher cross sections observed for reactions involving the excited states can be understood by the increased number of open product channels.

3.4 $O^+(^4S, ^2D, ^2P) + O_2$

We have also obtained absolute total state-selected charge transfer cross

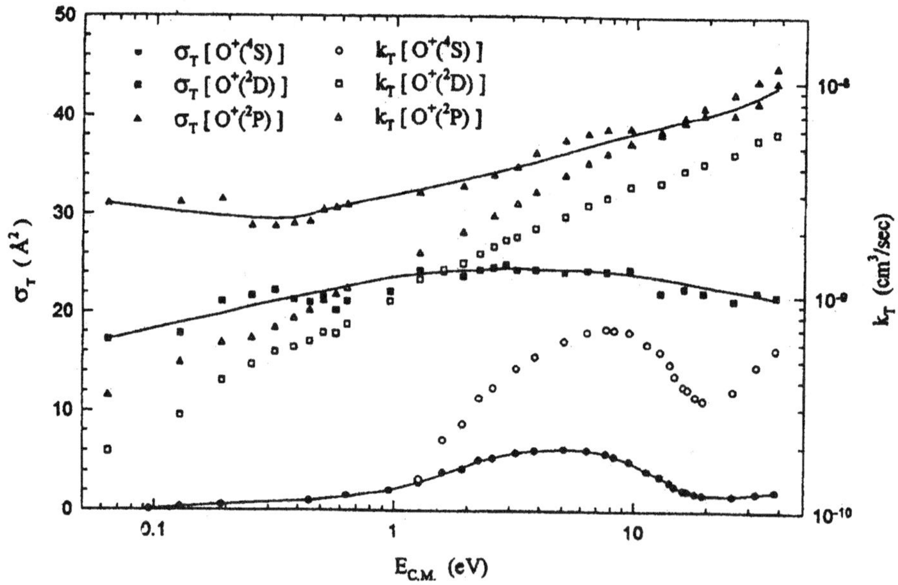

Figure 8. Comparison of σ_T's for the reactions $O^+(^4S) + N_2$ (solid circles), $O^+(^2D) + N_2$ (solid squares), and $O^+(^2P) + N_2$ (solid triangles). Here $\sigma_T = \sigma(N_2^+ + NO^+ + N^+)$. The total rate constants (k_T's) corresponding to σ_T's are also included. (hollow circles): $k_T[O^+(^4S)]$; (hollow squares): $k_T[O^+(^2D)]$; and (hollow triangles): $k_T[O^+(^2P)]$.[44]

sections for the $O^+(^4S, ^2D, ^2P) + O_2$ reactions in the $E_{c.m.}$ range of 0.05-40 eV.[45] Charge transfer product O_2^+ ions formed in the reaction of $O^+ + O_2$ are known to undergo rapid dissociative recombination reactions with electrons, giving rise to excited oxygen atoms.[46] The emission from $O(^1S)$ atoms thus formed is the source of the 5577 Å green line,[47] which is a prominent spectral feature in the night sky and is extensively used by aeronomers for probing the chemistry and dynamics of the ionosphere. The absolute total cross sections for O_2^+ from $O^+(^4S) + O_2$ in the $E_{c.m.}$ range of 0.05-3 eV obtained in our experiment are in the range of 0.5-4.5 $Å^2$ and are consistent with previous reaction rate measurements.[45] The charge transfer cross sections for $O^+(^2P) + O_2$ are found to be ≈3 fold greater than those for $O^+(^2D) + O_2$, which in turn are ≈2-20 fold greater than those for $O^+(^4S) + O_2$.[45] Since $O^+(^4S)$, $O^+(^2D)$, and $O^+(^2P)$ are produced nearly in equal intensities by solar UV dissociation and photoionization in the ionosphere, this result indicates that O_2^+ formed in the collisions of $O^+(^2D, ^2P) + O_2$ is overwhelmingly the main source of charge transfer product O_2^+ ions.

4. CONCLUSIONS

We have developed a novel method for the preparation of $O^+(^2D)$ and $O^+(^2P)$ with high purity. This method has made possible, for the first time, the cross section measurements of a series of ion-molecule reactions involving state selected $O^+(^2D, ^2P)$ ions. As expected, the cross sections observed in this study indicate that the reactions involving the excited $O^+(^2D, ^2P)$ ions are comparable or higher than those for the corresponding $O^+(^4S)$ reaction and thus the $O^+(^2D, ^2P)$ reactions cannot be ignored in a realistic modeling of the chemistry in planetary ionospheres.

ACKNOWLEDGEMENTS

The author is indebted to postdoctoral fellows and students, X. Li, Y.-L. Huang, and G. D. Flesch, who performed the original experiments described in this article. This work was supported by the National Science Foundation Grants ATM-9200785, ATM-9521558, and ATM-001646.

REFERENCES

1 *State-Selected and State-to-State Ion-Molecule Reaction Dynamics I: Experiment*, ed. by C. Y. Ng and M. Baer, Adv. Chem. Phys., **82** (Wiley, New York, 1992).
2 *State-Selected and State-to-State Ion-Molecule Reaction Dynamics II: Theory*, ed. by C. Y. Ng and M. Baer, Adv. Chem. Phys., **82** (Wiley, New York, 1992).
3 *Unimolecular and Bimolecular Ion-Molecule Reaction Dynamics*, ed. by C. Y. Ng, T. Baer and I. Powis (Wiley, Chichester, 1994).
4 P. B. Armentrout, in *Gas Phase Inorganic Chemistry*, ed. by D. H. Russell (Plenum, New York, 1989), p. 1.
5 P. M. Banks and G. Kockarts, *Aeronomy*, Part A (Academic, New York, 1973), pp. 14, 257.
6 M. R. Torr and D. G. Torr, Rev. Geophys. Space Phys. **20**, 91 (1982).
7 A. Dalgarno and M. B. McElroy, Planet. Space Sci. **13**, 947 (1965).
8 R. J. Henry, Ap. J. **161**, 1153 (1970).
9 J. L. Kohl, G. P. Lafyatis, H. P. Palenius, and W. H. Parkinson, Phys. Rev. A **18**, 571 (1978).
10 D. G. Torr and N. Orsini, Planet. Space Sci. **25**, 1171 (1977).
11 D. G. Torr and N. Orsini, Geophys. Res. Lett. **5**, 657 (1980).
12 M. R. Torr and D. G. Torr, Geophys. Res. Lett. **7**, 103 (1980).
13 E. E. Ferguson, F. C. Fehsenfeld, and D. L. Albritton, in *Gas Phase Ion Chemistry*, M. T. Bowers, ed., (Academic Press, New York, 1979), Vol. 1, p. 45.
14 A. Dalgarno and J. L. Fox, in *Unimolecular and Bimolecular Ion-Molecule Reaction Dynamics*, ed. by C. Y. Ng, T. Baer, and I. Powis (Wiley, Chichester, 1994) p. 1.
15 G. D. Flesch and C. Y. Ng, J. Chem. Phys. **92**, 3235 (1990); *ibid.* **94**, 2372 (1991) and references therein.
16 H. M. Rosenstock, K. Draxl, B. W. Steiner, and J. T. Herron, J. Phys. and Chem. Ref. Data **6**, Suppl. 1 (1977).
17 A. Dalgarno, Adv. At. Mol. Phys. **15**, 37 (1979).
18 D. R. Bates and H. S. W. Massey, Proc. R. Soc. London Ser. A **192**, 1 (1947).
19 D. G. Torr and N. Orsini, Planet. Space Sci. **25**, 1171 (1977).

20 D. G. Torr and N. Orsini, Geophys. Res. Lett. **5**, 657 (1978).
21 M. R. Torr and D. G. Torr, Geophys. Res. Lett. **7**, 103 (1980).
22 M. Mendillo, G. S. Hawkins, and J. A. Klobuchar, J. Geophys. Res. **80**, 2217 (1975).
23 W. D. Watson, Accounts Chem. Res. **10**, 221 (1977).
24 W. W. Duley and D. A. Williams, *Interstellar Chemistry* (Academic Press, New York, 1984).
25 R. A. Dressler and E. Murad, in *Unimolecular and Bimolecular Ion-Molecule Reaction Dynamics*, ed. by C. Y. Ng, T. Baer, and I. Powis, (Wiley, Chichester, 1994), Chapter 2, p. 87.
26 J. M. Grebowsky, H. A. Taylor, Jr., M. W. Pharo III, and N. Reese, Planet. Space Sci. **35**, 501 (1987); ibid. **35**, 1463 (1987).
27 D. E. Hastings, N. A. Gatsonis, and T. Mogstad, J. Geophys. Res. **A93**, 1961 (1988).
28 R. A. Dressler, J. A. Gardner, D. L. Cooke, and E. Murad, J. Geophys. Res. **96**, 13, 795 (1991).
29 E. Murad, Planet. Space Sci. **33**, 421 (1985).
30 G. E. Caledonia, J. C. Person, and D. E. Hastings, J. Geophys. Res. **92**, 273 (1987).
31 D. E. Hunton and J. M. Calo, Planet. Space Sci. **33**, 945 (1985).
32 X. Li, Y.-L. Huang, G. D. Flesch and C. Y. Ng, Rev. Sci. Instrum. **66**, 2871(1995)
33 D. Gerlich, in *"State-Selected and State-to-State Ion-Molecule Reaction Dynamics: I. Experiment"*, ed. by C. Y. Ng and M. Baer, *Adv. Chem. Phys.* **82** (Wiley, New York, 1992), p. 1.
34 G. Bischof and F. Linder, Z. Phys. **D1**, 303 (1986).
35 G. D. Flesch, S. Nourbakhsh, and C. Y. Ng, J. Chem. Phys. **92**, 3590 (1990).
36 C. Y. Ng in *State-Selected and State-to-State Ion-Molecule Reaction Dynamics I: Experiment*, ed. by C. Y. Ng and M. Baer, Adv. Chem. Phys. **82** (Wiley, New York, 1992). p.401-500.
37 X. Li, Y.-L. Huang, G. D. Flesch and C. Y. Ng, Rev. Sci. Instrum. **65**, 3724 (1994).
38 R. E. Fox, W. H. Hickam, D. J. Grove, and T. Kjeldaas, Jr., Rev. Sci. Instrum. **26**, 1101 (1955).
39 G. D. Flesch and C. Y. Ng, J. Chem. Phys. **94**, 2372 (1991).
40 X. Li, Y.-L. Huang, G. D. Flesch, and C. Y. Ng, J. Chem. Phys. **106**, 564 (1997).
41 G. Gioumousis and D. P. Stevenson, J. Chem. Phys. **29**, 294 (1958).
42 X. Li, Y.-L. Huang, G. D. Flesch, and C. Y. Ng, J. Chem. Phys. **102**, 5100 (1995).
43 X. Li, Y.-L. Huang, G. D. Flesch, and C. Y. Ng, J. Chem. Phys. **106**, 928 (1997).
44 X. Li, Y.-L. Huang, G. D. Flesch, and C. Y. Ng, J. Chem. Phys. **106**, 1373 (1997).
45 X. Li, Y.-L. Huang, G. D. Flesch, and C. Y. Ng, to be published.
46 S. L. Guberman, Science **278**, 1276 (1997); D. Kella, L. Vejby-Christensen, P. J. Johnson, H. B. Pedersen, and L. H. Anderson, Science **276**, 1530 (1997); ibid. **277**, 167 (1997).
47 W. W. Campbell, Ap. J. **2**, 162 (1895).

Calculation of Electron Affinities
A Roadmap

Thom H. Dunning, Jr.,[1] Kirk A. Peterson,[2] Tanja Van Mourik[3]
[1]North Carolina Supercomputing Center, 3021 Cornwallis Road, Research Triangle Park, NC 27709 and Department of Chemistry, University of North Carolina, Chapel Hill, NC 27799, Current Address: Oak Ridge National Laboratory, MS-6414, Post Office Box 2008, Oak Ridge, TN 37831; [2]Environmental Molecular Sciences Laboratory, Pacific Northwest National Laboratory, Richland, WA 99352 and Department of Chemistry, Washington State University, Richland, WA 99352; [3]Chemistry Department, University College London, 20 Gordon Street, London WC1H 0AJ, United Kingdom
dunningthjr@ornl.gov

1. INTRODUCTION

Molecular anions are important in many gas- and solution-phase chemical processes; examples include the classic Walden inversion reaction in organic chemistry and the complex set of anion-molecule reactions that establish the equilibrium concentrations of negatively-charged species in the upper atmosphere. Molecular anions also play an important role in the scattering or capture of electrons by molecules, including dissociative attachment processes—a subject of this symposium. However, little definitive experimental information exists on the structure and properties of molecular anions, other than electron affinities, which can be accurately measured by photodetachment spectroscopy.[1]

Information on anions can, of course, be obtained from solution of the Schrödinger equation. However, accurate calculation of the properties of molecular anions, especially electron affinities, is very demanding. Although the "extra" electron in an anion may be bound by as much as 4 eV, *e.g.*, CN^- (3.86 eV) and C_2^- (3.27 eV), for many anions the additional electron is bound by just a few tenths or hundredths of an eV, *e.g.*, O_2^- (0.45 eV) and NO^- (0.026 eV). The presence of a weakly bound electron means that the electronic structure of anions differs substantially from that of the corresponding neutral molecule. This difference places a substantial burden on the techniques used to calculate the energies, structures and properties of molecular anions.

In this paper we will examine the calculation of electron affinities with the intent of providing a clear roadmap for the reliable determination of electron affinities. This approach should provide a general guide for computational studies of molecular anions, for calculations which yield accurate values of the electron affinity are also expected to yield accurate values for other anion properties.

2. COMPUTATIONAL METHODS

Although the Hartree-Fock (HF) method provides a useful, qualitative description of the electronic structure of anions, accurate predictions of electron affinities and anion properties cannot be made without explicit inclusion of the effects of electron correlation. For example, HF calculations on O_2^- and CN^- with basis sets near the complete basis set limit predict electron affinities (EA_0) of –0.65 and 3.27 eV. This is to be compared to measured values of 0.45[2] and 3.86[3] eV. Thus, for O_2^-, HF calculations do not even predict the anion to be stable relative to $O_2 + e$.

If both the neutral and anion are well described in zero order by the HF wave function, the effects of electron correlation can be included in the calculations using configuration interaction[4] (CISD, CISDT, ...), perturbation theory[5] (MP2, MP3, MP4, ...), or coupled cluster theory[6] (CCSD, CCSDT,...).

Table 1. General characteristics of configuration interaction, perturbation theory, and coupled cluster methods. N is the number of basis functions; m is the number of iterations.

Method	Computational Cost	Size Extensive?
Configuration Interaction		
CISD	mN^6	No
Perturbation Theory		
MP2	N^5	Yes
MP3	N^6	Yes
MP4	N^7	Yes
Coupled Cluster		
CCSD	mN^6	Yes
CCSD(T)	$mN^6 + N^7$	Yes
CCSDT	mN^7	Yes

The computational characteristics of these three methods are summarized in Table 1. All three techniques have found, and continue to find, use in computational studies of the electronic structure of molecules.

If the zero-order wave functions are not adequately described by a single HF wave function, multi-reference wave functions must be used. MRCI calculations are well established,[7] but the cost of these calculations increases dramatically with the size of the molecule. Multireference perturbation theories have recently been implemented[8] but have not yet been used to compute electron affinities. Work is still underway on multireference generalizations of coupled cluster approaches.[9]

We will restrict our analysis here to techniques based on a HF reference wave function. The extension of this analysis to multireference methods is straightforward.

2.1 Configuration Interaction

Configuration interaction calculations including all single and double excitations from the HF wave function, CISD, have a computational cost that scales as mN^6, where N is the number of basis functions and m is the number of diagonalization iterations. However, its lack of size extensivity,[10] *i.e.*, its improper scaling with the number of electrons, limits its accuracy in molecular calculations. Only a full CI within a given basis set is size extensive. Size extensivity is most important when there are changes in the number of electrons, as is the case in the calculation of electron affinities (although the difference is just one).

2.2 Perturbation Theory

Although Møller-Plesset perturbation theory is size extensive, the cost of the calculation increases significantly with the order of perturbation theory, rising from N^5 for MP2 to N^7 for MP4. Although many computational studies continue to be based on perturbation theory methods, especially the relatively low cost MP2 method, recent studies have shown that the perturbation series expansion is either slowly convergent or non-convergent.[11] Convergence difficulties are not limited to molecules that are poorly described by a HF wave function but afflict all molecules, especially those with several electron pairs localized in a confined region of space. In addition, for binding energies, it has been shown that, *at the complete basis set limit*, the MP2-MP4 methods are incapable of providing chemical accuracy for a wide range of molecules.[12] We shall also find this to be true for electron affinities.

2.3 Coupled Cluster Theory

Coupled cluster theory[13] is a latecomer to the toolkit used by quantum chemists to describe the electronic structure of molecules. Nonetheless, it is increasingly popular because it provides a theoretically sound and accurate means of solving the electronic Schrödinger equation. In fact, the CCSD(T) method,[14] *i.e.*, coupled cluster theory including all single and double excitation clusters plus a perturbative treatment of connected triple excitation clusters, is able to predict many properties, including molecular binding energies, to an accuracy that rivals that of all but the most sophisticated experimental techniques. Coupled cluster methods are also size extensive and have the added appeal that they explicitly recognize the dominant role played by electron pairs in describing the electronic structure of molecules. Unlike the perturbation theory equations, the coupled cluster equations must be solved iteratively. In the CCSD method, the cost of each iteration scales as N^6, while in the CCSD(T) method a final N^7 calculation must be performed to calculate the (T) correction.

2.4 Basis Sets

Solution of the electronic Schrödinger equation usually involves an expansion of the electronic wave function in a set of basis functions. In the 1980s a thorough study of correlation effects in atoms led to the development of the correlation consistent basis sets[15]: cc-pV*n*Z (or v*n*z), where n = D(ouble) (2), T(riple) (3), Q(uadruple) (4), *etc*. These sets systematically extend the coverage of both the angular and radial coordinates in an atom and, as the cardinal number increases, approach a complete basis set (*CBS*). In fact, in many cases it is possible to extrapolate the properties computed with the cc-sets to the *CBS* limit. For example, the correlation energy is well described by a simple inverse power law:[16]

$$\mathrm{E}_{corr}(n) = \mathrm{E}_{corr}(\infty) + \frac{A}{n_{\max}^3} \tag{1}$$

where A is a constant and $n_{\max}$ is the maximum cardinal number in the basis set. The number of functions in the cc-sets increases with increasing cardinal number and is given in Table 2.

Augmented correlation consistent basis sets,[17] aug-cc-pV*n*Z (or av*n*z), have been developed to describe negative ions and other molecular systems where it is important to properly describe the long range behavior of the wave function. The singly augmented sets are obtained from the standard sets by addition of a single diffuse function to each angular symmetry (*s*, *p*, *d*, *f*, ...)

Table 2. Scaling of correlation consistent basis sets. N is the number of basis functions.

Basis Set (n)	Scaling (N)
cc-pVnZ	$\frac{1}{3}(n+1)(n+\frac{2}{3})(n+2)$
aug-cc-pVnZ	$+(n+1)^2$
d-aug-cc-pVnZ	$+2(n+1)^2$
cc-pCVnZ	$+n^2$
cc-pwCVnZ	$+n^2$

present in the standard basis set. The exponents for the additional functions are obtained from calculations on the atomic anions (or interpolated in the case of nitrogen). Additional sets of even more diffuse functions are added in the doubly and triply augmented sets (dav*n*z or tav*n*z).

Basis sets for all-electron calculations have also been developed.[18] Two types of core-valence basis sets have been put forward for all-electron calculations. The first, the cc-pCV*n*Z sets (cv*n*z), focus on recovery of all correlation effects (core, core-valence, and valence). In the second, the weighted core-valence or cc-pwCV*n*Z sets (wcv*n*z), the extra functions are concentrated in the core-valence regions. They take advantage of the fact that many molecular properties are little influenced by correlation of the core electrons. Such properties tend to converge more rapidly with *n* for the weighted core valence sets.

For a more complete discussion of correlation consistent basis sets, see Ref. 19.

2.5 Details of the Calculations

The equilibrium electron affinities were computed as the difference in the electronic equilibrium energies of the corresponding neutral and anionic molecules, i.e.,

$$EA_e = E_e(X) - E_e(X^-) \tag{2}$$

The equilibrium energies were obtained from sixth-order polynomial fits of the diatomic energies, computed at seven internuclear separations in the range $-0.3a_0 \leq (r - r_e^{expt'l}) \leq 0.5a_0$.

The "experimental" EA_e's were determined by subtracting from the measured EA_0's the differences in vibrational frequencies obtained from valence-electron CCSD(T)/av6z calculations. A standard Dunham analysis was used to derive the harmonic frequencies and anharmonicities from the calculated energies. The calculated harmonic frequencies and anharmonicities are 1604.1/1136.3 and 10.93/8.61 cm^{-1} for O_2/O_2^- and 2067.2/2065.6 and 13.37/12.21 cm^{-1} for CN/CN^-. These values compare well with the measured values for the neutral molecules of 1580.2 and 12.0 cm^{-1} for O_2[20] and 2068.6 and 13.1 for CN.[20] The "experimental" equilibrium electron affinities so obtained are:

$$EA_e(O_2) = 0.422 \pm 0.007 \text{ eV} \quad \text{and} \quad EA_e(CN) = 3.862 \pm 0.004 \text{ eV}$$

All calculations were performed using the MOLPRO program package,[21] except the restricted MPn calculations, which were done with the ACES II package.[22] The calculations were carried out at PNNL on a SGI Origin 2000 and an HP J5600 and at UCL on a 1.7 GHz PC running Linux.

The coupled cluster calculations on the $X^2\Pi$ state of O_2^- used state-averaged Hartree-Fock orbitals ($2\pi_x$, $2\pi_y$).

3. TWO CASE STUDIES: O_2^- AND CN^-

In this section we will develop guidelines for the calculation of electron affinities. We will consider two prototype anions, O_2^- and CN^-. The "extra" electron in O_2^- is only weakly bound, whereas the "extra" electron in CN^- is strongly bound. Thus, these species represent two ends of the spectrum of electron affinities and should provide a severe test of the techniques used to compute electron affinities.

3.1 Basis Sets

The first decision to be made is the selection of a basis set *family* for calculations on both the neutral and negatively charged molecules. In Fig. 1, we plot the EA_e's from CCSD(T) calculations with three families of basis sets: v*n*z, av*n*z, and dav*n*z for $n = 2$-6. The inadequacy of the standard correlation consistent sets (v*n*z) is clearly evident in the Figure. For O_2, the double and triple zeta sets predict negative electron affinities and even for the sextuple zeta set, there is a 0.07 eV difference between the EA_e computed with the v6z set and those with the av6z/dav6z sets. Similar problems are found in CN where it is seen that the avtz set (92 functions) gives a result closer to that obtained with the av6z set than does the much larger v5z set (182 functions).

For O_2, in which the "extra" electron is very weakly bound, addition of an extra set of diffuse functions to the avdz set increases the EA_e by 0.014 eV. However, the extra binding decreases with increasing *n*, being just 0.004 eV with the avqz set and 0.002 eV with the av6z set. It is always much better to use the av(*n*+1)z set than the dav*n*z set, even though the dav*n*z set has slightly

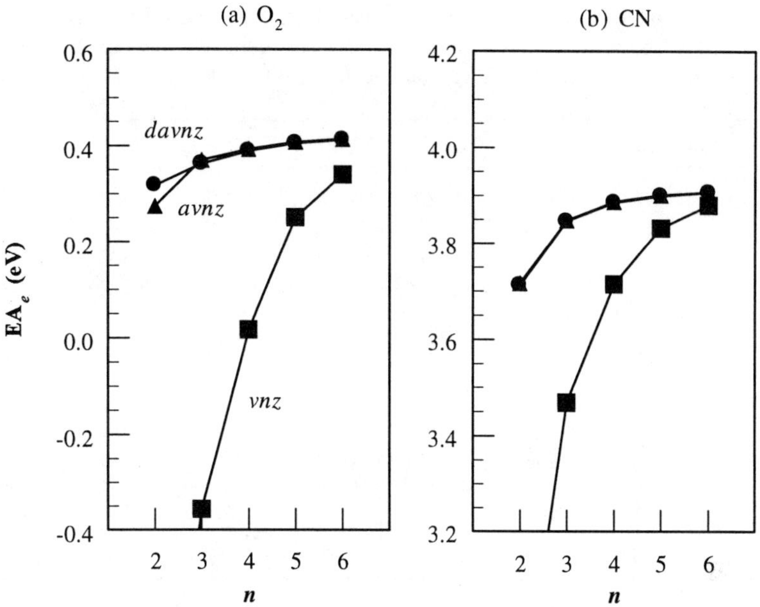

Figure 1. Calculated electron affinities, EA_e, of O_2 (a) and CN (b) using the v*n*z, av*n*z and dav*n*z basis set families.

fewer functions. For CN, use of the dav*n*z sets never has a large effect (<0.004 eV) and results in a decrease in EA_e.

3.2 Electron Correlation Methods (Valence Electron)

Having selected the augmented correlation consistent sets, av*n*z, for the calculation of EA_e's, we can now assess the suitability of the various electron correlation methods [CISD, MP2-4, CCSD, CCSD(T)] for computing electron affinities. These results are summarized in Table 3. The row labeled *CBS* is an estimate of EA_e at the *complete basis set limit*. This estimate was obtained by adding $E_{corr}(\infty)$ from Eq. (1) and the results from the largest two basis sets listed to the HF energy from the av6z set.

CISD. Although CISD calculations give a very accurate value for EA_e(CN), its performance for $EA_e(O_2)$ is abysmal. For O_2, the computed electron affinity, *at the estimated CBS limit*, is in error by 0.34 eV—an error of over 80%! Clearly, the CISD method does not provide a consistent description of the electron affinities of these two molecules.

MPn. The perturbation theory expansions have markedly different behaviors for EA_e for O_2 and CN. In O_2, all of the MPn calculations underestimate the experimentally derived EA_e even at the *CBS* limit—by 0.22 eV for MP2, 0.096 eV for MP3, and 0.044 eV for MP4. For CN, on the other hand, MP2-MP4 calculations at the *CBS* limit overestimate the experimentally-derived EA_e—by 0.20 eV for MP2, 0.24 eV for MP3, and

Table 3. Calculated electron affinities (eV), EA_e, for O_2 and CN. All calculations use the aug-cc-pV*n*Z basis sets. *CBS* refers to the estimated complete basis set limit; see the text. The derived experimental EA_e values are 0.422 (O_2) and 3.862 (CN).

Molecule	*n*	CISD	MP2	MP3	MP4	CCSD	CCSD(T)
O_2	2	-0.005	-0.017	0.156	0.227	0.252	0.317
	3	0.043	0.100	0.260	0.317	0.285	0.362
	4	0.055	0.141	0.289	0.345	0.308	0.389
	5	0.067	0.168	0.306	0.360	0.322	0.405
	6	0.072				0.328	0.412
	CBS	*0.081*	*0.199*	*0.326*	*0.378*	*0.338*	*0.423*
CN	2	3.716	3.815	3.924	3.762	3.728	3.712
	3	3.827	3.964	4.046	3.906	3.857	3.846
	4	3.858	4.012	4.078	3.946	3.894	3.885
	5	3.870	4.033	4.090	3.961	3.908	3.899
	6	3.876				3.915	3.906
	CBS	*3.884*	*4.057*	*4.104*	*3.977*	*3.924*	*3.915*

0.12 eV for MP4. For CN, even the MP4 method does not achieve chemical accuracy (<0.04 eV).

In all cases, the calculated EA_e's increase with increasing *n*. Because the MPn calculations overshoot EA_e for CN, the errors for a particular choice of method and basis set may be smaller than those at the *CBS* limit. However, it is the latter, which is referred to as the *intrinsic error* that is indicative of the true accuracy of the method.

CCn. The CCSD(T) method predicts the EA_e's of O_2 and CN within or close to chemical accuracy, namely, +0.001 and +0.053 eV, respectively. For the avdz sets, the basis set convergence errors, *i.e.*, the errors relative to the CBS limit are -0.11 and -0.20 eV for O_2 and CN. This error drops by a factor of 2-3 for the avtz set (to -0.061 and -0.069 eV, respectively) and by another factor of 2 for the avqz set (-0.034 and -0.030 eV, respectively).

The intrinsic error for the CCSD method is relatively small for CN, just +0.062 eV. However, it is larger (and of opposite sign) for O_2, -0.084 eV. It is interesting to note that the difference between the EA_e's calculated for CN with the CCSD and CCSD(T) methods converge rapidly—to within 0.009 eV for the avqz basis set. This means that it would be possible to estimate the result of a CCSD(T) calculation with a large basis set by combining the results of a CCSD calculation with that large basis set with a CCSD(T) calculation with a much smaller basis set (avtz or avqz). For a similar approach using perturbation theory to estimate the results of large basis set CCSD(T) calculations, see Ref. 23.

3.3 Higher Order Correlation Corrections

In the previous section it was found that only the coupled cluster method, specifically the CCSD(T) method, provided accurate, reliable predictions of molecular electron affinities. There are two major corrections to be considered for CCSD(T) calculations—the effect of core-valence correlation and the effect of full inclusion of the triple excitation clusters. In this section, we consider each of these corrections separately.

3.3.1 Core-Valence Correlation Effects

It is important to calculate core-valence correlation corrections using exactly the same basis set for the valence- and all-electron calculations. Otherwise, the magnitudes of the computed core-valence corrections will include contributions from the differences in the basis sets. In Table 4 we report valence- and all-electron calculations of the EA_e's of O_2 and CN using the weighted core-valence correlation consistent basis sets.

As can be seen in the table, the core-valence correlation corrections to EA_e converge rapidly with *n* for the weighted basis sets. For O_2, the value of the correction is essentially converged with the awcvdz set (-0.002 eV). The convergence is somewhat slower for CN, but even here, calculations with the awcvtz set are within 0.001 eV of the converged value (+0.009 eV).

The core-valence correction is larger for CN than for O_2 because the orbital in which the "extra" electron is placed in CN has considerable 2*s*-character. In O_2, the "extra" electron occupies a 2*p*-like orbital. It has been found that interaction between the electrons in the 1*s* core orbital and the valence orbitals is significantly enhanced by 2*s*-character in the valence orbitals.

Table 4. Core-valence correlation corrections, EA_e^{all}-$EA_e^{valence}$, to the valence electron affinities, EA_e's, of O_2 and CN. The CCSD(T)/RCCSD(T) method with augmented weighted core-valence correlation consistent basis sets (awcv*n*z) was used in all calculations. In eVs.

	$EA_e(O_2)$			$EA_e(CN)$		
	valence	all	Δ	valence	all	Δ
2	0.259	0.257	-0.002	3.726	3.730	+0.004
3	0.368	0.366	-0.002	3.851	3.859	+0.008
4	0.391	0.389	-0.002	3.887	3.896	+0.009
5	0.406	0.404	-0.002	3.900	3.909	+0.009

Table 5. Full triples correlation corrections, EA_e^{UCCSDT}-$EA_e^{UCCSD(T)}$, to the electron affinities, EA_e's, of O_2 and CN. The augmented valence correlation consistent basis sets (av*n*z) were used in all calculations. In eVs.

	$EA_e(O_2)$			$EA_e(CN)$		
	CCSD(T)	CCSDT	Δ	CCSD(T)	CCSDT	Δ
2	0.255	0.261	+0.005	3.743	3.668	-0.075
3	0.354	0.355	+0.001	3.869	3.797	-0.072
4	0.381	0.380	-0.001	3.905	3.835	-0.070

3.3.2 Full Triples Corrections

Sordo[24] has recently reported unrestricted CCSDT calculations on both O_2/O_2^- and CN/CN^- using the augmented correlation consistent basis sets and compared these results to the corresponding unrestricted CCSD(T) calculations. Although it was only possible to carry out CCSDT calculations with the avdz-avqz sets, this is sufficient to reliably estimate the correction due to the full inclusion of triple excitation clusters.

The results of the UCCSDT/UCCSD(T) calculations are summarized in Table 5. The full triples correction for O_2 decreases with increasing basis set size, decreasing the CCSD(T) value by just 0.001 eV with the avqz set. For CN the triples correction decreases the CCSD(T) values of EA_e by ~0.07 eV, clearly a significant effect. This is likely a result of the fact that CN is more poorly described by a HF wave function than is CN^-.

The relatively rapid convergence of the full triples correction with basis set is encouraging. Even for CN, for which is the correction is large, the full triples correction appears to be converged to approximately 0.002 eV for the aug-cc-pVTZ set.

3.3.3 Best Estimates of $EA_e(O_2)$ and $EA_e(CN)$

If we estimate the CCSDT all-electron complete basis set limit by adding (1) the full triples correction computed with the avqz set and (2) the core-valence correction computed with the awcvqz set to the estimated CCSD(T) *CBS* limit, we predict:

$$EA_e(O_2) \approx 0.420 \text{ eV}$$

$$EA_e(CN) \approx 3.854 \text{ eV}$$

These values, which define the estimated intrinsic errors for the all-electron CCSDT method for the electron affinities of O_2 and CN, differ from the measured values by just 0.002–0.008 eV, much better than chemical accuracy (~0.04 eV).

4. CONCLUSIONS

Although the two molecules studied in detail here, O_2^- and CN^-, are together expected to be representative of a broad range of molecular anions, the conclusions drawn must be considered tentative. More definitive insights into the accuracy and reliability of the three methods considered here (configuration interaction, perturbation theory, and coupled cluster theory) will require a sufficiently large sample of benchmark calculations to permit a statistical analysis of the predicted electron affinities.[25] Nonetheless, the current study yields important insights into the selection of basis sets and electronic structure methods for computational studies of negatively charged molecules.

Basis Sets. Benchmark calculations on O_2/O_2^- and CN/CN^- using the standard correlation consistent basis sets (v*n*z) and singly and doubly augmented basis sets (av*n*z and dav*n*z, respectively) clearly show the importance of augmenting the standard basis sets. Calculations with the v*n*z sets yield very poor results, *e.g.*, $EA_e(O_2)$ computed with the avdz set is closer to the complete basis set (*CBS*) limit than the electron affinity computed with the v5z set and a similar situation is found for EA_e(CN). On the other hand, use of the dav*n*z set offers only modest improvements in EA_e over those obtained with the av*n*z sets, despite the fact that there are $(n+1)^2$ more functions in the dav*n*z sets. As expected, EA_e(dav*n*z)-EA_e(av*n*z) decreases with increasing *n*.

Although molecular electron affinities converge slowly with basis set, *i.e.*, with *n*, a common trend for properties sensitive to electron correlation effects, calculations with the avqz basis set are converged to better than 0.03 eV.

Electronic Structure Methods. Although valence electron calculations with the CISD method predict a very accurate value of the electron affinity of CN, *at the CBS limit* $EA_e(O_2)$ is just 0.08 eV compared to the experimentally-derived value of 0.422 eV—an 80% error! Clearly, the CISD method does not provide a consistent description of the electron affinities of O_2 and CN.

The (valence electron) MPn methods underestimate the EA_e's of O_2 and overestimate the EA_e's of CN. The error in the MP2 method *at the CBS limit* ranges from -0.22 eV for O_2 to +0.20 eV for CN. The error decreases significantly for the MP4 method: to –0.04 eV for O_2 and +0.12 eV for CN. For CN, the MP4 value for EA_e is not within chemical accuracy (~0.04 eV).

Valence-electron calculations with the CCSD(T) method predict $EA_e(O_2)$

at the CBS limit to within chemical accuracy and EA_e(CN) to near chemical accuracy. All-electron calculations increase the errors slightly (by 0.009 eV for CN). It is estimated that the all-electron CCSDT method will yield EA_e's in error by just 0.002–0.008 eV.

The errors for the (valence-electron) CCSD method are over 0.04 eV for both O_2 (-0.08 eV) and CN (+0.06 eV). However, the difference in EA_e between the CCSD and CCSD(T) method converges relatively rapidly with basis set. Calculations with the avqz set predict the (T) correction to within 0.004 eV.

ACKNOWLEDGEMENTS

This work was supported by the Division of Chemical Sciences in the Office of Basic Energy Sciences of the U.S. Department of Energy at Pacific Northwest National Laboratory, a multiprogram national laboratory operated by Battelle Memorial Institute, under Contract No. DE-AC06-76RLO 1830. Computer resources were provided by the Division of Chemical Sciences.

REFERENCES

1 For a recent survey of atomic and molecular electron affinities determined using photoelectron techniques, see J. C. Rienstra-Kiracofe, G. S. Tschumper, H. F. Schaefer III, S. Nandi, and G. B. Ellison, Chem. Rev. (to be published).
2 M. J. Travers, D. C. Cowles, and G. B. Ellison, Chem. Phys. Lett. **164**, 449 (1989).
3 S. E. Bradforth, E. H. Kim, D. W. Arnold, and D. M. Neumark, J. Chem. Phys. **98**, 800 (1993).
4 I. Shavitt, Mol. Phys. **94**, 3 (1998).
5 D. Cremer, "Møller-Plesset Perturbation Theory," in *Encyclopedia of Computational Chemistry*, ed. by P. R. v. Schleyer, (John Wiley & Sons Ltd., New York, 1998).
6 J. Gauss, "Coupled-cluster Theory," in *Encyclopedia of Computational Chemistry*, ed. by P. R. v. Schleyer, (John Wiley & Sons Ltd., New York, 1998).
7 M. W. Schmidt and M. S. Gordon, Ann. Rev. Phys. Chem. **49**, 233 (1998).
8 K. Andersson, "Complete Active Space Self-consistent Field (CASSCF) Second Order Perturbation Theory (CASPT2)," in *Encyclopedia of Computational Chemistry*, ed. by P. R. v. Schleyer, (John Wiley & Sons Ltd., New York, 1998).
9 S. A. Kucharski and R. J. Bartlett, J. Chem. Phys. **95**, 8227 (1991); X. Li and J. Paldus, J. Chem. Phys. **107**, 6257 (1997).
10 For a discussion of size extensivity, see J. A. Pople, J. S. Binkley, and R. Seeger, Int. J. Quantum Chem. Symp. **10**, 1 (1976); R. J. Bartlett and G. D. Purvis, Int. J. Quantum Chem. **14**, 561 (1978).
11 J. Olsen, O. Christiansen, J. Olsen, and P. Jørgensen, J. Chem. Phys. **105**, 5082 (1996); O. Christiansen, J. Olsen, P. Jørgensen, H. Koch, and P.-A. Malmqvist, Chem. Phys. Lett. **261**, 369 (1996); J. Olsen, P. Jørgensen, T. Helgaker, and O. Christiansen, J. Chem. Phys.

112, 9736 (2000); T. H. Dunning, Jr. and K. A. Peterson, J. Chem. Phys. **108**, 4761 (1998); M. L. Leininger, W. D. Allen, H. F. Schaefer III, and D.C. Sherrill, J. Chem. Phys. **112**, 9213 (2000).

12 See, *e.g.*, T. H. Dunning, Jr., J. Phys. Chem. *A* **104**, 9062 (2000) and references therein.

13 See, *e.g.*, R. J. Bartlett, J. Phys. Chem. **93**, 1697 (1989); T. J. Lee and G. E. Scuseria,, in *Quantum Mechanical Calculations with Chemical Accuracy*, ed. by S. R. Langhoff, (Kluwer Academic Publishers, Dordrecht, 1997).

14 K. Raghavachari, G. W. Trucks, J. A. Pople, and M. Head-Gordon, Chem. Phys. Lett. **157**, 479 (1989); K. Raghavarchi, J. A. Pople, E. S. Replogle, and M. Head-Gordon, J. Phys. Chem. **94**, 5579 (1990).

15 T. H. Dunning, Jr., J. Chem. Phys. **90**, 1002 (1989).

16 A. Halkier, T. Helgaker, P. Jørgensen, W. Klopper, H. Koch, J. Olsen, and A. K. Wilson, Chem. Phys. Lett. **286**, 243 (1998).

17 R. A. Kendall, T. H. Dunning, Jr., and R. J. Harrison, J. Chem. Phys. **96**, 6796 (1992); D. E. Woon and T. H. Dunning, Jr., J. Chem. Phys. **100**, 2975 (1994).

18 D. E. Woon and T. H. Dunning, Jr., J. Chem. Phys. **103**, 4572 (1995); K. A. Peterson and T. H. Dunning, Jr. (unpublished).

19 T. H. Dunning, Jr., K. A. Peterson, and D. E. Woon, "Basis Sets: Correlation Consistent," in *Encyclopedia of Computational Chemistry*, ed. by P. R. v. Schleyer, (John Wiley & Sons Ltd., New York, 1998).

20 K. P. Huber and G. Herzberg, *Molecular Spectra and Molecular Structure. IV. Constants of Diatomic Molecules*, (Van Nostrand, Princeton, 1979).

21 H.-J. Werner, P. J. Knowles, R. D. Amos, A. Berning, D. L. Cooper, M. J. O. Deegan, A. J. Dobbyn, F. Eckert, C. Hampel, G. Hetzer, T. Leininger, R. Lindh, A. W. Lloyd, W. Meyer, M. E. Mura, A. Nicklass, P. Palmieri, K. A. Peterson, R. M. Pitzer, P. Pulay, G. Rauhut, M. Schütz, H. Stoll, A. J. Stone, and T. Thorsteinsson, *MOLPRO Quantum Chemistry Package 2000.1*, **2000**. See also, C. Hampel, K. A. Peterson, and H.-J. Werner, Chem. Phys. Lett. **190**, 1 (1992); P. J. Knowles, C. Hampel, and H.-J. Werner, J. Chem. Phys. **99**, 5219 (1994).

22 R. J. Bartlett, in *Encyclopedia of Computational Chemistry*, ed. by P. R. v. Schleyer, (John Wiley & Sons Ltd., New York, 1998). See also, W. J. Lauderdale, J. F. Stanton, J. Gauss, J. D. Watts, and R. J. Bartlett, Chem. Phys. Lett. **187**, 21 (1991).

23 T. H. Dunning, Jr. and K. A. Peterson, J. Chem. Phys. **113**, 7799 (2000).

24 J. A. Sordo, J. Chem. Phys. **114**, 1974 (2001).

25 For a corresponding statistical analysis of atomization energies, see K. L. Bak, P. Jørgensen, J. Olsen, T. Helgaker, and W. Klopper, J. Chem. Phys. **112**, 9229 (2000).

Dissociative Electron Attachment in Gas and Condensed Phases

I. I. Fabrikant
Department of Physics and Astronomy University of Nebraska-Lincoln
iif@unlserve.unl.edu

1. INTRODUCTION

The resonance theory of electron-molecule collisions developed by O'Malley[1] and Bardsley[2] has become an efficient tool for *ab initio* calculations of dissociative attachment (DA) processes for diatomic molecules. However, theoretical calculations still cannot be performed routinely in the area of electron-atom and electron-molecule (nonreactive) collisions. Semiempirical methods are still widely used in studies of electron attachment to polyatomics. The resonance R-matrix theory[3] is particularly useful for this purpose.

The present paper discusses recent theoretical and experimental developments in physics of electron attachment to polyatomic molecules and clusters. We will concentrate on the phenomenon of vibrational Feshbach resonances[4,5] observed recently in DA to methyl iodide,[6] the influence of the cluster environment on these resonances, and the condensed-matter effects in DA.

2. VIBRATIONAL FESHBACH RESONANCES

Vibrational (or "nuclear-excited"[4,5]) Feshbach resonances (VFR) in DA are similar to Rydberg resonances in dissociative recombination (so-called

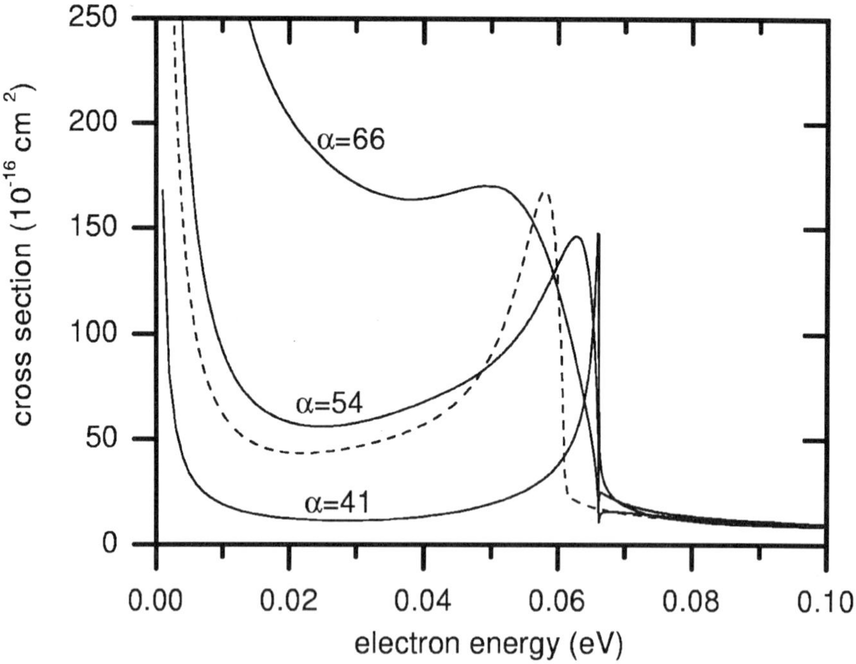

Figure 1. Calculated cross sections for DA to methyl iodide molecules with fixed dipole moments and different polarizability, α. Dashed curve: DA to CD_3I with the actual polarizability, α = 54 a.u.

indirect mechanism[2]) and are associated with a temporary capture of the electron by the long-range field of the target molecule with a simultaneous vibrational excitation of the target. Since the long-range (dipolar plus polarization) field of a neutral molecule is substantially weaker than the Coulomb field of a molecular ion, VFR are typically not as pronounced as Rydberg resonances in dissociative recombination.[7] Although some indication of VFR was found in studies of vibrational excitation of hydrogen halides,[8,9] there was no clear experimental evidence of VFR until the recent discovery[6] of a pronounced peak in DA to methyl iodide below the threshold for the symmetric C-I stretch ($\nu_3 = 1$) vibrational excitation. The higher vibrational onsets are free of such resonances. As detailed calculations,[6,10] based on the resonance R matrix theory, have shown, the effect is caused by a combined action of the dipolar and polarization interactions.

In Fig. 1 we present the DA cross section for CH_3I obtained with different choice of polarizability. For a weaker long-range interaction, VFR moves towards the vibrational excitation threshold and turns eventually into a virtual state. The latter manifests itself in a pronounced upward cusp at the $\nu_3 = 1$ threshold. On the other hand, for a stronger long range interaction the resonance becomes very broad and looks more like a shoulder in the DA cross section. Fig. 1 also presents the isotope effect by comparing cross sections for CH_3I and CD_3I. A higher reduced mass makes the DA cross section typically lower. However, the VFR becomes sharper for CD_3I, and this makes the cross section higher in the resonance region.

For other methyl halides VFRs appear at higher thresholds,[11] although the calculated cross sections for DA from the ground vibrational state are substantially lower. For CH_3Cl, they are too low for experimental observations. As demonstrated in Fig. 2, sharp virtual-state peaks appear at the $\nu_3 = 7$ and 8 thresholds for CH_3Cl and at the $\nu_3 = 8$ and 9 thresholds for its isotopic substitute, CD_3CI, but the absolute values of the cross section at the highest peak is about $3x10^{-23}$ cm^2 and $5x10^{-24}$ cm^2 for CH_3Cl and CD_3Cl respectively. However, for DA from the $CH_3Cl(\nu_3 = 6)$ state the cross section at the $\nu_3 = 8$ threshold exceeds 10^{-15} cm^2.[11] This explains the strong temperature effect in CH_3Cl observed experimentally[12] and described in our previous calculations.[11,12]

These results can be interpreted[11] in terms of the potential curve behavior for methyl iodide and methyl chloride. It is well known that the DA cross sections for individual vibrational levels rise dramatically with ν below the energy corresponding to the crossing between the neutral and the *diabatic* negative-ion curve. For methyl halides, the *adiabatic* anion curve does not cross the neutral curve down to distances close to the equilibrium internuclear separation because of the long-range electron-molecule interaction. Therefore, well below the crossing point the vibrational negative-ion states supported by the adiabatic curve are very close to those supported by the neutral curve. The VFR in DA occurs when the electron is captured into one of the negative-ion states and then predissociates into the repulsive resonance state.

The crossing between the diabatic anion curve and the neutral curve occurs in the vicinity of $\nu_3 = 1$ for CH_3I whereas for CH_3Cl it occurs in the vicinity of $\nu_3 = 8$. Therefore we observe a dramatic increase in the DA cross section in CH_3Cl when going from the ground vibrational state to $\nu_3 = 8$. The vibrational enhancement for CH_3I is much weaker. Only a small increase is observed when going from $\nu_3 = 0$ to $\nu_3 = 1$.[6]

The VFR width is small for the negative-ion states lying below the crossing point. For states lying close and above the crossing point the width is large due to the large predissociation rates. Therefore VFR in CH_3I is observed only at the $\nu_3 = 1$ threshold, whereas for the $\nu_3 = 2$ threshold a pronounced cusp structure is seen.[6] In contrast, for CH_3Cl, the most pronounced resonance is seen at the $\nu_3 = 8$ threshold. At lower thresholds the resonances become extremely narrow.

The increase of the long-range electron-molecule interaction shifts the adiabatic negative-ion curve downwards and increases the VFR's width. On

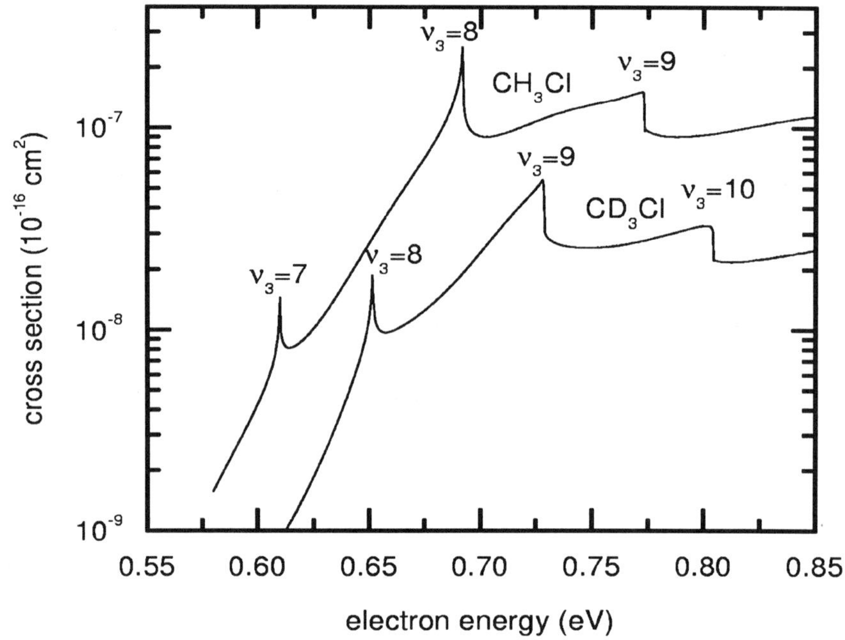

Figure 2. Cross sections for DA to CH_3Cl and CD_3Cl from the ground vibrational state. v_3 = 7,8,... indicate excitation thresholds for symmetric-stretch vibrations.

the other hand, with the decrease of the strength of the long-range interaction, a negative ion state moves into the continuum and becomes a virtual state. This explains the behavior of the cross sections calculated for different values of the dipole moment[10] and polarizability as in Fig. 1.

3. VIBRATIONAL FESHBACH RESONANCES IN ELECTRON ATTACHMENT TO CLUSTERS

The above observations explain the major features of recently observed electron attachment to Van der Waals clusters: CH_3I,[13] CO_2,[14] and N_2O[15,16] clusters. VFR's in these clusters are usually observed close to a vibrational excitation threshold of a *monomer.* Therefore, in our theoretical description of electron attachment to a cluster, we assume that the electron excites vibrational motion of an individual molecule, and the remaining part of the cluster serves as an environment producing a shift in the energy of the negative-ion state, the so-called solvation energy. The solvation energy produces different effects depending on the nature of a cluster. For the methyl iodide clusters, the VFR is rather broad even for monomers. The solvation energy effect leads to further broadening of the resonance[13] similar to that caused by the increasing polarizability of the target, indicated in Fig. 1. Qualitatively this corresponds to observations of electron attachment to the methyl iodide dimers[13]: the resonance is almost destroyed at the corresponding solvation energy 0.39 eV. However, the cusp structure associated with the threshold for vibrational excitation of the symmetric C-I stretch is still clearly seen in calculations, whereas no structure has been detected in the experiment.

For a more quantitative description of experimental data we assume that the most likely DA process in electron collisions with $(CH_3I)_N$ clusters has the form

$$(CH_3I)_{N+1} + e^- \rightarrow (CH_3I)_N \bullet I^- + CH_3 \qquad (1)$$

We note that this is not necessarily the only possible scenario for the cluster dissociation since it ignores the evaporation process,[17,18] i.e. the release of the excess energy in the form of neutral clusters $(CH_3I)_q$. On the other hand, due to the mass ratio of the DA products (cluster ion and CH_3 radical) the excess energy most likely will reside on the light methyl radical, therefore monomer evaporation seems unlikely in this case.[13] In developing a model for electron attachment to the dimer, we include the interaction of the C-I stretch mode with other modes in the complex. These modes are typically very soft because of the relatively weak interaction between monomers. Thoss and Domcke[19] used the displaced oscillator model[20] to describe intramolecular vibrational relaxation in low-energy electron scattering and photoionization of large molecules. We used a similar approach[13] to describe the coupling between the C-I stretch and soft modes in DA to $(CH_3I)_2$ dimers.

The theory[13] confirms that the solvation and polarization effects destroy VFR and lead to almost complete disappearance of the threshold cusp due to coupling with the bath modes. On the other hand, the theoretical attachment rate drops with energy too fast as compared to the experimental observation. We should stress that the present theory contains too many model assumptions to expect quantitative agreement.

The case of CO_2 and N_2O clusters is quite different. In these systems the long-range electron-target interactions are insufficient to promote a VFR state for a single molecule. They are, however, sufficiently strong for clusters above a certain size to give rise to VFRs. Moreover, the occurrence of VFRs is less sensitive to changes in the long range forces as compared to the case of methyl iodide clusters. In particular the VFR width does not change noticeably with the cluster size. For each vibrational series, the experimental resonance position corresponds to the binding energy of the captured electron in the $[M_{N-1}M(\nu)]^-$ anion state relative to the energy of the neutral $[M_{N-1}M(\nu)]$ cluster,[14] where M stands for CO_2 or N_2O.

The binding energy can be estimated using a simple model[14,16] incorporating the polarization interaction, $V_{pol} = -Ne^2\alpha/2r^4$, outside the cluster radius R_N, where α is the polarizability of the monomer. Within the cluster, at $r < R_N$, the electron-cluster interaction is assumed to be a constant, U_N, for a given N. Ab initio calculations[21] show that the N-dependence of U_N can be approximated as

$$U_N = aN^{-1/3} + b \tag{2}$$

where a and b are constants. The constant b has the meaning of the energy of the excess electron in the bulk and can be estimated from the effective interaction of the electron with the monomer and dielectric properties of the bulk. The cluster radius R_N is proportional to $R_{WS}N^{1/3}$, where R_{WS} is the Wigner-Seitz radius. For CO_2 and N_2O clusters, we estimate R_N as $R_N = 3.43N^{1/3}$ a.u.,[14,16] and consider a and b as empirical parameters.

In Fig. 3 we present the binding energies as functions of N and compare them with experimental data on VFR red shifts (relative to vibrational excitation thresholds) in CO_2 and N_2O clusters. A good fit for binding energies is obtained by using $a = 0.70$ eV, $b = -0.866$ eV for CO_2,[14] and $a = 0$, $b = 0.2$ eV for N_2O.[16] Larger red shifts for CO_2 clusters are due to differences in the short-range interactions between the electron and respective monomers, since polarizabilities of CO_2 and N_2O are very close.

This conclusion is confirmed by analysis of experimental and theoretical data for e-CO_2 and e-N_2O scattering. Cross sections for low-energy electron scattering by CO_2 are rapidly increasing towards lower energies[22-24] leading to a large negative scattering length of about -6.3 to -7.2 a.u. (corresponding to the zero-energy cross section between 140 and 183 A^2). This behavior is dominated by the isotropic polarization interaction while the quadrupolar and anisotropic polarization interactions play a minor role.[25] Therefore, the polarization model for clusters described above can also describe electron scattering by monomers.

On the other hand, experimental[26] and theoretical[27,28] data on electron scattering by the N_2O molecule show that the cross section in this case is much smaller and does not exceed 8 A^2 below 1 eV. At ultralow energies, the cross section for N_2O should increase due to the dipole moment. However, the dipole moment of N_2O is very small (about 0.16 D), and this effect should not be important for clusters anyway. The observed behavior of e-N_2O scattering cross section can be reproduced by modelling the e-N_2O interaction potential as a repulsive barrier with a polarization tail which is consistent with our findings for N_2O clusters.[14, 16]

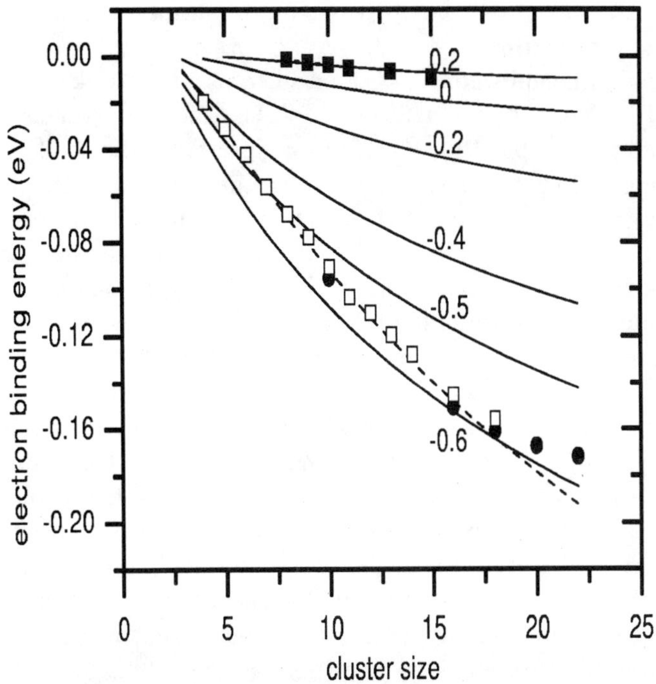

Figure 3. Binding energies of electron in a Van der Waals cluster as a function of the cluster size N for different values of the short-range potential U_0. Dashed curve presents calculations with a variable depth given by Eq. (2). Experimental data[14,16] for VFR red shifts: open squares, CO_2, average of the two (020) and (I00) series; full circles, CO_2, average of the two (030) and (110) series; full squares, N_2O, (I00) series.

4. CONDENSED-MATTER EFFECTS IN DISSOCIATIVE ATTACHMENT

In a variety of applications, such as plasma chemistry, radiation chemistry, biology, photon-induced surface chemistry, and dielectric aging (see references in Ref. 29), low-energy electron interaction with molecules occurs under complex collision conditions. Typical examples are ozone depletion in polar stratospheric clouds[30,31] and environmental remediation.[32] In both examples the DA reaction occurs on surfaces, and the surface environment influences strongly the reaction rate. In particular it increases the DA cross section for methyl chloride by about five orders of magnitude.[33] This strong influence can be explained qualitatively by the polarization interaction between the intermediate negative ion state and the surface. However, the quantitative theory of the DA processes on the surface and in a bulk medium is still at its very preliminary stage.

Most experimental studies of DA on surfaces and in bulk media have been performed with the rare gas films.[29,33-36] The molecules under study were physisorbed on (typically) Kr film or inserted in a bulk medium by sandwiching them between two films. There are two major effects that should be incorporated in the theory of DA on a surface or in a bulk medium. First, we have to take into account the interaction between the negative-ion state and the medium. In the framework of the resonance R-matrix theory this can be accomplished by shifting the position of the R matrix pole by the amount corresponding to the polarization energy.[33] In terms of the standard resonance theory this corresponds to a uniform downward shift of the diabatic negative - ion state. Secondly, we have to include interaction of the continuum electron with the medium. Both effects influence the resonance width, the most important parameter of the resonance theory. However, the second effect may not be so significant since the average interaction of the free electron with the medium can be characterized by the position of the bottom of the conduction band, which is only slightly shifted with respect to the vacuum level (e.g., by -0.25 eV for solid Kr). Accordingly, in our surface calculations[33,37] we included only the first effect. For a more accurate treatment of the surface effects one has to incorporate the wave functions calculated from the low-energy electron diffraction (LEED) theory.

Calculations of DA in a medium were carried out with the inclusion of the second effect in the continuum medium approximation[35] whereby the electron wave function is described by a plane wave with a modified dispersion relation between momentum and energy

$$E=E_o+p^2/2m^* \tag{3}$$

where E_0 is the bottom of the conduction band and m^* is the effective electron mass.

Fig. 4 presents the results for Cl^- formation for 0.1 monolayers (ML) of CF_3Cl condensed on a 13 ML Kr surface and 0.1 ML of CF_3Cl sandwiched between two 13 ML Kr films.[29] Whereas the theory describes the surface results quite well, the situation for attachment to CF_3Cl embedded in a medium is more complicated. If we don't include the electron-medium interaction at all, the theory underestimates the cross section by a factor of two, and the position of the peak is shifted towards lower energies as compared to experimental data. On the other hand, the inclusion of the electron-medium interaction in the effective mass approximation overestimates the cross section by a factor of three. Note that the structure at vibrational excitation thresholds disappears when the electron-medium interaction is included.

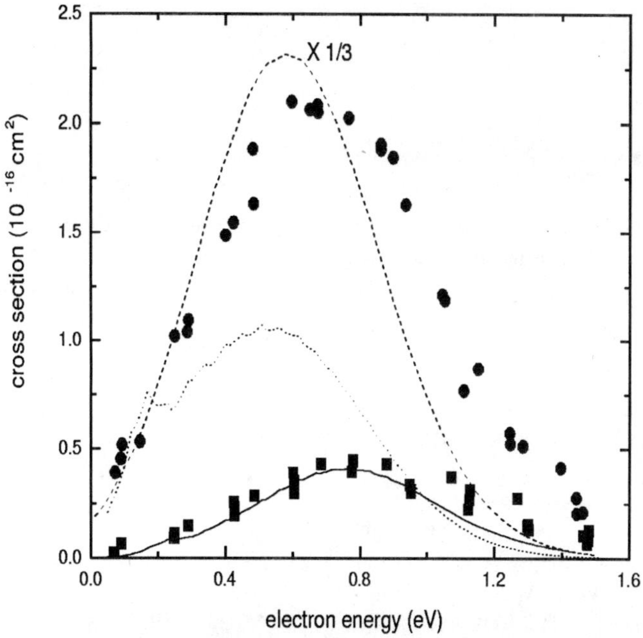

Figure 4. DA cross section for CF_3Cl condensed on a Kr surface (squares) and embedded into a Kr film (circles). Theory: solid curve, calculations including the surface polarization energy, $E_{pol} = 0.72$ eV; dotted curve, calculations including the bulk polarization energy, E_{pol} = I.I5eV eV; dashed curve, calculations including the electron-medium interaction.

5. CONCLUSION

The past few years have been very fruitful in terms of our understanding of the physics of DA. The first clear evidence of VFR in electron attachment to molecules and clusters was obtained, and a theoretical description of this phenomenon within the framework of the resonance R-matrix theory was developed. These studies indicate many similarities between DA and dissociative recombination with its indirect mechanism. On the other hand, in contrast to Rydberg resonances in dissociative recombination, VFRs in DA are very fragile and are easily destroyed by a slight modification of the long-range electron-molecule interaction. This has been demonstrated by studying DA to methyl iodide clusters.

In the area of DA on a surface and in a bulk, we have achieved a basic understanding of mechanisms responsible for the change of the cross sections as compared to those in the gas phase. On the other hand, a more advanced theory, incorporating LEED wave functions for the projectile electron, should be developed, in order to achieve good quantitative agreement between theory and experiment.

ACKNOWLEDGEMENTS

This work became possible due to contribution of graduate students, R. S. Wilde and Y. Xu, and due to a fruitful collaboration with G. A. Gallup, H. Hotop, M.-W. Ruf, L. Sanche and K. Nagesha. Support from the National Science Foundation is gratefully acknowledged.

REFERENCES

1 T. F. O'Malley, Phys. Rev. **150**, 14 (I966).
2 J. N. Bardsley, J. Phys. B **1**, 349 (I968).
3 I.I. Fabrikant, Comments At. Mol. Phys. **24**, 37 (1990).
4 N. Bardsley and F. Mandl, Rep. Prog. Phys. **31**, 47I (I968)
5 J. P. Gauyacq and A. Herzenberg, Phys. Rev. A **25**, 2959 (I982).
6 A. Schramm, I. I. Fabrikant, J. M. Weber, E. Leber, M.-W. Ruf, and H.Hotop, J. Phys. B **32**, 2I53 (I999).
7 I. F. Schneider, O. Dulieu, and A. Giusti-Suzor, J. Phys. B **24**, L289 (1991).
8 G. Knoth, M. Gote, M. Radle, K. Jung, and H. Ehrhardt, Phys. Rev. Lett. **62**, I735 (I989).
9 A.-Ch. Sergenton, L. Jungo, and M. Allan, Phys. Rev. A **61**, 06272 (2000).
10 E. Leber, I. I. Fabrikant, J. M. Weber, M.-W. Ruf, and H. Hotop, In *Dissociative Recombination: Theory, Experiment, and Applications IV,* ed. M. Larsson, J. B. A. Mitchell, and I. F. Schneider. World Scientific, Singapore (2000), p. 69.
11 R. S. Wilde, G. A. Gallup, and I. I. Fabrikant, J. Phys. B **33**, 5479 (2000).
12 D. M. Pearl, P. D. Burrow, I. I. Fabrikant, and G. A. Gallup, J. Chem. Phys. **115**, 9876 (I995).
13 J. M. Weber, I. I. Fabrikant, E. Leber, M.-W. Ruf, and H. Hotop, Eur. Phys. J. D **11**, 247 (2000).
14 E. Leber, S. Barsotti, I. I. Fabrikant, J. M. Weber, M.-W. Ruf, and H. Hotop, Eur. Phys. J. D **12**, I25 (2000).
15 J. M. Weber, E. Leber, M.-W. Ruf, and H. Hotop, Phys. Rev. Lett. **82**, 5I6 (I999).
16 E. Leber, S. Barsotti, J. Bommels, J. M. Weber, I. I. Fabrikant, M.-W. Ruf, and H. Hotop, Chem. Phys. Lett. **325**, 345 (2000).
17 T. Kondow, J. Phys. Chem. **91**, I307 (I987).
18 F. Misaizu, K. Mitsuke, T. Kondow, and K. Kuchitsu, J. Chem. Phys. **94**, 243 (1991).
19 M. Thoss and W. Domcke, J. Chem. Phys. **109**, 6577 (I998).
20 W. Domcke and L. S. Cederbaum, Phys. Rev. A **16**, I465 (I977).
21 P. Stampfli, Phys. Rep. **255**, 1 (I995).
22 J. J. Lowke, A. V. Phelps, and B. W. Irwin, J. Appl. Phys. **44**, 4664 (I973).
23 M. A. Morrison, N. F. Lane, and L. A. Collins, Phys. Rev. A **15**, 2I86 (I977).
24 D. Field, N. C. Jones, S. Lunt, and J.-P. Ziesel, Phys. Rev. A **64**, 022708 (200I).
25 I. I. Fabrikant, J. Phys. B **17**, 4223 (I984).

26 C. Szmytkowski, G. Karwasz, and K. Maciag, Chem. Phys. Lett. **107**, 481 (1984).
27 B. K. Sarpal, K. Pfingst, B. M. Nestmann, and S. D. Peyerimhoff, J. Phys. B **29**, 857 (1996); *ibid.* **29**, 1877 (1996).
28 C. Winstead and V. McKoy, Phys. Rev. A **57**, 3589 (1998).
29 K. Nagesha, I. I. Fabrikant, and L. Sanche, J. Chem. Phys. **114**, 4934 (2001).
30 Q.-B. Lu and Th. E. Madey, Phys. Rev. Lett. **82**, 4122 (1999).
31 Q.-B. Lu and L. Sanche, Phys. Rev. Lett. **87**, 078501 (2001).
32 P. D. Burrow, K. Aflatooni, and G. A. Gallup, Environ. Sci. Technol. **34**, 3368 (2000).
33 L. Sanche, A. D. Bass, P. Ayotte and I. I. Fabrikant, Phys. Rev. Lett. **75**, 3568 (1995).
34 K. Nagesha and L. Sanche, Phys. Rev. Lett. **78**, 4725 (1997).
35 I. Fabrikant, K. Nagesha, R. Wilde, and L. Sanche, Phys. Rev. B **56**, R5725 (1997).
36 K. Nagesha and L. Sanche, J. Appl. Phys. **88**, 5211 (2000).
37 P. Ayotte, J. Gamache, A. D. Bass, I. I. Fabrikant, and L. Sanche, J. Chem. Phys. **106**, 749 (1997).

Electron Collisions with C_4^- : Detachment Process and Resonant Structure

A. Le Padellec[1], F. Rabilloud[1], D. Pegg[2], K. Andersson[3], D. Hanstorp[3], A. Neau[4], M. Larsson[4], F. Hellberg[4], and R. Thomas[4]
[1]LCAR UMR 5589, Université Paul Sabatier-Toulouse III, 118 route de Narbonne, Bât. III R1B4, 31062 Toulouse Cedex 4, France; [2]Department of Physics, University of Tennessee, Knoxville, Tennessee 37996, USA; [3]Department of Physics, Chalmers University of Technology/Göteborg University, SE- 412 96 Göteborg, Sweden; [4]Department of Physics, Stockholm University, Box 6730, S-113 85 Stockholm, Sweden
arnaud.lepadellec@irsamc.ups-tlse.fr; Franck.rabilloud@lasim.univ-lyon1.fr

1. INTRODUCTION

A detailed knowledge of the carbon chain's spectroscopy is of considerable importance in astrophysics. More than two decades ago, Douglas[1] put forward the suggestion that the unknown absorbing species responsible for over 150 diffuse interstellar bands, might be long carbon chain molecules, C_n. More recently, Tulej et al.[2] measured the gas phase electronic spectrum of the C_7^- chain and found that the 5 narrow absorption bands in the lowest energy transition correlate quite well with the observed wavelengths of narrow diffuse interstellar bands. CRYRING was used to study collision processes between an electron and a negative ion cluster C_4^-. We made use of a grid to separate four neutral channels over the five that contribute to the total neutral 4C production. Only the detachment cross sections, leading to the production of C_4, are presented here. Although the doubly charged negative ion C_4^{-2} has received little previous attention, a resonance observed in the present work in the threshold region of the detachment cross section is associated with a short lived C_4^{-2} state.

Dissociative Recombination of Molecular Ions with Electrons
Edited by S. Guberman, Kluwer Academic/Plenum Publishers, 2003

2. THEORETICAL RESULTS

2.1 Computational Methods

We present a study of the structure of the C_4^- and C_4^{-2} ions. All calculations are carried out using the GAUSSIAN 98 package[3] and are performed in two steps: The geometries of the negative ions are first optimized using the coupled cluster (CC) method involving double substitutions (CCD) in the Hartree-Fock determinant.[4] Restricted Hartree-Fock (RHF) reference wave functions are used for singlet states, and unrestricted Hartree-Fock (UHF) functions for states with higher multiplicities. Various electronic states are considered and several geometrical structures (a $D_{\infty h}$ linear structure, a D_{2h} rhomboid and a C_{2v} three-membered ring) are tested and optimized using a gradient algorithm. In the case of the CCD determined geometries, additional coupled cluster calculations are performed to obtain the total energy of the systems. The CC method involves single and double substitutions, and non-iterative triple excitations (CCSD(T)).[5] In order to reduce the computation time, carbon atoms are represented by relativistic effective pseudo-potentials. In the latter treatment, the 1s electrons are included in the core but only the 2s and 2p electrons are active. Calculations were performed using the linear combination of atomic orbitals (LCAO) scheme. The Gaussian basis set was 5s6p1d. With this basis, the electron affinity of the carbon atom is found to be 0.92 eV. The electron affinity of the C_2 dimer is found to be 2.81 eV, in good agreement with the previous theoretical studies.[6]

2.2 The C_4^- Ion

The optimized geometries and relative energies of the C_4^- anion are displayed in Fig. 1. The lowest energy isomer A has a linear structure and a $^2\Pi_g$ electronic state. We calculate the length of the central bond to be 1.343 Å, while those of the terminal bonds are 1.277 Å each. These bond lengths compare well with prior calculations.[7-9] Moreover, three other isomers are identified. Isomer B is a $^2\Sigma_g$ linear structure and lies 1.082 eV above isomer A. Isomer C is a planar three membered ring of C_{2v} symmetry and isomer D has a D_{2h} rhombus shape. We find these two latter structures at 1.211 eV and 1.241 eV above A, respectively. For each isomer, we examine the localization of the negative charge using a Mulliken population analysis. The negative charge is localized and shared by the two terminal atoms in isomers A and B. The excess charge is mostly on the terminal carbon atom of the isomer C "tail." For isomer D, the additional negative charge is localized on the two atoms situated on opposite sides of the long diagonal of the rhombus.

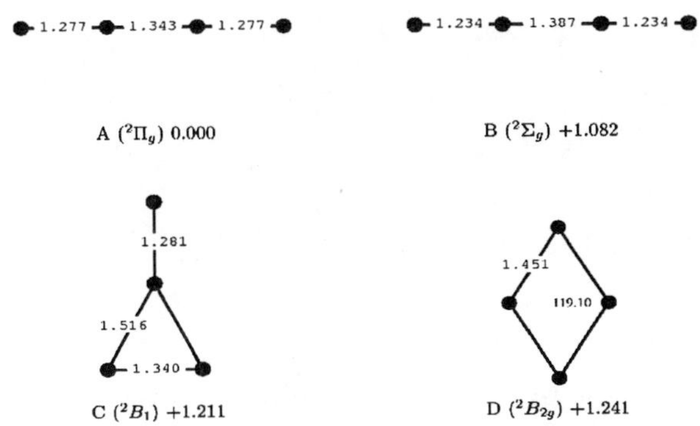

Figure 1. Calculated isomeric structures of the C_4^- ion. Shown are the electronic states, the relative energies (eV) and the atomic separations (Å).

2.3 The C_4^{-2} ion

The quantum chemistry calculations are extended to C_4^{-2}. In particular, we focus upon the stability of the C_4^{-2} ion with respect to the linear, rhomboidal or three-membered ring isomeric structures. The geometry of the isomers, relative to a minimum in the energy surface, are displayed in Fig. 2. The ground state has a linear structure and a $^1\Pi_g^+$ electronic state. The bond lengths (1.411 Å for the central one and 1.256 Å for the two outer bonds), are close in value to those obtained by a prior ROHF study.[10] The effect on the geometry of C_4^{-2} that results from adding one electron to the C_4^- ion, is an increase in the difference between the central bond and the outer bonds (from 0.066 Å to 0.155 Å). This result, obtained here with correlated calculations, confirms the results from the previous ROHF study,[10] in which the authors suggest a representation of the doubly charged ion as $^-$C≡C-C≡C$^-$. Moreover, the population analysis is in agreement with the intuitive picture since it indicates that the negative charges are mostly localized on the two terminal atoms. A Mulliken analysis indicates -0.8e on each one. A second isomer is found at 1.947 eV above the first one. It has a three-membered ring structure, and a $^1A_1$ ground state. According to a Mulliken analysis, the charge on the terminal carbon of the "tail" is close to -2e. For the rhombus shape, it is found that a ^1A_g state could perhaps be stable, but the single determinant wave function that is used here is not a satisfactory zero order description. Therefore, we do not report it in the present work since it is not expected to be one of the low energy isomers. For completeness, we mention that a few triplet states are considered but no local minima are found in the energy surfaces.

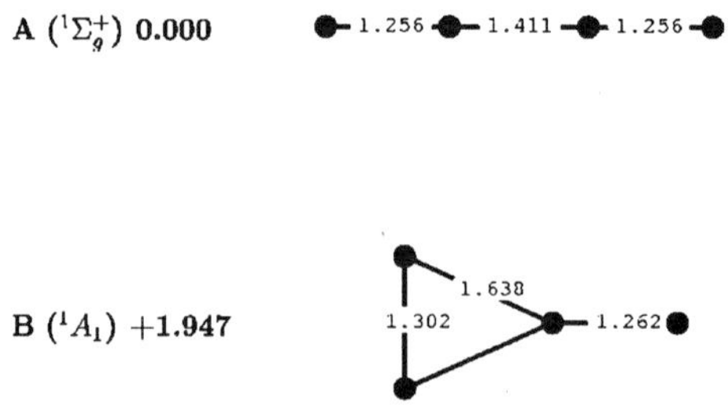

Figure 2. Calculated structures of C_4^{-2}. Shown are the electronic ground state, the relative energy (eV) and the atomic separations (Å).

3. EXPERIMENTAL RESULTS

The experiment was performed at the heavy ion storage ring, CRYRING, located at the Manne Siegbahn Institute, University of Stockholm, Sweden. C_4^- anions are produced in a sputter ion source.[11] The isomeric composition of the negative ion beam is indeed an important issue in this experiment. Zajfman et al.[12] investigated the photodetachment of negative carbon clusters C_n^-, using a similar sputter ion source. They observed the detachment threshold at about 2.1 eV, which agrees very well with the predicted electron affinity of the rhombus isomer 2.192 eV, rather than with that of 3.75 eV, which would have been expected for a target beam comprised exclusively of the ground state isomer A. On the other hand, Bhardwaj et al.[13] also produced C_4^- projectiles with a sputter ion source but they claim that their ions were predominantly in the A configuration. This is rather puzzling and contradictory evidence.

The following fact led us to adopt Zaijman's findings. Indeed, in the work by Bhardwaj et al., the main goal was not to study the internal excitation of the C_4^- ions as was the case in the Zajfman reference but rather to study collisional dissociation patterns. Moreover, photodetachment experiments (with subsequent threshold characterizations) constitute the sharpest and most reliable test for the evaluation of the internal excitation of a given target. Unfortunately, the drawback is that photodetachment studies do not allow evaluation of the relative proportions of the isomers leaving a given ion source. Therefore, we conclude that the proportions in the various C_4^- isomers remain unknown in the present experiment. Moreover, the relaxation of the C and D isomers to the A configuration prior to any kind of measurement is inefficient, since these isomers are "trapped" in local minima.

We do not present here the operational details of the ring and the analysis procedure since they are included in a companion paper in this book. We give here a few essential numbers concerning the C_4^- target ion beam and the electron projectiles: the ion beam current after acceleration was typically 770 nA; the ion revolution frequency, often called the Schottky frequency, was 54.42 kHz; the electron beam current and electron density at the 24.77 eV cooling energy, were 0.305 mA and 3.4×10^5 cm^{-3}, respectively.

A detailed description of the reduction of the branching fractions is also included in the companion paper and is not reproduced here. We shall only discuss a few points that are specific to the present study. *Only neutral fragments* resulting from the C_4^- electron interaction were detected in our experiment. The number of counts from the energy sensitive surface barrier detector is plotted against the energy of the given neutral fragment, in a so-called MCA spectrum. The energy of each fragment is a known fraction of the ion beam energy. Thus the four peaks in the MCA spectrum correspond, in increasing order, to C, 2C, 3C and 4C "neutral events". For the electron energy range studied here, 5.0-32.9 eV, the following channels are taken into consideration:

- $C_3^- + C + e - 6.4eV$, $C_2^- + C^- + C - 10.9eV$, $C_3^+ + C + 3e - 19.7eV$, $C_2^- + C + C^+ + 2e - 23.0eV$, $C_2^+ + C + C^- + 2e - 25.5eV$
- $C_2^- + C_2 + e - 6.2eV$, $C_2^- + 2C + e - 12.1eV$, $C_2 + 2C^- - 12.9eV$, $2C + 2C^- - 18.8eV$, $C_2 + C_2^+ + 3e - 20.9eV$, $C_2 + C^+ + C^- + 2e - 25.0eV$, $C_2^+ + 2C + 3e - 26.8eV$
- $C_3 + C^- + e - 7.2eV$, $C_2 + C + C^- + e - 14.1eV$, $C_3 + C^+ + 3e - 19.3eV$, $3C + C^- + e - 20.0eV$, $C_2 + C + C^+ + 3e - 26.2eV$
- $C_4 + 2e - 3.9eV$, $C_3 + C + 2e - 8.4eV$, $2C_2 + 2e - 9.5eV$, $C_2 + 2C2e - 15.4eV$, $4C + 2e - 21.3eV$,

where the four groups of channels contained in the four peaks correspond to the C, 2C, 3C and 4C "neutral events", respectively. The reaction energies are calculated assuming all particles are in their lowest states. Moreover, the estimates do not include contributions from the Coulomb repulsion between the interacting partners in the entrance channel. Fragments associated with several of the above channels appeared in the same peak in the MCA spectrum, and are therefore not distinguishable from each other. In order to determine their relative contributions, i.e. the branching fractions, a grid is inserted in front of the detector. In the present case, the major difference in the analysis of branching fractions compared to the analysis routinely used in dissociative recombination measurements concerns the "redistributed" signal. Here, it does not originate exclusively from the full energy peak 4C, but also from the 3C and 2C peaks. The analysis is, therefore, more complicated and in practice the problem is not fully solvable. Without the grid, we find that the total cross sections for the channels containing at least one charged carbon fragment, are at least thirty times lower than those for the sum of the five channels for which only neutral carbon atoms (molecules) are produced. We therefore make the simplifying assumption that when the grid is inserted, the C, 2C and 3C peaks grow only at the exclusive expense of the 4C peak.

The cross section curve for the pure detachment channel (C_4 + 2e) is displayed in Fig. 3. The CM energy resolution is about 80 meV at the 6 eV threshold. The error bars on the data points represent only the statistical uncertainties on the cross section measurements, but do not include a 16% systematic uncertainty. Also in Fig. 3, an inset displays the near threshold cross sections. The dashed-dotted line shows a fit to the data using a model cross section, $\sigma=\sigma_0(1-E_{th}^{exp}/E)$, with $E_{th}^{exp}=6.0eV$. The dotted line represents the fit of a Lorentzian function, $\sigma=\sigma_0/\left[(E-E_r)^2+(\Gamma/2)^2\right]$, to the measured data after the dashed-dotted line has been subtracted from it. For this structure, the fit parameters are found to be $E_r=6.75eV$ and $\Gamma=0.93eV$. We believe that this residual structure in the neutral C_4 channel is attributed to a resonance associated with the short-lived doubly charged negative ion, C_4^{-2}. This point is developed below. At 15 eV, the following "channels" are open : C_4 (pure detachment), C_3+C, $2C_2$, C_2+2C, C_3 (Total) and C_2 (Total). They represent 87.1, 2.4, 1.0, 2.6, 4.7 and 2.2 % of the overall flux, respectively. The global dissociation events account for 12.9 % compared with 87.1% for detachment only. Therefore, it appears that one electron is easily detached from the target ion and in most cases, this may be the least tightly bound electron. The detachment only channel dominates overwhelmingly over the numerous fragmentation processes.

The threshold occurs at about 6.0 eV. Part of this value arises from the Coulomb repulsion. We stated that our ion beam was an unknown mixture of the three lowest energy isomers of C_4^-, labeled A, C and D. In that situation, the threshold must be associated with the isomer with the lowest binding energy, the rhombic structure D. Its electron binding energy is 2.2 eV and therefore the pure Coulomb contribution is 3.8 eV (= 6.0-2.2eV). The cross section reaches a maximum value of roughly $2x10^{-16}$ cm^2. This can be explained from a purely geometrical model. Indeed, one obtains a cross sectional area for the target electron in the ground state linear isomer of $3.3x10^{-16}$ cm^2. This value might be considered as an upper limit since it is known that other, more compact isomers, such as C and D, populate our ion beam.

The resonance in the threshold region of the pure detachment cross section of Fig. 3 is attributed to a doubly charged negative ion, C_4^{-2}. However, it is difficult to know the isomeric structure of this ion. Nevertheless, we carried out a theoretical search for possible "stable" isomeric structures (see above). The measured width yields a lifetime of 0.7 fs and this value compares very well with the 0.3 fs lifetime that was evaluated[14-15] for C_2^{-2}. The two extra electrons are even closer in this latter case and therefore the repulsion should be stronger. Therefore, it makes sense that the C_2^{-2} lifetime should be shorter that that for C_4^{-2}. We shall consider the stability of the C_4^{-2} ion with respect to detachment. To do this, one has to

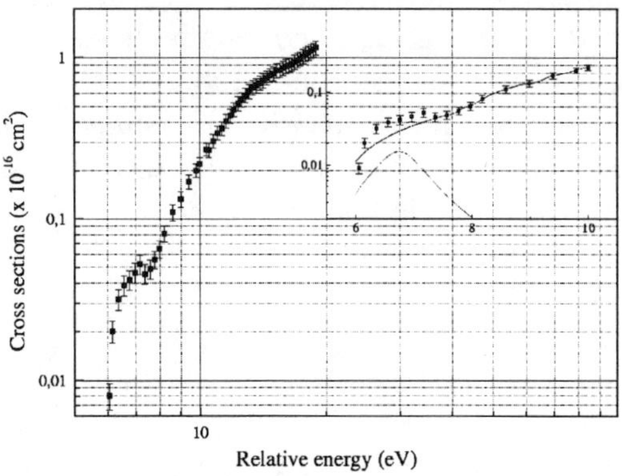

Figure 3. The full squares represent the absolute cross-sections for the pure detachment into C_4 from threshold up to 19 eV. The inset displays the near threshold cross section for pure C_4 detachment. The dashed-dotted line displays a model cross section. The dotted line represents a fit to the data obtained by subtraction of the model cross section fit from the raw experimental data. The fit to this residual structure has the Lorentzian form (see text).

consider the adiabatic electron affinity of the C_4^- ion, which is defined as the energy released when an extra electron is added to C_4^-, the geometry being allowed to relax. The calculations are performed using the CCSD(T) method. For the calculation on the C_4^{-2} dianion, we checked that the *extra electron did not localized on the more diffuse basis functions* and so the electron was really bound to the ion. The electron affinity of the linear C_4^- ion is found to be negative (-2.16 eV), so that the doubly charged negative ion is metastable with respect to the system consisting of the anion plus an infinitely separated electron. In the same way, we find that the adiabatic electron affinity of the three-membered ring C_4^- ($^2B_1$) to the three-membered ring C_4^{-2} is -2.90 eV. Being metastable (short lived) with respect to single detachment, it is reasonable for the C_4^{-2} resonance to decay via single detachment, but we do not have any experimental evidence for this. On the other hand, we do observe that the decay of the dianionic state occurs via double detachment. This indicates that the relaxation might involve more rearrangement within the cluster. Indeed, there might also be the dissociation pathways observed by Mathur et al.[16] Our experimental arrangement do not allow us, however, to intercept the negative fragments necessary to study the production, in coincidence, of $C_2^- + C_2^-$ or/and $C_3^- + C^-$. Mathur et al. carried out a systematic search for doubly charged negative ions of small carbon clusters formed by laser ablation of graphite. A multicoincidence time-of-flight technique was used to monitor the fragment ion pairs produced by the decay of the doubly charged ions. The two covariance maps that display the formation of these time-correlated $C_2^- + C_2^-$ and $C_3^- + C^-$ ion pairs clearly indicate that C_4^{-2} breaks up preferentially into $C_2^- + C_2^-$. This observation supports our finding concerning the bond lengths in the linear dianion. We found 1.411 Å for the central one and 1.256 Å for the outer two. This, together with the results from the ROHF study,[10] that suggests a representation of the dianion as $^-$C≡C-C≡C$^-$, explains why the fragmentation is symmetrical. The central single bond is easier to break than the two outer triple bonds.

4. CONCLUSION

The results of a study of electron impact detachment of C_4^- are presented. A theoretical approach is used to calculate the geometries of C_4^- and C_4^{-2}. Three different isomers with linear, "three-membered ring" and rhombus structures are associated with the target ion beam but it is not possible to determine their relative populations. Absolute cross sections for pure detachment are presented, with a maximum cross section of 2×10^{-16} cm^2. The corresponding threshold location is found to be approximately 6.0 eV. A resonance with a width of 0.93 eV (0.7 fs lifetime) is observed in the near threshold region. It is attributed to a short lived state of C_4^{-2}. Decay mechanisms for this state are discussed.

ACKNOWLEDGEMENTS

We thank the staff of the Manne Siegbahn Laboratory for their invaluable help and use of the heavy ion storage ring facility.

REFERENCES

1 A. E. Douglas, Nature **269**, 130 (1977).
2 M. Tulej, D.A. Kirkwood, M. Pachkov and J.P. Maier, Ap. J, **506**, L69 (1998).
3 M. J. Frisch *et al*, Gaussian 98, Revision A.6 (Gaussian Inc., Pittsburgh, 1998).
4 J. A. Pople, R. Krishnan, H. B. Schlegel and J. S. Binkley, Int. J. Quant. Chem. **14,** 545 (1978).
5 K. Raghavachari, G. W. Trucks, J. A. Pople and M. Head-Gorden, Chem. Phys. Lett. **157**, 479 (1989).
6 J. D. Watts and R. J. Bartlett, J. Chem. Phys. **96**, 6073 (1992).

7 J. D. Watts, J. Gauss, J. F. Stanton and R. J. Bartlett, J. Chem. Phys. **97**, 8372 (1992).
8 A. K. Ray and B. K. Rao, Z. Phys. D. **33**, 197 (1995).
9 S. Schmatz and P. Botschwina, Int. J. Mass Spectrom. Ion Proc, **149/150**, 621 (1995).
10 J. D. Watts and R. J. Bartlett, J. Chem. Phys. **97**, 3445 (1992).
11 Peabody Scientific, Peabody, Massachusetts, USA
12 Z. Vager, H. Feldman, D. Kella, E. Malkin, E. Miklazky, D. Zajfman and R. Naaman, Z. Phys. D - At. Mol. Cl. **19**, 413 (1991).
13 V. R. Bhardwaj, C.P. Safvan, D. Mathur, and A.K. Gupta, Int. J. Mass Spec. Ion. Phys.**163**, 141 (1997).
14 H.B. Pedersen, N. Djuric, M.J. Jensen, D. Kella, C.P. Safvan, L. Vejby-Christensen and L. H. Andersen, Phys. Rev. Lett. **81,** 5302 (1998).
15 L. H. Andersen, P. Hvelplund, D. Kella, P. H. Mokler, H. B. Pedersen, H. T. Schmidt and L. Vejby-Christensen, J. Phys. B **29,** L643 (1996).
16 D. Mathur, V. R. Bhardwaj, F. A. Rajgara and C. P. Safvan, Chem. Phys. Lett. **277,** 558 (1997).

Dissociative Electron Attachment at Low Temperature to Clusters

J.-L. Le Garrec, T. Speck, B. R. Rowe, and J. B. A. Mitchell
Equipe "Astrochimie Expérimentale", Laboratoire P.A.L.M.S., U.M.R. 6627 du C.N.R.S., Université de Rennes 1, 35042 Rennes Cedex, France
Bertrand.Rowe@univ-rennes1.fr

1. INTRODUCTION

Dissociative electron attachment to neutral molecules is a process very similar to the dissociative recombination of molecular ions with electrons. It is known to occur very efficiently with neutral clusters for a variety of molecules,[1] although very little is known about the absolute value of the cross sections (or rate coefficients) involved. This is in contrast with the dissociative recombination of cluster ions, where the rate coefficients have been measured and found to be very large.[2] These large values are believed to be due to super-dissociative recombination.[3]

Electron attachment to clusters could be an important process in a variety of environments, depending on the ability of a neutral monomer to form dimers and higher oligomers. For example, the industrially important gas, SF_6, used in electrical switching and in plasma processing of semiconductors has a rich cluster attachment.[4] In interstellar space, although the search for dimers has not yet been successful,[5] clusters could provide a route for the formation of anions in the very low temperature environment of dense clouds, and these would be effective in destroying the free electrons.

In our laboratory at Rennes, we have recently conducted several studies of electron attachment with various molecules at very low temperatures (down to 50 K),[6,7,8] using a uniform supersonic flow as a flow reactor i.e. a CRESU (Cinétique de Réaction en Ecoulement Supersonique Uniforme) apparatus. In an attempt to check the thermalization of electrons in this kind of flow, we decided to study some endothermic electron attachment reactions involving HCl and HBr. It was found that the rate coefficients for electron attachment by these species showed an anomalous inverse temperature dependence, which was attributed to the attachment to neutral clusters formed at low temperatures. Several experimental checks were performed that showed unambiguously that this was the case and allowed us to derive a lower limit for the rate coefficient of electron attachment (hereafter β) with HBr clusters.[6] In this paper we describe the experimental technique as well as the results obtained respectively with HCl, HBr, C_3F_6, H_2O and benzene.

2. EXPERIMENTAL

2.1 Apparatus

The CRESU apparatus used for electron attachment studies has been described in detail elsewhere[6,7] and only the essential features will be summarized below.

Fig. 1 shows a general sketch of the apparatus. It consists of a large chamber (0.5 m diameter, 3 m long) pumped by large Roots blower (24 000 m^3/h). In the chamber, various Laval nozzles can be attached to a flange, mounted on a moveable reservoir in which a buffer gas (helium or nitrogen in the present case) can be mixed with a reactant gas. Expansion of the gas through the nozzle yields a uniform supersonic flow where the thermodynamic conditions are very well defined. Different temperature and pressure conditions can be achieved using various Laval nozzles and different buffer gases. An electron beam of 12.5 keV crosses the flow at 90° downstream of the nozzle exit. This results in the creation of a weakly ionized plasma with an initial electron density in the range 10^8-10^9 electrons/cm^3. Further downstream, a moveable mass spectrometer with a Langmuir probe attached to its outer casing allows the ion density and the electron density to be measured

2.2 Measurement of the Attachment Rate Coefficient

The rate coefficient for electron attachment is obtained using standard flow reactor techniques. Under our experimental conditions, diffusion and dissociative recombination with positive ions are negligible and therefore the equation governing the variation of the electron density, n_e, reduces to:

$$v\frac{dn_e}{dx} = -\beta[A]n_e \qquad (1)$$

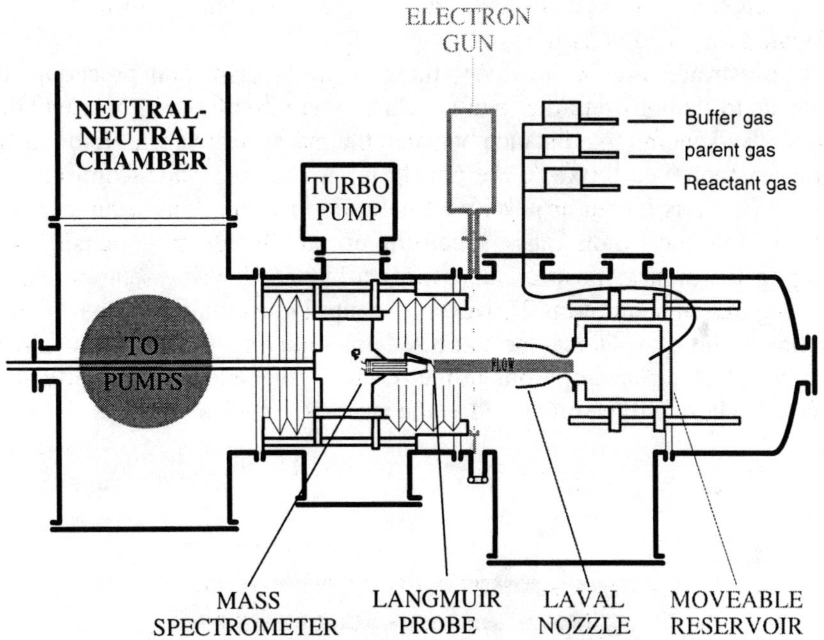

Figure 1. CRESU apparatus.

where v, x and [A] are the flow speed, the distance along the flow axis, and the attaching gas density (which is usually much larger than the electron density) respectively.

Since [A] >> n_e, the value of [A] remains constant along the flow and integrating equation (1) leads to :

$$\ln\left(\frac{n_e}{n_{e0}}\right) = -\beta\frac{[A]}{v}x \qquad (2)$$

Plotting $\ln(n_e)$ as a function of x, therefore, yields a straight line whose slope is $p_1 = -\frac{\beta[A]}{v}$.

By repeating this measurement for several values of [A] and plotting p_1 as a function of [A] one obtains the value of β.

2.3 Procedure Demonstrating Attachment to Neutral Clusters

The presence of masses corresponding to cluster ions in the anion mass spectrum as well as their variation versus distance between the electron beam and the mass spectrometer is not always sufficient to demonstrate an effect of neutral clusters since such ions can be very efficiently formed by ion-molecule termolecular association.

As illustrated Fig. 2, however, there is an experimental procedure that allows us to demonstrate that neutral clusters are involved in the attachment process. By keeping the distance between the mass spectrometer and electron beam constant (i.e. L fixed), the nozzle is moved backward from its initial position (i.e. z is increased). Without neutral clustering, this cannot have an influence on the anion mass spectrum or on the electron density since changing the nozzle position does not influence the flow conditions and attaching gas concentration. However, if neutral clustering plays a role in the attachment process, it can be expected that moving the nozzle backward enhances cluster concentrations in the reaction zone and resulting effects are expected. These will be presented in the next sections.

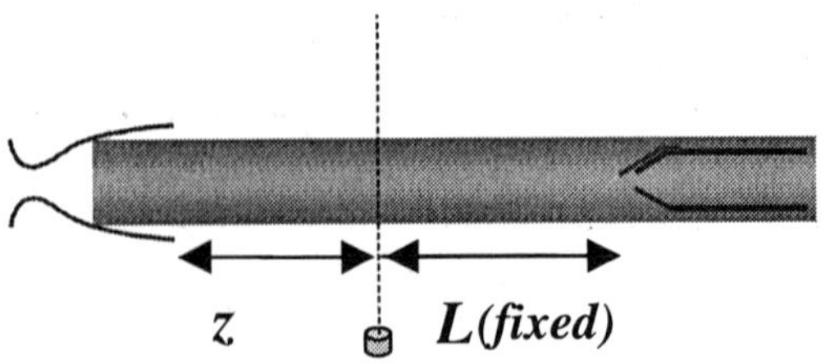

Figure 2. Test for formation of dimers in the neutral jet.

3. RESULTS AND DISCUSSION

3.1 Attachment with HCl and HBr

Typical plots of electron density versus distance along the flow axis x are shown in Fig. 3 for various flow rates (i.e. concentrations) of HBr. The slopes, p_1, of the lines shown in this Figure are then plotted as a function of HBr density (see Fig. 4) and the rate coefficient β is obtained as explained in Section 2.2.

Although the reaction

$$e + HBr \longrightarrow Br^- + H \qquad (3)$$

is endothermic (ΔH = 0.4 eV), the measured values of β for attachment with HBr exhibits a strong inverse temperature dependence as shown in Fig. 5. At 170 K only an upper limit of β can be obtained. This behavior cannot be explained by electron thermalization problems since the gas density at the lowest temperatures are an order of magnitude higher than at 170 K (see

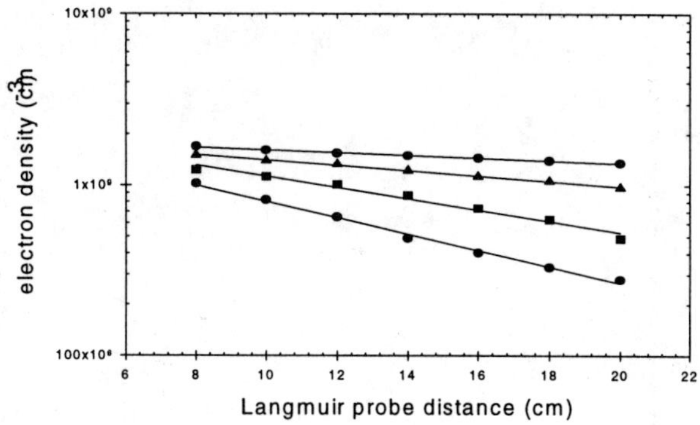

Figure 3. Electron density as a function of the Langmuir probe position in the plasma. Each decrease leads to the determination of the slope p_1 for a given reactant flow rate (HBr at 48 K).

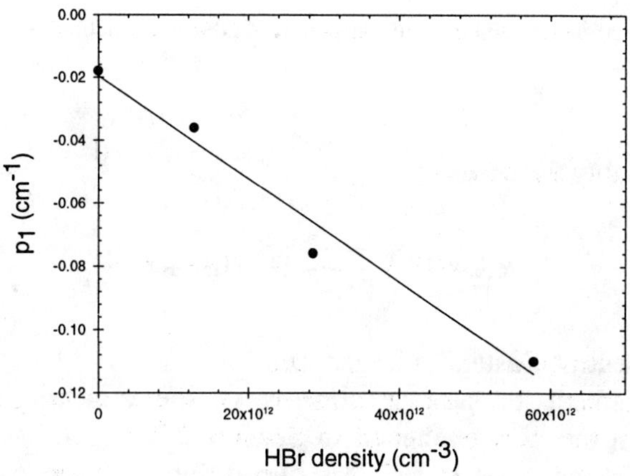

Figure 4. Plot of slopes p_1 as a function of the reactant density in the jet. The slope yields the electron attachment rate coefficient (HBr at 48 K).

Table 1 in Ref. 6) for a very similar hydrodynamic time.

The same results were obtained for attachment in HCl. Here again the reaction:

$$e + HCl \longrightarrow Cl^- + H \tag{4}$$

is endothermic ($\Delta H = 1.0$ eV). We were, therefore, led to the conclusion that in both cases, species other than the monomers were attaching electrons. Taking into account the known electron affinity of Cl·HCl (4.896 eV)[9] we estimate that $\Delta H \approx -0.3$ eV for the reaction :

$$e + (HCl)_2 \longrightarrow Cl^- \cdot HCl + H \tag{5}$$

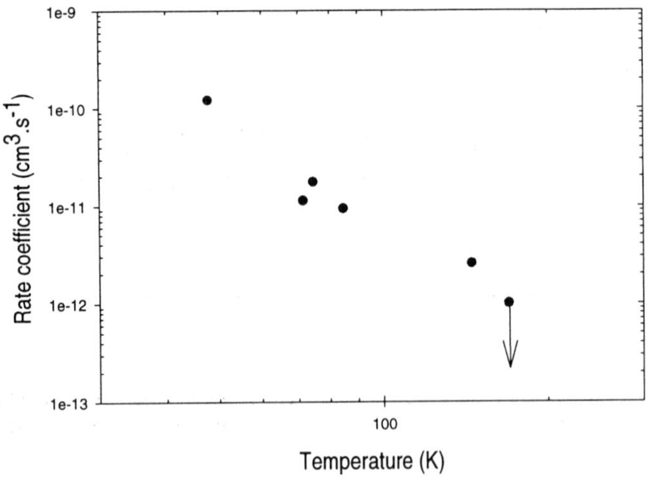

Figure 5. Electron attachment Rate coefficient of HBr as a function of temperature.

and this is probably the case for :

$$e + (HBr)_2 \longrightarrow Br^{-}\cdot HBr + H \tag{6}$$

as well as for neutral clusters of larger size.

The involvement of neutral clusters in the attaching process was demonstrated in the way explained in Section 2.3. Fig. 6 shows two mass spectra taken in the case of HCl for two different distances between the electron beam and the nozzle exit. The cluster ion signal increases strongly with this distance. The peak of Cl^- can be attributed to the endothermic Reaction (4), which certainly occurs in the thermalization zone close to the electron beam.

The first order dependence of β with HCl and HBr concentration seems at first sight surprising but can be explained if the cluster concentrations are in steady-state with that of the monomer.[6]

In the case of HBr it was possible to obtain an upper limit for the density of the various neutral clusters from a study of the neutral reaction:

$$OH + HBr \longrightarrow H_2O + Br \tag{7}$$

which exhibits a departure from pseudo-first order behavior at the onset of measurable clustering.

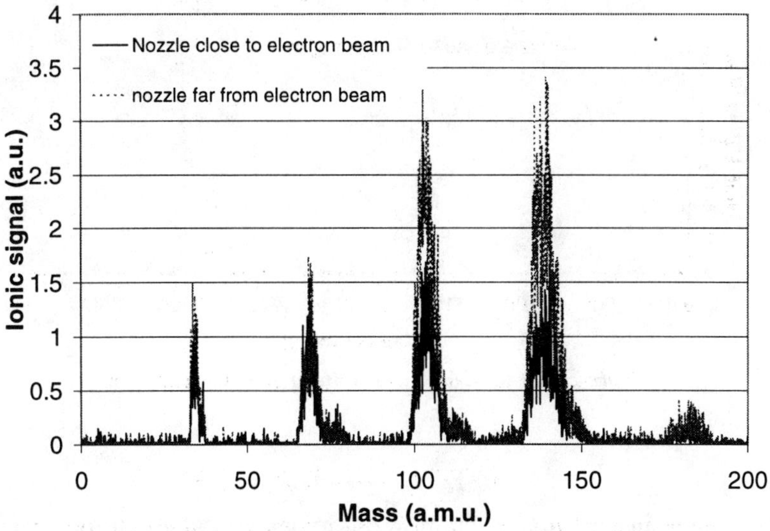

Figure 6. Mass spectrum taken in the case of HCl for two different nozzle positions from the electron beam. When the nozzle is moved backward to increase the neutral jet length, more clusters are generated in the jet and the negative ion signal increases.

This allowed the estimation of a lower limit of ~10^{-8} cm^3s^{-1} for electron attachment to HCl clusters at low temperatures (48 K).

3.2 Electron Attachment with C_3F_6

The attachment of an electron to C_3F_6 has been studied by various authors[10,11] and an anomalous pressure dependence of the rate coefficient has been found. Both ions $(C_3F_6)_2^-$ and $C_6F_{11}^-$ have been observed in the mass spectrum by Hunter et al.[10] and by Jarvis et al.[11] Both authors have suggested that the first step of attachment is :

$$e + C_3F_6 \longrightarrow (C_3F_6^-)^* \tag{8}$$

followed by :

$$(C_3F_6^-)^* + C_3F_6 \longrightarrow (C_3F_6)_2^{-*} \xrightarrow{+M} (C_3F_6)_2^- \tag{9}$$

or

$$(C_3F_6^-)^* + C_3F_6 \longrightarrow C_6F_{11}^- + F. \tag{10}$$

An ion such as $C_6F_{11}^-$ could be easily formed however by reaction with neutral dimers of C_3F_6:

$$(C_3F_6)_2 + e \longrightarrow (C_3F_6)_2^{-*} \tag{11}$$

followed by

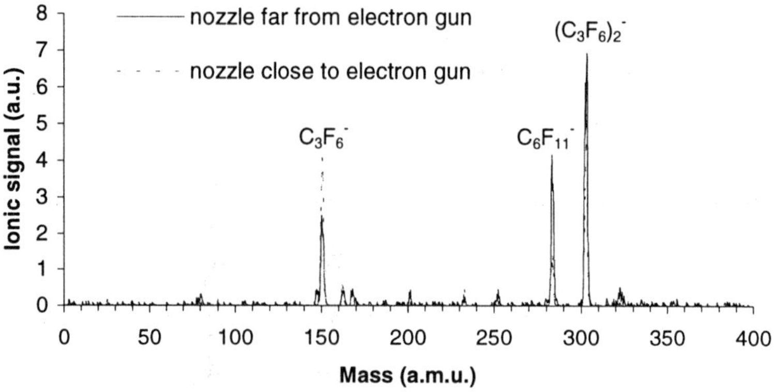

Figure 7. Test for C_3F_6 dimers formation in the neutral jet.

$$(C_3F_6^-)_2^{-*} \longrightarrow C_6F_{11}^- + F. \tag{12}$$

The importance of this reaction in our experimental conditions was again checked by varying the position of the nozzle exit and the resulting spectra are shown in Fig. 7. The $C_6F_{11}^-$ anion peak increases strongly when the nozzle is moved backward thus giving more time for dimer formation. The electron density is also observed to drop sharply during this operation as shown in Fig. 8. Since the concentration of these species will change from a given experiment to another depending upon the thermodynamic conditions, this makes a comparison of β values between various techniques unreliable as long as the importance of Reactions (11) and (12) is not addressed.

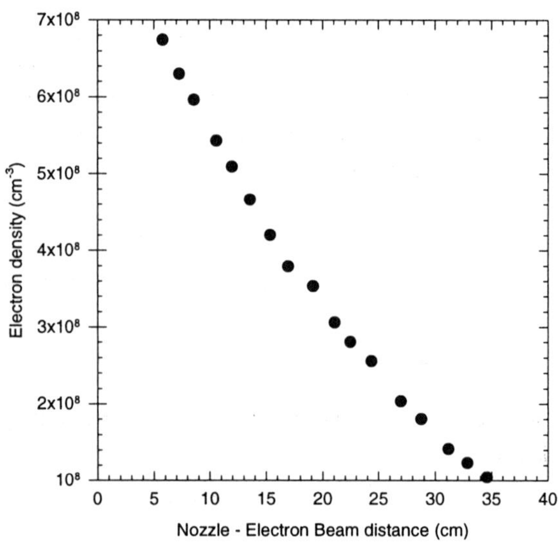

Figure 8. Electron density as a function of nozzle-electron beam distance at 48 K.

3.3 Preliminary Results with Water and Benzene

Water and benzene monomers do not attach electrons and any attachment noticed in these gases is due either to impurities or to neutral clusters. At sufficiently large flow rates of H_2O (or C_6H_6), we have observed a strong decrease of electron density in the supersonic flow. Under these conditions the mass spectrometer exhibits a rich spectrum (Fig. 9), which shows a variety of cluster anions that can only be attributed to attachment to neutral clusters.

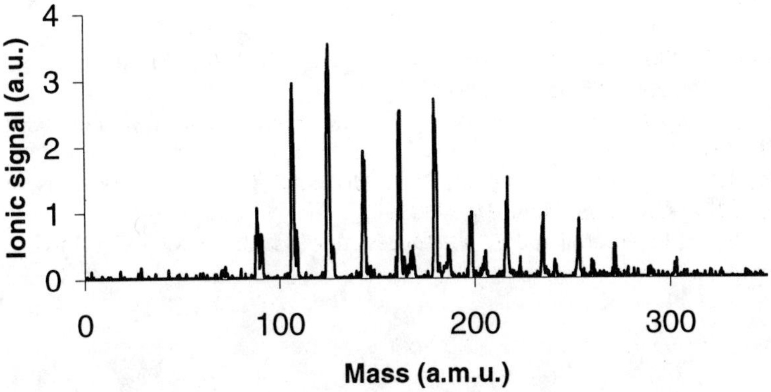

Figure 9. Negative ions observed for the electron attachment process on water at 22.9 K.

4. CONCLUSIONS

Although the present study has not permitted a definitive demonstration of electron thermalization in the CRESU experiments, the importance of electron attachment to neutral clusters at low temperatures has been highlighted. Such processes could be important in a variety of environments such as planetary atmospheres or interstellar clouds.

ACKNOWLEDGMENTS

We are grateful for funding from the "Programme National de Physique et Chimie du Milieu Interstellaire" (PCMI) and from the "Programme National de Planétologie" (PNP). We are grateful to the European Commission through a TMR Research Network grant, contract FMRX-CT97-0132 (DG12-MIHT), for research on Astrophysical Chemistry.

REFERENCES

1 T. D. Märk, Intern. J. Mass Spec. Ion. Proc. **107**, 143 (1991).
2 J. B. A. Mitchell and C. Rebrion-Rowe, Int. Rev. Phys. Chem. **16**, 201 (1997).
3 D. R. Bates, J. Phys. B: At. Mol. Opt. Phys. **25**, 3067 (1992).
4 T. Kraft, M. W. Ruf and H. Hotop, *Proceedings of the XVII ICPEAC*, ed. by W. R. MacGillivray, I. E. McCarthy and M. C. Standage (Adam Hilger, Bristol, UK, 1992), p. 599.
5 R. J. Allen, L. Loinard, A. R. W. McKellar and J. Lequeux, Astrophys. J. **489**, 102 (1997).
6 T. Speck, J-L. Le Garrec, S. Le Picard, A. Canosa, J. B. A. Mitchell and B.R. Rowe, J. Chem. Phys. **114,** 8303 (2001).
7 J-L. Le Garrec, O. Sidko, J-L Queffelec, S. Hamon, J. B. A. Mitchell and B. R. Rowe, J. Chem. Phys., **107**, 54 (1997).
8 T. Speck, T. Mostefaoui, C. Rebrion-Rowe, J. B. A. Mitchell and B. R. Rowe, J. Phys. B: At. Mol. Opt. Phys. **33**, 3575 (2000).
9 *Handbook of Chemistry and Physics*, 74th Edition, ed. by David R. Lide (CRC Press, 1993-1994).
10 S. R. Hunter, L. G. Christophorou, D. L. McCorkle, I. Sauers, H. W. Ellis and D. R. James, J. Phys. D: Appl. Phys., **16**, 573 (1983).
11 G. K. Jarvis, R. Peverall and C. A. Mayhew, J. Phys. B: At. Mol. Opt. Phys. **29**, L713 (1996).

Large Enhancements in Dissociative Electron Attachment to Chlorine Containing Molecules Adsorbed on H_2O Ice

Implications for Atmospheric Ozone Depletion

Q.-B. Lu and L. Sanche

Group of the Canadian Institutes of Health Research in the Radiation Sciences, Faculty of Medicine, University of Sherbrooke, Sherbrooke, Canada, J1H 5N4

qblu@caltech.edu

1. INTRODUCTION

The study of chlorine-containing molecules, especially chlorofluorocarbons (CFCs), has received continued interest mainly because of their well-known association with the ozone depletion in the Earth's stratosphere. It was first proposed[1] that chlorine atoms are produced by photolysis of such species as CF_2Cl_2 (CFC-12) and $CFCl_3$ (CFC-11). For the former species, the reaction is: $CF_2Cl_2 + h\nu \rightarrow Cl + CF_2Cl$. The resultant Cl atom destroys ozone via a (Cl, ClO) reaction chain. The significance of CFCs to the ozone depletion has been well recognized since the discovery of the Antarctic ozone hole.[2] Nevertheless, the above photodissociation model predicts a maximum rate of Cl atom production in the upper stratosphere at altitudes of 30~40 km and *a negligible rate below 20 km.*[1] However, the ozone hole is found at the lower polar stratosphere at an altitude of ~15 km. It is known that the creation of the ozone hole is closely related to the existence of polar stratospheric clouds (PSCs) that form in the Antarctic stratosphere due to the extreme cold temperature; these PSCs consist mainly of condensed-phase water ice and nitric acid/ice.[3,4] The formation of the ozone hole has been attributed to heterogeneous reactions occurring on surfaces of PSC ice particles, which convert chlorine from the inactive compounds (HCl and $ClONO_2$) into reactive Cl_2: $HCl + ClONO_2 \rightarrow Cl_2 + HNO_3$.[3,4] Upon photolysis, Cl_2 releases chlorine to destroy O_3.

Dissociative electron attachment (DEA) of CFC molecules with low-energy electrons near 0 eV is an extremely efficient process in the gas phase:

Dissociative Recombination of Molecular Ions with Electrons
Edited by S. Guberman, Kluwer Academic/Plenum Publishers, 2003

$e^- + CF_2Cl_2 \rightarrow Cl^- + CF_2Cl$.[5,6] This DEA cross section is four orders of magnitude higher than the photodissociation cross section.[5] In the stratosphere, the major source producing atmospheric ionization is cosmic rays.[7] However, the detected density of free electrons is very low, since most of the low-energy electrons produced by cosmic ray ionization are captured by stratospheric molecules to produce negative ions, mainly O_2^-.[7] Thus, DEA processes are not believed to be a significant sink for CFCs in the stratosphere.[8] However, the case may be different in the winter polar stratosphere with the presence of PSCs. Low-energy electrons can be self-trapped (solvated) in ices with large trapping cross sections.[9-11] It has also been observed that negative ion yields in electron-stimulated desorption of CF_2Cl_2 are greatly enhanced by factors of 10^2 and 10^4 when CF_2Cl_2 is coadsorbed with polar water and ammonia ices on a surface, respectively.[12] These enhancements have been identified as due to the transfer of ~0 eV electrons temporarily localized at precursors of the fully solvated state in polar molecular ice to CFC molecules that dissociate via DEA into Cl^- or F^-.[12,13] The Cl^- ions can create Cl atoms by photodetachment or react with other species on the ice surface to produce Cl_2 that dissociates to destroy ozone after dawn.[14] From field measurements (satellites, balloons, etc.), we have found that there exists a strong correlation between ozone depletion (CFC destruction) and cosmic ray flux in both the latitude and altitude distributions, as well as in calendar year variations.[15] Here, we present laboratory measurements on absolute DEA cross sections of CF_2Cl_2, $CFCl_3$ and HCl, adsorbed on surfaces of H_2O ice films. Implications of these measurements for atmospheric ozone depletion are discussed below.

2. EXPERIMENTS

Experimental details for measuring the absolute cross section for the dissociation of CFCs on water ice have been given elsewhere.[13] The experiments were conducted in an ultrahigh vacuum chamber with a base pressure of $\sim 5\times10^{-11}$ Torr. A Pt foil is used as a substrate; it can be cooled to ~20 K with a closed-cycle helium refrigerator. High-purity Kr, H_2O and chlorine-containing gases (CF_2Cl_2 / $CFCl_3$ / HCl) are in sequence condensed onto the Pt surface with two separate directional dosers. Their coverages are determined from temperature programmed desorption (TPD) spectra, where one monolayer (ML) of H_2O refers to a bilayer with a density of $\sim 1.0\times10^{15}$ cm^{-2}. Electron trapping is measured by the low energy electron transmission method.[16] The trapping coefficient $A_s(E)$ is defined as $d(\Delta V)/dt$ at $t=0$, where ΔV is the surface electrostatic potential change caused by electron trapping. For charging by a submonolayer of molecules on the H_2O film, the trapping cross section $\sigma(E)$ is obtained from[13]

$$A_s(E) = \frac{\sigma(E)\rho o J o}{\varepsilon o}\left(\frac{L_1}{\varepsilon_1} + \frac{L_2}{\varepsilon_2}\right) \quad (1)$$

where E is the incident electron energy, ρ_0 is the surface density of electron-trapping molecules, J_0 is the incident electron density and ε_0 is the vacuum permittivity. ε_1=1.91 and ε_2=3.3 are the dielectric constants of Kr and H_2O, respectively. L_1=32.6 Å (L_2=13.0 Å) is the film thickness for 10 ML Kr (5 ML H_2O).

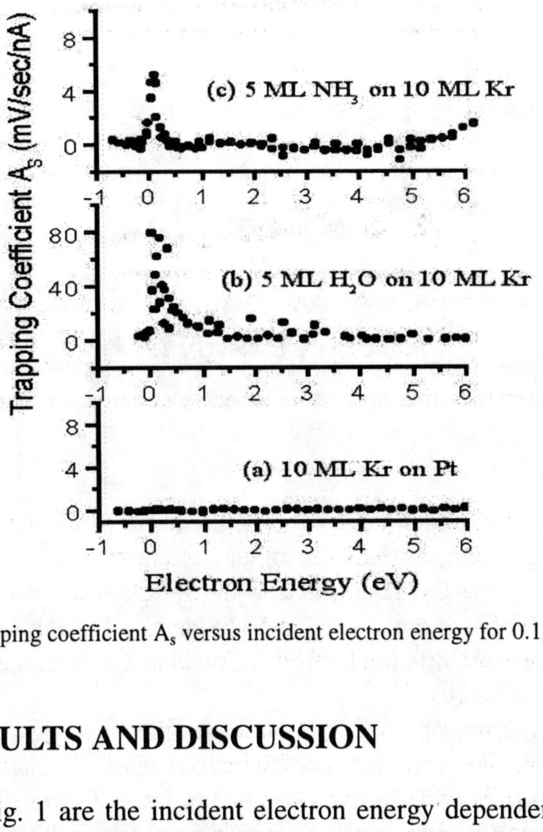

Figure 1. Trapping coefficient A_s versus incident electron energy for 0.1 ML CF_2Cl_2.

3. RESULTS AND DISCUSSION

Shown in Fig. 1 are the incident electron energy dependencies of $A_s(E)$ for (a) a 10 ML pure Kr, (b) a 5 ML H_2O and (c) a 5 ML NH_3 film condensed on top of the Kr film. Electron trapping near 0 eV is observed for both the H_2O and NH_3 film deposited on Kr, but the trapping coefficient for the H_2O is larger than that for the NH_3 film.

Shown in Fig. 2 are the $A_s(E)$ spectra for 0.1 ML CF_2Cl_2 adsorbed on a 10 ML Kr, a 5 ML H_2O and on 5 ML NH_3 film. From these data, a trapping cross section, $\sigma=1.4\times10^{-15}$ cm^2 at nearly 0 eV, is derived for CF_2Cl_2 on Kr. Gaseous CF_2Cl_2 has a maximum DEA cross section of ~1×10^{-16} cm^2 at ~0 eV.[5] The larger σ value (by about one order of magnitude) for the condensed

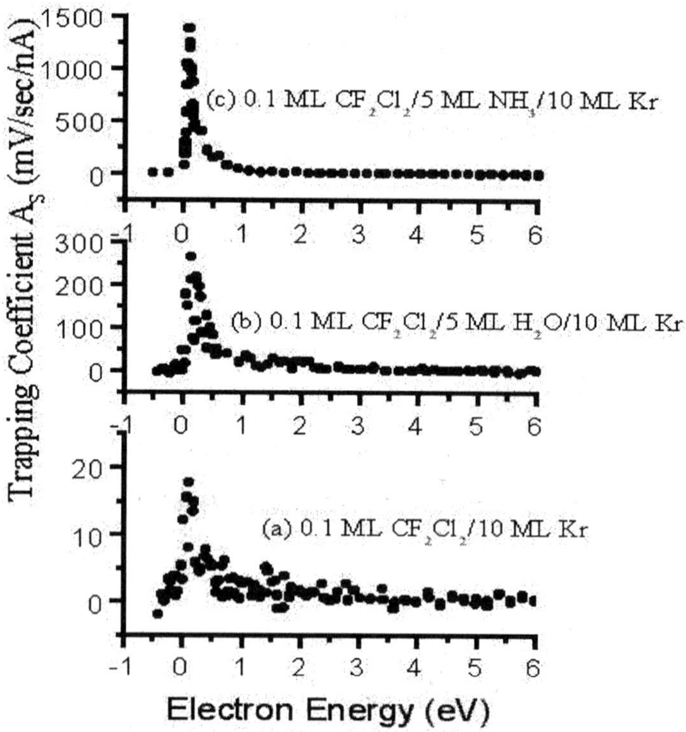

Figure 2. Trapping coefficient A_s versus incident electron energy for 0.1 ML CF_2Cl_2.

phase is reasonably explained by the polarization of the Kr film induced by the dissociating anion, as seen for other halomethanes.[17, 18] Due to the low kinetic energy of the Cl^- fragment and the image force, most Cl^- ions from the DEA at ~ 0 eV cannot desorb and are captured at the surface. Thus, the measured charges are due to electrons trapped as Cl^- ions, i.e., σ corresponds to the DEA cross section.[17, 18]

Evidently, the presence of H_2O increases the trapping cross section near 0 eV. However, the trapping coefficient in Fig. 2b should include the contribution from bound electrons trapped in the H_2O film shown in Fig. 1b. After subtracting the data for A_s in Fig.1b from that in Fig. 2b, we obtained from Eq.(1) a trapping cross section $\sigma=1.3\times10^{-14}$ cm^2 at ~ 0 eV for 0.1 ML CF_2Cl_2 on 5 ML H_2O, which is about two orders higher than the gaseous cross section. Most significantly, NH_3 causes a giant enhancement of DEA to CF_2Cl_2, as seen in Fig. 2c. The cross section σ at ~ 0 eV reaches ~9.7×10^{-14} cm^2 for 0.1 ML CF_2Cl_2 on 5ML NH_3, which is nearly three orders of magnitude larger than the corresponding gaseous value.

Shown in Fig. 3 are the $A_s(E)$ for 0.2 ML HCl adsorbed on a 10 ML Kr surface with and without the presence of 5ML H_2O, from which trapping

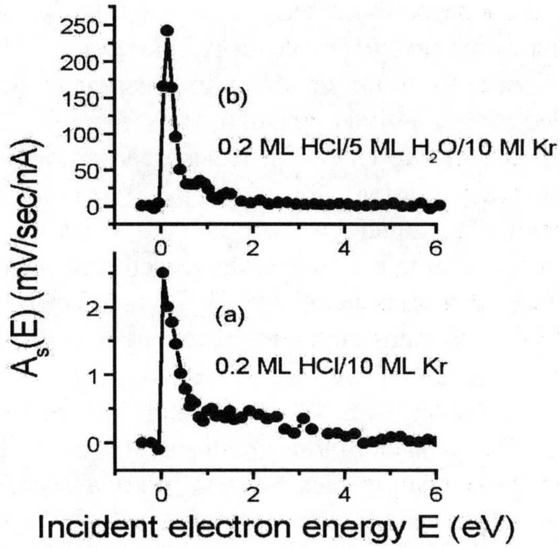

Figure 3. Trapping coefficient A_s versus incident electron energy for HCl.

cross sections of $4.0\text{x}10^{-15}$ and $1.1\text{x}10^{-16}$ cm^2 at ~0 eV are derived, respectively. The former is two orders of magnitude higher than the gaseous σ.[19]

Similarly, we have measured that the trapping cross sections σ for 0.1 ML $CFCl_3$ adsorbed on a 10 ML Kr surface with and without the presence of 5ML H_2O are $1.1\text{x}10^{-13}$ and $8.9\text{x}10^{-15}$ cm^2 at ~0 eV, respectively. The former is about one order of magnitude larger than the corresponding gaseous value.

At first glance, one might attribute the enhancements in σ to an effect of the polar molecular film on the direct DEA to CF_2Cl_2 (i.e., without electron transfer). However, it has been shown that direct DEA is significantly reduced for molecules deposited on H_2O and NH_3 films.[20, 21] Thus, the present observation of a large enhancement in the DEA cross section for HCl on ice must involve another mechanism. An excess electron in a polar medium like H_2O becomes trapped (solvated) due to the interaction of the electron with the dipole moment of surrounding polar molecules.[9] Before it becomes fully solvated, the electron first thermalizes and reaches a preexisting trapping site, i.e., a precursor of the fully solvated state.[22, 23] It has been predicted that electrons in a precursor state are highly reactive due to their large mobility; these electrons will react rapidly with electron attaching molecules, if the latter are present in the medium.[9] For DEA to CFCs (HCl) via electron transfer from fully solvated traps, we can take as an upper limit of A_s the value in the pure polar film plus that due to direct DEA to CFCs (HCl) on Kr. However, the measured values of A_s(0 eV) for CFCs (HCl) on H_2O / NH_3 are much larger than the upper limit. Thus, we must conclude that transfer of electrons from precursor states to CFCs (HCl) is the major mechanism responsible for the observed increase in σ near 0 eV. The lifetime of electrons localized at the precursor state of H_2O ice below 200 K should lie in the range of μs.[13] Because of the small binding energy and the relatively long lifetime of an electron (e_p^-) in the precursor state, there is a large probability for transfer of the electron to CFCs (HCl) adsorbed on the ice surface. As a result, the DEA cross sections for CFCs (HCl) on ice is significantly enhanced, as seen in Figs. 2 and 3.

With the data for cosmic ray flux in the lower stratosphere, we estimate that electron-induced dissociation rates of CF_2Cl_2 and $CFCl_3$ on the surface of ice particles in PSCs under cosmic ray radiation are 4.3×10^{-5} s^{-1} and 3.6×10^{-4} s^{-1}, respectively.[15]

According to our findings, the Cl^- yield (and thus the Cl yield) depends not only on the CFC concentrations in the stratosphere, but also on the densities of other molecular species, especially polar species such as H_2O. The greenhouse effect causes the concentration of H_2O in the lower stratosphere (16-20 km) to increase.[24] This increase may lead to more severe ozone depletion and a significant delay in the eventual recovery of the ozone hole.

4. SUMMARY

We have shown that transfer of near 0 eV electrons, temporarily localized in the precursor of the fully solvated state of H_2O/NH_3 ice, to CFCs (HCl), leads to large enhancements of the absolute cross sections for dissociative electron attachment to these molecules. The absolute DEA cross sections for these molecules adsorbed on a H_2O film grown on Kr at ~20 K are about two orders of magnitude larger than those in the gas phase. These findings indicate that electron-induced dissociation of CFCs in PSCs due to cosmic ray radiation is a very efficient process for destruction of CFCs in the lower polar stratosphere and should play a significant role in causing the ozone hole(s).

ACKNOWLEDGEMENTS

This work is supported by the Canadian Institutes of Health Research (CIHR). One of us (Q.-B. Lu) acknowledges financial support via a CIHR Senior Research Fellowship.

REFERENCES

1 M. J. Molina and F. S. Rowland, Nature **249**, 810 (1974).
2 J. C., Farman, B. G. Gardiner and J. D. Shanklin, Nature **315**, 207 (1985).
3 S. Solomon, R. R. Garcia, F. S. Rowland and D.-J. Wuebbles, Nature **321**, 755 (1986).
4 M. J. Molina, T.-L. Tso, L. T. Molina and F. C.-Y. Wang, Science **238**, 1253 (1987).
5 E. Illenberger, H.-U. Scheunemann and H. Baumgärtel, Chem. Phys. **37**, 21 (1979).
6 L. G. Christophorou, J. K. Olthoff, and Y. Wang, J. Phys. Chem. Ref. Data **26**, 1205(1997).
7 D. Smith and N. G. Adams, Top. Curr. Chem. **89**, 1 (1980).
8 F. C. Fehsenfeld, P. J. Crutzen, A. L. Schmeltekopf, C. J. Howard, D. L. Albritton, E. E Ferguson, J. A. Davidson, H. I. Schiff, J. Geophys. Res. **81**, 4454 (1976).
9 N. V. Klassen, in *Radiation Chemistry: Principles and Applications,* ed. by Farhataziz and M. A. Rodgers, (VCH, New Year, 1987), chap.2.
10 A. D. Bass and L. Sanche, J. Chem. Phys. **95**, 2910 (1991).
11 W. C. Simpson, T. M. Orlando, L. Parenteau,; K. Nagesha, L. J. Sanche, Chem. Phys. **108,** 5027 (1998).
12 Q.-B. Lu and T. E. Madey, Phys. Rev. Lett. **82**, 4122(1999); ibid., Surf. Sci. **451**, 238 (2000); ibid., J. Phys. Chem. B**105**, 2779 (2001).
13 Q.-B. Lu and L. Sanche, Phys. Rev. B**63**, 153403 (2001).
14 Q.-B. Lu and T. E. Madey, J. Chem. Phys. **111**, 2861 (1999).
15 Q.-B. Lu and L. Sanche, Phys. Rev. Lett. **87,** 078501 (2001).
16 R. Marsolais, M. Deschênes and L. Sanche, Rev. Sci. Instrum. **60**, 2724 (1989); K. Nagesha, J. Gamache, A. D. Bass and L. Sanche, Rev. Sci. Instrum. **68**, 3883 (1997).
17 P. Ayotte, J. Gamache, A. D. Bass, I. I. Fabrikant, and L. Sanche, J. Chem. Phys.**106**, 749 (1996).
18 A. D. Bass, J. Gamache, P. Ayotte, L. Sanche, J. Chem. Phys. **104**, 4258 (1996).
19 Q.-B. Lu and L. Sanche, J. Chem. Phys. **115**, 5711 (2001).
20 M. A. Huels, L. Parenteau, and L. Sanche, J. Chem. Phys. **100**, 3940 (1994).
21 R. Azria, Y. Le Coat, M. Lachgar, M. Tronc, L. Parenteau and L. Sanche, Surf. Sci. **451**, 91 (2000).
22 A. Migus, Y. Gauduel, J. L. Martin, and A. Antonetti, Phys. Rev. Lett. **58,** 1559 (1987).
23 R. Laenen, T. Roth, and A. Laubereau, Phys. Rev. Lett. **85,** 50 (2000).
24 S. J. Oltmans and D. J. Hofmann, Nature **374**, 146 (1995).

INDEX